Engineering Materials

This series provides topical information on innovative, structural and functional materials and composites with applications in optical, electrical, mechanical, civil, aeronautical, medical, bio- and nano-engineering. The individual volumes are complete, comprehensive monographs covering the structure, properties, manufacturing process and applications of these materials. This multidisciplinary series is devoted to professionals, students and all those interested in the latest developments in the Materials Science field, that look for a carefully selected collection of high quality review articles on their respective field of expertise.

More information about this series at http://www.springer.com/series/4288

Juan Carlos Moreno-Piraján ·
Liliana Giraldo-Gutierrez ·
Fernando Gómez-Granados
Editors

Porous Materials

Theory and Its Application for Environmental Remediation

Springer

Editors
Juan Carlos Moreno-Piraján
Universidad de Los Andes
Bogotá, Colombia

Liliana Giraldo-Gutierrez
Universidad Nacional de Colombia
Bogotá, Colombia

Fernando Gómez-Granados
Universidad Nacional de Colombia
Bogotá, Colombia

ISSN 1612-1317 ISSN 1868-1212 (electronic)
Engineering Materials
ISBN 978-3-030-65990-5 ISBN 978-3-030-65991-2 (eBook)
https://doi.org/10.1007/978-3-030-65991-2

This Springer imprint is published by the registered company Springer Nature Switzerland AG
The registered company address is: Gewerbestrasse 11, 6330 Cham, Switzerland

Preface

When we came up with the idea of writing this book, the authors told Prof. Rodríguez-Reinoso that we were working on this idea, which he loved, feeling honored. Later during the development of this work, he unfortunately passed away.

This book has been written in honor of the memory of Doctor Francisco Rodríguez-Reinoso (R.I.P.), Chemist, Doctor of Chemistry, Professor and Emeritus Professor at the University of Alicante.

Professor Reinoso was and will continue to be a world leader in the area of adsorption and porous materials. Anyone who has worked in this area has surely come across his countless publications in high-impact books and journals, where he published the results of his research. He was Editor of CARBON, the most important journal in the area of carbonaceous materials. He received several awards that recognized his work as a researcher, including the American Carbon Society (2010) and the Japan Carbon Society Prize (Japan Carbon Society, 2008).

His generous personality led him to collaborate with scientists around the world, transmitting his knowledge. Today, many research groups owe the basis of their work to him, such as in Latin America.

The industrial development of recent decades, seeking to obtain technological advances, has led to altered ecosystems breaking the existing balance, with serious consequences for the subsistence of living things. Scientists have strived to propose solutions to eliminate a great diversity of pollutants that are present both in water and in the atmosphere.

Worldwide, environmental regulation standards are established day by day, which requires different industries to decrease their limits for the generation of waste from their industrial processes and greater control over said waste.

Initially, the regulations sought to control emissions once produced to correct their polluting effects. However, in many countries, these regulations are not respected, particularly in developing countries. Consequently, pollutants from an industrial process are not adequately treated within industries and become a pollutant that eventually begins to accumulate, generating a problem in the environment that can be increasingly dangerous over time.

That is why the development of new materials to reduce pollutants has now become one of the priorities of science, which, together with industry, have made progress in this important area.

This book presents advances in the area of porous materials at the nanomaterial level, ranging from the detailed description of their preparation, characterization, mathematical modeling and application, to the removal of contaminants both in aqueous solution and in the vapor phase.

The book *Porous Materials: Theory and its Application for Environmental* is divided into 13 chapters.

- The first chapter was written by Profs. Tanay Kundu, Leisan Gilmanova, Wai Fen Yong, Stefan Kaskel entitled "Metal-Organic Frameworks for Environmental Applications", who very clearly illustrate this type of modern materials and their use in environmental applications. The authors show a comparison between functionalized porous materials such as activated carbons and zeolites. In the chapter is discuss MOFs proved their selectivity or preference to adsorptive removal of various toxic components. High surface area, varying geometries and various functional groups result in a high degree of functionality. Increasing the stability of MOFs for the environmental application purpose remains an actual concern. Studies in this regard showed that appropriate approach to MOFs design helps to solve this problem and allows to synthesize materials applicable in real-world conditions.
- The second chapter is a contribution by Prof. Teresa Bandoz, entitled "Porous Carbons as Oxygen Reduction Electrocatalysts", where it shows the development of "flat" heteroatom-doped carbons for oxygen reduction electrocatalysis, followed by a discussion of the role of defects as a general driving force, through porous carbons doped with heteroatoms. The emphasis is on the importance of porosity.
- The third chapter written by Profs. Lumeng Liu, Shiliang (Johnathan) Tan and D. D. Do presents an interesting topic entitled "Computer Simulation and Experimental Studies of Various Environmental Gases (NH_3, CH_2O, SO_2, H_2S,

Benzene, Water) on Carbon Materials", where they present a detailed analysis of adsorption studies of a series of gases as well as performing computer simulations.

- The fourth chapter was written by Profs. Artur P. Terzyk, Monika Zięba, Stanisław Koter, Emil Korczeniewski, Wojciech Zięba, Piotr Kowalczyk, Joanna Kujawa, who present in this book a chapter entitled "Recent Developments in the Electrophoretic Deposition of Carbon Nanomaterials", who describe in this chapter basic concepts of the electrophoretic deposition process (EPD) in relation to carbon nanomaterials and analyze recent developments in the deposition of carbon nanomaterials, mainly graphene, nanotubes, nanohorns, fullerenes and nanodiamonds.
- The fifth chapter is presented by Profs. S. Reljic, E.O. Jardim, C. Cuadrado-Collados, M. Bayona, M. Martinez-Escandell, J. Silvestre-Albero, F. Rodríguez-Reinoso, entitled "CO_2 Adsorption in Activated Carbon Materials", where they present the selective capture of CO_2 in different types of industrial streams to reduce actual emissions to the atmosphere and minimize environmental concerns. They focus on activated carbon materials and the effect of porous structure and surface chemistry on selective CO_2 adsorption.
- The sixth chapter is written by Profs. Shuwen Wang and Katsumi Kaneko, entitled "Graphene-Based Carbons of Tuned Nanoporosity and Crystallinity", who present an interesting review on the adjustment of porosity and crystallinity of graphene-based porous carbons through a series of methods such as the activation of chemical reagents (with KOH, $ZnCl_2$ and H_3PO_4), physical activation (with CO_2), compression and heat treatment.
- The seventh chapter was written by Profs. Nuno S. Graça, Ana M. Ribeiro, Alexandre Ferreira and Alírio E. Rodrigues, entitled "Application of Adsorption Processes for the Treatment of Diluted Industrial Effluents". It presents an analysis of how the use of a cyclic adsorption process, such as parametric pumping, where the mobile phase is filtered through the adsorption bed alternately up and down, associated with a simultaneous change in a thermodynamic variable (temperature, pressure, pH, etc.), enables the continuous purification of a contaminated stream avoiding the use of chemical regenerants and allows the recycling of the concentrated stream back to the industrial process.
- The eight chapter is written by Profs. Sebastião Mardonio Pereira de Lucena, Daniel Vasconcelos Gonçalves, José Carlos Alexandre de Oliveira, Moises Bastos-Neto, Célio Loureiro Cavalcante Jr. and Diana Cristina Silva de Azevedo, who shows their contribution entitled "Activated Carbons for H_2S Capture", which describes the main variables to guide the choice or synthesis of a suitable activated carbon for H_2S retention. The most recent studies of textural characterization and learning about the structure–property relationship in the capture of H_2S capture are presented. They also illustrate the promising molecular simulation technique, still weakly explored for this specific application, detailing how it can be useful.

- The ninth chapter ("The Immersion Calorimetry as a Tool to Study of the Adsorbate-Adsorbent Interactions on the Adsorption of Emerging Pollutants onto Activated Carbon from Water: Case Methylparaben and Paracetamol") is a contribution by Profs. Juan Carlos Moreno-Piraján, Liliana Giraldo, Fernando Gómez-Granados and Valentina Bernal who present the use of immersion calorimetry to determine the affinity between activated carbon and a pharmaceutical compound (emerging pollutant) and its correlation with adsorbate uptake. They present a thermodynamic study of adsorbate–adsorbent interactions by calculating two enthalpies: the enthalpy of immersion and interaction.
- The tenth chapter is presented by Profs. M. R. Gonzalez, A. M. Pereyra, E. I. Basaldella, entitled "Conversion of Nonconventional Aluminosiliceous Sources into Microporous Adsorbents for Water Remediation", who present in this writing the reuse of two distinctive industrial wastes: the discarded catalysts from the cracking units used in the oil industry and the fly ash generated in the production of energy from coal. Both materials are high-volume waste produced by the industrial energy sector. Its use as a negative cost raw material in the synthesis processes of microporous materials containing zeolites and its possible applications as adsorbents.
- The eleventh chapter is a contribution of the Profs. Nahum Andres Medellin-Castillo, Miguel Mauricio Aguilera-Flores, Bridinette Thiodjio Sendja entitled "Removal of Pollutants from Water by Adsorbents Prepared from Animal Bone Wastes" where the professors present an analysis on the importance of water, the types of pollutants present in the water and their sources of contamination. Likewise, they highlight the importance of adsorption processes as an easy-to-implement and low-cost technology in water treatment that uses various adsorbent materials such as bone carbon.
- The twelfth chapter is a contribution of the Profs. Kaihang Shi, Erik E. Santiso and Keith E. Gubbins, entitled "Current Advances in Characterization of Nanoporous Materials: Pore Size Distribution and Surface Area", who present methods for characterizing the most important textural properties of porous materials; particularly, pore size distribution (PSD) and specific surface area, using theoretical models to analyze standard experimental physisorption isotherms, are reviewed. For PSD analysis, we explain the basics of the classical methods, including those based on the Kelvin equation and adsorption potential theory, as well as the modern molecular methods based on the density functional theory (DFT) and molecular simulation.
- The thirteenth chapter closes this book with a contribution of the Profs. Roberto Leyva-Ramos, Aracely Jacobo-Azuara, Jesus Ivan Martínez-Costa, entitled "Organoclays. Fundamentals and Applications for Removing Toxic Pollutants from Water Solution"; they analyze adsorption capacity of the organoclay is highly dependent upon the properties of the precursor clay, the surfactant loading of the organoclay and the physicochemical properties of the surfactant, such as the size. The authors discuss the physicochemical properties of the pollutant considerably affect the adsorption capacity of the organoclays. Also, the organoclays have successfully applied for removing organic compounds,

metallic cations and anions, inorganic anions and pharmaceutical compounds from water solution. In this chapter, the adsorption capacity of the organoclay can be influenced by the operating conditions such as temperature and solution pH. The degree and tendency of the variation depend on the interactions between the organoclay and the contaminant in the solution.

This completes a material that will be very useful to both students and researchers in the area of materials for environmental protection. This book is the product of the collaboration of renowned scientists from different parts of the world, who decided to make a valuable contribution to pay tribute not only to the scientist and the professor, but also to the friend.

Bogotá, Colombia

Liliana Giraldo-Gutierrez
Juan Carlos Moreno-Piraján
Fernando Gómez-Granados

Contents

Editors and Contributors

About the Editors

Juan Carlos Moreno-Pirajún is a Full Professor of Chemistry at Universidad de Los Andes (Colombia). He received his Ph.D. from Universidad Nacional de Colombia. For his doctoral dissertation, he was awarded "distinción meritoria" (*Magna cum Laude*). He was a Postdoctoral Fellow at Instituto Nacional del Carbón (INCAR), Oviedo (Spain), under the supervision of Professor Juan Manuel Diez-Tascón. He has been invited to several universities, among them: Universidad de San Luis (Argentina), Universidad Nacional del Litoral (Argentina), Universidad Autónoma de San Luis Potosí (Mexico), Universidad Federal do Ceará (Brazil), Università Degli Studi di Napoli Federico II (Italy) and Universidad de Alicante (Spain). He works in the areas of thermodynamics, porous materials, particularly nanomaterials, special materials such as MOFs, mathematical modeling and design, construction and application of calorimeters to the characterization of materials. He has published over 200 papers in peer-reviewed journals, having 2650 citations and a 27 h-index (ISI Web). He is the director of the Porus Solids and Calorimetry Research Group.

Liliana Giraldo-Gutierrez is a Full Professor at Universidad Nacional de Colombia (Bogotá). She received her Ph.D. from Universidad de Colombia with "distinción meritoria" (*Magna cum Laude*). She works in porous materials, particularly with nanomaterials, special materials such as MOFs, modeling, design, and applying calorimeters to the characterization of materials and thermodynamic studies of interactions solute–solute, solute–solvent and adsorbate–adsorbent. She has been an advisor of over 50 dissertations, both for undergraduates as graduate students. She has been invited to Universidad Nacional de San Luis (Argentina) and Universidad de Alicante (Spain). He has published over 200 papers in peer-reviewed journals. She is the director of the Calorimetry Research Group.

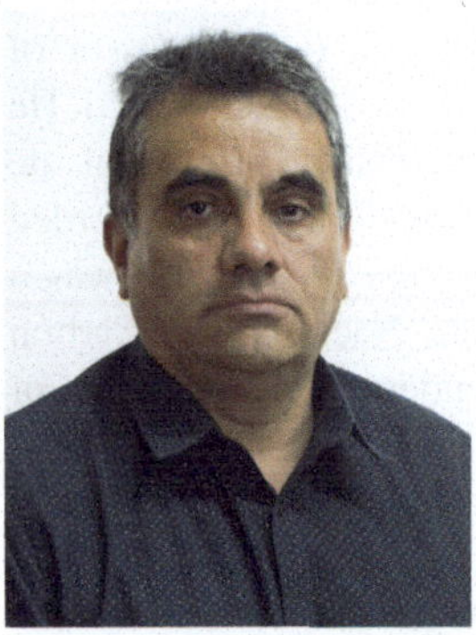

Dr. Fernando Gómez-Granados is a Chemist graduated from the Universidad Nacional de Colombia, a specialist in Geographic Information Systems from the Instituto Geográfico Agustín Codazzi—Universidad Distrital, Ph.D. in Analytical Chemistry of the Environment and Contamination from the Universitat de Barcelona with research in analytical methodology for the determination of antibiotic residues in environmental samples. Currently, he is an Associate Professor in the Department of Chemistry at the Universidad Nacional de Colombia, teaching in analytical chemistry. Has held academic–administrative positions in the Department of Chemistry (academic coordination) and the Curricular Area of Chemistry (director). He has worked in coal exploitation and characterization research, including petrographic analysis. He also worked in the formulation of mixtures and production of metallurgical coke, as well as in the design of databases and implementation of geographic information systems of the coal resource. His interest is currently focused on the study of adsorption processes of different analytes with activated carbon prepared, characterized and modified from carbon materials.

Contributors

Miguel Mauricio Aguilera-Flores Multidisciplinary Graduate Program in Environmental Sciences, Autonomous University of San Luis Potosi, San Luis Potosi, Mexico;
Interdisciplinary Professional Unit of Engineering Zacatecas Campus, Instituto Politécnico Nacional, Bioengineering Academy, Zacatecas, Mexico

Teresa J. Bandosz Department of Chemistry and Biochemistry, The City College of New York, New York, NY, USA

E. I. Basaldella Centro de Investigación y Desarrollo en Ciencias Aplicadas Dr. J.J. Ronco (CINDECA) (CONICET-CIC-UNLP), La Plata, Argentina

Moises Bastos-Neto Grupo de Pesquisas em Separações Por Adsorção (GPSA), Departamento de Engenharia Química (DEQ), Universidade Federal do Ceará (UFC), Fortaleza, CE, Brazil

M. Bayona Laboratorio de Materiales Avanzados, Departamento de Química Inorgánica-IUMA, Universidad de Alicante, Alicante, Spain

Valentina Bernal Grupo de Investigación en Sólidos Porosos y Calorimetría, Departamento de Química, Universidad de Los Andes, Bogotá, Colombia

Célio Loureiro Cavalcante Jr. Grupo de Pesquisas em Separações Por Adsorção (GPSA), Departamento de Engenharia Química (DEQ), Universidade Federal do Ceará (UFC), Fortaleza, CE, Brazil

C. Cuadrado-Collados Laboratorio de Materiales Avanzados, Departamento de Química Inorgánica-IUMA, Universidad de Alicante, Alicante, Spain

Diana Cristina Silva de Azevedo Grupo de Pesquisas em Separações Por Adsorção (GPSA), Departamento de Engenharia Química (DEQ), Universidade Federal do Ceará (UFC), Fortaleza, CE, Brazil

Sebastião Mardonio Pereira de Lucena Grupo de Pesquisas em Separações Por Adsorção (GPSA), Departamento de Engenharia Química (DEQ), Universidade Federal do Ceará (UFC), Fortaleza, CE, Brazil

José Carlos Alexandre de Oliveira Grupo de Pesquisas em Separações Por Adsorção (GPSA), Departamento de Engenharia Química (DEQ), Universidade Federal do Ceará (UFC), Fortaleza, CE, Brazil

D. D. Do School of Chemical Engineering, University of Queensland, St Lucia, Australia

Alexandre Ferreira Faculty of Engineering, University of Porto, Porto, Portugal

Leisan Gilmanova Technische Universität Dresden, Institut für Anorganische Chemie I, Dresden, Germany

Liliana Giraldo-Gutierrez Grupo de Calorimetría, Departamento de Química, Universidad Nacional de Colombia, Bogotá, Colombia

M. R. Gonzalez Centro de Investigación y Desarrollo en Ciencias Aplicadas Dr. J.J. Ronco (CINDECA) (CONICET-CIC-UNLP), La Plata, Argentina

Daniel Vasconcelos Gonçalves Grupo de Pesquisas em Separações Por Adsorção (GPSA), Departamento de Engenharia Química (DEQ), Universidade Federal do Ceará (UFC), Fortaleza, CE, Brazil

Nuno S. Graça Faculty of Engineering, University of Porto, Porto, Portugal

Keith E. Gubbins Department of Chemical and Biomolecular Engineering, North Carolina State University, Raleigh, NC, USA

Fernando Gómez-Granados Grupo de Calorimetría, Departamento de Química, Universidad Nacional de Colombia, Bogotá, Colombia

Araceli Jacobo-Azuara División de Ciencias Naturales y Exactas, Departamento de Química, Universidad de Guanajuato, Guanajuato, GTO, Mexico

E. O. Jardim Laboratorio de Materiales Avanzados, Departamento de Química Inorgánica-IUMA, Universidad de Alicante, Alicante, Spain

Katsumi Kaneko Research Initiative for Supra-Materials, Shinshu University, Nagano, Japan

Stefan Kaskel Technische Universität Dresden, Institut für Anorganische Chemie I, Dresden, Germany

Emil Korczeniewski Faculty of Chemistry, Physicochemistry of Carbon Materials Research Group, Nicolaus Copernicus University in Toruń, Toruń, Poland

Stanisław Koter Faculty of Chemistry, Chair of Physical Chemistry and Physiochemistry of Polymers, Nicolaus Copernicus University in Toruń, Toruń, Poland

Piotr Kowalczyk College of Science, Health, Engineering and Education, Murdoch University, Perth, WA, Australia

Joanna Kujawa Faculty of Chemistry, Chair of Physical Chemistry and Physiochemistry of Polymers, Nicolaus Copernicus University in Toruń, Toruń, Poland

Tanay Kundu Technische Universität Dresden, Institut für Anorganische Chemie I, Dresden, Germany

Roberto Leyva-Ramos Centro de Investigación y Estudios de Posgrado, FCQ, Universidad Autonoma de San Luis Potosi, San Luis Potosi, SLP, Mexico

Lumeng Liu School of Civil Engineering, Chongqing University, Chongqing, China

M. Martinez-Escandell Laboratorio de Materiales Avanzados, Departamento de Química Inorgánica-IUMA, Universidad de Alicante, Alicante, Spain

Jesus Ivan Martínez-Costa Centro de Investigación y Estudios de Posgrado, FCQ, Universidad Autonoma de San Luis Potosi, San Luis Potosi, SLP, Mexico

Nahum Andres Medellin-Castillo Multidisciplinary Graduate Program in Environmental Sciences, San Luis Potosi, Mexico;
Faculty of Engineering, Graduate Studies and Research Center, Autonomous University of San Luis Potosi, San Luis Potosi, Mexico

Juan Carlos Moreno-Piraján Grupo de Investigación en Sólidos Porosos y Calorimetría, Departamento de Química, Universidad de Los Andes, Bogotá, Colombia

A. M. Pereyra Centro de Investigación y Desarrollo en Ciencias Aplicadas Dr. J.J. Ronco (CINDECA) (CONICET-CIC-UNLP), La Plata, Argentina

S. Reljic Laboratorio de Materiales Avanzados, Departamento de Química Inorgánica-IUMA, Universidad de Alicante, Alicante, Spain

Ana M. Ribeiro Faculty of Engineering, University of Porto, Porto, Portugal

Alírio E. Rodrigues Faculty of Engineering, University of Porto, Porto, Portugal

F. Rodríguez-Reinoso Laboratorio de Materiales Avanzados, Departamento de Química Inorgánica-IUMA, Universidad de Alicante, Alicante, Spain

Erik E. Santiso Department of Chemical and Biomolecular Engineering, North Carolina State University, Raleigh, NC, USA

Bridinette Thiodjio Sendja Department of Mathematic and Physical Science, National Advanced School of Engineering, University of Yaounde I, Yaounde, Cameroon

Kaihang Shi Department of Chemical and Biomolecular Engineering, North Carolina State University, Raleigh, NC, USA;
Department of Chemical and Biological Engineering, Northwestern University, Evanston, IL, USA

J. Silvestre-Albero Laboratorio de Materiales Avanzados, Departamento de Química Inorgánica-IUMA, Universidad de Alicante, Alicante, Spain

Shiliang (Johnathan) Tan School of Chemical and Biomedical Engineering, Nanyang Technological University, Singapore, Singapore

Artur P. Terzyk Faculty of Chemistry, Physicochemistry of Carbon Materials Research Group, Nicolaus Copernicus University in Toruń, Toruń, Poland

Shuwen Wang Research Initiative for Supra-Materials, Shinshu University, Nagano, Japan

Wai Fen Yong Technische Universität Dresden, Institut für Anorganische Chemie I, Dresden, Germany;
School of Energy and Chemical Engineering, Xiamen University Malaysia, Selangor Darul Ehsan, Malaysia;
College of Chemistry and Chemical Engineering, Xiamen University, Xiamen, Fujian, China

Monika Zięba Faculty of Chemistry, Physicochemistry of Carbon Materials Research Group, Nicolaus Copernicus University in Toruń, Toruń, Poland

Wojciech Zięba Faculty of Chemistry, Physicochemistry of Carbon Materials Research Group, Nicolaus Copernicus University in Toruń, Toruń, Poland

Metal-Organic Frameworks for Environmental Applications

Tanay Kundu, Leisan Gilmanova, Wai Fen Yong, and Stefan Kaskel

Abstract Environmental remediation is a worldwide concern to reverse the impact of rapid urbanization. Outstanding porosity, tunable pore structure and unique flexibility render metal-organic frameworks (MOF) and MOF based separation technologies as promising for air and water filtration. In this chapter, we summarize the design of MOF based adsorbents for pollutant removal. A special emphasis is given on their stability, membrane and fixed bed forming ability, role, performance and mechanism. In addition, limitations and further improvement scopes are discussed in comparison with the current state of the art systems.

1 Introduction

The rapid urbanization and industrial revolution mark the unprecedented growth of human civilization. On the other hand, ever-growing release of toxic pollutants into the environment poses a worldwide risk on flora and fauna. Excessive use of fossil fuels, chemical processes and industrial discharges are major sources for the contamination of air and water [1, 2]. For example, gases of NO_x and SO_x family cause respiratory diseases and when combined with water create inorganic acids that corrode and disrupt soil pH balance. Water pollution is mostly caused by discharge of dyes and personal care products [3]. As a result, air filtration and water purification became one of the major concerns and research highlights for global scientific communities.

T. Kundu · L. Gilmanova · W. F. Yong · S. Kaskel (✉)
Technische Universität Dresden, Institut für Anorganische Chemie I, Dresden, Germany
e-mail: stefan.kaskel@tu-dresden.de

W. F. Yong
School of Energy and Chemical Engineering, Xiamen University Malaysia, Selangor Darul Ehsan 43900, Malaysia

College of Chemistry and Chemical Engineering, Xiamen University, Xiamen 361005, Fujian, China

J. C. Moreno-Piraján et al. (eds.), *Porous Materials*, Engineering Materials,
https://doi.org/10.1007/978-3-030-65991-2_1

Currently, the state of the art air and water purification technologies are incineration, oxidative degradation and membrane/fixed-bed separation [4]. Among them, separation is the most environmentally benign option, with the advantage of low cost of operation, low energy penalty and no secondary product generation. The adsorptive capture and removal process of harmful chemicals rely on the total uptake, size/shape selectivity and polarity of the adsorbent bed. Based on the interaction energy, the process can be classified as physisorption (mainly van der Waals force) or chemisorption (chemical bond formation between the adsorbate and adsorbent). The physisorption capacity is dependent on the surface area, pore size and aperture, whereas the chemisorption depends on the surface functionality of the adsorbent [5]. Till date, activated carbon, porous silica and zeolites are the main porous materials that have been used in industry as adsorbents for pollutants owing to their inexpensiveness, availability and good capacity.

Metal-organic frameworks (MOF) are a relatively new class of hybrid porous materials, made of inorganic secondary building units (SBUs, can be ions or clusters) coordinatively joined by organic ligands. They feature ultrahigh porosity, controllable pore size/shape and homogeneously dispersed functional groups and active sites for suitable applications. Such advantages make MOFs very attractive candidates for uptake and separation of toxic chemicals, which have outperformed the traditional porous materials in terms of adsorption capacity and selectivity [6]. Both physisorption and chemisorption interactions have been utilized for designing next generation MOF-based adsorbents for environmental remediation [5].

This chapter deals with the different MOF designs and their processing as membranes and fixed bed for pollutant removal. Firstly, several necessary features of MOFs such as open metal sites and specific host guest-interaction have been introduced and their use in different chemical capture in pure or competitive environments has been discussed. Then, fixed bed design criteria and breakthrough measurements are introduced and various shaping technologies that facilitate the industrial transition of MOF research for lab-scale have been outlined in the light of air and water purification capabilities. Finally, MOF based mixed matrix membranes has been discussed in detail for water and gas treatment. At the end, we conclude the findings and future design perspectives of MOF based adsorbent for pollutant removal.

2 Design Strategies of MOFs

2.1 Structural MOF Building Blocks

Linker design, diversity of metal nodes and topologies allow to create MOFs with predetermined functions. For the environmental application of MOFs, such as sensing, filtration, adsorption and degradation of toxic gases and vapours, suitable materials have to be explored thoroughly. The most important features of the adsorbent for those tasks are pore size, capacity, selectivity and stability in ambient conditions.

Economic and ecologic synthesis of target MOFs and their easy scalability are highly important for the industry concerning applications such as adsorption and degradation of hazardous gases [7]. Therefore, using inexpensive and stable systems is reasonable. Thus, a number of MOFs with commercially available linkers (benzenedicarboxylate, biphenyldicarboxylate, terephthalic acid, dihydroxyterephthalic acid etc.) and various metal sites was examined for efficient and selective toxicants adsorption [1, 7, 8].

Specific interactions of the absorbent with pore walls (π-stacking, Van der Waals interactions, hydrogen bonding, Lewis acid-base complexes etc.) enhance adsorption efficiency. Such interplays are provided by MOFs, containing coordinatively unsaturated sites (CUS) or functionalised pore surface. The latter can be achieved using linkers with additional functionalities such as amino or hydroxyl groups [9, 10]. More efficient and widespread environmental application found MOFs with coordinatively unsaturated secondary building units (so called open metal sites). Unsaturated metal sites enhance physisorption properties by the attractive forces occurring due to the partial positive charged on metal sites [11].

Active site creation and characterization were described in detail in a very recent review by Hall and Bollini on the example of three prototypical MOFs with open-metal sites—Cu-BTC, UiO-66, MIL-100/101 (Fig. 1) [12].

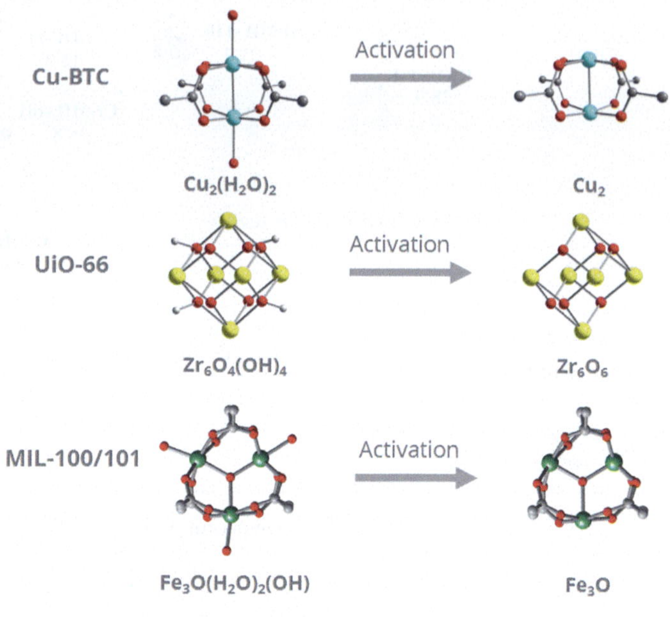

Fig. 1 Creation of open-metal sites in Cu-BTC, UiO-66 and MIL-100/101 [12]

2.2 Stability Under Operation Conditions

It is common knowledge that poor stability of MOFs brings limitations for their application. It remains a sophisticated task and challenge to synthesize MOFs with resistance to degradation under thermal and humid conditions, and chemically and mechanically stable MOFs [13]. De novo synthesis and post-synthetic modification are promising approaches to improve the stability of MOFs. Latest advances in the design and synthesis using these methods were discussed in the review of Ding et al. [14]. Stability tests against air, water, acid/base were reported and reviewed within the last decade to reveal highly stable MOFs such as MIL-53(Al, Cr), Al-fum, MIL-96(Al), MIL-100(Cr, Fe), MIL-101(Cr), ZIF-8, UiO-66, MOF-74, NOTT-300 and some others [3, 15–20] (Fig. 2).

Water stability of MOFs is perhaps the main concern for the real-world environmental applications among all other dimensions. Weak coordination between the metal and organic linkers in MOFs lead to easy interaction of water molecules with metal clusters through different degradation mechanisms [3]. Using high-oxidation metal ions, water-protective groups on ligands [20] and hydrophobicity enhancement [21] improve the stability of MOFs in aqueous environment and expand the range of their application [22–24].

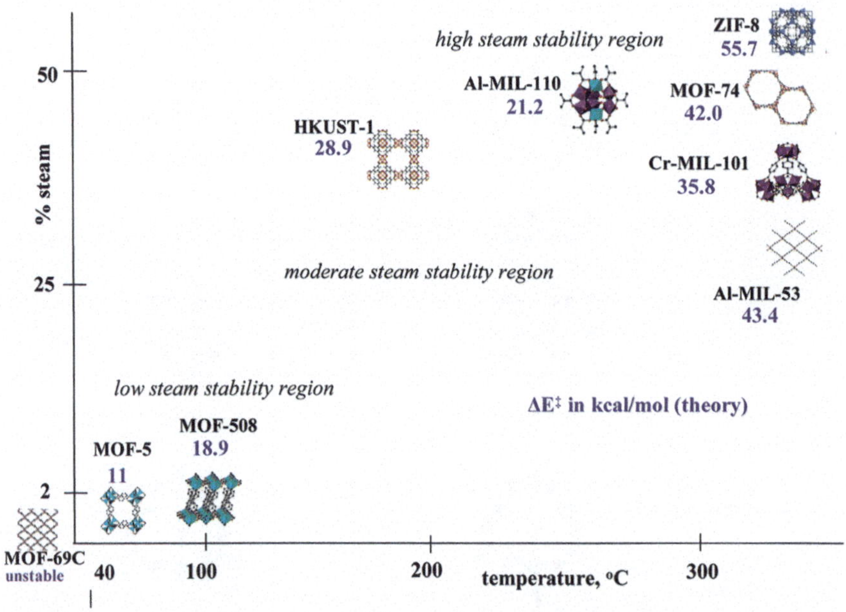

Fig. 2 Steam stability map for some MOFs. The position of the structure for a given MOF represents its maximum structural stability by XRD measurement, while the activation energy for ligand displacement by a water molecule, determined by molecular modelling, is represented by the magenta number in kcal/mol. Reprinted with permission from [21]

Another matter of concern is reusability. Some MOFs show great pollutant adsorption and degradation, but are only usable in a very limited number of cycles. Thus, the fresh HKUST-1 shows 92% of CEES degradation in the first cycle, but this value is decreased to 45% in the 4th cycle. It cannot be reactivated fully afterwards due to the strong host-guest interactions and the partial irreversible collapse of the structure [11]. Despite regeneration of adsorbents is not necessarily required for respiratory filters, a general drawback of MOF adsorbents is poor recyclability as compared to activated carbons. Probably, only chemical recycling (recovery of linker and metal) is reasonable.

2.3 *Specific Host-Guest Interactions and Selective Adsorption of Toxicants*

Specific interactions between the hazardous guests and the host are essential for adsorption selectivity. Adsorption of certain pollutants and groups of toxicants was discussed comprehensively in reviews by Woellner et al. and Bobbit et al. Due to the configurable reactivity of pore walls of MOFs, both physisorption and chemisorption are possible for these materials [1, 7, 25].

The importance of open metal sites (OMS) in MOF structures for adsorption properties was mentioned above. OMS act as Lewis acid and bind base molecules such as NH_3 and PH_3. Ammonia adsorption was studied extensively on a number of MOFs, since it is a common toxic industrial chemical (TIC) [26, 27]. MOF-74 was found to adsorb ammonia very effectively due to the high density of Cu (II) sites [28]. However, loss of crystallinity in humid conditions limits reusability of the material after one cycle. The same MOF exhibits high phosphine capacity at low pressure regimes [29]. MOF-74(Mn) in particular can be recycled and reused after mild room temperature evacuation. The presence of OMS in a coordination polymer has a particularly pronounced effect on the NO adsorption [30]. Nitric oxide adsorption was tested on several MOFs (HKUST-1, MOF-74(Zn, Mg, Co, Ni) [31, 32]. M(II)-NO complex formation was spectroscopically observed and binding energies were calculated [33]. Sun et al. reported the removal of SO_2, NO_x, and CO_2 from flue gases. Mg-MOF-74 was recommended as the only potential material for this separation due to its high density of open-metal sites which provide strong interactions with dipole and quadruple molecules [34].

The availability of open metal sites is not always the one and only circumstance for efficient toxicant adsorption. Additional amino-group functionalities on linkers show great results in terms of hazardous gas adsorption. Hereby UiO-66-NH_2 removes nitrogen dioxide at extraordinary levels in both – dry and humid conditions, by diazonium ion formation [4]. Also, UiO-66-NH_2 improves pyridine adsorption parameters in both vapor and liquid phases in comparison with unfunctionalized UiO-66, because of hydrogen bonding interaction between the pyridine and the amino group of the MOF [35]. NH_2-MIL-53(Al) was reported recently showing high adsorption efficiency and good selectivity towards toxic Hg^{2+} due to the strong coordination between amino groups and Hg^{2+} [36]. MFM-300(In) was reported as highly selective MOF material with respect to SO_2 in a presence of CO_2, CH_4 and N_2 under ambient conditions (i.e., 50:50 mixture at 1 bar and 298 K) [10]. The MFM-300(In) structure consists of $[InO_4(OH)_2]$ chains and tetracarboxylate linkers. A study revealed, that adsorbed SO_2 forms multiple specific supramolecular interactions with the free hydroxyl groups and aromatic rings on the pore surface of this material. Much higher strength of those interactions in comparison with CO_2 and N_2 gives observed selectivity.

For volatile organic compound (VOC) adsorption, for example aromatics or toxic dyes [37] π–π interactions are significant. Selective or preferential adsorption of benzene in comparison to water was found for HKUST-1 [38]. Higher isosteric heat and activation energies of desorption of benzene and stronger interactions with surface led to reported results.

2.4 Water Adsorption and Competitive Adsorption

Since interaction of the coordination polymer with water is a matter of concern for environmental application, water adsorption was studied on a high number of MOFs, including stable, promising ones such as MOF-74, UiO-66 and its analogues MOF-801 series, MIL-100 and MIL-101 [17, 18, 39–44]. Burtch et al. concluded that OMS play a crucial role in water adsorption for microporous rigid MOFs, attracting water molecules in lower pressures and giving type I isotherms [17]. Microporous MOFs without OMS behave slightly hydrophobic, filling up pores with water in the pressure range of 0.2–0.4 p/p_0. Macroporous rigid MOFs undergo capillary condensation in a different pressure range depending on pore size and result in a type-V isotherm. Water adsorption studies help to get better understanding which MOFs can be used in aqueous environment adsorption and which MOFs are suitable for water competitive adsorption. For example, for VOC adsorption, the sorbent should have low water uptake ate least below 60–70% RH. Otherwise, the capacity for VOC under ambient conditions will be quite limited. Competitive multicomponent adsorption was discussed in detail by Burtch et al. [17]. Carbon dioxide gas adsorption is the most studied pollutant, since it is a major source of global warming. In cabin air, CO_2 regulation significantly improves the productivity and human concentration ability. Interestingly, water vapour presence can influence CO_2 adsorption not only in negative, but also negligible or in particularly cases positive way [45]. Thus, MOFs having OMS show high affinity to water, may show reduced CO_2 capacity due to competitive adsorption with water vapour. However, carbon dioxide uptake increasing phenomena after water pre-adsorption was observed for HKUST-1 (Fig. 3) [46]. It was concluded from GCMS simulations, that electrostatic interactions between the quadrupole moment of CO_2 molecules and the electric field gradient of the sorbent increases, after occupation of Cu(II) OMS by water molecules.

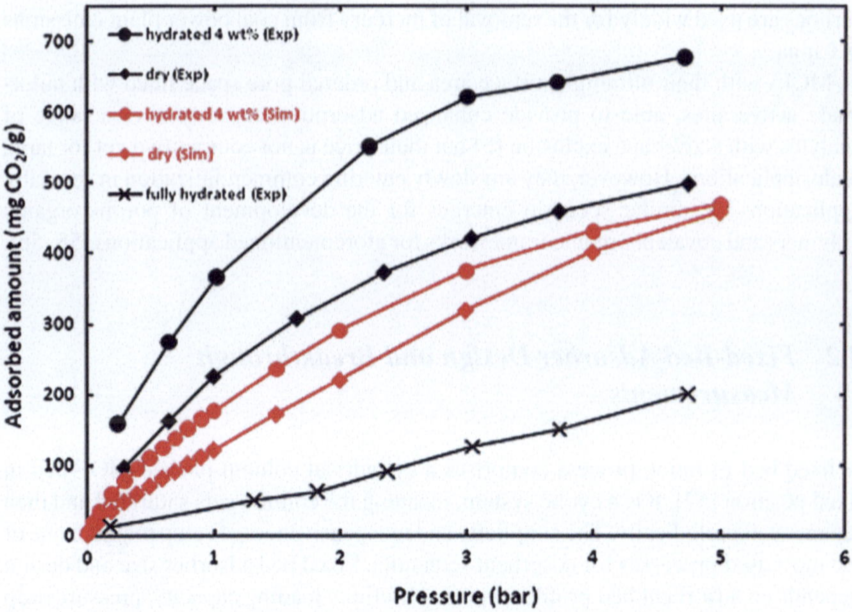

Fig. 3 Simulated and experimental adsorption isotherms for CO_2 at 298 K in Cu-BTC with varying water contents. Reprinted with permission from American Chemical Society [46]

3 MOFs in Fixed Bed Filter Applications

3.1 Background

Adsorptive removal of analytes in large scale by fixed bed filters attract increasingly worldwide attention for challenging yet relevant systems, like polluted air and wastewater treatment [2, 47–49]. Conventional adsorbents such as porous carbon, mesoporous silica and zeolites offer easy synthesis, processing, stability and good performance for air and water purification. For example, natural zeolites (e.g. mordenite, chabazite, laterite), single and binary oxides (e.g. alumina, Fe–Mn oxides) can capture harmful gases from air and heavy metals from wastewater, by adsorption and/or ion-exchange mechanism [50–52]. Porous carbons have also been extensively explored, especially derived from natural resources (e.g. coconut shell, wood, fruit stones, rice husk, dry leaf, saw dust), providing inexpensive yet effective solutions for thriving economies [53]. Mechanical properties of these adsorbents have been utilized for different shaping (granular beads, tablets etc.), suitable for application needs [54]. Modern purification requirements demand filtration down to ppb level, long lifetime, able to trap multiple analytes with different polarity, selectivity and recyclability, which is difficult to obtain by traditional adsorbents. Wood derived carbons are used widely for the removal of mercury from coal power plant emissions in China.

MOFs with their ultrahigh surface area and ordered pore space lined with tailor-made active sites, able to provide enhanced adsorption capacity over a range of analytes with size/shape exclusion [5] but their price is not competitive yet for large scale applications. However, they are slowly entering commercialization in specialty applications. A similar scenario emerges for the development of porous organic polymers and covalent organic frameworks for aforementioned applications [55, 56].

3.2 Fixed-Bed Adsorber Design and Breakthrough Measurements

A fixed bed filtration process comprises a cylindrical column-packed MOF bed in fixed position [57]. It is a cyclic system, meaning the column gets saturated and then regenerated periodically. The simplicity and inexpensiveness of setup makes it one of the most used processes for adsorbent screening. Fixed bed adsorber size and design depends on adsorbent bed profile, expected lifetime, loading capacity, pressure drop and reaction kinetics. Three regions define the profile of a fixed bed, viz. equilibrium zone (part of the column in equilibrium with feed flow), mass transfer zone (where active adsorption takes place) and active zone (where bed is untouched by feed). Regeneration of MOF bed can be performed either by heating (temperature swing adsorption) or pressure change (pressure swing adsorption), while the former is more effective.

Breakthrough experiments are essential for evaluating the adsorption efficiency of MOFs. Typically, packed adsorbent beds or columns of MOFs are eluted with analytes, and the efficacy is determined by the time the analyte reaches a certain limit at the other end, termed as breakthrough time (T_b). Adsorbents for air purification are evaluated against their total capacity (measured from adsorption isotherm using analyte) and T_b.

When a gas stream containing analyte is passed through the adsorbent column, it first interacts with peripheral layers (stage 1, Fig. 4). Then, the mass transfer zone moves forward (stage 2, Fig. 4) until breakthrough is detected. Further gas flow results in gradual exit of the analyte through bed (stage 3, Fig. 4), and after complete saturation reaches an exhaustion limit (stage 4, Fig. 4). The adsorption capacity is dependent on gas flow rate, temperature, concentration and shaping of the MOF, while the gas diffusion rate being the rate determining step [58]. Breakthrough time can be calculated from modified Wheeler equation,

$$T_b = \left(\frac{W.C_a}{Q.C_0}\right) - \left(\frac{D.C_a}{R.C_0}\right) \ln\left(\frac{C_a - C_1}{C_0}\right) \quad (1)$$

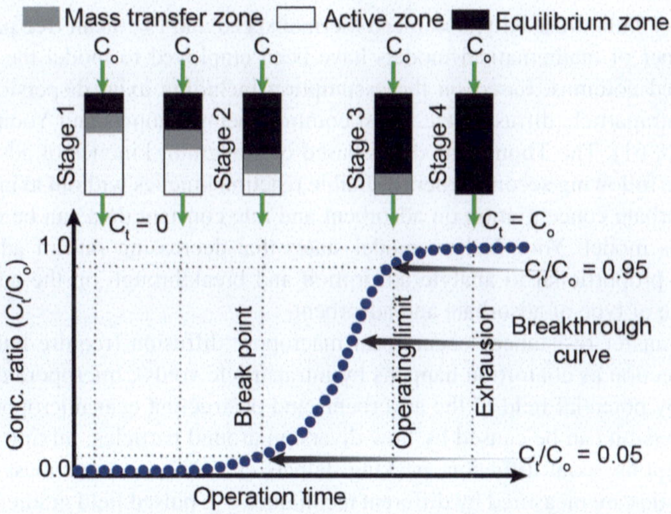

Fig. 4 Schematic representation of processes inside a breakthrough apparatus

W = weight of the MOF, C_a = analyte adsorption capacity of the MOF, C_o = analyte inlet concentration, Q = flow rate, D = MOF density, R = adsorption rate coefficient, C_1 = breakthrough concentration.

For trace gas toxic gas adsorption channelling is also a critical issue. In this case the concentration before breakthrough reaches a critical threshold value. Channelling may occur if the bed is inhomogeneous, the flow distribution deviates significantly from ideal plug flow (ill defined front), or adsorption is partially kinetically hindered.

3.3 *Transport Mechanism*

In MOFs, the rate of mass transfer within different pores and with the outer space are the key factors for determining transport mechanism. The diffusion or transport phenomena in MOFs can be modelled by molecular dynamic simulation studies and quasi-elastic neutron scattering experiments [59]. These measurements can take weeks for example in case of sorption hysteresis dynamics over femtosecond scale. The main phenomena of mass transfer occur first from bulk to the solid surface by diffusion or convection and then gets adsorbed in the inner surface-by-surface diffusion with heat release. The flow rate may not be the same throughout the MOF bed, due to temperature gradient generated by adsorption and role of bed-wall interphase. An estimation of mass transfer rate through fixed-bed of MOF can be obtained from the following equation,

$$D_b = \frac{1}{3}\overline{u}.l \tag{2}$$

Where D_b = bulk diffusivity, $\bar{u}$ = mean thermal speed and l = mean free path.

A number of mathematical models have been employed to model the efficacy of fixed bed columns, based on the assumption including axial dispersion, mass transfer, intraparticle diffusion etc. Most common being Thomas and Yoon-Nelson models [60, 61]. The Thomas model is based on Langmuir kinetics of adsorption-desorption, following second-order reversible reaction kinetics without axial dispersion. Adsorbate concentration on adsorbent and rate constant data can be obtained by Thomas model. Yoon-Nelson model states that decreasing rate of adsorption is directly proportional to analyte adsorption and breakthrough on the adsorbent, irrespective of type of adsorbate and adsorbent.

Mass transfer resistances occur from macropore diffusion (require bulk diffusivity correction as contortion happens by intraparticle voids), micropore diffusion (induced by potential field of the adsorbent) and obstruction near micropore openings. Dispersion can be caused by flow diversion around particles, mixing in interparticle regions, axial diffusion, packing defects etc. Diffusion and mass transfer characteristics are measured by different techniques e.g. pulsed field gradient NMR, quasi-elastic neutron scattering, microimaging by IR (interference microscopy) and frequency response [62]. Smaller particle sizes overcome transport limitations by offering higher molecular exchange rate. Hierarchical MOFs are also emerging as they can be considered as fused crystallites or pelletized MOFs with a network of transporting pore spaces to enhance mass transfer [63].

3.4 MOF Granulation and Shaping

Conventional large-scale MOF synthesis results in fluffy low-density powder after activation, unsuitable for engineering processes due to pressure drops in packed beds, abrasion etc. Thus, MOFs must be pre-processed into granules, pellets or monoliths for standard industrial operations (Materials Center, 2020). However, such mechanical treatment and binder additives lead to trade-off regarding porosity, as most MOFs show 20–30% loss in uptake capacity upon compaction, along with changes in pore volumes, crucial for separation applications [64]. Moreover, use of binder in most cases impart another level of complexity in the resulting properties of the composite. Thus, optimisation of such treatments must be performed on a case to case basis. For example, separation applications require good crushing strength and hardness, while devices and sensors need high wear resistance and Young's modulus.

In this section, we will briefly discuss different parameters for MOF shaping and a few examples with reported values. A detailed discussion can be found in recent review articles [65, 66]. Although shaping is suitable for many industrial applications, we will limit our discussion to the separation aspect based on the relevance of this chapter.

3.4.1 Mechanical Properties

Elastic Moduli

Based on the direction of elastic force and stress-strain curve, three most important parameters in MOF research are Young's, bulk and shear modulus. Young's modulus (Y) is defined as the ratio of applied stress and generated strain along a defined axis. It is expressed as, Y = (F/A)(dL/L), where F = applied force, A = stress area, dL = length change and L = original length. Bulk modulus (B) is the ratio of volumetric strain and stress, defined as B = V * (dP/dV) where V = initial volume, dP = differential pressure change and dV = differential volume change. Shear modulus (S) is similar to Young's modulus applied parallel to the surface, and defined as S = (F/A)(dx/L) where F = shear force, A = area parallel to shear force, dx = displacement and L = unit length of sample. In addition, strength or deformation resistance of MOF pellets are measured in terms of ultimate tensile strength, which comprise the highest value in the stress-strain curve. In industry the most practical parameter is the crushing strength. It is determined by placing a defined load on a spherical granule. Typical crushing strength values range from 30 to 70 N for zeolites and 10–30 N for MOFs.

Attrition

The attrition resistance is the measure of stability of MOF pellets or granules against different forces in packed bed columns. Shaped MOFs are rolled in a cylinder at fixed rpm and the breakage/deformations are recorded over a certain time.

$$Attriction(A) = \frac{Initial\ weight - weight\ recovered}{Initial\ weight} \times 100\% \tag{3}$$

3.4.2 Different Techniques for Shaping of MOFs

Shaping techniques depend on the specific requirement of adsorption process, such as mass transfer, diffusion kinetics and pressure drop. Granulation process can be either wet or dry (presence or absence of solvent during the compaction process). Wet granulation typically done in shear mixer or extruder, while dry granulation is achieved using tablet or pellet press. Extruded monoliths were first reported by Küsgens (2010). Both granulation methods are compared using three prototypical MOFs, viz. CPO-27(Ni), MIL-100(Fe), and HKUST-1 for ammonia capture in respiratory filters, where wet granulation shows higher attrition resistances [67]. Regarding tensile strength, a linear relationship with pellet density can be observed in common MOFs below 80% relative density, then decreases readily with decrease

Fig. 5 Photographs of shaped granules of MOFs with 5 wt% alumina binder: **a** MIL-100(Fe), **b** MIL-101(Cr), **c** UiO-66(Zr), and **d** UiO-66(Zr)-NH_2. Reproduced with permission from [68]

in specific surface area. For example, 2–2.5 mm MIL-100(Fe), MIL-101(Cr), UiO-66(Zr), and UiO-66(Zr)-NH_2 granules with 5% mesoporous alumina binder was prepared using wet granulation (Fig. 2), showing good NH_3 and CO_2 adsorption performance (Fig. 5) [68].

3.5 Application of MOFs Fixed Bed for Air Purification

Maintaining air quality is a critical aspect in the modern world. Continuous release of pollutants [NO_x, SO_x, CO_x, H_2S, NH_3, volatile organic compounds (VOCs) and polycyclic aromatic compounds (PAHs)] in airstream leads to respiratory illness, lower immunity, neurological damages and death in many cases [69, 70]. The versatility of MOFs such as regular adsorption sites, selectivity and high thermochemical stability made it a promising choice for air purification application.

Carbon monoxide is a lethal gas, mostly generated from incomplete combustion. Iron based MOF, Fe-BTTri shows high CO uptake and IAST selectivity of 7400 for CO/H_2, due to unique spin-transition of Fe that is not possible with another gas [71]. For 0.05 mol fraction of CO in N_2 and CO_2, the IAST selectivity reaches 1750 and 193 for 1 bar, respectively. The CO/N_2 selectivity of Cu(I) loaded MIL-100(Fe), denoted as 0.8Cu(I)@MIL-100(Fe) was 169 at 1 bar, 112 times higher than pristine MIL-100(Fe) [72].

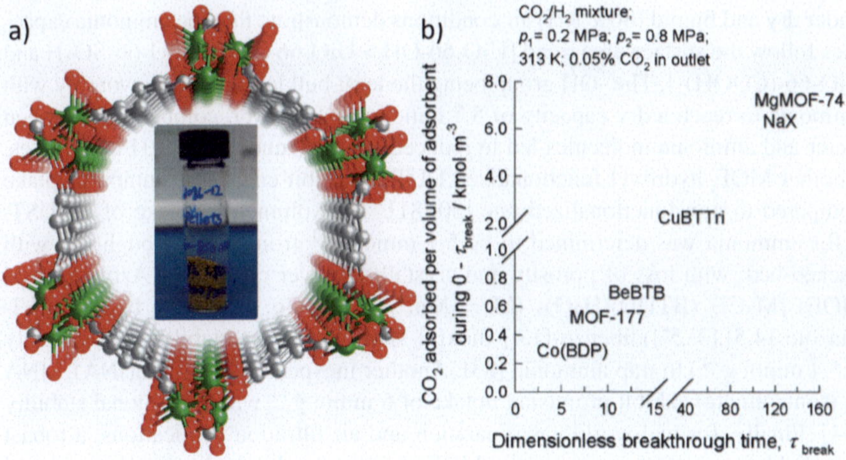

Fig. 6 **a** MOF-74 structure with unsaturated metal sites (inset photograph of the pellets of Ni-MOF-74). **b** Breakthrough measurement in low concentration of CO_2 to assay CO_2 affinity of different state-of-the-art MOFs and carbon, demonstrating highest affinity for Mg-MOF-74. Permission obtained from [57]

Modern industrialization led to extensive use of fossil fuels, elevating CO_2 levels in air that spark CO_2 capture research. Adsorptive removal of CO_2 from air is conventionally done by amine scrubbing, which attracts high regeneration cost. MOF relies mainly on unsaturated metal sites, and polar interactions for high CO_2 uptake [73]. Unsaturated Mg-sites endow Mg-MOF-74 a high CO_2 uptake capacity of 8.9 wt% (Fig. 6) and low regeneration temperature of 80 °C [6, 57]. In order to capture ultra low concentration of CO_2 from air (407 ppm) or flue gas, diamine incorporation inside Mg-MOF-74 leads to steep uptake at very low partial pressure of CO_2 at 1 bar [74]. Analogous diamine loaded Mg-DOBPDC exhibit 13.7 wt% CO_2 uptake at 1 bar air with 400 ppm CO_2 [75]. Carbamic acid formation and cooperativity within the appended diamines are responsible for such high affinity [76]. Similar high uptakes were also observed for two azolate based Cu-MOF, viz. Cu-BTTri and Cu-BTTri(en) [77]. Optimizations are being carried out to tune heat of adsorption in between physisorption and chemisorption (30–50 kJ mol^{-1}) to balance affinity and regeneration temperature [57]. Another MOF, Ni-MOF-74 demonstrated excellent pellet formation (Fig. 6a inset) and useful characteristics in fixed-bed breakthrough studies [78]. High CO_2 uptake, selectivity (CO_2/N_2 and CO_2/CH_4) and performance under competing agents (3 wt% water) was also tested.

For ammonia capture, acidic functional groups (–OH, –COOH) and metal ions (Cu(II), Ag(I) etc.) are advantageous for high uptake. Zr-based MOFs (UiO-66) are widely tested for air purification, due to their high thermal, chemical and mechanical stability. Functional varieties of UiO-66 [e.g. UiO-66-OH, UiO-66-$(OH)_2$, UiO-66-NO_2, UiO-66-NH_2, UiO-66-SO_3H and UiO-66-$(COOH)_2$] were probed for ammonia removal from air [79]. Ammonia breakthrough measurements under dry and humid (80% RH) air conditions demonstrate that the ammonia capacities follow the surface area trend [UiO-66-OH > UiO-66-NH_2 > UiO-66-SO_3H and UiO-66-$(COOH)_2$]. The -OH group being the least bulkier, interacts favorably with ammonia to reach a dry capacity of 5.7 mmol g^{-1}. However, competition between water and ammonia molecules led to reduced capacity under 80% RH in all cases. Another MOF, hydroxyl functionalized DUT-6, exhibit enhanced ammonia uptake compared to non-functionalized one [80, 81]. The volumetric uptake of HKUST-1 for ammonia was determined to be 5.1 mmol g^{-1} from breakthrough test with packed-bed, with loss of porosity and crystallinity over cycle [82]. Azolate based MOFs [M_2Cl_2 (BTDD)$(H_2O)_2$ (M = Mn, Ni, and Co; BTDD = bis(1H-1,2,3-triazolo [4,5],[4',5'])dibenzo-[1,4] dioxin] show necessary stability and capacity (15.4 mmol g^{-1}) to trap ammonia [83]. Another inexpensive MOF Zn$(INA)_2$ (INA = isonicotinate) exhibit promising uptake of 6 mmol g^{-1} with operational stability [84]. Finally, for real-world gas separation and air filtration applications, a robust yet pliable support is necessary for MOFs incorporated, which was demonstrated by Kaskel and co-workers [85]. They have used chitin networks derived from marine sponge as a support material for MOF impregnation. A continuous deposition of HKUST-1 particles onto the network was observed (Fig. 7a, b), without loss in crystallinity and specific surface area. Ammonia breakthrough measurements showed NH_3 capacity of the composite being 39.3 mg g^{-1} (Fig. 7c). In another instance, HKUST-1 coated on polypropylene mat shows 36 times higher NH_3 capacity compared to pristine polypropylene [86].

Nitrogen oxides are major pollutants, causing acid rain, smog and ozone depletion. NO_2 removal potential from air has been measured through breakthrough setup. Among BPL carbon, UiO-66 and UiO-66-NH_2, later show the longest elution time and saturation capacity of 0.9 g g^{-1}, which further increases to 1.4 g g^{-1} in 80%

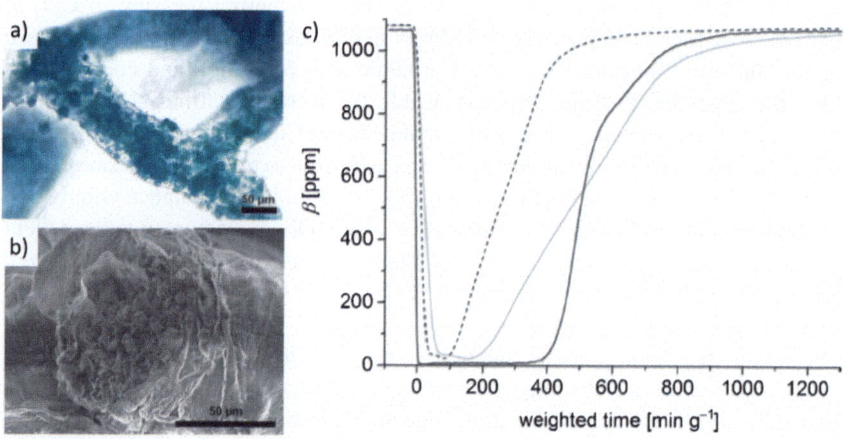

Fig. 7 **a** Microscopic and **b** SEM images of HKUST-1 coated on chitin network. **c** Breakthrough measurement with NH_3 for pure HKUST-1 (black), composite (dashed gray), and for composite normalized with the amount of HKUST-1 (gray). Reproduced with permission from [85]

humidity due to reaction with the framework [9]. Urea modified UiO-66 afforded high NO_2 adsorption capacity of 101 mg g^{-1}, which further increases to 154 mg g^{-1} in humid condition [87]. Oxalic acid modification in UiO-66 also doubled NO_2 uptake (8.4 mmol g^{-1}) compared to unmodified UiO-66 [88].

SO_2 removal from air is important, as it is a major component for acid rain, corrosion and respiratory illness. Schröder group synthesized In-based MOF, NOTT-202a for high SO_2 uptake of 13.6 mmol g^{-1} at 1 bar and 268 K followed by irreversible phase change to NOTT-202b [89]. Another MOF, MFM-300(In), shows SO_2 uptake of 8.28 mmol g^{-1} at 298 K and 1 bar [10]. The IAST selectivity of MFM-300(In) for 1:1 binary mixture of SO_2/CO_2, SO_2/N_2, and SO_2/CH_4 are 60, 5000 and 425, respectively. The free hydroxyl groups and aromatic rings interact with SO_2 to afford molecular level selectivity and framework robustness. Another important candidate is SIFSIX-1-Cu, showing high SO_2 capacity of 11 mmol g^{-1} is achieved at 298 K and 1 bar, and high uptake at very low partial pressure due to electrostatic dipole-dipole interaction [90]. SIFSIX-2-Cu-i shows high SO_2/CO_2 IAST selectivity of 86 within 0.1–0.9 mol fraction of SO_2, while the SO_2/N_2 selectivity of SIFSIX-1-Cu is 2510 [90].

H_2S is corrosive and flammable gas, and can be poisonous even at ppm level. Gallium based MOF, Ga-soc-MOF shows highly selective H_2S retention (40 min) over CO_2 (5 min) and CH_4 (<1 min), evident by breakthrough tests [91]. Another rare earth MOF, Y-1,4-NDC-fcu-MOF showed a much higher retention time for H_2S (70 min) over CO_2 (11 min) due to smaller pore aperture [92]. Similarly, a tetrazolate MOF kag-MOF-1, shows high uptake of H_2S, CO_2 and H_2O, and rejects organic molecules [93].

3.6 *Application of MOFs Fixed Bed for Wastewater Treatment*

Drinking water quality maintenance is one of the major challenges for modern society. Affordable water purification technology requires easy setup, economic operation and good filtration capability [4]. Arsenic, iron, lead, cadmium and fluoride are main inorganic contaminants, while pharmaceuticals and personal care products (PPCPs), dyes and pesticides are common organic pollutants [2, 3]. Reverse osmosis, membrane filtration, precipitation and adsorption are most common techniques for water purification. Among them, adsorption is the most commonly used, in the form of zeolite and activated carbon adsorbents.

MOFs with their uniform pore window, high surface area, structural and functional tunability, offer a unique and efficient design platform for adsorptive removal of diverse contaminants [2]. MOFs have greater promise for trapping different contaminants than conventional adsorbents due to multiple interaction possibilities (e.g. electrostatic, hydrophobic, ionic, π-π and hydrogen bonding interactions) (Fig. 8d) [94]. For example, ZIF-8 (surfactant assisted synthesized) exhibits adsorption capacity

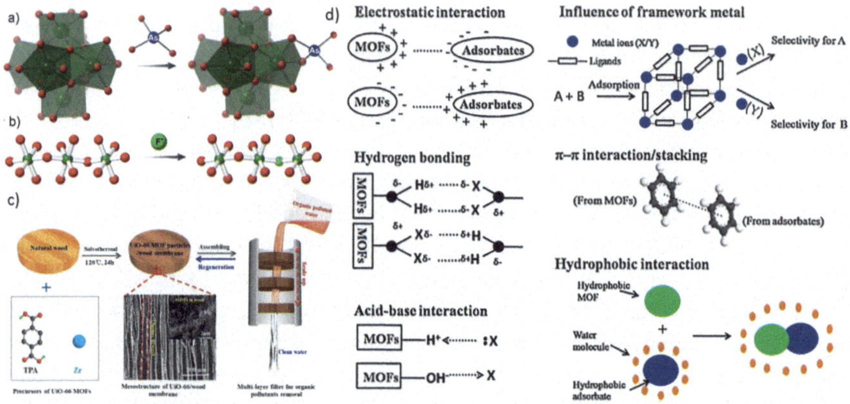

Fig. 8 **a** Proposed mechanism of **a** Arsenate removal by substitution at Zr-oxo/hydroxido clusters and **b** fluoride containment by substitution at hydroxo-bridged Al-chains in MOFs. **c** A wood/UiO-66 composite for column purification of water. **d** Schematic diagram of different interactions operating for pollutant removal involving MOFs. Permission obtained from [94, 115]

up to 91 mg g^{-1} for As(V), exceeding activated carbon (3 mg g^{-1}) and zeolite (13 mg g^{-1}) filters [95]. Similarly, MIL-53 (Al), MIL-88A, MIL-88B and Fe/Mg MIL-88B shows good uptake of 106, 145, 156 and 303 mg g^{-1} respectively for As(V) from aqueous solution [96]. Zn-MOF-74 shows adsorption capacity up to 325 and 211 mg g^{-1}, while mixed-metal MOF-74 viz. Fe_2Co-MOF-74 exhibit 266 and 292 mg g^{-1}, respectively for As(V) and As(III) [97]. $CoFe_2O_4$@MIL-100(Fe) composite can remove As(III) and As(V) (144 and 115 mg g^{-1}, respectively) by forming Fe-O-As complexes via -OH groups [98]. Zr-based MOF (UiO-66) demonstrates excellent stability and adsorption capacity of 205 and 68 mg g^{-1} for As(III) and As(V), respectively employing similar mechanisms (Fig. 8a) [99].

Fluoride is another common contaminant present in ground water and difficult to remove [100]. The main mechanism of fluoride removal by MOF is replacement of framework anions or hydroxyl groups (Fig. 8b). MOF-801, a Zr-based MOF, shows moderate adsorption capacity of 40 mg g^{-1}, with excellent pH tolerance, temperature range of 293–323 K and effective in presence of other counteranions [101]. UiO-66(Zr) shows similar performance of 41 mg g^{-1} [102]. Aluminium-fumarate MOF shows excellent adsorption capacity of 600 mg g^{-1}, but small particle size of the MOF hinders fixed-bed operational perspective [103]. MIL-96, another Al-MOF, demonstrates 31 mg g^{-1} removal of fluoride [104].

Iron and other inorganic species (Al, Hg, Cd etc.) are also present in trace level, and can be toxic above a certain level upon daily usage. Co-TATAB, a MOF having both carboxylate and secondary amine site, shows common metal ion impurity removal [Al(III), Hg(II), Fe(III), Cd(II)], a wide pH range (2–6) and enhanced capacity for Al(III) and Fe(III) at elevated exposure time and pH [105]. Another Cu-terephthalate MOF composite with graphene oxide and mordenite zeolite also shows heavy metal removal with similar pH dependence as Co-TATAB [106].

Organic pollutants are an increasing menace to the environment and water due to their functional diversity and limited knowledge about proper treatment. Various interaction modes of MOFs enable adsorption of a wide range of contaminants from water. For example, diamine functionalized MIL-101(Cr) can adsorb naproxen and clofibric acid by acid-base interaction, and can be regenerated by ethanol wash [107]. Tetracycline, a potent antibiotic, can be trapped by an In-based MOF with tetrafluoroterephthalate linker via π-π interaction [108]. Another mesoporous MOF, PCN-222(Fe) can exhibit uptake capacities of 854, 812 and 417 mg g^{-1} for brilliant green, crystal violet, and acid red dyes, respectively, which is much higher than conventional activated carbons [109]. Ion exchange is another common mechanism, where anionic MOFs prefer cationic dyes and vice versa. For example, a dimethylammonium ion containing Cu-MOF, FJU-C2 shows high methylene blue capture of 1.3 g g^{-1} [110]. ZIF-8 (Zinc imidazolate), another well studied MOF, shows high congo red adsorption of 1.25 g g^{-1}, while its Co counterpart ZIF-67 adsorb 3.9 g g^{-1} [111, 112]. UiO-66 exhibits 400 mg g^{-1} uptake for alizarin S [113]. Finally, particle size tuning in MIL-68(In) shows four times higher congo red uptake by nanosized particles as compared to microsized samples [114]. A good example of MOF composite is wood-UiO-66 composite, where UiO-66 particles grow inside wood bulk and exhibit efficient removal of water contaminants upon filtration via the composite column (Fig. 8c) [115].

3.7 Limitations

Above mentioned examples demonstrate excellent future promise for MOF based fixed-bed adsorbents as an environmental remedy. Till date, there are few stumbling blocks for realizing these potential candidates into the market. Firstly, among the vast majority of MOFs, extended water stability is of prime importance for environmental applications, and only limited MOFs can surpass these stringent criteria. Secondly, MOF's production cost is still high compared to conventional adsorbents, posing threat to future widespread use. While activated carbons are available for few EUR/kg, zeolites and specialty adsorbents reach prices 10–100 EUR/kg depending on composition and application. The market price of most MOFs is still above that value (100–1000 EUR/kg) however mostly, because of the small scale of production (typically <50 kg). Finally, large scale fixed-bed testing is still scarce for MOFs (batch screening still prevails), and further optimization of cost and setup is still at infancy. However, companies, university spin-offs and start-ups worldwide have been successful to commercialize MOFs by large-scale production (which eventually lowers cost) and find interesting real-world special applications to catalyse the economic boost for future research and development (MOFs for Industry, 2020).

4 MOFs in Membranes

4.1 Background

Membrane separation has received increasing attention as one of the most competitive alternative separation technologies for energy and environmental applications including wastewater treatment, CO_2 capture and storage because of its low footprint, continuous process and energy efficiency without the requirement of any phase change or addition of chemicals during separation [116]. The membrane is a key component in the separation system. It acts as a semipermeable thin film that allows selective components to permeate through it from a feed stream when a driving force for instance, pressure difference, concentration difference, temperature difference, the difference in gravity or magnetic fields is applied across the membrane [117]. A number of continuous and self-supporting membranes made of porous materials have been developed [118]. It is challenging in preparing mechanical robust and defect-free continuous membranes from porous materials. Polymeric membranes are attractive because of their advantages of cost effectiveness and ease of processability. However, there is a well-known trade-off between permeability and selectivity where a membrane with high permeability comes with low selectivity and vice versa. The trade-off for a gas separation membrane is depicted as the Robeson Upper Bound [119, 120].

To overcome this limitation, mixed matrix membranes (MMMs) containing inorganic or organic fillers as a dispersed phase in a polymer matrix have been substantially developed and reported in numerous reviews [121–126]. A broad range of nanoporous fillers including carbon molecular sieves, zeolites, silica, polyhedral oligomeric silsesquioxane (POSS), graphene oxides (GOs), carbon nanotubes (CNTs), covalent organic frameworks (COFs) and metal-organic frameworks (MOFs) have been explored [121, 118]. Nevertheless, the poor compatibility, especially between the polymer and the inorganic filler, leads to the evolution of non-selective voids between the interface of polymer and filler, thus decreasing the selectivity of the membrane. Also, the aggregations of fillers remain as one of the current challenges in MMMs. Among the aforementioned fillers, MOFs have tremendous potential as one of the promising fillers in the fabrication of MMMs owing to the large surface areas, large pore volume, selective adsorption of gases, regular and tuneable pore structures, and customizable chemical functionalities, which allow MOFs possess better compatibility and dispersity in the polymer matrix. In addition, MOFs have a greater impact on the membrane separation properties in a given MOFs loading because of the higher pore volumes and lower density in MOFs than that of zeolites [127]. The selection of MOFs possessing suitable pore size, shape and functional groups is important to tailor with the polymer to form appropriate membranes for efficient separation.

4.2 Transport Mechanism for Membrane Gas Separation

Dense polymeric membranes are commonly used in commercial gas and vapor separation. The separation through a dense membrane occurs by a solution-diffusion model where the gas penetrants dissolve in the membrane and then diffuse through it at a concentration gradient (Fig. 9). In the solution-diffusion model, it is assumed that there are no permanent pores that appear in the dense selective layer of the membrane where the separation is achieved through the differences in the amount of gas penetrants which dissolve in the membrane (e.g., gas solubility) and the rate of the gas penetrants that diffuse through the membrane (e.g., gas diffusivity) [128].

Fick's first law was derived from the studies of gas transport through the non-porous membrane [129]. The amount of gas transport through a dense membrane at a cross-section area and a given time is expressed as:

$$J = -D\frac{dc}{dx} \tag{4}$$

where J is the flux (mol/cm^2.s), D is the gas diffusion coefficient (cm^2/s), dc/dx is the concentration gradient, c is the concentration (mol/cm^3) and x is the position (cm).

The concentration is proportional to the applied pressure in the gas phase. A solubility coefficient is a constant which is defined by Henry's law is introduced in the equation.

$$c = S \times P_i \tag{5}$$

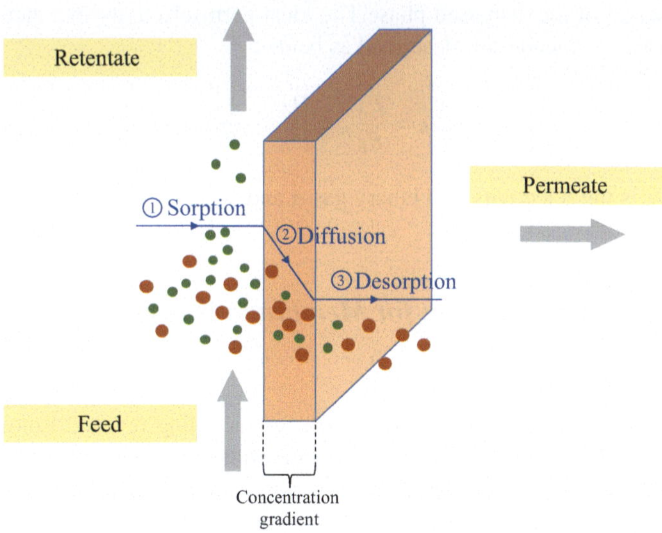

Fig. 9 A schematic diagram for the solution-diffusion mechanism in a dense membrane

where S is the solubility coefficient (cm^3(STP)/(cm^3.atm)) and P_i is the applied pressure (atm).Thus, Eq. (4) is rearranged as follows:

$$J = -DS\frac{\Delta P}{l} \tag{6}$$

where ΔP is the pressure difference applied across the membrane and l is the membrane thickness (cm).

Generally, permeability and selectivity are the important gas separation performance indicators for a membrane. The permeability or the rate of gases permeating through the membrane is a product of diffusion coefficient and solubility coefficient.

$$P = D \times S \tag{7}$$

where P is the gas permeability expressed as Barrer (1 Barrer = 1×10^{-10} cm^3(STP)cm/(cm^2s.cmHg)), D is the gas diffusion coefficient (cm^2/s) and S is the solubility coefficient (cm^3(STP)/(cm^3.atm)). The permeability for MMMs is predicted using the Maxwell model by considering the volume fraction of fillers into the polymer matrix, permeabilities in the continuous phase and dispersed phase [130]. The Maxwell model for MMMs is expressed according to the following equation:

$$P_{eff} = P_c\left[\frac{P_d + 2P_c - 2\emptyset_d(P_c - P_d)}{P_d + 2P_c + \emptyset_d(P_c - P_d)}\right] \tag{8}$$

where P_{eff} is the effective permeability (Barrer), P_c is the permeability in the continuous phase (Barrer), P_d is the permeability in the dispersed phase (Barrer), $\emptyset_d$ is the volume fraction of the dispersed phase.The ideal permselectivity in a membrane is the ratio of gas permeabilities described as below:

$$\propto_{A/B} = \frac{P_A}{P_B} = \frac{D_A}{D_B} \times \frac{S_A}{S_B} \tag{9}$$

where $\propto_{A/B}$ is the selectivity in a binary gas A and B.

4.3 Transport Mechanism for Membrane Wastewater Treatment

Pressure-driven membrane separation processes including reverse osmosis (RO), nanofiltration (NF), ultrafiltration (UF) and microfiltration (MF) are widely applied in wastewater treatment to effectively remove suspended particles, bacteria, viruses, multivalent ions and monovalent ions [131]. The membrane pore size is tailored accordingly for removing a wide range of components towards targeted water quality economically. Generally, the membranes used in pressure-driven wastewater treatment are asymmetric which consist of a porous support and an ultra-thin selective layer. Water flux and salt rejection are the two main parameters in analyzing the membrane performance for wastewater treatment.

The water flux across a membrane is determined based on the following equation.

$$J_w = \frac{V}{A \cdot \Delta t} \tag{10}$$

where J_w is the water flux (L/m^2 h), V is the volume of permeate (L), A is the effective membrane area (m^2), and Δt is the testing duration (h). The separation efficiency or solute rejection is determined as follows:

$$R = \left(1 - \frac{C_p}{C_f}\right) \times 100\% \tag{11}$$

where R is the solute rejection (%), C_p and C_f are the permeate and feed solution concentrations, respectively. Other than the water flux and solute rejection, the membranes should possess good antifouling properties. Therefore, various modification strategies on materials have been employed in tailoring the desirable membranes with high water flux, solute rejection and antifouling properties.

4.4 MOFs for Membrane Preparation

Thin film composite (TFC) membrane is a typical structure membrane favourable in industries for gas and liquid separation. Figure 10 shows the TFC structure where the bottom layer is a porous structure acting as the mechanical support and the top layer is an ultra-thin selective layer. A gutter layer is coated optionally to enhance the compatibility between the selective layer and support layer which is usually made

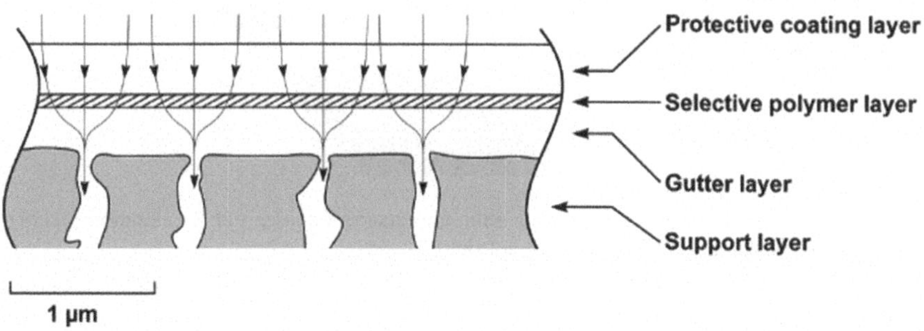

Fig. 10 Schematic diagram of the cross-section of thin film composite membrane [123]. Copyright 2017. Reproduced with permission from American Chemical Society

from the highly permeable polymer. For gas separation using polymeric membrane, a dense defect-free selective layer is needed to achieve high separation properties. Thus, a protective coating layer which consists of highly permeable polymer is applied on top of the selective layer if there are any defects on the selective layer.

On the other hand, the TFC membranes used in liquid separation applied the idea of interfacial polymerization, which was initially reported by Morgan in 1965 [132]. The interfacial polymerization process occurs when the reactive monomers such as aliphatic or aromatic diamine in aqueous phase contact with the acid chloride monomers in the organic phase and thus form a polyamide layer. The commonly used diamines are m-phenylenediamine (MPD) and piperazine (PIP), while the acid chloride monomer is trimesoyl chloride (TMC). Nanofillers such as MOFs could be incorporated into the polyamide layer for liquid separation to improve the water flux and/or solute rejection. Depending on the hydrophilic or hydrophobic nature of the MOFs, they could be added in the aqueous phase or organic phase. The TFC membranes incorporated with nanofillers is known as thin film nanocomposite (TFN) membranes. Since the thickness of a desirable selective layer is about 0.1–1.0 μm, thus the size of the MOF particles should be less than 100 nm [133]. Jeong et al. developed TFN membranes by incorporating molecular sieve nanoparticles in the polyamide layer for RO [134]. Inspired by this invention, Sorribas et al. were the first to embed the MOFs in TFN membranes [135]. Figure 11 illustrates a typical TFN fabrication process consists of UiO-66 in the polyamide selective layer on top of a polysulfone (PSF) support.

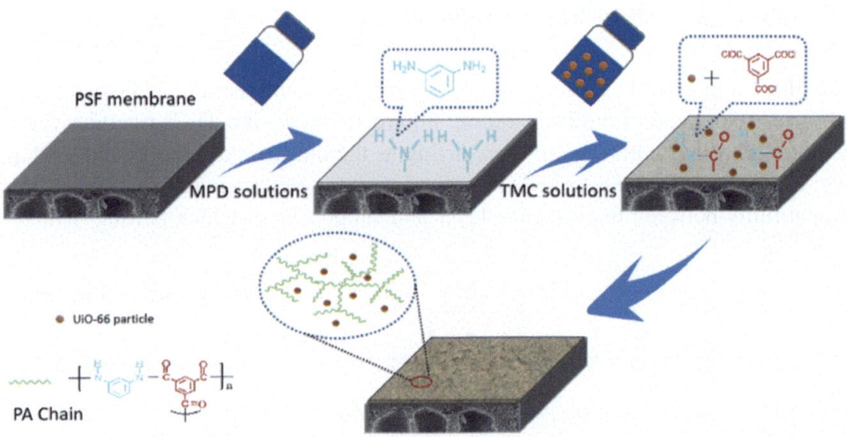

Fig. 11 Schematic diagram of fabricating thin film nanocomposite (TFN) membrane [124]. Copyright 2019. Reproduced with permission from Elsevier

4.5 Application of MOFs Membrane for Gas Purification

Among the four main gas separation areas including air purification, hydrogen recovery, CO_2 capture and propylene separation [136, 137], the potential of MOF-based membranes for CO_2 removal from natural gas and biogas have emerged to mitigate the challenges of global warming and climate change. Higher MOFs loading could be incorporated in polymers to form MMMs because of the better interaction between MOFs and polymer. In contrast, the optimal loading of inorganic fillers such as zeolite in MMMs is less than 10 wt% [138]. A variety of MOFs such as HKUST-1 (i.e., $Cu_3(BTC)_2$ or MOF-199) [139–141], manganese (II) formate ($Mn(HCOO)_2$) [140], Cu–4,4-bipyridine–hexafluorosilicate (Cu–BPY–HFS) [142], MOF-5 [143], MOF-74 (i.e., Mg_2dobdc) [144], copper and terephthalic acid (CuTPA) [127], MIL-53 [139], Fe(BTC) [145, 146], MIL-101(Cr) [141] and many other MOFs have been incorporated in polymer to form MMMs.

To enhance the compatibility between MOFs and polymer chains, MOFs with various functional groups have been tailored. Couck et al. functionalized the NH_2-MIL-53(Al) MOFs with amino groups and investigated the adsorption and separation of CO_2 [147]. The CO_2 adsorption and CO_2/CH_4 selectivity of the NH_2-MIL-53(Al) were increased significantly. This was attributed to the presence of the additional –NH_2 functional groups together with –OH groups increased the CO_2/CH_4 selectivity. Inspired by this work, Zornoza et al. synthesized NH_2-MIL-53(Al) using microwave and incorporated them in polysulfone (PSF) Udel® P-3500 [138]. MMMs with homogeneous dispersion of fillers were observed even at a high NH_2-MIL-53(Al) loading up to 40 wt%. An increase in the loading of NH_2-MIL-53(Al) led to an improvement in CO_2 permeability and CO_2/CH_4 selectivity of the PSF/NH_2-MIL-53(Al) MMMs. A two-fold increment in CO_2/CH_4 selectivity was observed when the NH_2-MIL-53(Al) was at 25 wt%. This synergistic effect was ascribed to the hydrogen bonding between the amine of NH_2-MIL-53(Al) and sulfone groups of PSF which promote the interaction between fillers and polymer. It is worth mentioning that the CO_2/CH_4 selectivity of the PSF/NH_2-MIL-53(Al) MMMs increased with pressure, which was noticeably different from the conventional MMMs. This was mainly due to the introduction of the amino groups that enhanced the flexibility in the NH_2-MIL-53(Al) compared to the MIL-53(Al). However, a further increase in the CO_2 permeability and a dramatic decrease in the CO_2/CH_4 selectivity was observed then the NH_2-MIL-53(Al) loading was at 40 wt%. Thus, the results suggested that there was an optimal loading of MOFs in the polymer matrix.

In another similar study, Rodenas et al. reported the NH_2-MIL-53(Al)-based MMMs, using Matrimid as a continuous phase for CO_2 capture [148]. The synthesized MMMs showed homogenous dispersion of fillers at 25 wt% NH_2-MIL-53(Al) loading but with the presence of interphase voids. Figure 12 depicts the influence of MOFs loading on the gas transport properties of the Matrimid/NH_2-MIL-53(Al) MMMs. The CO_2/CH_4 selectivity of Matrimid/NH_2-MIL-53(Al) MMMs increased marginally with increasing the MOFs loading. However, it was noticed that low NH_2-MIL-53(Al) loadings of 8 and 15 wt% reduced the CO_2 permeability compared to the

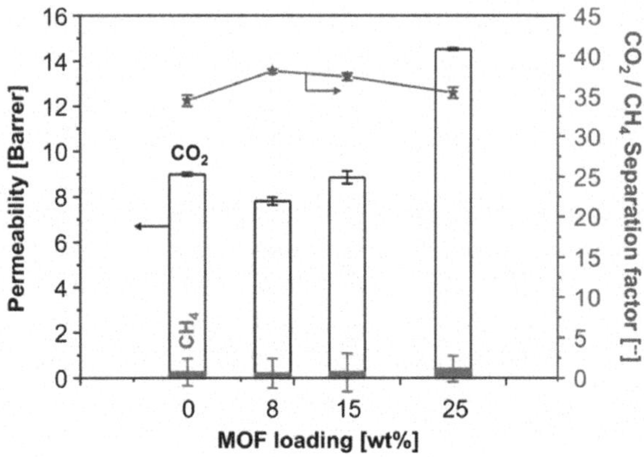

Fig. 12 CO_2/CH_4 separation performance as a function of MOF loading of Matrimid/NH_2-MIL-53(Al) MMMs at 35°C in a binary CO_2:CH_4 = 1:1 mixture [148]. Copyright 2014. Reproduced with permission from John Wiley and Sons

pristine Matrimid. Similar to the PSF/NH_2-MIL-53(Al) MMMs [138], the MMMs with a 25 wt% NH_2-MIL-53(Al) content doubled the CO_2 permeability.

Venna et al. reported an extensive study on surface-functionalized the MOFs to promote a better interaction between polymer-fillers [149]. Among the four types of surface functionalized UiO-66-NH_2 (e.g., polar, non-polar or aromatic functional groups), the phenyl acetyl group modified UiO-66-NH_2 in Matrimid MMMs showed the best separation performance with CO_2 permeability and CO_2/N_2 selectivity of MMMs increased by 200 and 25%, respectively. The synergistic improvement in the MMMs was attributed to the formation of π–π interactions and hydrogen bonding between imide groups of Matrimid and phenyl acetyl groups of the surface-modified UiO-66-NH_2. Also, the CO_2/N_2 selectivity of all the MMMs incorporated with 23 wt% loading of MOFs improved, suggesting the presence of a defect-free interface between polymer-filler.

Notably, a subclass of MOFs namely zeolitic imidazolate frameworks (ZIFs) exhibit exceptional thermal and chemical stability, as well as zeolite chemistry, which the transition reviews on the ZIF-based membranes for gas separation, pervaporation and fuel cells have been reported [150]. A diversity of ZIFs such as ZIF-7 [151–157], ZIF-8 [154, 158–169], ZIF-11 [170], ZIF-67 [171], ZIF-71 [172–175], ZIF-90 [127, 160, 162] and ZIF-301 [176] have been added into polymer to fabricate MMMs for gas separation. The MMMs consisted of ZIFs demonstrated a remarkable improvement in permeability and/or selectivity.

Bae et al. synthesized submicrometric ZIF-90 using the nonsolvent-induced crystallization method and incorporated them in polyimide to develop MMMs [127]. They crystallized ZIF-90 particles at ambient temperature by supersaturating the solution with the addition of nonsolvents (e.g., methanol and deionized water) to the reagent solution (e.g., *N*,*N*-dimethylformamide (DMF)). The submicrometric ZIF-90 demonstrated good compatibility with polyimide in the MMMs. Interestingly, the ZIF-90 embedded MMMs showed promising CO_2/CH_4 separation performance, which was believed to be accredited to the pore windows of 0.35 nm of ZIF-90 acted as size exclusion of CH_4 from the gas mixtures [177].

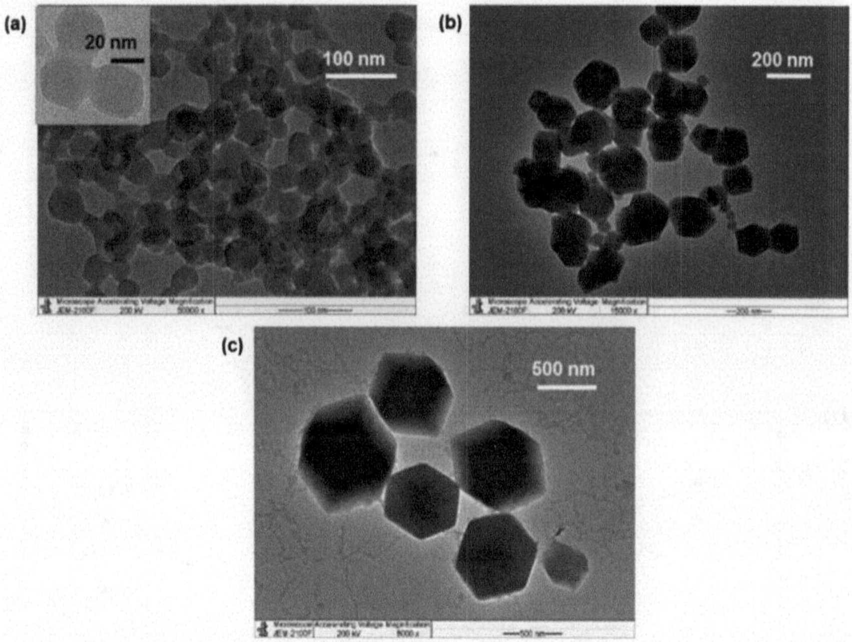

Fig. 13 Morphology of ZIF-71 particles synthesized in different volumetric ratios of DMF/MeOH with various particles **a** 30 nm, **b** 200 nm and **c** 600 nm. Copyright 2016. Reproduced with permission from American Chemical Society

Besides that, the effects of particle sizes of MOFs namely ZIF-71 on the N_2 adsorption and gas separation of MMMs have been investigated by Japip et al. [174]. ZIF-71 with three distinct particle sizes (i.e., 30, 200 and 600 nm) were synthesized using varying volume ratios of MeOH and DMF solvents (Fig. 13). The adsorption of these ZIF-71 particles in different sizes followed the typical type I isotherm pattern of microporous materials. It is interesting to note that the smallest particle size of ZIF-71 (i.e., 30 nm) showed the lowest surface area, total pore volume and volume of adsorbed N_2, which was attributed to the particles had the highest outer surface area to volume ratio and they were at their nucleation stage. In addition, the incorporation of ZIF-71 in 6FDA-Durene MMMs increased the gas diffusion coefficient significantly. Noticeably, there was a significant gas separation performance enhancement in MMMs consisting of ZIF-71 particle sizes of $\leq$200 nm compared to that of the MMMs with the particle size of >200 nm.

In contrast to the obvious increment of permeability of MMMs at high MOFs loading, Yong et al. observed a significant increment in the CO_2 permeability even at a low loading (e.g., less than 10 wt%) of ZIF-71 in a rubbery polymer namely poly(ethylene oxide) terephthalate and poly(butylene terephthalate) (PEOT/PBT) MMMs [175]. The ZIF-71 nanoparticles were synthesized at room temperature with the particle sizes less than 100 nm. Interestingly, the CO_2 permeability increased by 48% at an optimal loading of MMMs composed of 1 wt% ZIF-71 (Fig. 14a). This

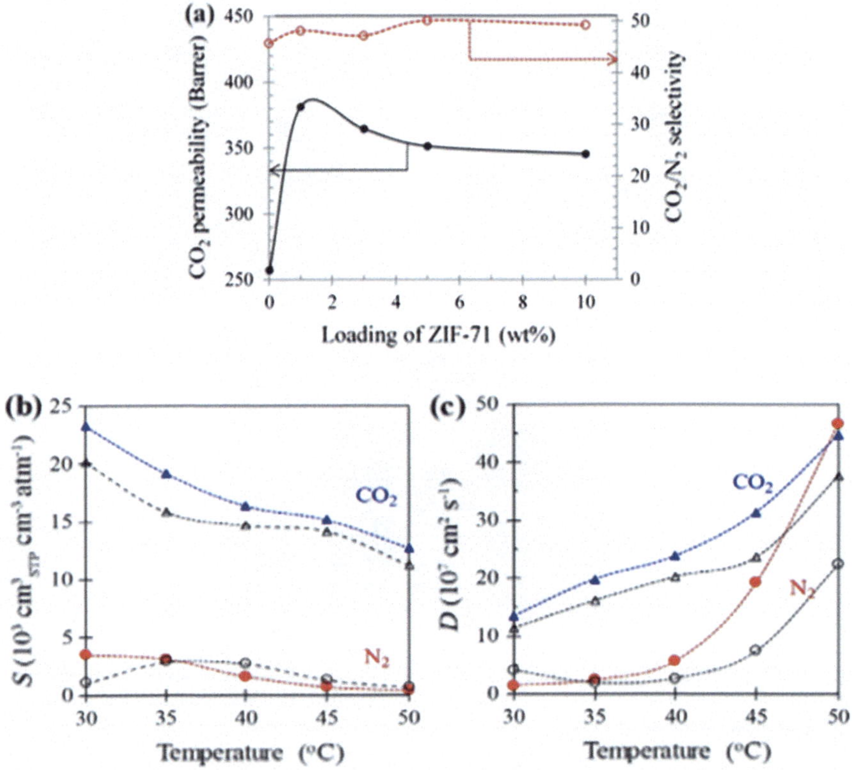

Fig. 14 **a** CO_2 permeability and CO_2/N_2 selectivity of PEOT/PBT/ZIF-71 MMMs as a function of ZIF-71 content. **b** Solubility coefficients and **c** diffusion coefficients of CO_2 and N_2 of PEOT/PBT (open symbols) and PEOT/PBT composed of 1 wt% ZIF-71 (filled symbols). Copyright 2018. Reproduced with permission from John Wiley and Sons

was mainly due to the enhancement of ZIF-71 on the fractional free volume of the PEOT/PBT matrix, as evidenced by the positron annihilation lifetime spectroscopy (PALS). Moreover, the resultant MMMs showed improvement in the CO_2 diffusion and solubility coefficients with the incorporation of ZIF-71 (Fig. 14b–c). It is believed that the altered chain packing caused an increase in CO_2 diffusion coefficient, while the high CO_2 affinity toward ZIF-71 increased the CO_2 solubility coefficient.

4.6 Application of MOFs Membrane for Wastewater Treatment

The incorporation of MOFs in membranes for wastewater treatment has been enthusiastically explored. Since the membranes will be in contact with a liquid solution, the important characteristic of the MOFs membranes for wastewater treatment is their stability in water. In other words, the structure of MOFs should not decompose, and do not experience losses in overall porosity and crystallinity when exposed to water moisture [178]. Generally, water molecules favour to transport through hydrophilic MOFs such as MILs and UiO-66 [178–181].

Among various studies on MOFs, Zr-based MOFs exhibited the highest stability in aqueous solutions and has emerged as a potent candidate for water-based applications [182–184]. UiO-66 is one of zirconium (IV)-carboxylate MOFs with pore size of about 6.1 Å, possesses excellent chemical and thermal stabilities. UiO-66 is able to facilitate water permeation while selectively excluding hydrated cations attributed to the intrinsic sub-nanometer pores [124, 137, 184–188]. Owing to its superior properties, UiO-66 has been employed as nanofiller in the TFN for a wide range of wastewater treatments including UF [188], NF [186], forward osmosis [184], and RO [187]. For example, Chung and co-workers developed TFN membranes comprising UiO-66 to remove toxic components such has selenium (Se) and arsenic (As) from the wastewater [186]. The developed TFN comprising of UiO-66 displayed excellent water flux and rejections. This is because the aperture size of UiO-66 is identical to those of the diameters of Se and As [189]. Moreover, high solute rejections were also reported in the mixed ions feed of Se and As.

Apart from the UiO-66 nanoparticles, a chromium based porous MOFs material named MIL-101(Cr) with large surface area and pore size of 30–35 Å, which able to provide broad water channels have gained much attention [190]. MIL-101(Cr) has been embedded in the TFN membranes for organic solvent nanofiltration (OSN) [135, 191, 192], NF [193, 194] and RO [195]. Other than the pristine MIL-101(Cr), the amino modification on MIL-101 namely NH_2-MIL-101 has been conducted and added in TFN membranes for NF [196]. Navorro et al. embedded a monolayer of MIL-101 on top of the cross-linked polyimide support by employing the Langmuir-Schaefer (LS) method [193]. Subsequently, the polyamide layer was formed on top of the MIL-101. The resultant TFN membranes showed a homogenous dispersion of MIL-101 and a superior NF performance. The exceptional high separation performance in the TFN membranes compared to TFC membranes was due to the higher roughness in the former which lead to an increased in the effective surface area of the membrane. Moreover, it is noteworthy that the LS method required only a very mere loading of MOFs (e.g., 3.8 $\mu g/cm^2$) to form a defect-free MOF layer without any loss of filler.

Different from the hydrophilic MOFs, hydrophobic MOFs such as ZIF-8 could be incorporated in the selective layer of TFN for RO membranes [197–199]. Wang et al. studied the effect of particle sizes of ZIF-8 on the water desalination performance of TFN membranes [199]. Among three particle sizes such as 50, 150 and 400 nm, the TFN membranes incorporated with the particle sizes of 50 nm showed the most promising water flux and salt rejection properties, suggesting the importance of particle sizes on promoting the dispersion in the polyamide layer.

To fabricate large scale defect-free membranes with high surface area, MOFs can be incorporated into the bulk polymer matrix or the selective layer in the typical membrane configurations including tubular, hollow fiber and spiral-wound modules.

Tubular modules are made of ceramic as the support while a pure-MOF layer can be deposited on top of the support as the selective layer. Hollow fiber membranes are normally formed by polymers and fabricated using solution or wet spinning. In the hollow fiber membranes, MOFs can be incorporated onto the polymer matrix as MMMs in the bulk structure or in the TFC as the selective layer. Spiral-wound modules consist of flat membrane and space in membrane envelopes that wound around a perforated tube. MOFs can be embedded onto the polymer matrix as MMMs in spiral-wound modules.

4.7 Limitations

Excellent MMMs and TFN comprising of MOFs with unprecedented high separation performance have been developed. Despite these significant advances, there are still critical challenges that need to be addressed for the further development of MOFs based membranes. To be technologically attractive, the MMMs of TFN containing 2D or 3D MOFs should be processable, reproducible and mechanically robust in actual industrial conditions. Besides, approaches including improving interface compatibility should be seriously taken into consideration. The selection of appropriate pairs of polymer and MOFs in MMMs remains imperatively important in designing next-generation defect-free MMMs to cope with the vast and diverse applications. As stated by Merkel et al., ultrahigh membrane selectivity is not necessary for economical and large-scale CO_2 capture from natural gas or flue gas because the process is pressure-ratio-limited, where the downstream of the more permeable gas plateaus when the selectivity continues to rise [136]. Thus, a membrane with ultrahigh permeability and modest selectivity may be efficient and practical for the industry in handling enormous volumetric flow rate of the feed gas. Other than having exceptional separation properties, the MOFs should also be stable in moisture or water especially when used in membranes for water application.

5 Conclusion and Future Perspectives

In comparison to other functional porous materials such as activated carbons and zeolites MOFs proved their selectivity or preference to adsorptive removal of various toxic components. High surface area, varying geometries and various functional groups result in a high degree of functionality. Increasing the stability of MOFs for the environmental application purpose remains an actual concern. Studies in this regard showed, that appropriate approach to MOFs design help to solve this problem and allow to synthesize materials applicable in real-world conditions.

MOF based fixed bed adsorbents may play a vital role in air and wastewater purification in future, and lead to a paradigm shift from batch method to larger scale breakthrough setup highlight the great potential of MOFs for as industrial standards. The stability, regeneration and mechanical properties of MOFs need to be further explored for widespread trials with different separation applications. The tradeoff between density and surface area of the shaped MOFs is crucial for the development of better MOF based adsorber. Future MOFs need to combine ultrahigh porosity, high packing density, operational longevity and functional diversity to cater the commercial development of the field. In current technologies they should not be regarded as competitors to zeolites or activated carbon but used as functional additives for increased removal of small polar molecules that can not be effectively treated with traditional adsorbents. Promising targets are aldehydes and other TICs.

MOF-based membranes have great potential in diverse energy and environmental applications. Many efforts have been dedicated to overcome the trade-off between the permeability and selectivity as well as the interfacial compatibility of MOFs based membranes. However, the interfacial chemistry of polymer and MOFs in membrane applications have not been fully understood. The correlations of the molecular level properties of the MOFs on the separation properties and feasibility of membranes scale-up are important to commercialize the MOFs-based membranes. Taking a step further towards sustainable development, green MOFs and membranes synthesis methods which are environmentally benign, recyclable and low cost should be one of the important milestones for MOFs membranes.

References

1. Bobbitt, N.S., Mendonca, M.L., Howarth, A.J., Islamoglu, T., Hupp, J.T., Farha, O.K., Snurr, R.Q.: Metal-organic frameworks for the removal of toxic industrial chemicals and chemical warfare agents. Chem. Soc. Rev. **46**, 3357–3385 (2017)
2. Rojas, S., Horcajada, P.: Metal-organic frameworks for the removal of emerging organic contaminants in water. Chem. Rev. **120**, 8378–8415 (2020)
3. Dhaka, S., Kumar, R., Deep, A., Kurade, M.B., Ji, S.W., Jeon, B.H.: Metal–organic frameworks (MOFs) for the removal of emerging contaminants from aquatic environments. Coord. Chem. Rev. **380**, 330–352 (2019)
4. Johnson, D.M., Hokanson, D.R., Zhang, Q., Czupinski, K.D., Tang, J.: Feasibility of water purification technology in rural areas of developing countries. J. Environ. Manage. **88**, 416–427 (2008)
5. Furukawa, H., Cordova, K.E., O'Keeffe, M., Yaghi, O.M.: The chemistry and applications of metal-organic frameworks. Science **80**, 341 (2013)
6. Britt, D., Furukawa, H., Wang, B., Glover, T.G., Yaghi, O.M.: Highly efficient separation of carbon dioxide by a metal-organic framework replete with open metal sites. Proc. Natl. Acad. Sci. U. S. A. **106**, 20637–20640 (2009)
7. Woellner, M., Hausdorf, S., Klein, N., Mueller, P., Smith, M.W., Kaskel, S.: Adsorption and detection of hazardous trace gases by metal-organic frameworks. Adv. Mater. **30**, 1704679 (2018)
8. Barea, E., Montoro, C., Navarro, J.A.R.: Toxic gas removal-metal-organic frameworks for the capture and degradation of toxic gases and vapours. Chem. Soc. Rev. **43**, 5419–5430 (2014)
9. Peterson, G.W., Mahle, J.J., DeCoste, J.B., Gordon, W.O., Rossin, J.A.: Extraordinary NO_2 removal by the metal-organic framework UiO-66-NH_2. Angew. Chemie Int. Ed. **55**, 6235–6238 (2016)
10. Savage, M., Cheng, Y., Easun, T.L., Eyley, J.E., Argent, S.P., Warren, M.R., Lewis, W., Murray, C., Tang, C.C., Frogley, M.D., Cinque, G., Sun, J., Rudić, S., Murden, R.T., Benham, M.J.,

Fitch, A.N., Blake, A.J., Ramirez-Cuesta, A.J., Yang, S., Schröder, M.: Selective adsorption of sulfur dioxide in a robust metal-organic framework material. Adv. Mater. **28**, 8705–8711 (2016)
11. Roy, A., Srivastava, A.K., Singh, B., Shah, D., Mahato, T.H., Srivastava, A.: Kinetics of degradation of sulfur mustard and sarin simulants on HKUST-1 metal organic framework. Dalt. Trans. **41**, 12346–12348 (2012)
12. Hall, J.N., Bollini, P.: Structure, characterization, and catalytic properties of open-metal sites in metal organic frameworks. React. Chem. Eng. **4**, 207–222 (2019)
13. Mouchaham, G., Wang, S., Serre, C.: The stability of metal-organic frameworks. In: Metal-Organic Frameworks. pp. 1–28. Wiley-VCH Verlag GmbH & Co. KGaA, Weinheim, Germany (2018)
14. Ding, M., Cai, X., Jiang, H.L.: Improving MOF stability: approaches and applications. Chem. Sci. **10**, 10209–10230 (2019)
15. Leus, K., Bogaerts, T., De Decker, J., Depauw, H., Hendrickx, K., Vrielinck, H., Van Speybroeck, V., Van Der Voort, P.: Systematic study of the chemical and hydrothermal stability of selected "stable" Metal Organic Frameworks. Microporous Mesoporous Mater. **226**, 110–116 (2016)
16. McHugh, L.N., McPherson, M.J., McCormick, L.J., Morris, S.A., Wheatley, P.S., Teat, S.J., McKay, D., Dawson, D.M., Sansome, C.E.F., Ashbrook, S.E., Stone, C.A., Smith, M.W., Morris, R.E.: Hydrolytic stability in hemilabile metal–organic frameworks. Nat. Chem. **10**, 1096–1102 (2018)
17. Burtch, N.C., Jasuja, H., Walton, K.S.: Water stability and adsorption in metal-organic frameworks. Chem. Rev. **114**, 10575–10612 (2014)
18. Tan, K., Nijem, N., Gao, Y., Zuluaga, S., Li, J., Thonhauser, T., Chabal, Y.J.: Water interactions in metal organic frameworks. Cryst. Eng. Comm. **17**, 247–260 (2015)
19. Jiao, Y., Morelock, C.R., Burtch, N.C., Mounfield, W.P., Hungerford, J.T., Walton, K.S.: Tuning the kinetic water stability and adsorption interactions of Mg-MOF-74 by partial substitution with Co or Ni. Ind. Eng. Chem. Res. **54**, 12408–12414 (2015)
20. Canivet, J., Fateeva, A., Guo, Y., Coasne, B., Farrusseng, D.: Water adsorption in MOFs: fundamentals and applications. Chem. Soc. Rev. **43**, 5594–5617 (2014)
21. Low, J.J., Benin, A.I., Jakubczak, P., Abrahamian, J.F., Faheem, S.A., Willis, R.R.: Virtual high throughput screening confirmed experimentally: Porous coordination polymer hydration. J. Am. Chem. Soc. **131**, 15834–15842 (2009)
22. Wang, S., Lee, J.S., Wahiduzzaman, M., Park, J., Muschi, M., Martineau-Corcos, C., Tissot, A., Cho, K.H., Marrot, J., Shepard, W., Maurin, G., Chang, J.S., Serre, C.: A robust large-pore zirconium carboxylate metal–organic framework for energy-efficient water-sorption-driven refrigeration. Nat. Energy. **3**, 985–993 (2018)
23. Padial, N.M., Quartapelle Procopio, E., Montoro, C., López, E., Oltra, J.E., Colombo, V., Maspero, A., Masciocchi, N., Galli, S., Senkovska, I., Kaskel, S., Barea, E., Navarro, J.A.R.: Highly Hydrophobic Isoreticular Porous Metal-Organic Frameworks for the Capture of Harmful Volatile Organic Compounds. Angew. Chemie Int. Ed. **52**, 8290–8294 (2013)
24. Drache, F., Bon, V., Senkovska, I., Marschelke, C., Synytska, A., Kaskel, S.: Postsynthetic inner-surface functionalization of the highly stable zirconium-based metal-organic framework DUT-67. Inorg. Chem. **55**, 7206–7213 (2016)
25. Khan, N.A., Hasan, Z., Jhung, S.H.: Adsorptive removal of hazardous materials using metal-organic frameworks (MOFs): a review. J. Hazard. Mater. **244–245**, 444–456 (2013)
26. Khabzina, Y., Farrusseng, D.: Unravelling ammonia adsorption mechanisms of adsorbents in humid conditions. Microporous. Mesoporous. Mater. **265**, 143–148 (2018)
27. Vikrant, K., Kumar, V., Kim, K.H., Kukkar, D.: Metal-organic frameworks (MOFs): potential and challenges for capture and abatement of ammonia. J. Mater. Chem. A. **5**, 22877–22896 (2017)
28. Katz, M.J., Howarth, A.J., Moghadam, P.Z., DeCoste, J.B., Snurr, R.Q., Hupp, J.T., Farha, O.K.: High volumetric uptake of ammonia using Cu-MOF-74/Cu-CPO-27. Dalt. Trans. **45**, 4150–4153 (2016)

29. Weston, M.H., Morris, W., Siu, P.W., Hoover, W.J., Cho, D., Richardson, R.K., Farha, O.K.: Phosphine gas adsorption in a series of metal-organic frameworks. Inorg. Chem. **54**, 8162–8164 (2015)
30. Hinks, N.J., McKinlay, A.C., Xiao, B., Wheatley, P.S., Morris, R.E.: Metal organic frameworks as NO delivery materials for biological applications. Microporous Mesoporous Mater. **129**, 330–334 (2010)
31. Cattaneo, D., Warrender, S.J., Duncan, M.J., Kelsall, C.J., Doherty, M.K., Whitfield, P.D., Megson, I.L., Morris, R.E.: Tuning the nitric oxide release from CPO-27 MOFs. RSC Adv. **6**, 14059–14067 (2016)
32. McKinlay, A.C., Xiao, B., Wragg, D.S., Wheatley, P.S., Megson, I.L., Morris, R.E.: Exceptional behavior over the whole adsorption-storage-delivery cycle for NO in porous metal organic frameworks. J. Am. Chem. Soc. **130**, 10440–10444 (2008)
33. Xiao, B., Wheatley, P.S., Zhao, X., Fletcher, A.J., Fox, S., Rossi, A.G., Megson, I.L., Bordiga, S., Regli, L., Thomas, K.M., Morris, R.E.: High-capacity hydrogen and nitric oxide adsorption and storage in a metal-organic framework. J. Am. Chem. Soc. **129**, 1203–1209 (2007)
34. Sun, W., Lin, L.-C., Peng, X., Smit, B.: Computational screening of porous metal-organic frameworks and zeolites for the removal of SO_2 and NO_x from flue gases. AIChE J. **60**, 2314–2323 (2014)
35. Hasan, Z., Tong, M., Jung, B.K., Ahmed, I., Zhong, C., Jhung, S.H.: Adsorption of pyridine over amino-functionalized metal-organic frameworks: attraction via hydrogen bonding versus base-base repulsion. J. Phys. Chem. C **118**, 21049–21056 (2014)
36. Zhang, L., Wang, J., Du, T., Zhang, W., Zhu, W., Yang, C., Yue, T., Sun, J., Li, T., Wang, J.: NH2-MIL-53(Al) metal-organic framework as the smart platform for simultaneous high-performance detection and removal of Hg_{2+}. Inorg. Chem. **58**, 12573–12581 (2019)
37. Li, L., Shi, Z., Zhu, H., Hong, W., Xie, F., Sun, K.: Adsorption of azo dyes from aqueous solution by the hybrid MOFs/GO. Water Sci. Technol. **73**, 1728–1737 (2016)
38. Zhao, Z., Wang, S., Yang, Y., Li, X., Li, J., Li, Z.: Competitive adsorption and selectivity of benzene and water vapor on the microporous metal organic frameworks (HKUST-1). Chem. Eng. J. **259**, 79–89 (2015)
39. Li, Y., Wang, X., Xu, D., Chung, J.D., Kaviany, M., Huang, B.: H_2O adsorption/desorption in MOF-74: Ab initio molecular dynamics and experiments. J. Phys. Chem. C **119**, 13021–13031 (2015)
40. Ghosh, P., Colón, Y.J., Snurr, R.Q.: Water adsorption in UiO-66: the importance of defects. Chem. Commun. **50**, 11329–11331 (2014)
41. Furukawa, H., Gándara, F., Zhang, Y.B., Jiang, J., Queen, W.L., Hudson, M.R., Yaghi, O.M.: Water adsorption in porous metal-organic frameworks and related materials. J. Am. Chem. Soc. **136**, 4369–4381 (2014)
42. Küsgens, P., Rose, M., Senkovska, I., Fröde, H., Henschel, A., Siegle, S., Kaskel, S.: Characterization of metal-organic frameworks by water adsorption. Microporous Mesoporous Mater. **120**, 325–330 (2009)
43. Akiyama, G., Matsuda, R., Kitagawa, S.: Highly porous and stable coordination polymers as water sorption materials. Chem. Lett. **39**, 360–361 (2010)
44. Ehrenmann, J., Henninger, S.K., Janiak, C.: Water adsorption characteristics of MIL-101 for heat-transformation applications of MOFs. Eur. J. Inorg. Chem. **2011**, 471–474 (2011)
45. Li, P., Chen, J., Zhang, J., Wang, X.: Water stability and competition effects toward CO_2 adsorption on metal organic frameworks. Sep. Purif. Rev. **44**, 19–27 (2015)
46. Yazaydin, A.Ö., Benin, A.I., Faheem, S.A., Jakubczak, P., Low, J.J., Richard, R.W., Snurr, R.Q.: Enhanced CO_2 adsorption in metal-organic frameworks via occupation of open-metal sites by coordinated water molecules. Chem. Mater. **21**, 1425–1430 (2009)
47. Decoste, J.B., Peterson, G.W.: Metal-organic frameworks for air purification of toxic chemicals. Chem. Rev. **114**, 5695–5727 (2014)
48. Kumar, P., Kim, K.H., Kwon, E.E., Szulejko, J.E.: Metal-organic frameworks for the control and management of air quality: advances and future direction. J. Mater. Chem. A. **4**, 345–361 (2015)

49. Dias, E.M., Petit, C.: Towards the use of metal-organic frameworks for water reuse: a review of the recent advances in the field of organic pollutants removal and degradation and the next steps in the field. J. Mater. Chem. A. **3**, 22484–22506 (2015)
50. Ackley, M.W., Yang, R.T.: Adsorption characteristics of high-exchange clinoptilolites. Ind. Eng. Chem. Res. **30**, 2523–2530 (1991)
51. Wang, S., Peng, Y.: Natural zeolites as effective adsorbents in water and wastewater treatment. Chem. Eng. J. **156**, 11–24 (2010)
52. Kirchner, A., Brown, I.W.M., Bowden, M.E., Kemmitt, T., Smith, G.: Preparation and high-temperature characterisation of nanostructured alumina ceramic membranes for gas purification. Curr. Appl. Phys. **8**, 451–454 (2008)
53. Gupta, V.K., Saleh, T.A.: Sorption of pollutants by porous carbon, carbon nanotubes and fullerene—an overview. Environ. Sci. Pollut. Res. **20**, 2828–2843 (2013)
54. Permyakova, A., Skrylnyk, O., Courbon, E., Affram, M., Wang, S., Lee, U.H., Valekar, A.H., Nouar, F., Mouchaham, G., Devic, T., De Weireld, G., Chang, J.S., Steunou, N., Frère, M., Serre, C.: Synthesis optimization, shaping, and heat reallocation evaluation of the hydrophilic metal-organic framework MIL-160(Al). Chemsuschem **10**, 1419–1426 (2017)
55. Weston, M.H., Peterson, G.W., Browe, M.A., Jones, P., Farha, O.K., Hupp, J.T., Nguyen, S.B.T.: Removal of airborne toxic chemicals by porous organic polymers containing metal-catecholates. Chem. Commun. **49**, 2995–2997 (2013)
56. Kuehl, V.A., Yin, J., Duong, P.H.H., Mastorovich, B., Newell, B., Li-Oakey, K.D., Parkinson, B.A., Hoberg, J.O.: A highly ordered nanoporous, two-dimensional covalent organic framework with modifiable pores, and its application in water purification and ion sieving. J. Am. Chem. Soc. **140**, 18200–18207 (2018)
57. Krishna, R., Long, J.R.: Screening metal-organic frameworks by analysis of transient breakthrough of gas mixtures in a fixed bed adsorber. J. Phys. Chem. C **115**, 12941–12950 (2011)
58. Belmabkhout, Y., Pirngruber, G., Jolimaitre, E., Methivier, A.: A complete experimental approach for synthesis gas separation studies using static gravimetric and column breakthrough experiments. Adsorption **13**, 341–349 (2007)
59. Beerdsen, E., Dubbeldam, D., Smit, B.: Understanding diffusion in nanoporous materials. Phys. Rev. Lett. **96**, 044501 (2006)
60. Shafeeyan, M.S., Wan Daud, W.M.A., Shamiri, A.: A review of mathematical modeling of fixed-bed columns for carbon dioxide adsorption (2014)
61. Sotelo, J.L., Ovejero, G., Rodríguez, A., Álvarez, S., García, J.: Adsorption of carbamazepine in fixed bed columns: experimental and modeling studies. Sep. Sci. Technol. **48**, 2626–2637 (2013)
62. Kärger, J., Ruthven, D.M.: Diffusion in nanoporous materials: fundamental principles, insights and challenges. New J. Chem. **40**, 4027–4048 (2016)
63. Kabtamu, D.M., Wu, Y., Li, F.: Hierarchically porous metal–organic frameworks: synthesis strategies, structure(s), and emerging applications in decontamination. J. Hazard. Mater. 122765 (2020)
64. Peng, Y., Krungleviciute, V., Eryazici, I., Hupp, J.T., Farha, O.K., Yildirim, T.: Methane storage in metal-organic frameworks: current records, surprise findings, and challenges. J. Am. Chem. Soc. **135**, 11887–11894 (2013)
65. Shah, B.B., Kundu, T., Zhao, D.: Mechanical properties of shaped metal-organic frameworks. Top. Curr. Chem. **377**, 1–34 (2019)
66. Valizadeh, B., Nguyen, T.N., Stylianou, K.C.: Shape engineering of metal–organic frameworks. Polyhedron **145**, 1–15 (2018)
67. Hindocha, S., Poulston, S.: Study of the scale-up, formulation, ageing and ammonia adsorption capacity of MIL-100(Fe), Cu-BTC and CPO-27(Ni) for use in respiratory protection filters. Faraday Discuss. **201**, 113–125 (2017)
68. Valekar, A.H., Lee, S.G., Cho, K.H., Lee, U.H., Lee, J.S., Yoon, J.W., Hwang, Y.K., Cho, S.J., Chang, J.S.: Shaping of porous metal-organic framework granules using mesoporous ρ-alumina as a binder. RSC Adv. **7**, 55767–55777 (2017)

69. Manisalidis, I., Stavropoulou, E., Stavropoulos, A., Bezirtzoglou, E.: Environmental and health impacts of air pollution: a review. Front. Public Heal. **8**, 14 (2020)
70. Mannucci, P.M., Franchini, M.: Health effects of ambient air pollution in developing countries. Int. J. Environ. Res. Public Health. **14** (2017)
71. Reed, D.A., Xiao, D.J., Gonzalez, M.I., Darago, L.E., Herm, Z.R., Grandjean, F., Long, J.R.: Reversible CO scavenging via adsorbate-dependent spin state transitions in an iron(II)-triazolate metal-organic framework. J. Am. Chem. Soc. **138**, 5594–5602 (2016)
72. Peng, J., Xian, S., Xiao, J., Huang, Y., Xia, Q., Wang, H., Li, Z.: A supported Cu(I)@MIL-100(Fe) adsorbent with high CO adsorption capacity and CO/N2 selectivity. Chem. Eng. J. **270**, 282–289 (2015)
73. Seoane, B., Coronas, J., Gascon, I., Benavides, M.E., Karvan, O., Caro, J., Kapteijn, F., Gascon, J.: Metal-organic framework based mixed matrix membranes: A solution for highly efficient CO2 capture? Chem. Soc. Rev. **44**, 2421–2454 (2015)
74. Choi, S., Watanabe, T., Bae, T.H., Sholl, D.S., Jones, C.W.: Modification of the Mg/DOBDC MOF with amines to enhance CO 2 adsorption from ultradilute gases. J. Phys. Chem. Lett. **3**, 1136–1141 (2012)
75. McDonald, T.M., Lee, W.R., Mason, J.A., Wiers, B.M., Hong, C.S., Long, J.R.: Capture of carbon dioxide from air and flue gas in the alkylamine-appended metal-organic framework mmen-Mg 2(dobpdc). J. Am. Chem. Soc. **134**, 7056–7065 (2012)
76. Siegelman, R.L., McDonald, T.M., Gonzalez, M.I., Martell, J.D., Milner, P.J., Mason, J.A., Berger, A.H., Bhown, A.S., Long, J.R.: Controlling cooperative CO_2 adsorption in diamine-appended Mg_2(dobpdc) metal-organic frameworks. J. Am. Chem. Soc. **139**, 10526–10538 (2017)
77. Demessence, A., D'Alessandro, D.M., Foo, M.L., Long, J.R.: Strong CO2 binding in a water-stable, triazolate-bridged metal-organic framework functionalized with ethylenediamine. J. Am. Chem. Soc. **131**, 8784–8786 (2009)
78. Liu, J., Tian, J., Thallapally, P.K., McGrail, B.P.: Selective CO 2 capture from flue gas using metal-organic frameworks-a fixed bed study. J. Phys. Chem. C **116**, 9575–9581 (2012)
79. Jasuja, H., Peterson, G.W., Decoste, J.B., Browe, M.A., Walton, K.S.: Evaluation of MOFs for air purification and air quality control applications: ammonia removal from air. Chem. Eng. Sci. **124**, 118–124 (2015)
80. Klein, N., Senkovska, I., Gedrich, K., Stoeck, U., Henschel, A., Mueller, U., Kaskel, S.: A Mesoporous Metala Organic Framework. Angew. Chemie Int. Ed. **48**, 9954–9957 (2009)
81. Spanopoulos, I., Xydias, P., Malliakas, C.D., Trikalitis, P.N.: A straight forward route for the development of metal-organic frameworks functionalized with aromatic-OH groups: synthesis, characterization, and gas (N_2, Ar, H_2, CO_2, CH_4, NH_3) sorption properties. Inorg. Chem. **52**, 855–862 (2013)
82. Peterson, G.W., Rossin, J.A., Decoste, J.B., Killops, K.L., Browe, M., Valdes, E., Jones, P.: Zirconium hydroxide-metal-organic framework composites for toxic chemical removal. Ind. Eng. Chem. Res. **52**, 5462–5469 (2013)
83. Rieth, A.J., Tulchinsky, Y., Dincă, M.: High and reversible ammonia uptake in mesoporous azolate metal-organic frameworks with open Mn Co, and Ni sites. J. Am. Chem. Soc. **138**, 9401–9404 (2016)
84. Chen, Y., Yang, C., Wang, X., Yang, J., Ouyang, K., Li, J.: Kinetically controlled ammonia vapor diffusion synthesis of a Zn(II) MOF and its H2O/NH3 adsorption properties. J. Mater. Chem. A. **4**, 10345–10351 (2016)
85. Wisser, D., Wisser, F.M., Raschke, S., Klein, N., Leistner, M., Grothe, J., Brunner, E., Kaskel, S.: Biological Chitin-MOF composites with hierarchical pore systems for air-filtration applications. Angew. Chemie Int. Ed. **54**, 12588–12591 (2015)
86. Zhao, J., Losego, M.D., Lemaire, P.C., Williams, P.S., Gong, B., Atanasov, S.E., Blevins, T.M., Oldham, C.J., Walls, H.J., Shepherd, S.D., Browe, M.A., Peterson, G.W., Parsons, G.N.: Highly adsorptive, MOF-functionalized nonwoven fiber mats for hazardous gas capture enabled by atomic layer deposition. Adv. Mater. Interfaces. **1**, 1400040 (2014)

87. Ebrahim, A.M., Bandosz, T.J.: Effect of amine modification on the properties of zirconium-carboxylic acid based materials and their applications as NO2 adsorbents at ambient conditions. Microporous Mesoporous Mater. **188**, 149–162 (2014)
88. DeCoste, J.B., Demasky, T.J., Katz, M.J., Farha, O.K., Hupp, J.T.: A UiO-66 analogue with uncoordinated carboxylic acids for the broad-spectrum removal of toxic chemicals. New J. Chem. **39**, 2396–2399 (2015)
89. Yang, S., Liu, L., Sun, J., Thomas, K.M., Davies, A.J., George, M.W., Blake, A.J., Hill, A.H., Fitch, A.N., Tang, C.C., Schröder, M.: Irreversible network transformation in a dynamic porous host catalyzed by sulfur dioxide. J. Am. Chem. Soc. **135**, 4954–4957 (2013)
90. Cui, X., Yang, Q., Yang, L., Krishna, R., Zhang, Z., Bao, Z., Wu, H., Ren, Q., Zhou, W., Chen, B., Xing, H.: Ultrahigh and selective SO_2 uptake in inorganic anion-pillared hybrid porous materials. Adv. Mater. **29**, 1606929 (2017)
91. Belmabkhout, Y., Pillai, R.S., Alezi, D., Shekhah, O., Bhatt, P.M., Chen, Z., Adil, K., Vaesen, S., De Weireld, G., Pang, M., Suetin, M., Cairns, A.J., Solovyeva, V., Shkurenko, A., El Tall, O., Maurin, G., Eddaoudi, M.: Metal-organic frameworks to satisfy gas upgrading demands: fine-tuning the soc-MOF platform for the operative removal of H_2S. J. Mater. Chem. A. **5**, 3293–3303 (2017)
92. Bhatt, P.M., Belmabkhout, Y., Assen, A.H., Weseliński, Ł.J., Jiang, H., Cadiau, A., Xue, D.X., Eddaoudi, M.: Isoreticular rare earth fcu-MOFs for the selective removal of H_2S from CO_2 containing gases. Chem. Eng. J. **324**, 392–396 (2017)
93. Mohideen, M.I.H., Pillai, R.S., Adil, K., Bhatt, P.M., Belmabkhout, Y., Shkurenko, A., Maurin, G., Eddaoudi, M.: A fine-tuned MOF for gas and vapor separation: a multipurpose adsorbent for acid gas removal, dehydration, and BTX sieving. Chem **3**, 822–833 (2017)
94. Hasan, Z., Jhung, S.H.: Removal of hazardous organics from water using metal-organic frameworks (MOFs): plausible mechanisms for selective adsorptions. J. Hazard. Mater. **283**, 329–339 (2015)
95. Massoudinejad, M., Ghaderpoori, M., Shahsavani, A., Jafari, A., Kamarehie, B., Ghaderpoury, A., Amini, M.M.: Ethylenediamine-functionalized cubic ZIF-8 for arsenic adsorption from aqueous solution: Modeling, isotherms, kinetics and thermodynamics. J. Mol. Liq. **255**, 263–268 (2018)
96. Wang, C., Luan, J., Wu, C.: Metal-organic frameworks for aquatic arsenic removal. Water Res. **158**, 370–382 (2019)
97. Sun, J., Zhang, X., Zhang, A., Liao, C.: Preparation of Fe–Co based MOF-74 and its effective adsorption of arsenic from aqueous solution. J. Environ. Sci. (China) **80**, 197–207 (2019)
98. Yang, J.C., Yin, X.B.: $CoFe_2O_4$ @MIL-100(Fe) hybrid magnetic nanoparticles exhibit fast and selective adsorption of arsenic with high adsorption capacity. Sci. Rep. **7**, 1–15 (2017)
99. Wang, C., Liu, X., Chen, J.P., Li, K.: Superior removal of arsenic from water with zirconium metal-organic framework UiO-66. Sci. Rep. **5**, 1–10 (2015)
100. Bhatnagar, A., Kumar, E., Sillanpää, M.: Fluoride removal from water by adsorption—a review. Chem. Eng. J. **171**, 811–840 (2011)
101. Zhu, X.H., Yang, C.X., Yan, X.P.: Metal-organic framework-801 for efficient removal of fluoride from water. Microporous Mesoporous Mater. **259**, 163–170 (2018)
102. Massoudinejad, M., Ghaderpoori, M., Shahsavani, A., Amini, M.M.: Adsorption of fluoride over a metal organic framework Uio-66 functionalized with amine groups and optimization with response surface methodology. J. Mol. Liq. **221**, 279–286 (2016)
103. karmakar, S., Dechnik, J., Janiak, C., De, S.: Aluminium fumarate metal-organic framework: a super adsorbent for fluoride from water. J. Hazard. Mater. 303, 10–20 (2016)
104. Wang, X., Zhu, H., Sun, T., Liu, Y., Han, T., Lu, J., Dai, H., Zhai, L.: Synthesis and study of an efficient metal-organic framework adsorbent (MIL-96(Al)) for fluoride removal from water. J. Nanomater. **2019**, 3128179 (2019)
105. Abbasi, A., Moradpour, T., Van Hecke, K.: A new 3D cobalt (II) metal-organic framework nanostructure for heavy metal adsorption. Inorganica Chim. Acta. **430**, 261–267 (2015)
106. Rahimi, E., Mohaghegh, N.: Removal of toxic metal ions from sungun acid rock drainage using mordenite zeolite, graphene nanosheets, and a novel metal-organic framework. Mine Water Environ 35

107. Luo, X., Ding, L., Luo, J.: Adsorptive removal of Pb(II) ions from aqueous samples with amino-functionalization of metal-organic frameworks MIL-101(Cr). J. Chem. Eng. Data **60**, 1732–1743 (2015)
108. Defuria, M.D., Zeller, M., Genna, D.T.: Removal of pharmaceuticals from water via π-π stacking interactions in perfluorinated metal-organic frameworks. Cryst. Growth Des. **16**, 3530–3534 (2016)
109. Sarker, M., Shin, S., Jeong, J.H., Jhung, S.H.: Mesoporous metal-organic framework PCN-222(Fe): promising adsorbent for removal of big anionic and cationic dyes from water. Chem. Eng. J. **371**, 252–259 (2019)
110. Wang, X.S., Liang, J., Li, L., Lin, Z.J., Bag, P.P., Gao, S.Y., Huang, Y.B., Cao, R.: An anion metal-organic framework with lewis basic sites-rich toward charge-exclusive cationic dyes separation and size-selective catalytic reaction. Inorg. Chem. **55**, 2641–2649 (2016)
111. Jiang, C., Fu, B., Cai, H., Cai, T.: Efficient adsorptive removal of Congo red from aqueous solution by synthesized zeolitic imidazolate framework-8. Chem. Speciat. Bioavailab. **28**, 199–208 (2016)
112. Thanh Tu, N.T., Thien, T.V., Du, P.D., Thanh Chau, V.T., Mau, T.X., Khieu, D.Q.: Adsorptive removal of Congo red from aqueous solution using zeolitic imidazolate framework-67. J. Environ. Chem. Eng. **6**, 2269–2280 (2018)
113. Embaby, M.S., Elwany, S.D., Setyaningsih, W., Saber, M.R.: The adsorptive properties of UiO-66 towards organic dyes: a record adsorption capacity for the anionic dye Alizarin Red S. Chinese J. Chem. Eng. **26**, 731–739 (2018)
114. Jin, L.N., Qian, X.Y., Wang, J.G., Aslan, H., Dong, M.: MIL-68 (In) nano-rods for the removal of Congo red dye from aqueous solution. J. Colloid Interface Sci. **453**, 270–275 (2015)
115. Guo, R., Cai, X., Liu, H., Yang, Z., Meng, Y., Chen, F., Li, Y., Wang, B.: In situ growth of metal-organic frameworks in three-dimensional aligned lumen arrays of wood for rapid and highly efficient organic pollutant removal. Environ. Sci. Technol. **53**, 2705–2712 (2019)
116. Koros, W.J., Zhang, C.: Materials for next-generation molecularly selective synthetic membranes. Nat. Mater. **16**, 289–297 (2017)
117. Baker, R.W.: Membrane Technology and Applications 3rd edition. Jon Wiley & Son. (2012)
118. Rangnekar, N., Mittal, N., Elyassi, B., Caro, J., Tsapatsis, M.: Zeolite membranes—a review and comparison with MOFs. Chem. Soc. Rev. **44**, 7128–7154 (2015)
119. Comesaña-Gándara, B., Chen, J., Bezzu, C.G., Carta, M., Rose, I., Ferrari, M.C., Esposito, E., Fuoco, A., Jansen, J.C., McKeown, N.B.: Redefining the Robeson upper bounds for CO2/CH4 and CO2/N2 separations using a series of ultrapermeable benzotriptycene-based polymers of intrinsic microporosity. Energy Environ. Sci. **12**, 2733–2740 (2019)
120. Robeson, L.M.: The upper bound revisited. J. Memb. Sci. **320**, 390–400 (2008)
121. Chung, T.S., Jiang, L.Y., Li, Y., Kulprathipanja, S.: Mixed matrix membranes (MMMs) comprising organic polymers with dispersed inorganic fillers for gas separation. Prog. Polym. Sci. **32**, 483–507 (2007)
122. Denny, M.S., Moreton, J.C., Benz, L., Cohen, S.M.: Metal-organic frameworks for membrane-based separations. Nat. Rev. Mater. **1**, 1–17 (2016)
123. Galizia, M., Chi, W.S., Smith, Z.P., Merkel, T.C., Baker, R.W., Freeman, B.D.: 50th anniversary perspective: polymers and mixed matrix membranes for gas and vapor separation: a review and prospective opportunities. Macromolecules **50**, 7809–7843 (2017)
124. Liu, L., Xie, X., Qi, S., Li, R., Zhang, X., Song, X., Gao, C.: Thin film nanocomposite reverse osmosis membrane incorporated with UiO-66 nanoparticles for enhanced boron removal. J. Memb. Sci. **580**, 101–109 (2019)
125. Li, W., Zhang, Y., Li, Q., Zhang, G.: Metal-organic framework composite membranes: synthesis and separation applications. Chem. Eng. Sci. **135**, 232–257 (2015)
126. Ma, L., Svec, F., Lv, Y., Tan, T.: Engineering of the filler/polymer interface in metal–organic framework-based mixed-matrix membranes to enhance gas separation. Chem. An Asian J. **14**, 3502–3514 (2019)
127. Bae, T.H., Lee, J.S., Qiu, W., Koros, W.J., Jones, C.W., Nair, S.: A high-performance gas-separation membrane containing submicrometer-sized metal-organic framework crystals. Angew. Chemie Int. Ed. **49**, 9863–9866 (2010)

128. Wijmans, J.G., Baker, R.W.: The solution-diffusion model: a review. J. Memb. Sci. **107**, 1–21 (1995)
129. Fick, A.: Ueber diffusion. Ann. Phys. **170**, 59–86 (1855)
130. Maxwell, J.C.: A treatise on electricity and magnetism. Dover Publications Inc, New York (1954)
131. Van Der Bruggen, B., Vandecasteele, C., Van Gestel, T., Doyen, W., Leysen, R.: A review of pressure-driven membrane processes in wastewater treatment and drinking water production. Environ. Prog. **22**, 46–56 (2003)
132. Von Morgan, P.W.: Condensation Polymers: By Interfacial and Solution Methods. Interscience Publishers, Geneva (1965)
133. Baker, R.W., Low, B.T.: Gas separation membrane materials: a perspective. Macromolecules **47**, 6999–7013 (2014)
134. Jeong, B.H., Hoek, E.M.V., Yan, Y., Subramani, A., Huang, X., Hurwitz, G., Ghosh, A.K., Jawor, A.: Interfacial polymerization of thin film nanocomposites: a new concept for reverse osmosis membranes. J. Memb. Sci. **294**, 1–7 (2007)
135. Sorribas, S., Gorgojo, P., Téllez, C., Coronas, J., Livingston, A.G.: High flux thin film nanocomposite membranes based on metal-organic frameworks for organic solvent nanofiltration. J. Am. Chem. Soc. **135**, 15201–15208 (2013)
136. Merkel, T.C., Lin, H., Wei, X., Baker, R.: Power plant post-combustion carbon dioxide capture: an opportunity for membranes. J. Memb. Sci. **359**, 126–139 (2010)
137. Yong, W.F., Chung, T.S., Weber, M., Maletzko, C.: New polyethersulfone (PESU) hollow fiber membranes for CO_2 capture. J. Memb. Sci. **552**, 305–314 (2018)
138. Zornoza, B., Martinez-Joaristi, A., Serra-Crespo, P., Tellez, C., Coronas, J., Gascon, J., Kapteijn, F.: Functionalized flexible MOFs as fillers in mixed matrix membranes for highly selective separation of CO2 from CH4 at elevated pressures. Chem. Commun. **47**, 9522–9524 (2011)
139. Basu, S., Cano-Odena, A., Vankelecom, I.F.J.: MOF-containing mixed-matrix membranes for CO_2/CH_4 and CO_2/N_2 binary gas mixture separations. Sep. Purif. Technol. **81**, 31–40 (2011)
140. Car, A., Stropnik, C., Peinemann, K.V.: Hybrid membrane materials with different metal-organic frameworks (MOFs) for gas separation. Desalination **200**, 424–426 (2006)
141. Nuhnen, A., Klopotowski, M., Tanh Jeazet, H.B., Sorribas, S., Zornoza, B., Téllez, C., Coronas, J., Janiak, C.: High performance MIL-101(Cr)@6FDA-: M PD and MOF-199@6FDA- m PD mixed-matrix membranes for CO2/CH4 separation. Dalt. Trans. **49**, 1822–1829 (2020)
142. Zhang, Y., Musselman, I.H., Ferraris, J.P., Balkus, K.J.: Gas permeability properties of Matrimid® membranes containing the metal-organic framework Cu-BPY-HFS. J. Memb. Sci. **313**, 170–181 (2008)
143. Perez, E.V., Balkus, K.J., Ferraris, J.P., Musselman, I.H.: Mixed-matrix membranes containing MOF-5 for gas separations. J. Memb. Sci. **328**, 165–173 (2009)
144. Bae, T.H., Long, J.R.: CO2/N_2 separations with mixed-matrix membranes containing Mg_2(dobdc) nanocrystals. Energy Environ. Sci. **6**, 3565–3569 (2013)
145. Dorosti, F., Alizadehdakhel, A.: Fabrication and investigation of PEBAX/Fe-BTC, a high permeable and CO2 selective mixed matrix membrane. Chem. Eng. Res. Des. **136**, 119–128 (2018)
146. Nabais, A.R., Ribeiro, R.P.P.L., Mota, J.P.B., Alves, V.D., Esteves, I.A.A.C., Neves, L.A.: CO_2/N_2 gas separation using Fe(BTC)-based mixed matrix membranes: a view on the adsorptive and filler properties of metal-organic frameworks. Sep. Purif. Technol. **202**, 174–184 (2018)
147. Couck, S., Denayer, J.F.M., Baron, G.V., Rémy, T., Gascon, J., Kapteijn, F.: An amine-functionalized MIL-53 metal-organic framework with large separation power for CO_2 and CH4. J. Am. Chem. Soc. **131**, 6326–6327 (2009)
148. Rodenas, T., Luz, I., Prieto, G., Seoane, B., Miro, H., Corma, A., Kapteijn, F., Llabrés I Xamena, F.X., Gascon, J.: Metal-organic framework nanosheets in polymer composite materials for gas separation. Nat. Mater. **14**, 48–55 (2015)

149. Venna, S.R., Lartey, M., Li, T., Spore, A., Kumar, S., Nulwala, H.B., Luebke, D.R., Rosi, N.L., Albenze, E.: Fabrication of MMMs with improved gas separation properties using externally-functionalized MOF particles. J. Mater. Chem. A. **3**, 5014–5022 (2015)
150. Aceituno Melgar, V.M., Kim, J., Othman, M.R.: Zeolitic imidazolate framework membranes for gas separation: a review of synthesis methods and gas separation performance. J. Ind. Eng. Chem. **28**, 1–15 (2015)
151. Al-Maythalony, B.A., Alloush, A.M., Faizan, M., Dafallah, H., Elgzoly, M.A.A., Seliman, A.A.A., Al-Ahmed, A., Yamani, Z.H., Habib, M.A.M., Cordova, K.E., Yaghi, O.M.: Tuning the interplay between selectivity and permeability of ZIF-7 mixed matrix membranes. ACS Appl. Mater. Interfaces. **9**, 33401–33407 (2017)
152. Li, T., Pan, Y., Peinemann, K.V., Lai, Z.: Carbon dioxide selective mixed matrix composite membrane containing ZIF-7 nano-fillers. J. Memb. Sci. **425–426**, 235–242 (2013)
153. Li, Y., Liang, F., Bux, H., Yang, W., Caro, J.: Zeolitic imidazolate framework ZIF-7 based molecular sieve membrane for hydrogen separation. J. Memb. Sci. **354**, 48–54 (2010)
154. Wijenayake, S.N., Panapitiya, N.P., Versteeg, S.H., Nguyen, C.N., Goel, S., Balkus, K.J., Musselman, I.H., Ferraris, J.P.: Surface cross-linking of ZIF-8/polyimide mixed matrix membranes (MMMs) for gas separation. Ind. Eng. Chem. Res. **52**, 6991–7001 (2013)
155. Xiang, L., Sheng, L., Wang, C., Zhang, L., Pan, Y., Li, Y.: Amino-functionalized ZIF-7 nanocrystals: improved intrinsic separation ability and interfacial compatibility in mixed-matrix membranes for CO_2/CH_4 separation. Adv. Mater. **29** (2017)
156. Yang, T., Shi, G.M., Chung, T.S.C.: Symmetric and asymmetric zeolitic imidazolate frameworks (ZIFs)/Polybenzimidazole (PBI) nanocomposite membranes for hydrogen purifi cation at high temperatures. Adv. Energy Mater. **2**, 1358–1367 (2012)
157. Yang, T., Xiao, Y., Chung, T.S.: Poly-/metal-benzimidazole nano-composite membranes for hydrogen purification. Energy Environ. Sci. **4**, 4171–4180 (2011)
158. Askari, M., Chung, T.S.: Natural gas purification and olefin/paraffin separation using thermal cross-linkable co-polyimide/ZIF-8 mixed matrix membranes. J. Memb. Sci. **444**, 173–183 (2013)
159. Díaz, K., López-González, M., Del Castillo, L.F., Riande, E.: Effect of zeolitic imidazolate frameworks on the gas transport performance of ZIF8-poly(1,4-phenylene ether-ether-sulfone) hybrid membranes. J. Memb. Sci. **383**, 206–213 (2011)
160. Diestel, L., Wang, N., Schwiedland, B., Steinbach, F., Giese, U., Caro, J.: MOF based MMMs with enhanced selectivity due to hindered linker distortion. J. Memb. Sci. **492**, 181–186 (2015)
161. Hao, L., Li, P., Yang, T., Chung, T.S.: Room temperature ionic liquid/ZIF-8 mixed-matrix membranes for natural gas sweetening and post-combustion CO_2 capture. J. Memb. Sci. **436**, 221–231 (2013)
162. Yang, T., Chung, T.S.: Room-temperature synthesis of ZIF-90 nanocrystals and the derived nano-composite membranes for hydrogen separation. J. Mater. Chem. A. **1**, 6081–6090 (2013)
163. Lively, R.P., Dose, M.E., Xu, L., Vaughn, J.T., Johnson, J.R., Thompson, J.A., Zhang, K., Lydon, M.E., Lee, J.S., Liu, L., Hu, Z., Karvan, O., Realff, M.J., Koros, W.J.: A high-flux polyimide hollow fiber membrane to minimize footprint and energy penalty for CO 2 recovery from flue gas. J. Memb. Sci. **423–424**, 302–313 (2012)
164. Ordoñez, M.J.C., Balkus, K.J., Ferraris, J.P., Musselman, I.H.: Molecular sieving realized with ZIF-8/Matrimid® mixed-matrix membranes. J. Memb. Sci. **361**, 28–37 (2010)
165. Song, Q., Nataraj, S.K., Roussenova, M.V., Tan, J.C., Hughes, D.J., Li, W., Bourgoin, P., Alam, M.A., Cheetham, A.K., Al-Muhtaseb, S.A., Sivaniah, E.: Zeolitic imidazolate framework (ZIF-8) based polymer nanocomposite membranes for gas separation. Energy Environ. Sci. **5**, 8359–8369 (2012)
166. Thompson, J.A., Chapman, K.W., Koros, W.J., Jones, C.W., Nair, S.: Sonication-induced Ostwald ripening of ZIF-8 nanoparticles and formation of ZIF-8/polymer composite membranes. Microporous Mesoporous Mater. **158**, 292–299 (2012)
167. Yang, T., Chung, T.S.: High performance ZIF-8/PBI nano-composite membranes for high temperature hydrogen separation consisting of carbon monoxide and water vapor. Int. J. Hydrogen Energy **38**, 229–239 (2013)

168. Zhang, C., Dai, Y., Johnson, J.R., Karvan, O., Koros, W.J.: High performance ZIF-8/6FDA-DAM mixed matrix membrane for propylene/propane separations. J. Memb. Sci. **389**, 34–42 (2012)
169. Carlin, D.J., Naujokas, M.F., Bradham, K.D., Cowden, J., Heacock, M., Henry, H.F., Lee, J.S., Thomas, D.J., Thompson, C., Tokar, E.J., Waalkes, M.P., Birnbaum, L.S., Suk, W.A.: Arsenic and environmental health: state of the science and future research opportunities. Environ. Health Perspect. **124**, 890–899 (2016)
170. Safak Boroglu, M., Yumru, A.B.: Gas separation performance of 6FDA-DAM-ZIF-11 mixed-matrix membranes for H_2/CH_4 and CO_2/CH_4 separation. Sep. Purif. Technol. **173**, 269–279 (2017)
171. Wu, X., Liu, W., Wu, H., Zong, X., Yang, L., Wu, Y., Ren, Y., Shi, C., Wang, S., Jiang, Z.: Nanoporous ZIF-67 embedded polymers of intrinsic microporosity membranes with enhanced gas separation performance. J. Memb. Sci. **548**, 309–318 (2018)
172. Japip, S., Liao, K.S., Xiao, Y., Chung, T.S.: Enhancement of molecular-sieving properties by constructing surface nano-metric layer via vapor cross-linking. J. Memb. Sci. **497**, 248–258 (2016)
173. Japip, S., Wang, H., Xiao, Y., Chung, T.S.: Highly permeable zeolitic imidazolate framework (ZIF)-71 nano-particles enhanced polyimide membranes for gas separation. J. Memb. Sci. **467**, 162–174 (2014)
174. Japip, S., Xiao, Y., Chung, T.S.: Particle-size effects on gas transport properties of 6FDA-Durene/ZIF-71 mixed matrix membranes. Ind. Eng. Chem. Res. **55**, 9507–9517 (2016)
175. Yong, W.F., Ho, Y.X., Chung, T.-S.: Nanoparticles embedded in amphiphilic membranes for carbon dioxide separation and dehumidification. Chemsuschem **10**, 4046–4055 (2017)
176. Sarfraz, M., Ba-Shammakh, M.: Synergistic effect of adding graphene oxide and ZIF-301 to polysulfone to develop high performance mixed matrix membranes for selective carbon dioxide separation from post combustion flue gas. J. Memb. Sci. **514**, 35–43 (2016)
177. Morris, W., Doonan, C.J., Furukawa, H., Banerjee, R., Yaghi, O.M.: Crystals as molecules: postsynthesis covalent functionalization of zeolitic imidazolate frameworks. J. Am. Chem. Soc. **130**, 12626–12627 (2008)
178. Wang, C., Liu, X., Keser Demir, N., Chen, J.P., Li, K.: Applications of water stable metal-organic frameworks. Chem. Soc. Rev. **45**, 5107–5134 (2016)
179. Elrasheedy, A., Nady, N., Bassyouni, M., El-Shazly, A.: Metal organic framework based polymer mixed matrix membranes: review on applications in water purification. Membranes (Basel) **9** 88 (2019)
180. Férey, C., Mellot-Draznieks, C., Serre, C., Millange, F., Dutour, J., Surblé, S., Margiolaki, I.: Chemistry: A chromium terephthalate-based solid with unusually large pore volumes and surface area. Science **309**(80), 2040–2042 (2005)
181. Serre, C., Millange, F., Thouvenot, C., Noguès, M., Marsolier, G., Louër, D., Férey, G.: Very large breathing effect in the first nanoporous chromium(III)-based solids: MIL-53 or CrIII(OH)·{O2C-C6H4- CO2}·{HO2C-C6H4 -CO2H}x·H2Oy. J. Am. Chem. Soc. **124**, 13519–13526 (2002)
182. Guillerm, V., Ragon, F., Dan-Hardi, M., Devic, T., Vishnuvarthan, M., Campo, B., Vimont, A., Clet, G., Yang, Q., Maurin, G., Férey, G., Vittadini, A., Gross, S., Serre, C.: A series of isoreticular, highly stable, porous zirconium oxide based metal-organic frameworks. Angew. Chemie - Int. Ed. **51**, 9267–9271 (2012)
183. Kadhom, M., Hu, W., Deng, B.: Thin film nanocomposite membrane filled with metal-organic frameworks UiO-66 and MIL-125 nanoparticles for water desalination. Membranes (Basel). **7**, 31 (2017)
184. Ma, D., Peh, S.B., Han, G., Chen, S.B.: Thin-film nanocomposite (TFN) membranes incorporated with super-hydrophilic metal-organic framework (MOF) UiO-66: toward enhancement of water flux and salt rejection. ACS Appl. Mater. Interfaces. **9**, 7523–7534 (2017)
185. Cavka, J.H., Jakobsen, S., Olsbye, U., Guillou, N., Lamberti, C., Bordiga, S., Lillerud, K.P.: A new zirconium inorganic building brick forming metal organic frameworks with exceptional stability. J. Am. Chem. Soc. **130**, 13850–13851 (2008)

186. He, Y., Tang, Y.P., Ma, D., Chung, T.S.: UiO-66 incorporated thin-film nanocomposite membranes for efficient selenium and arsenic removal. J. Memb. Sci. **541**, 262–270 (2017)
187. Liu, X., Demir, N.K., Wu, Z., Li, K.: Highly water-stable zirconium metal-organic framework UiO-66 membranes supported on alumina hollow fibers for desalination. J. Am. Chem. Soc. **137**, 6999–7002 (2015)
188. Ma, J., Guo, X., Ying, Y., Liu, D., Zhong, C.: Composite ultrafiltration membrane tailored by MOF@GO with highly improved water purification performance. Chem. Eng. J. **313**, 890–898 (2017)
189. Howarth, A.J., Katz, M.J., Wang, T.C., Platero-Prats, A.E., Chapman, K.W., Hupp, J.T., Farha, O.K.: High efficiency adsorption and removal of selenate and selenite from water using metal-organic frameworks. J. Am. Chem. Soc. **137**, 7488–7494 (2015)
190. Férey, G., Mellot-Draznieks, C., Serre, C., Millange, F.: Crystallized frameworks with giant pores: are there limits to the possible? Acc. Chem. Res. **38**, 217–225 (2005)
191. Echaide-Górriz, C., Navarro, M., Téllez, C., Coronas, J.: Simultaneous use of MOFs MIL-101(Cr) and ZIF-11 in thin film nanocomposite membranes for organic solvent nanofiltration. Dalt. Trans. **46**, 6244–6252 (2017)
192. Echaide-Górriz, C., Sorribas, S., Téllez, C., Coronas, J.: MOF nanoparticles of MIL-68(Al), MIL-101(Cr) and ZIF-11 for thin film nanocomposite organic solvent nanofiltration membranes. RSC Adv. **6**, 90417–90426 (2016)
193. Navarro, M., Benito, J., Paseta, L., Gascón, I., Coronas, J., Téllez, C.: Thin-film nanocomposite membrane with the minimum amount of MOF by the langmuir-schaefer technique for nanofiltration. ACS Appl. Mater. Interfaces. **10**, 1278–1287 (2018)
194. Xu, Y., Gao, X., Wang, Q., Wang, X., Ji, Z., Gao, C.: Highly stable MIL-101(Cr) doped water permeable thin film nanocomposite membranes for water treatment. RSC Adv. **6**, 82669–82675 (2016)
195. Xu, Y., Gao, X., Wang, X., Wang, Q., Ji, Z., Wang, X., Wu, T., Gao, C.: Highly and stably water permeable thin film nanocomposite membranes doped with MIL-101 (Cr) nanoparticles for reverse osmosis application. Materials (Basel) **9** (2016)
196. Ma, X.H., Yang, Z., Yao, Z.K., Xu, Z.L., Tang, C.Y.: A facile preparation of novel positively charged MOF/chitosan nanofiltration membranes. J. Memb. Sci. **525**, 269–276 (2017)
197. Aljundi, I.H.: Desalination characteristics of TFN-RO membrane incorporated with ZIF-8 nanoparticles. Desalination **420**, 12–20 (2017)
198. Hu, Z., Chen, Y., Jiang, J.: Zeolitic imidazolate framework-8 as a reverse osmosis membrane for water desalination: Insight from molecular simulation. J. Chem. Phys. **134**, 134705 (2011)
199. Wang, F., Zheng, T., Xiong, R., Wang, P., Ma, J.: Strong improvement of reverse osmosis polyamide membrane performance by addition of ZIF-8 nanoparticles: Effect of particle size and dispersion in selective layer. Chemosphere **233**, 524–531 (2019)

Porous Carbons as Oxygen Reduction Electrocatalysts

Teresa J. Bandosz

Abstract In search for new alternative sources of energy carbon materials placed themselves as a valuable alternative to replace efficient noble metal- or transition metal-based catalysts. Even though the search for this kind of efficient catalysts started from the discovery and wonders of graphene, soon it has been established that to bring a catalytic activity for oxygen reduction to graphene an introduction of heteroatoms to its matrix or defect engineering is a must. The intent of this review is to present a brief journey on the development of heteroatom -doped "flat" carbons for an oxygen reduction electrocatalysis, followed by an indication of the role of defects as a general driving force, through heteroatom-doped porous carbons to heteroatom-free porous carbons. The emphasis is on importance of porosity, no necessary that 3-D engineered. The development of that concept is presented in the chronological order. Since to summarize all works on this topic sounds as a rather impossible task, and numerous reviews on the specific aspects of carbon-based metal-free ORR catalysts have been published recently, our intention is to stress the works where other factors than an alteration in the electronic/structure affect or could affect the ORR efficiency. These other factors include oxygen accessibility to small pores and its adsorption there. Even though the development of pores certainly is associated with an increase in the level of defect, the pore- influenced oxygen reduction mechanisms addressed here differs from that on the electronic defects-based catalytic sites.

Keywords Porous carbons · Oxygen reduction reaction (ORR) · Oxygen adsorption · Surface chemistry · Heteroatoms · Microporosity

1 Introduction

One of the major tasks facing a contemporary society is the development of renewable energy sources being able to replace fossil fuels. There are some specific requirements concerning the feasibility of these new sources/technology implementations. Thus,

---

T. J. Bandosz (✉)
Department of Chemistry and Biochemistry, The City College of New York, New York, NY 10031, USA
e-mail: tbandosz@ccny.cuny.edu

J. C. Moreno-Piraján et al. (eds.), *Porous Materials*, Engineering Materials,
https://doi.org/10.1007/978-3-030-65991-2_2

their costs should be low and they should not cause any further damage to our environment. Therefore, there is an extensive search for inexpensive materials, which would be able to produce photocurrent in photovoltaic devices, split water to produce H_2 with minimum or zero fossil energy requirements, reduce oxygen for fuel cell applications, or convert CO_2 released from fossil fuel into hydrocarbons.

The efficiency of the processes mentioned above is based on the application of efficient catalysts promoting target reactions and enhancing their yield. Here noble metals are considered as the best catalysts. Even though they provide a high efficiency, their drawbacks are scarcity and high costs. Therefore, as a natural path of a research development, transition metals as catalysts in energy related applications have started to be extensively explored. In their case detrimental environmental effects are an disadvantage. Considering the above, the efforts of materials scientists have focused on searching for non-metal-based catalysts. The obvious candidates are those carbon-based [1, 2], graphitic carbon nitride [3–5] or, quite recently, boron nitride [6].

The usage of the carbon materials as direct catalysts was boosted by the discovery of carbon nanotubes and graphene. The advantageous assets of these materials are a high electrical conductivity and easiness of surface modifications. Extensive efforts on those modifications and the discovery of an enhanced activity owing to heteroatom doping directed the attention of scientists to carbon nitride [3–5] or boron nitrides [6]. Nevertheless, they need carbon doping in order to increase an electrical conductivity. Since, in catalysis, a surface area is an important factor, there are also extensive efforts towards modifications of these 2-D materials leading to the development of porosity [7] and 3-D structures. All of these efforts result in some cost increase for materials, technology and thus for consumers.

In many strategically important applications of nanoporous carbons, such as an energy storage, the porosity of supercapacitor electrodes is very important. It provides space for an electric double layer, which is the main mechanism for a charge storage. For this, not only the volume of pores but also their sizes are of critical importance [8–10]. It was found that pores similar in size to electrolyte ions enhance the capacitance in an "abnormal way" [11]. This directed scientists to look for reasons explaining this behavior. One of the hypotheses that have been recently taken into consideration is that ions in pores behave similarly to the adsorbed phase on the surface of solids [12, 13]. The molecular dynamic simulations suggested that to better explain the properties of an electrolyte-solid interface physical adsorption forces should be also taken into account [12], besides Coulombic interactions.

The discovery of graphene changed carbon science, and opened other, unforeseen applications, mainly in energy-related fields, such as optics and electronics. This, combined with growing interest in an importance of an interface for various processes and with advancements in instrumentation, especially in the resolution of microscopes, directed carbon scientists to take another look at nanoporous carbons and to reach *beyond adsorption* to better understand and further explore the new applications of these materials [14, 15]. Recent studies have shown that indeed the pores of carbons are built of distorted graphene layers [16] and those "poor person-graphenes" can exhibit similar properties to those of graphene [14]. Moreover, a developed porosity is a very unique asset of porous carbon, either amorphous or with 3-D carbon structures. In fact, for many applications of graphene the specific porous structure has to be built, and often the target chemistry is the same chemistry common to all carbon-based solids [17].

Recently, owing to growing scientific interests, numerous review papers have been published summarizing the results on the oxygen reduction reaction (ORR) activity of various carbon-based materials [1, 18–25]. Therefore, the objective of this chapter is not reintroducing the well -described findings but rather to focus on an aspect, which we consider has not been addressed sufficiently yet, and which has been recently found as important for ORR—on the role of carbon pores, especially micropores where adsorption forces are the strongest ones. Even though one can argue that the oxygen reduction is a purely catalytic process, interactions of the oxygen molecule with a catalyst surface cannot be neglected (Scheme 1) [20, 26–29]. Only very recently, the review paper by Jorge et al. addressed the 3-D carbon structures as ORR catalysts but it rather focuses on more "exotic" carbons such as hard-and soft-templated, self-sacrificial templated, assembly of 3D carbon nanostructures, or 3D porous carbons macrostructures [24] than on amorphous and microporous carbons. Thus, in this review, ORR on "classical" platinum catalysts and the findings on "flat" 2-D carbon materials such as graphene are addressed very briefly as an introduction to the role of a surface chemistry-induced catalytic activity. Then, based on those findings, the role of pores, especially micropores in ORR is discussed in a detailed summary of very recent reports. Of course, one has to remember that the pores in carbons do not exist in isolation and the complexity of the system includes also surface chemistry/doping and defects. Therefore, as another objective, the excellent performance of porous carbons as electrocatalysts for ORR is linked to the combination of all these features with emphases on porosity, as a co-determining feature.

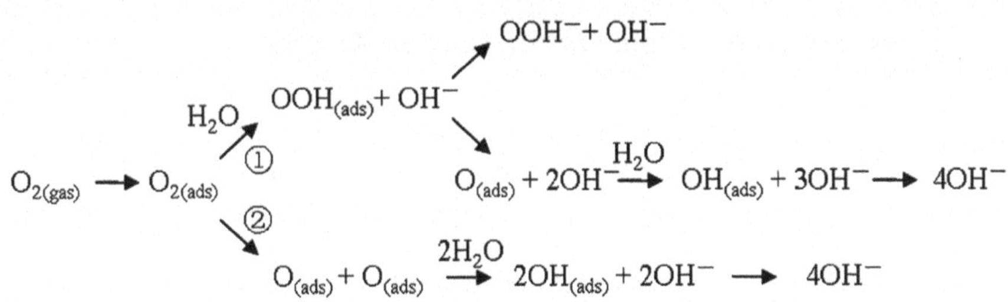

Scheme 1 The associative (1) and dissociative (2) ORR mechanism in alkaline electrolyte. Reprinted with permission from Ref. [20]. Copyright 2011, Elsevier

2 ORR on Pt Catalysts Supported on Carbon Materials

Important catalysts in energy fields are those promoting oxygen reduction in fuel cells. Fuel cells convert chemical energy from a fuel into electric energy [30]. The oxygen reduction reaction occurs on a cathode. The reaction is very slow and the application of a catalyst is a necessity. In aqueous solutions, the reaction goes through two different mechanisms, depending on the chemical nature of a medium/electrolyte [31]. The oxygen reduction through a $4e^-$ pathway, producing either water or OH−, in an acidic or basic electrolyte, respectively, is the desired reaction in the fuel cells. Besides a high energy production, it prevents the formation of sub-products, such as H_2O_2, that can be harmful to the cell and to the catalyst.

The most commonly used electrocatalyst is platinum supported on carbon materials [31–35]. Nowadays, all scientific efforts focus on enhancing several aspects of platinum-based catalysts, trying to increase the efficiency of platinum through formation of various alloys [34, 35], or using different facets in the platinum surfaces [36, 37]. The replacement of platinum in the ORR catalysts targets either the use of other metals of higher availability and lower costs, or the development of metal-free carbon materials doped with various heteroatoms and functional groups [1, 38]. The first approach proposes the use of metals such as iron, cobalt and copper. In recent years there has been a great development in non-noble metal or non-metal electrocatalysts, which could have a sufficient performance in ORR and could be used as replacement for Pt/C materials [19, 39–50].

3 Fundamentals of ORR on Nonporous/Flat and 2-D Carbon Materials

3.1 Effects of Heteroatoms in the Carbon Matrix

The most common carbon catalysts used in ORR are those based on nanoforms of carbons, such as CNT, graphite oxide or graphene [39–52]. To increase the efficiency of the catalytic processes, their surfaces have been modified with heteroatoms such as nitrogen [43, 51–55], sulfur [52], phosphorus [44, 45] or boron [44]. These heteroatoms have been indicated as providing catalytically active sites for oxygen electro-sorption and reduction [49, 50]. Either a direct doping or introduction of functional groups have been explored [41–47, 56–61].

So far, the most extensively studied metal-free ORR carbon-based catalysts have been those doped with nitrogen or co-doped with nitrogen and sulfur. Qiao et al. [50] did their experiments on materials considered as mesoporous graphene. They were a silica templated assembly of defectous graphene sheets doped with nitrogen and sulfur. As heteroatom precursors, melamine and benzyl disulfide were used, respectively. The co-doped materials contained 2% of S, 4.5% of N and 4.6% of O. Even though doping only with nitrogen resulted in a similar number of electron transfer as that on nonmodified graphene and equal to 3, a slight increase in a kinetic limited current upon doping was found. That effect was more pronounced for the sulfur doped sample on which the number of electron transfer increased to 3.3. On the other hand, the co-doped with N and S graphene sample exhibited the kinetic limited current higher than that on Pt/C (32 mA cm^{-2}) with the number of electron transfer reaching 3.6. To explain their findings, the authors used density functional theory (DFT) calculations. Thus, the activity on the N-doped sample was linked to the incorporation of N atom to the carbon matrix, which is more electronegative than the carbon atom is (3.04 vs. 2.55). It resulted in a high charge density of surrounding carbon atoms. The incorporation of sulfur, owing it its similar electronegativity to that of carbon (2.58) resulted in a very small effect on the charge transfer and the catalytic effect was linked to a mismatch of the outer orbitals of sulfur and carbon, which caused that sulfur was positively charged and thus was a catalytic center for oxygen reduction. On the other hand, doping with N and S resulted in asymmetrical spin and a positive charge density on surrounding carbon atoms, which are the catalytic centers for ORR [49, 50]. Thus, the introduction of heteroatoms to the carbon rings changed the electronic state of the carbon matrix, resulting in the formation of catalytic sites, on which the oxygen reduction took place. On such catalysts, the number of electrons transfer was close to 4 [48], and the kinetic current reached almost 30 mA/cm^2 [39]. Nevertheless, the onset potential was still much less positive that that on the platinum catalysts. Co-doping with S and N heteroatoms was reported to bring a very good activity for ORR. Such promising results as $n = 3.8e^-$ and $J_k = 27.0$ mA cm^{-2} at −0.70 V versus Ag/AgCl (0.28 V vs. RHE), or $n = 3.3e^-$ [41] and $J_k = 24.5$ mA cm^{-2} at −0.50 V versus Ag/AgCl (0.48 V vs. RHE) were recorded [50]. However, in the majority of studies on doped carbons as the ORR catalysts, the onset potentials or the number of electrons transfer, although close to $4e^-$, are not sufficient for fuel cell applications. Nevertheless, an important feature of the carbon-based catalysts is their high tolerance to methanol crossover, contrary to that of Pt/C [48, 49, 62].

Recently, the performance of N and B co-doped graphene or carbon nanotubes has been explored and found promising for ORR [44, 47, 63–69]. On these materials, reduction of oxygen occurs with the number of electron transfer, n, close to 4 [67]. Lee et al. [70] evaluated N and B co-doped carbon- based materials obtained from flue gas as the ORR catalysts. Both experiments and simulations were used. The content of N and B in their materials reached 9% and 2.4% respectively. The highest current density was 7 mA cm^{-2} and was larger than that on Pt/C and the number of electron transfer was 3.8. By DFT calculations the authors found that the Gibbs free energy on the co-doped carbon materials was favorable for the 4-electron transfer process. A summary of the effects of other dopants than nitrogen, including boron, phosphorus and sulfur on ORR is presented in the reviews by Paraknowitsch and Thomas [71] and by Zhao and Xia [72]. The latter presents the design principles for dual-element doped catalysts. Providing a descriptor Φ being a product of electronegativity of heteroatoms (E_X) and electron affinity (A_X) relative to those of carbon was an interesting step helping to design bifunctional catalysts (Fig. 1). Regarding phosphorus, even though an incorporation of this element to "flat" carbons has not been extensively studied, the published results show some advantages of those structures

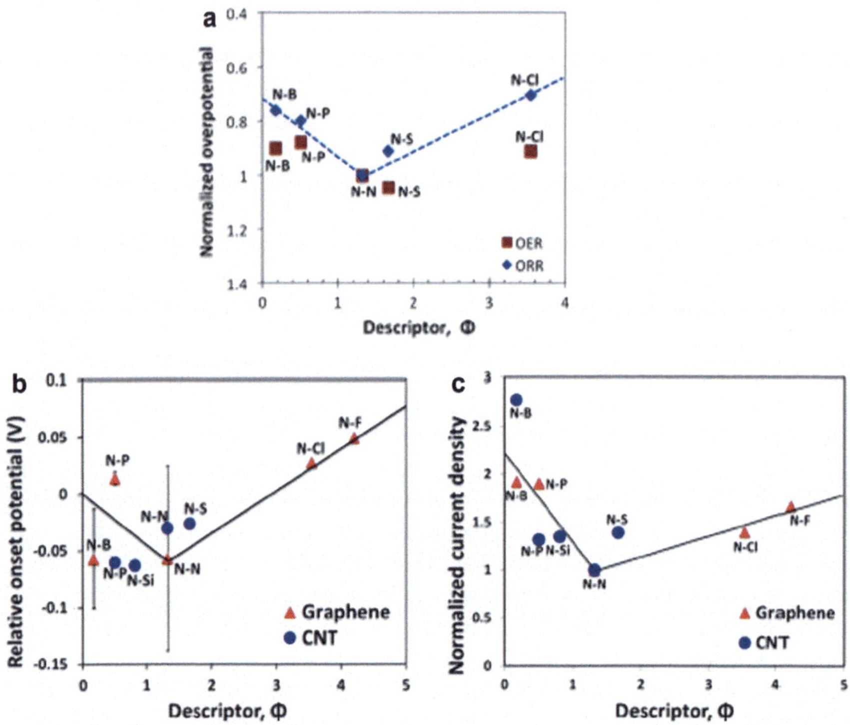

Fig. 1 **a** Predicted average ORR/OER overpotentials of the active sites on N–X codoped graphene (X = P, B, S, and Cl), normalized by the overpotential on the same positions in N–N graphene nanoribbons, as a function of the descriptor, Φ. **b** Average relative onset potential (onset potential of codoped graphene minus onset potential of Pt/C electrode in the same experiment). **c** Average measured limiting current density from the linear scan voltammogram (LSV) curves, normalized by N-doped carbon electrode density under the same conditions, as a function of description Φ for N–B, N–P, N–F, N–S am N–Cl codoped graphene, and for N–B, N–P, N–Si, and N–S codoped CNTs. Reprinted with permission from Ref. [72]. Copyright 2016, American Chemical Society

for ORR (Fig. 2) [41, 44, 72]. A very comprehensive review of the ORR process has been recently published by Ma et al. [22].

3.2 *Effect of Defects in the Carbon Matrix*

With the exception of pure graphene layers or ideal carbon nanotubes, carbons, either flat or 2-D, have some degree of complexity related to their defects. These defects are intrinsic and arise from the presence of "not a 6-member" carbon rings. Such imperfection affects the electronic density of a carbon matrix. The latter is also affected by the introduction of heteroatoms, as discussed above, and by the incorporation of single metal atoms to a carbon matrix.

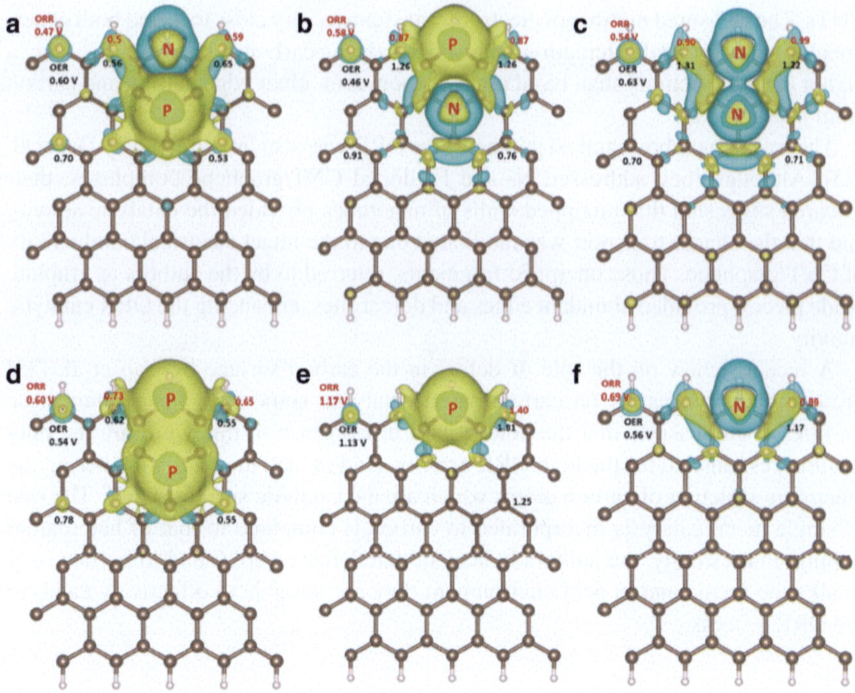

Fig. 2 Differential charge density (between codoped and undoped graphene) and overpotential distribution of **a** N–P codoped, **b** P–N codoped, **c** N–N-codoped, **d** P–P codoped, **e** single P doped, **f** single N-doped graphene nanoribbon with zig-zag edge. Brown and white balls refer to C and H atoms, respectively, Yellow and blue colors indicate the negative and positive values of electron quantities. The isosurface value is set to 0.0015. The ORR (in red) and OER (in black) overpotentials of the active sites ($U < 1.5$ V) are also listed on them. Compared with single element doping, the duel-element doping leads to more active sites for ORR and OER. Reprinted with permission from Ref. [72]. Copyright 2016, American Chemical Society

The presence of defects was used to explain the activity of pure carbon nanocages (heteroatom free) by Hu et al. [73]. Combining experiments with theoretical calculations, they found that intrinsic carbon defects, existing in a carbon matrix, contribute to the observed enhanced ORR activity, irrespective of the presence of nitrogen. They suggested that zigzag edge defects or pentagon defects provide the catalytic activity for ORR, and the number of these defects might be naturally increased in heteroatom-doped carbons. In their most active carbon, not only pores with sizes less than 1 nm were present but also a high volume of pores with sizes of about 10 nm was detected; however, to explain the activity the emphasis was placed only on the edge and pentagon intrinsic defects, which were linked to a high surface area. Nevertheless, the electron transfer number was only 2.9. The effect of the zing-zag carbon on ORR has been also recently addressed by Xue et al. [74]. They have constructed a cathode catalyst based on a composite of zigzag-edged graphene nanoribbons and

CNTs. The measured number of electrons transfer was very close to 4, and both experiments and theoretical calculations showed the zigzag carbon defects as possessing a higher catalytic activity than basal plane carbon, arm- chair edge carbon and carbon atom near a void.

The role of a carbon catalyst geometry for ORR was also investigated by Dai et al. [75]. Although they addressed N- and Fe-doped CNT/graphene complexes, their research suggested that unzipped walls of nanotubes provided the catalytic activity and that the charge transport was facilitated due to the intact electrical conductivity of CNT/graphene. Those unzipped fragments, referred to by the authors as graphite oxide pieces, provided abundant edges and defect sites, enhancing the ORR catalytic activity.

A recent review on the role of defect in the carbon surface by Yao et al. [25] presents the best results on carbon-based catalysts, some even better than those on Pt/C. It emphasizes that the defects in a broad sense of this word are the only features responsible for the high ORR activity. And in fact, the clue is in altering the electronic structure of carbon atoms which are the catalytic sites for ORR. The role of single metal catalysts incorporated to carbon is compared to that of heteroatom doping. Interestingly, the authors indicated, that defects were found important only in alkaline media and in acid environment they are not able to effectively catalyze the ORR process.

4 Doped Porous Carbons with Main Emphasis on Their Surface Chemistry

4.1 ORR on Nitrogen-Doped Porous Carbon

The findings on the high activity of nitrogen doped graphene directed the attention of scientists to investigate nitrogen modified porous carbons as the ORR catalysts. One of the first reports addressing this issue was published by Dai et al. [76]. They prepared ordered mesoporous carbons of the surface area reaching 2000 $m^2\ g^{-1}$, which were doped with nitrogen. The N content was between 3.6 and 6%. The nitrogen chemistry of all tested samples was similar and ~45% of nitrogen species was in pyridinic configurations and about 9%—in graphitic configurations. Interestingly, the highest number of electron transfer (3.72) was found on lowest nitrogen-content carbon of the highest surface area. An onset potential was more positive with an increase in the surface area. Since mesopores contributed to the surface area development, it was concluded that pyridinic and graphic nitrogen are active sites for ORR but the limited performance of the sample with the high content of nitrogen was linked it its high contribution of micropores which made the catalytic sites kinetically inaccessible to O_2. Generally, the dependence of the activity on the mesopore surface area was indicated. In fact, one could also link it to the high dispersion of the catalytically active centers.

Mentius et al. derived specific carbon nanoparticles from polyaniline [77]. The degree of microporosity was between 63 and 90% and the largest surface area reached 540 m^2 g^{-1}. The content of nitrogen on the surface was between 5.5 (for the sample with the largest surface area) and 7% (for the sample with the smallest surface area of 410 m^2 g^{-1}). The catalysts exhibited a good performance in ORR. The onset potentials were −0.05, −0.12 and −0.1 versus SCE and the number of electron transfer reached 3.5 (2–3.5). The kinetic current density was 2.25 mA cm^{-2}. The best performing sample was the one having the highest content of nitrogen and pyridinic nitrogen and the smallest surface area and lowest degree of microporosity. The authors suggested that the high degree of microporosity might hinder the electrochemical response by reducing the access of molecular O_2 to active sites in micropores. They suggested that ORR rather takes place on the surface of carbon particles and in mesopores owing to the slow access of O_2 to micropores.

Microporous nitrogen doped carbons as the ORR catalysts were studied by Yao et al. [78]. They directly carbonized aluminum-based metal organic framework (MIL-53) modified with amines. Resulting carbon had the surface area between 1000 and 1600 m^2 g^{-1} and an increase in the carbonization temperature increased the volume of mesopores. The content of nitrogen was between 1 and 6% and decreased with the increased carbonization temperature from 600 to 1000 °C. While the kinetic current density was about 4 mm^{-2} and close to that on Pt/C (5 mA cm^{-2}), the number of electron transfer reported was more than 4 (4.2 for Pt/C), which must have been a result of the calculation error. Nevertheless, the best performing sample having n close to that of Pt/C was the sample obtained at 700 °C and having the highest content of nitrogen and the relatively high surface area in micropores (1276 m^2 g^{-1}). The authors suggested that the high surface area in micropores combined with nitrogen doping might provide more catalytically active area and more active sites for ORR. Their observation on the role of microporosity was contradictory to that provided by Dai et al. [76].

Nitrogen doped carbon spheres of hierarchical porosity were synthesized in a mesoporous silica template by Chen et al. [79]. Methyl violet was their source of nitrogen. The carbon had micro/mesoporous structure with the BET surface area reaching 1400 m^2 g^{-1} and the very high total pore volume of 2.96 cm^3 g^{-1}. They were tested as the ORR catalysts in an alkaline medium. The total content of nitrogen was not reported but an N/C ratio decreased from 4.82% for the sample obtained at 600 °C to 2.25% for that one synthesized at 1000 °C. Nitrogen was present in pyridinic, pyrrolic, quaternary and pyridine-N-oxide configurations and an increase in the synthesis temperature increased the contribution of quaternary nitrogen from 12 to 50% and decreased the contribution of pyridinic nitrogen from 38 to 23% and that of pyrrolic one- from 40 to 18%. The highest kinetic current density was 4.6 mA cm^{-2} and it exceeded than that on Pt/C (3.5 mA cm^{-2}). The number of electron transfer, as in the work by Yao et al. [78] was higher than 4 (4.06) and also higher than that on Pt/C (4.01). The sample obtained at the highest temperature (1000 °C) was considered as the best performing one. It had the highest surface area and pore volume. The relative contribution of quaternary and pyridinic nitrogen was also the highest (73.6%) with the ratio of N/C was the lowest (2.5%). The good performance was linked to a three-dimensional interconnected hierarchical pore framework (with micropores) and a relatively high conductivity. The catalyst showed the high tolerance to methanol crossover and generally, it was considered as overperforming that of Pt/C.

Nitrogen doped microporous carbon spheres were also investigates as the ORR catalyst by Zhang and coworkers. Poly(o-methylaniline) was used as a source of carbon [80]. A high pyrolysis temperature of 900 °C resulted in the materials of the 727 $m^2 g^{-1}$ surface area and the N content varied from 4.5 to 1%. The latter decreased with an increase in the carbonization temperature between 700 and 1000 °C. The current density was 4 mA cm^{-2} and very close to that on Pt/C. The number of electron transfer was reported as 4. The catalysts were tolerant to methanol crossover. Four types of nitrogen species were detected on the surface with N/C between 1 and 5%. But in this case an increase in carbonization temperature to 1000 °C decreased the number of electron transfer from 4 (at 900 °C) to 3.4. That increase also drastically decreased the surface area from 727 to 202 $m^2 g^{-1}$, however the effect of porosity on ORR was not discussed. The onset potential for the best preforming sample was 0. 91 versus RHE and close to that on Pt/C (0.93 V). The authors believed that a synergistic effect between the high BET surface area, specific and "tuned" total nitrogen content with a high contribution of quaternary nitrogen was responsible for the good performance. It is important to mention that the performance of those microporous microspheres was similar to that of nitrogen doped carbon spheres of hierarchical porosity investigated by Chen et al. [79]. Both materials have similar contents and chemistry of nitrogen but the surface areas of the carbons investigated by Zhang et al. [80] were much similar to each other and their catalysts were predominantly microporous. This might suggest that hierarchical porosity is a crucial feature of the efficient ORR carbon-based catalysts.

The direct effect of a pore size distribution in nitrogen doped carbon on their activity in ORR was first mentioned by Titirici et al. [81]. They also tested carbon microspheres but their materials were obtained from pyrrole in porous silica as a template. The pore size in silica templates were 2.25 nm and 3.6 nm. While the former template led to microporous carbon with a 70% pore volume in pores smaller than 2 nm (predominant size was 1 nm) the latter- to mesoporous carbon of a bimodal pore size distribution centered at 3.2 and 13.7 nm. The content of nitrogen was ~ 8% and the mesoporous sample was richer in quaternary nitrogen than the microporous one, which had a higher contribution of pyridines on the surface. Interestingly, the surface areas were very similar and of about 1200 m^2/g. When tested as the ORR catalysts in the alkaline electrolyte, the kinetic current density on both samples was ~3.3 mA cm^{-2} and lower than that on Pt/C (4.2 mA m^{-2}). Similar were also the numbers of electron transfer and very close to 4. On the other hand, a limited current density was higher on mesoporous carbon and it exceeded that on Pt/C (6.5 mA m^{-2} vs. mA m^{-2}). The onset potential on the mesoporous sample as also higher than that on the microporous one. The good performance of the mesoporous sample was linked not only to its higher content of quaternary nitrogen, as more active catalytic centers than those of pyridinic origin, but also to an easily accessible porosity consisting of two interconnected systems of mesopores which facilitated the access of electrolyte to the active sites. Since, according to the authors, the main difference in the surface features was in the pore size distribution, this feature was indicated as affecting the ORR efficiency and large pores-as more beneficial for the catalytic reduction process. It is important to mention here that even though the electronic conductivity of both carbons was considered as similar, the mesoporous sample was more conductive than the microporous one (10 S m^{-1} vs. 2 S m^{-1}) and the conductivity of the carbon electrode affect the electron transfer processes.

The effect of a hierarchical porosity in nitrogen doped carbons on ORR was also noticed by Eisenberg et al. [82]. They synthesized N-doped self-templated carbon by carbonization of magnesium nitrilotriacetate between 600 and 1000 °C. The surface areas were between 400 and 1800 m^2 g^{-1}. Although their main objective was to study the evolution of porosity with an increase in the heat treatment temperature, they tested the synthesized carbons as the ORR catalysts in an alkaline environment. A marked increase in the porosity was found when the synthesis temperature increased to 900 °C and the surface areas ranged from 409 to 1831 m^2 g^{-1}. Interestingly, the content of nitrogen was affected rather slightly (decreased from 6.9–5.9%) and the distributions of nitrogen species (pyridines, quaternary and N-ox) were changed very little. In all samples quaternary nitrogen predominated (~50%). On the other hand, the treatment at 1000 °C decreased the porosity and the surface area (1519 m^2 g^{-1}). The nitrogen content also decreased to ~4%. The number of electron transfer on the sample obtained at 600 °C was 2, on those at 700–800 °C ~ 3, and on the samples heated at 900 and 1000 °C n was 3.64 and 3.79, respectively. Even though in those samples such properties as the surface area, graphitization degree, nitrogen concentration and distributions were affected by heat treatment temperatures, five activity trends related to the structure were identified. They were as follows: (1) microstructure plays the deciding role over the whole range of the temperature treatment. It affects the graphitization level and nitrogen content; (2) an increase in the surface area boosts the catalytic performance; (3) the benefits of the increased graphitization between 600 and 1000 °C outweigh the loss of the nitrogen-based catalytic centers; (4) the presence of mesopores is important; (5) the first step pyrolysis temperature determines the majority of the carbon properties. Interestingly, the main conclusion was that the ORR activity correlated best with microporosity, which emphasized the importance of active sites exposure.

Another group of interesting N-doped porous carbons evaluated as the ORR electrocatalysts was addressed by Cazorla-Amoros et al. [83]. Their carbons were obtained from PANI at 600 and 800 °C using either inert (N_2) or slightly oxidized atmosphere (0.3% O_2 in N_2). The carbons were microporous with the surface area between 340 and 630 m^2 g^{-1}. Their content of nitrogen was relatively high and ~12% in the samples obtained at 600 °C and ~7% in the samples obtained at 800 °C. Quaternary nitrogen was found only in the latter group of samples. The kinetic current density on the latter group was higher than that on the former one (~2 mA cm^{-2}) and reached 5 mA/cm^2. The number of electron transfer on the sample treated at 800 °C was 3 and the onset potential—0.75 V versus RHE (0.94 V was measured in Pt/C). Interestingly, the stability of the catalysts was higher than that of Pt/C. The best performing samples in the each temperature treatment category were those obtained at the atmosphere containing oxygen. Their activity was linked to surface chemistry and especially to N–C–O species as most active for ORR. In this configuration the carbon atom was found as having the highest positive charge. It was proposed that two N–C–O configurations presented in vicinity at the carbon plane edges (two pyridones) lead to the direct reduction of dioxygen to water (Fig. 3). Such chemistry was found on the samples obtained at 800 °C. Even though the effect of porosity was not analyzed, it is important to mention that the best performing carbons from both categories, besides having the high nitrogen content were also the most porous.

Following the previous line of research, Cazorla-Amoros et al. prepared N-doped carbon from aniline by the thermal treatment at 1100 and 1200 °C [84]. On those carbons the kinetic current density was similar to that of Pt/C but H_2O_2 yield was still between 0 and 40% in the potential range between 0.77 and 1 V versus RHE. The carbons had a very high purity and their surface areas were about 750 $m^2\ g^{-1}$. They had the low nitrogen content of about 1–1.8%, with a comparable contribution of pyrroles and quaternary nitrogen (~35% each). The authors noticed the change in the activity of the catalytic sites with the potential for those high temperature treated samples. While at the potential lower than 0.75 V 4 electrons were transferred, at the higher potential—only 2. Their results, along with the analysis of the model structures, suggested that quaternary nitrogen in the zig-zag configurations was the most active site in their samples. It was because in this configuration oxygen molecule can be bonded through the edges of the graphene plane with a minimum repulsion between O_2 and the N atoms and each oxygen atom of O_2 can be adsorbed in a bridge configuration on the two positively charged carbon atoms. On the other hand, internal quaternary N species were not active. The authors linked the activity of the high-temperature synthesized N-doped carbons to those quaternary species existing at the edge sites, which were formed from pyridinic configurations via thermal transformations. They also mentioned that the conductivity and porosity are important for ORR. However, the role of the latter was not analyzed in details. Nevertheless, it should be mentioned here that since those high temperature synthesized N-doped carbons had higher surface areas than those addressed in Ref. [83] (740 $m^2\ g^{-1}$ vs. 628 $m^2\ g^{-1}$ max for the sample obtained at 800 °C in the presence of oxygen [83]) and much less nitrogen, the porosity might play indeed an important role in their ORR activity.

Another view on the catalytic activity on N-doped porous carbon was proposed by Chen et al. [85]. They prepared their nitrogen doped carbons (NCF) by pyrolysis of a melamine foam at 1000 °C, which was followed by reduction with hydrazine. The obtained samples were micro/mesoporous with the surface areas between 730 and 810 $m^2\ g^{-1}$. The kinetic current density in the alkaline electrolyte reach 4 mA cm^{-2} and the highest number of electron transfer (3.8) was measured on the nonreduced sample. That sample had also the most positive onset potential (−0.05 V vs. Ag/AgCl). The content of nitrogen was 4.8 at.% for the NCF sample and decreased slightly upon reduction. The detailed XPS analysis of C 1 s, O 1 s and N 1 s core energy level spectra suggested that quinones on the surface of the best performing sample were responsible for its enhanced ORR activity. This was an interesting finding but worth to mention is also the highest surface area of that sample and, as

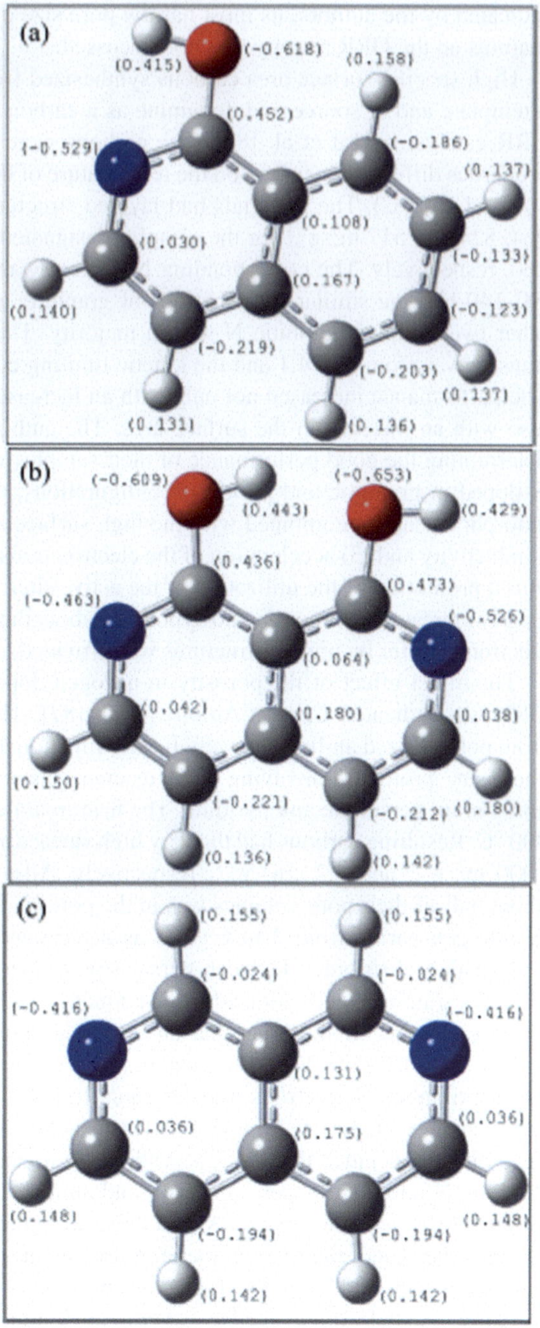

Fig. 3 Model structures for different heteroatom-containing molecules and the calculated effective charge of all atoms: **a** on pyridone groups, **b** two adjacent pyridones groups and **c** pyridone. H is white, C is grey, N is blue, and O is red. Reprinted with permission from Ref. [83]. Copyright 2017, Elsevier

indicated by the authors, its most narrow pore size distribution. The effect of these features on the ORR activity was not discussed.

High specific surface area carbons synthesized from g-C_3N_4, as simultaneously a template and N source and dopamine as a carbon source, were addressed as the ORR catalyst by Lu et al. [86]. The carbons were obtained at 900 °C and their properties differed depending on the temperature of the precursors preparation (120, 140 and 160 °C). The materials had layered structures and the surface areas were 964, 824 and 517 $m^2 g^{-1}$ for the samples designated as NC-120, NC-140 and NC-160, respectively. The corresponding N content was 7.71, 5.43 and 3.45%. While NC-140 had the similar contributions of graphitic and pyridinic nitrogen, for the other two samples graphitic N was in majority. The reported number of electron transfer was from 3 to 4.1 and the kinetic limiting current from 1.1 to 5 mA cm^{-2}. The performance increased not only with an increase in the content of nitrogen but also with an increase in the surface area. The authors concluded that the features determining the good performance of their samples were (1) layered structure with N-doped in graphitic and pyridinic configurations; (2) coexistence of graphic and amorphous carbon combined with the high surface area that provided the electron conductivity and (3) acceleration of the electron transfer by the porous structure and thus a promotion of the utilization of the active sites. Even though the porosity was clearly indicated as important, no hypothesis about the mechanism of the acceleration electron transfer by porous structures was provided.

The direct effect of the porosity in nitrogen doped carbon on the efficiency of ORR was indicated Cazorla-Amoros et al. [87]. Their carbons were synthesized from polymerized aniline and were further functionalized using the specific organic chemistry protocols involving such reactants as nitric acid, ammonium nitrate, ammonium persulfate and pyridine. The heat treatment temperatures were 600 and 800 °C. Resulting carbons had the very high surface areas and pore volume reaching 3000 $m^2 g^{-1}$ and 1.2 $cm^3 g^{-1}$, respectively. Adsorption of CO_2 suggested that about half of their pore volumes was in the pores smaller than 0.7 nm. The content of nitrogen varied from 3 to 6 at.%, as determined by XPS and nitrogen bonds were widely distributed between pyrroles/pyridones, pyridines/imines, quaternary N, amides/amines, and oxidized N. The kinetic current density was between 4 and 5 mA cm^{-2} and smaller than that on Pt/C. The most positive onset potential was 0.88 V (on Pt/C 0.98 V was measured). That particular sample exhibited also the highest number of electrons transfer equal to 3.4. Generally, the number of electron transfer varied from 2.5 to 3.4. After the detailed analysis of surface chemistry, the authors concluded that the low synthesis temperature led to carbon rich in non-catalytic N-based sites such as amines and amides and to a decrease in the populations of the catalytic edge sites, and those were the reasons for their low activity. On the other hand, the high temperature treatment generated more electrocatalytic nitrogen groups and led to the high concentration of N–C–O configurations, which were previously [83] indicated as governing the high ORR efficiency. An important finding, although without a deeper analysis, was that the pristine sample showed a remarkable photoactivity and that photoactivity was linked to a well-developed porosity.

Rich in micropores N-doped carbon aerogel was studied as the ORR catalysts by Zhang et al. [88]. Their material was obtained from graphene oxide and dopamine was used as both N-source and a cross-linking agent. The aerogels were of a very low density (0.004 g $cm^{-3)}$ and had a hierarchical pore structure with the surface area of 1631 m^2 g^{-1}. The ratio of the micropore to mesopore volumes was 80% and a pore size analysis indicated that pores with sizes 0.78, 2.2 and 30, 35 nm were present. The materials had also the high content of nitrogen (7.21 at.%). An XPS analysis indicated that nitrogen was present in pyridinic, pyrrolic and graphitic configurations, which contributed to 15.3%, 21.4% and 63.3%, respectively, to the total nitrogen content. Interestingly, the kinetic current density reported was higher than that on Pt/C (~8 eV vs. 4.6 eV). The number of electron transfer was 3.92 and less than 5% peroxide were formed in the potential range between 0.1 and 0.9 V. The Tafel plot was also analyzed and found to be almost equal to that on Pt/C (67 mV/dec). The authors linked the catalytic activity to graphitic nitrogen distributed in the rich micropore structure. The proposed relationship between the surface feature of N-rich aerogel and the ORR performance is illustrated in Fig. 4. The authors hypothesized that: (1) the synthesis conditions led to the dense slit-like microporous networks of the high N content; (2) graphitic N drew electrons from nearby carbon atoms and made those atoms active adsorption centers, on which strong C–O bonds were formed. Thus, by the presence of nitrogen, the adsorption energy of O_2 decreased facilitating its fast adsorption; (3) the presence of micro- and mesopores decreased the O_2 diffusion time and thus increased the transport rate of O_2 and KOH inside the pores. (4) the latter led to the high O_2 concentration on the surface and to the fast ORR kinetics. Step 2 in Fig. 4 was indicated as a rate limiting step and, in this step, active carbon atoms increased the electron transfer rate and accelerated the sequential transfer rate. In Step 4, C–N bonds were formed and they were indicated as contributing to a continuous electron transfer and thus to the high current density. This mechanism is a significant advancement in addressing the role of porosity. Nevertheless, it suggested that the efficient catalyst, besides porosity needs N groups on the surface since they are of paramount importance for electron transfer processes.

4.2 *Doping Porous Carbons Beyond Nitrogen for High ORR Performance: S, P, B Doping And Co-doping with Nitrogen*

Another popular heteroatom used to dope carbon matrices in order to affect their electric properties is sulfur. Early paper by Qiao and coworkers presented the change in the electronic structure of carbon upon doping [50] where difference in an electronegativity between heteroatoms and carbon brought a positive charge to the surface atoms and these sites were catalytic centers for oxygen reduction. In fact, to support their hypothesis, the authors used mesoporous graphene obtained in silica template. The size of mesopores ranged between 10 and 40 nm the content of nitrogen and

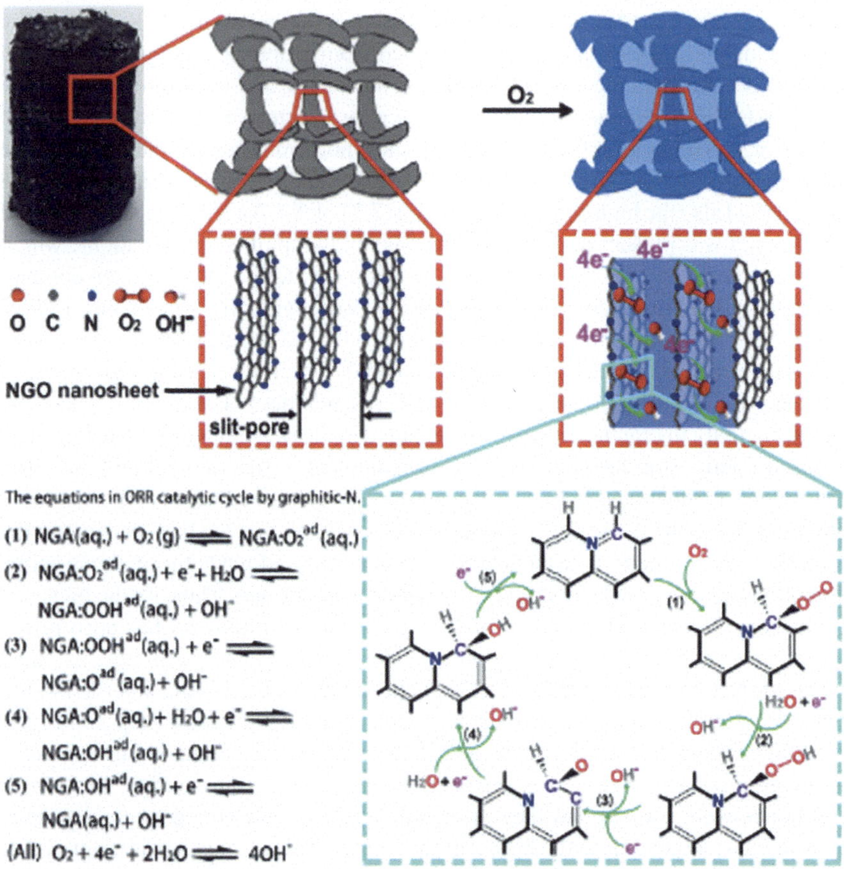

Fig. 4 Schematic relationship between the rich micropore structure in NGA and its effects on the high ORR performance. (Only three NGO nanosheets and four molecules are illustrated to make graphene skeleton and the pores of ORR legible). Reprinted with permission from Ref. [88]. Copyright 2019, Elsevier.

sulfur on the surface was 4.5 and 2 at.%, respectively. The kinetic current on S and N doped carbon reached that on Pt/C (10 mA cm^{-2}) and the number electron transfer was 3.6. Even though the authors linked the best performance of their catalyst to the existence of active centers formed owing to coexistence of S and N doped to the carbon matrix, they also concluded that mesoporosity was an important asset of these materials, although no details on its role were provided. Interestingly, only 1.3 at.% sulfur was found as advancing the ORR catalysis to greater extent than 5.1% nitrogen did, and besides resulting in the higher current density than that on the nitrogen-doped sample, it also led to the number of electron transfer 3.3 (on nitrogen doped sample n was 3). Still the kinetic limiting current density of the S-doped sample at –0.6 V was almost threes time less than that on the N and S-co-doped sample (13 vs. 30 mA cm^{-2}). The latter overperformed Pt/C with respect to this parameter. The co-doping also resulted in the most positive onset potential (−0.06 V vs. Ag/AgCl).

Dual S and N doped carbons of the high porosity were tested as the ORR catalyst by Yao et al. [89]. They were synthesized from carrageenan-urea aerogels pyrolyzed at temperatures between 600 and 1000 °C. An annealing at 1000 and 1110 °C was carried out to separate N-, S and O atoms and create defects. The catalysts were tested in acidic and alkaline electrolytes. Excellent results were obtained and the current density in KOH was up to 6 mA cm^{-2} and higher than that on Pt/C. The best performing sample had the number of electrons transfer 4. Based on extensive morphological and chemical analyses, the authors concluded that the high activity was the result of the defects engineering applied. The latter led to the formation of S- and N–S–C. The content of heteroatom in the carbons was not provided but the noisy XPS core energy level S 2p and N 1 s spectra suggest that their content was rather very small. Those specific bonds arrangements were indicated as the catalytic centers in both acidic and alkaline electrolytes (Fig. 5). The edge thiophene, graphitic N and pentagon defects were indicated as responsible for the formation of these centers. Since the carbons were porous of the high surface area (the best-performing sample had a surface of 1307 $m^2\ g^{-1}$), the authors noticed that the extent of the carbon surface promoted the high density of the exposed sites and mass transport. Even though the samples had a marked volume of micropores their effect on ORR was not analyzed.

That sulfur effect was also explored by Guo et al. [90]. They synthesized ordered mesoporous carbon (OMC) in SBA-15 silica and used dibenzyl sulfide as a source of sulfur. The samples had similar content of sulfur of about 1.5 wt% but differed in the contributions of oxidized and reduced sulfur. One sample had sulfur in predominantly –C–S–C configurations and it showed the best activity for ORR. The kinetic current measured on that catalyst was higher than that on Pt/C and the number of electron transfer reached 3.8. The surface areas were between 1099 and 1261 $m^2\ g^{-1}$and the main difference was in the volume of mesopores. The volume of micropores was very similar for all sample (~0.1 $cm^3\ g^{-1}$). Even though the effect of porosity was not analyzed, that sample had the highest degree of microporosity. The predominance of sulfur in C–S–C configuration led the authors to indicate the role of these catalytic centers as the most important for ORR. It was also mentioned that very small amount of sulfur was able to significantly affect the efficiency of ORR.

Effect of sulfur incorporated to the carbon matrix was also studied by Bandosz et al. [91]. In their first attempt, mesoporous carbon materials synthesized from graphite oxide in an amorphous silica template were used and the tests were run in a neutral electrolyte. Hydrogen sulfide was used to introduce sulfur to the surface. The porosity of the S-fee and S-doped samples was a replica of silica porosity and the surface area reached 190 $m^2\ g^{-1}$. Contrary to the results of Gao and coworkers in the alkaline medium, it was found that an introduction of sulfur by the H_2S treatment resulted in a decrease in the ORR activity. The number of electron transfer on the S-doped sample was 15–2.5 while on its unmodified counterpart 3–3.5. Since the porosity was practically the same, the alteration in surface chemistry upon the H_2S treatment was linked to the difference in the ORR activity. Even though upon the

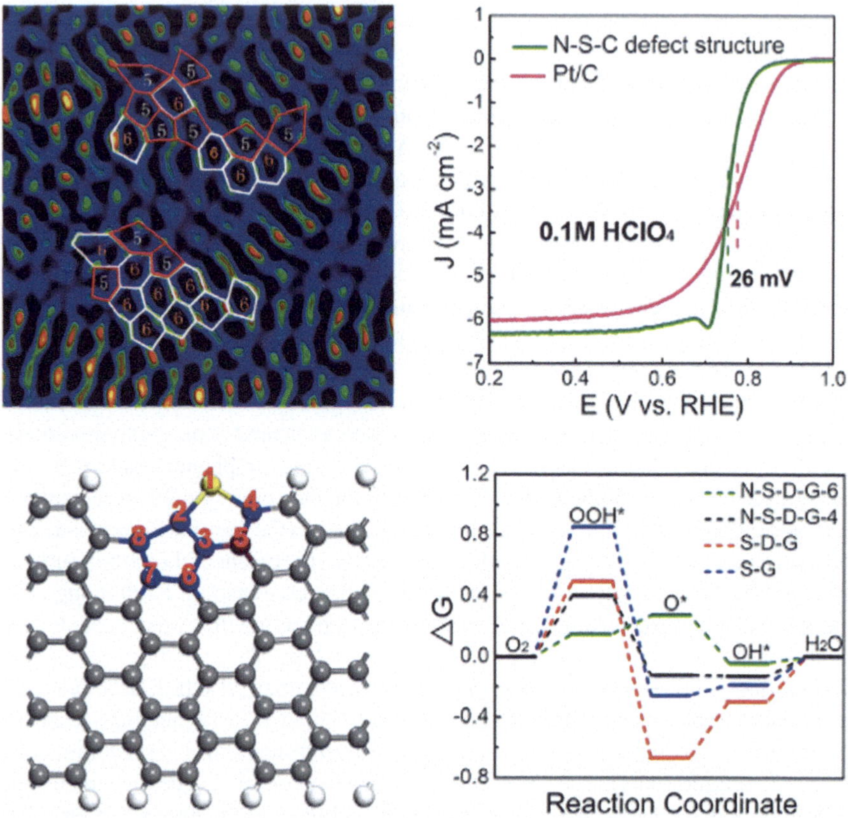

Fig. 5 Upper left: STEM filtered image of N and S codoped carbons; upper right: LSV curves; Bottom left: the optimized structure on N and S-codoped sample (grey, white, yellow, and claret balls represent C, H, S, and N atoms, respectively. The blue atoms represent S- or N-adjacent C atoms); Bottom right: Free energy diagram. Reprinted with permission from Ref. [89]. Copyright 2018, CellPress

modification 3 at.% sulfur was introduced, that sulfur was mainly in –SH configurations and only a small fraction was in –C–S–C bonds, which were indicated previously as the catalytic centers for ORR [90]. The authors concluded that the total content of O and S and their chemical configurations along with the hydrophobicity level of the surface played a role in the ORR activity in the neural electrolyte. Interestingly, when the same materials were used as the ORR catalysts in an alkaline electrolyte, the number of electron transfer on the S-doped sample was between 2 and 3.5 and the presence of OH- ions and their attraction to the sulfur -modified carbon matrix enhanced the pore space utilization, and this, along with hydrophobicity of sulfur species attracting oxygen, was indicated as enhancing the catalytic activity [92].

As a continuation of the study on the sulfur-doped ORR catalysts, Bandosz et al. evaluated carbon aerogel [60] and their composites with graphite oxide [59]. Micro/mesoporous aerogel was obtained from resorcinol–formaldehyde polymer aerogel and to introduce sulfur it was heat treated in the H_2S atmosphere at 650 or 800 °C. Difference in the posttreatment resulted in an increase in the surface area from 693 to 734 $m^2 g^{-1}$ and in a marked increase in the volume of ultramicropores (smaller than 0.7 nm) in the high temperature treated carbon. Interestingly, the sulfur contents in both samples were similar and ~1 at.% and similar were also the contributions of various sulfur species with thiophenes/bisulfides (~70%) and sulfides (14%) as the predominant species. While on the sulfur-free sample and on that treated with H_2S at 600 °C the number of electron transfer was about 2, it significantly increased for the sample treated at 800 °C and reached almost 4. On that sample the kinetic current density was similar to that on Pt/C (5 mA cm^{-2}). The kinetic limiting current density was as high as 20 mA cm^{-2}. The onset potentials were 0.780, 0.766 and 0.809 V on the initial sample and those treated at 600 and 800 °C, respectively. Treatment with sulfur also increased the stability of the catalyst. The excellent performance of the sample heated at 800 °C, besides its S-based catalytic centers, was linked to an increased level of hydrophobicity due to the reduction of surface oxygen groups, and to ultramicropores. It was hypothesized that both these features are important for the withdrawal of oxygen from electrolyte and its accumulation on the surface. For this, the mesoporosity was important since it positively affected the transport of electrolyte with dissolved oxygen to micropores.

To further enhance the electrocatalytic performance of S-doped aerogels, Bandosz et al. synthesized their composites with graphite oxide (GO) [59]. The objective of that modification was to increase the electrical conductivity beneficial for ORR. The composites were synthesized using a self-assembly method based on dispersive forces with the 1:1 weight ratio of the sulfur doped aerogels (obtained at 600 and 800 °C) and GO. As expected, an owing to the nonporous surface of GO, the porosity of the composites decreased, however, the small extent of that decrease indicated a synergistic effect of a new porosity formation, especially in the mesopore range. Interestingly, the content of sulfur on the surface, as determined by XPS, increased up to 1.8% and 2.5% for the samples treated at 600 and 800 °C, respectively. It was owing to sulfur present in GO alone. The addition of GO significantly improved the performance. The onset potential increased to 0.819 and 0.829 V versus RHE for 600 and 800 °C treated samples, respectively (on Pt/C 0.947 V was measured). Even though the number electron transfer for the latter sample was 4 and similar to that on aerogel alone [59] (it was not expected to increase over 4) on the initial sample and on the one heated at 600 °C n was 3.5 and significantly increased compared to those on aerogel alone. Since GO itself had sulfur in sulfones and sulfonic acid, its addition increased the contribution of those species in the composites, however reduced sulfur in thiols and thiophenes was still in majority. The sample heated at 800 °C had the higher contribution of those species than the one heated at 600 °C (66% vs. 59%). Besides to the catalytic centers based on thiophene, the good performance of these composites was also linked the hydrophilic mesopores having their origin in graphite oxide, which helped to supply oxygen to small pores in the aerogel phase and to the S-based catalytic centers.

Another heteroatom whose introduction to porous carbon has been investigated with an intention to increase the carbon activity in ORR was boron. However, its effects have been rather investigated in conjunction with that on nitrogen. Ozaki et al. prepared a series of carbons from furfuryl alcohol, which were doped with boron, nitrogen and codoped with these two elements [64]. BF_3-MeOH and melamine were used a source of boron and nitrogen, respectively. The samples were tested in an acidic electrolyte. Chemical modifications increased the surface area compared to that of the pristine sample. The sample doped with boron had a surface area of 150 $m^2 g^{-1}$ and that doped with nitrogen- 310 $m^2 g^-1$. NB-doped samples reached 180 $m^2 g^{-1}$. Interestingly, the modification decreased slightly the samples' electrical conductivity. The ratio of B/C in boron doped sample was 0.008 and in those co-doped-up to 0.033. On the other hand, a N/C ratio in the N-doped samples was 0.032 and on those codoped-up to 0.082. The variety of boron bonds was detected by XPS including BCO_2, BCO BN_2C/BC_3 and BC_2O. Nitrogen existed in typical five configurations and B–N–C moieties were formed from those of pyrroles/pyridine and quaternary nitrogen. While the addition of boron only slightly shifted the onset potential toward more positive, the coexistence of both boron and nitrogen resulted in the onset potential about 0.1 V more positive than that on the unmodified sample. The positive correlation between the kinetic current density at 0.4 V normalized per unit BET surface area and the ratio of edge nitrogen and B–N–C moieties to carbon was found. Even though the direct role of the porosity was not indicated or analyzed, that normalization per unit surface area suggests that the authors considered that parameter as important for the ORR activity.

Another group of B and N co-doped carbons as the ORR catalyst was investigated by Baik and Lee [93]. The carbon tested was a commercial activated carbon Darco (AC) with a surface area of 600 $m^2 g^{-1}$. Boric acid and urea were used as the sources of B and N, respectively. To provide a wide range of surface chemistry the carbons were heat -treated at three different temperatures up to 1000 °C. The content of boron in B-AC reached 1.86% and nitrogen in N-AC 4.6%. The N, B-co-doped samples had up to 8.13 and 4.6% of boron and nitrogen, respectively. Co-doping resulted in the higher current density than that on Pt/C (7.1 mA cm^{-2} vs. 5.5 mA cm^{-2}) and in the more positive onset potential than that on the initial AC. While the number of electron transfer on the B-doped sample was 3.16 and, on the N, -doped one-about 3.47, on the codoped counterpart n was about 3.81 when measured at −0.6 V vs. Ag/AgCl. An XPS analysis showed B–C, B–O and B–N bonds in the codoped samples. Detailed analysis of the data suggested that B–N–C groups were responsible for the high catalytic activity. Even though the chemistry and the microstructure of the catalysts were addressed in details, no information about the porosity of the samples was provided besides the surface area. Darco of 600 $m^2 g^{-1}$ is a well-known steam activated carbon (lignin origin) of a predominantly microporous structure and of a total pore volume 0.9 $cm^3 g^{-1}$. Since the treatment applied might rather shrink pore sizes than expand them, these results are very interesting from the point of view of the performance of highly microporous carbonaceous catalysts. However, these aspects were not analyzed by Baik and Lee [93].

Expanding the concept of the ORR catalytic activity of B-, N- co-doped carbon Ozaki and coworkers synthesized nanoshell carbon structures from polyfurfuryl alcohol, cobalt phthalocyanine, melamine and a trifuoroborane-methanol complex at 1000 °C [94]. The co-doped sample had over three times more nitrogen than boron. XPS analysis indicated that the former existed mainly in pyridine, B–N–C and pyrrole/pyridones and the latter in BC_2O an B–C–N. Even though the porosity of carbons was not studied in details, TEM images showed the existence of agglomerated nanoshell structures with wavy layers of about 50 nm in size between which some level of porosity could exist. The authors concluded that the graphic structure of the nanoshells was disturbed by BN doping and besides electrochemically active edge sites consisting B–C–N also defects were formed. All of this increased the ORR current eight to ten times compared to that on the non-doped catalyst. Even though the nanoshell structure was stressed out as a morphological feature affecting the ORR activity, the porosity was not reported and the surface accessibility for oxygen was not discussed.

Recent paper by Bandosz et al. addressed directly not only boron and nitrogen -based carbons as ORR catalysts but also analyzed the effect of porosity on the catalytic activity [28]. In fact, their carbons could be considered as multidoped since they had also sulfur in their structures. They were synthesizing from sulfur-containing commodity polymers (sodium-4-styrene sulfonates and 4-styrene sulfonic aid-co-maleic acid) at 1000 °C. NH_3 present in the carbonization atmosphere was a source of nitrogen and boric acid- a source of boron. The majority of samples had the surface areas between 400 and 800 $m^2\ g^{-1}$. Only one sample, not BN -doped, had the surface area almost 1600 $m^2\ g^{-1}$. All carbons were predominantly microporous. Interestingly, in the co-doped samples the contents of boron and nitrogen were similar and between 7 and 11%. In all samples sulfur was present in the amount of ~0.5%. The best samples were those co-doped and their performance was very similar to that of Pt/C with the number of electrons transfer very close to 4. Interestingly, the non-doped sample of the highest porosity also performed extremely well. That trend was also found for the kinetic current density. Doping with boron and nitrogen also visibly moved the onset potential to the more positive values. XPS and EDS analyzes showed a high dispersion of heteroatoms, and, as expected B–N and B–C bonds were present increasing the catalytic activity. One finding in this work desires the specific attention. It is that the good performance in ORR was not exclusively granted to the BN-doped carbons. The carbon sample without those elements but of the highest surface area and micropore volume exhibited similar behavior. It had also the highest volume of ultramicropores smaller than 0.7 nm. The authors proposed that the interactions of oxygen with both pore walls, owing to their proximity, promote the efficient ORR on that carbon. In such pores the adsorption of oxygen should be the strongest and they are expected to be hydrophobic and thus attract oxygen from the electrolyte. Based in the trend in the onset potential values, the results also suggested that the porosity was more important for governing the onset potential than was the population of the catalytic centers. Thus, a small volume of ultramicropores led to the less positive onset potential and the high numbers of B and N-based centers were not able to compensate this negative effect. Even though this was not mentioned in the paper, the dispersion of the catalytic center might play a role in determining the extent of the onset potential and their high dispersion might benefit ORR.

The effect of porous carbon network doping with phosphorus on ORR was investigated by Gao et al. [95]. They applied a dual chemical activation with $ZnCl_2$ and NaH_2PO_2 which led to porous carbon of the surface area of 943 $m^2\ g^{-1}$. Synergistic effects of the combined activating agents led to the formation of active sites and defects enhancing the ORR efficiency. The modified carbon catalyst showed an increased limiting current whose density reached 5.83 mA cm^{-2}.

Another interesting approach enhancing the ORR activity of porous carbon was co-doping with nitrogen and phosphorus [96]. The carbon samples were synthesized from N and P doped hydrogel obtained using P123, phytic acid and pyrrole. The gel was either only carbonized or activated with CO_2. The surface areas were 2850 or 1820 $m^2\ g^{-1}$ and higher for the activated sample. While the former sample was micro/mesoporous, the latter was predominantly microporous. The activated sample had 3.4% N and 0.6% P and non-activated- 3.8% N and 1.2% P. The elements were very homogeneously distributed on the surface. The evaluation of the ORR activity showed the largest current density for the activated sample at the maximum of the reduction humps (0.75 mA cm^{-2} vs. 0.65 mA cm^{-2} for the non-activated one). The same trend was found for the current densities which were between 2 and 2.5 mA cm^{-2} (on Pt/C ~ 3 mA cm^{-2} was measured). The high surface area sample had also the higher number of electron transfer. Generally, for both samples n was between 3.75 and 3.95. The smaller Tafel plot than that for 5% Pt/C suggested superior kinetics of ORR on the N, P dual doped samples. The detailed XPS analysis of the surface was also provided and phosphorus was found in P–C and P–O bonds. The authors linked the high activity of their samples to the dual doping and to the high content of pyridinic and graphitic nitrogen. The latter was larger in the activated sample. They clearly indicated the important role of micro and mesopores in providing the easy access of electrolyte to the heteroatom -based catalytic centers. The latter were considered also as defects. It is important to mention that the higher surface area sample of the marked activity had less heteroatoms thus the development of mesopores by activation and an increase in microporosity were providing additional assets beyond the chemistry of the active centers.

5 ORR Activity of Non-heteroatom Doped Carbons: Exploring the Effects of Porosity

Even though the heteroatom-doped carbon nanostructures or quite recently porous carbons have been investigated quite extensively as the metal-free ORR catalysts owing to the well established mechanism related to the alteration in the electronic structure of the carbon matrix, quite recently the researchers have started to observe, and thus to focus, on the enhanced ORR activity on heteroatom -free carbon structures. An interesting approach explaining the activity of pure carbon nanocages (heteroatom free) was proposed by Hu et al. [73]. Combining experiments with theoretical calculations, they found that intrinsic carbon defects, existing in the carbon matrix, contribute to the observed enhanced ORR activity, irrespective of the presence of nitrogen. They suggested that zigzag edge defects or pentagon defects provide the catalytic activity for ORR, and the number of these defects might be naturally increased in heteroatom-doped carbons. In their most- active carbon, not only pores with sizes less than 1 nm were present but also a high volume of pores with sizes of about 10 nm was detected; however, to explain the activity, the emphasis was placed only on the edge and pentagon intrinsic defects, which were linked to a high surface area. Nevertheless, the electron transfer number was only 2.9. The effect of the zing-zag carbon on ORR has been also recently addressed by Xu et al. [74]. They have constructed a cathode catalyst based on a composite of zigzag-edged graphene nanoribbons and CNTs. The measured number of electrons transferred was very close to 4, and both experiments and theoretical calculations showed the zigzag carbon defects as possessing the higher catalytic activity than those of the basal plane carbon atom, arm- chair edge carbon and carbon atom near a void.

The role of a carbon catalyst geometry for ORR was also investigated by Dai et al. [75]. Although they addressed N- and Fe-doped CNT/graphene complexes, their research suggested that unzipped walls of nanotubes provided the catalytic activity and that the charge transport was facilitated due to the intact electrical conductivity of CNT/graphene. Those unzipped fragments, referred to by the authors as graphite oxide pieces, provided abundant edges and defect sites, enhancing the ORR catalytic activity.

The mentioned above good ORR activity of heteroatom-free carbon materials led to the generalization of the active sites in carbon as those creating defects [25, 73, 74, 78, 89, 97]. The mentioned above studies were carried out on rather nonporous carbons. Indicated in the previous section, but not deeply studied, the role of porosity in heteroatom-doped carbons directed the attention of researchers to take a closer look at the effects of the pore structure on the ORR activity. That pore structure should not only be considered as providing the surface for a high distribution of heteroatom-based catalytic sites or edge defect. It might also affect the adsorption processes which are associated with carbon immersed in O_2 saturated electrolytes.

One of the first study of porous carbon not containing heteroatoms as ORR catalysts was carried out by Tammeveski et al. [98]. Besides with carbon blacks, their electrodes were also modified with carbide derived carbon, CDC. The surface areas of CDC were 177 and 684 $m^2\ g^{-1}$. The authors measured a high reduction current on the CDC materials and concluded that their high graphitization level and mesopores provided the fast charge transfer contributed to that effect. LSV curves showed two plateau and the number of electron transfer of the high surface area catalyst was close to 4 and on that of the smaller surface—about 2.5. Interestingly, the % of peroxide formation on all carbons was between 40 and 80%, and it seems unusually high with 4 electrons transferred in the reduction reactions. The catalytic activity was linked to the presence of quinones although it was also indicated that the structure and porosity of carbons might also affect the results.

An unusually high activity on "all-carbon" electrocatalysts in ORR was described by Wei et al. [99]. Their engineered materials were rich in an interface between graphene and curved carbon nanotubes. On them, the high kinetic current was measured and $n = 3.86$ with only 7% of peroxide formed. The authors speculated that the high activity of their G-CNT hybrids was caused by a localized charge separation at the interface of graphene and nanotubes. That Fermi level mismatch on the planar and curved sp^2 surfaces tuned the electron transfer. It is important to mention that the BET surface area of the hybrid was 470 $m^2\ g^{-2}$ however, its extent and the effect on the activity were not discussed.

A direct effect of the porosity on ORR influenced by oxygen adsorption was first indicated by Bandosz et al. [26, 27]. They studied three samples of Poly-HIPE carbons [26] of the micro/mesoporous structure, which had the same pore size distributions but differed in the volume of pores (mainly ultramicropores < 0.7 nm) and in the oxygen content and its chemical environment. The surface areas between 500 and 720 $m^2\ g^{-1}$ and the volume of ultramicropores- between 0.130 and 0.253 $cm^3\ g^{-1}$. The carbons had traces of nitrogen and their content of oxygen decreased with an increase in the surface area from 10.5 to 8.0%. Detailed analysis of surface chemistry indicated that the highly porous carbon was also least oxidized and most hydrophobic one. Interestingly, testing those carbons as the catalysts showed an increase in the activity with an increase in the volume of ultramicropores and in the hydrophobicity level. The kinetic current density on the most porous sample was higher than that on Pt/C (11 mA cm^{-2} vs. 6 mA cm^{-2}) and the number of electron transfer reached 4 at 0.2 V versus RHE. The kinetic current density at 0.18 V versus RHE was over 35 mA cm^{-2}. The carbon catalysts were also more stable that Pt/C and very resistant to methanol crossover. Since the activity could not be linked to the nitrogen-based centers or quinones, the authors suggested that the reduction process was adsorption -driven and ultramicropores worked as pseudo -catalytic centers for ORR promoting O_2 adsorption, which in turn facilitated the electron transfer. Since the heat of O_2 adsorption in pores similar to the size of the oxygen molecule can reach 24 kJ mol^{-1}, such strong adsorption energy can promote charge transfer leading to weakening of O=O bonds. An important asset was hydrophobicity in those pores and thus their affinity to withdraw the oxygen molecule from the electrolyte. Owing to the proximity of the water phase, formed OH- did not accumulate in pores but was attracted to the water phase and the pseudo- catalytic centers were free to accept other O_2 molecules and further promote ORR. For this process, the totally hydrophobic surface would hinder the reduction process, but ~8% oxygen was incorporated in larger pores (at pore entrances) and this could facilitate the transfer of electrolyte with dissolved oxygen to ultramicropores. It has been also mentioned that the higher porosity carbons are expected to have more defect sites which might also be the ORR catalytic centers but they are totally different in the mechanism than are the pores.

To further explore the effects of porosity on ORR and to support the hypothesis on the importance of ultramicropores as the ORR pseudo-catalytic centers, Bandosz and workers tested oxygen electroreduction on cellular vitreous carbon foams [27]. The initial foam had a broad distribution of pores in micro, meso- and macro-range and the surface area of 640 $m^2\ g^{-1}$. To modify both, pore sizes and surface chemistry, the composites with GO were synthesized using the self-assembly in the suspension method. The samples were further modified by the introduction of sulfur or nitrogen to the surface or by only a heat treatment imposing the surface reduction at 950 °C. Dicyandiamine and hydrogen sulfide were used a source of nitrogen and sulfur heteroatoms, respectively. The surface area of the modified samples was in the range between 393 (nitrogen -modified foam) and 655 $m^2\ g^{-1}$ (reduced foam/GO composite). The ratio of the micropore volume to that of the total pore volume was about 90% for the majority of samples. As expected, the introduction of GO increased the number of oxygen groups and thus the surface hydrophilicity while the heat treatment made the surface more hydrophobic. The content of nitrogen on the surface was maximum 3.7 at.% and that of sulfur-1.2 at.%. The majority of nitrogen was in pyridines and pyrroles/pyridones and sulfur- in bisulfide/thiophenes. The most positive onset potential was found for the initial foam sample and that modified with GO, while the introduction of heteroatoms resulted in less positive values. The number of electron transfer was also the highest for the heteroatom-free samples (3.94–3.96) and only about 3% of H_2O_2 were formed (Fig. 6). The samples were also resistant to methanol cross-over and showed the 80% stability after 1000 cycles. The good performance of the heteroatom-free samples was linked to the specific porosity. The foams had a high volume of micropores in the walls of macropores. Those large macropores with oxygen groups on the surface promoted the transport of the electrolyte to the micropores where, owing to the strong adsorption of oxygen in the pores similar to the size of the oxygen molecule, the splitting of O=O took place and 4 electrons were transferred. This conclusion was also based on the fact that the introduction of heteroatoms markedly decreased the volume of small pores and this changed the mechanism of ORR. Even though N and S-provided the catalytic centers, they were not able to compensate the effects of the decreased micropore volume.

These two different extents of the ORR mechanisms on carbons were also addressed by Bandosz and coworkers in their study of ordered mesoporous carbons as

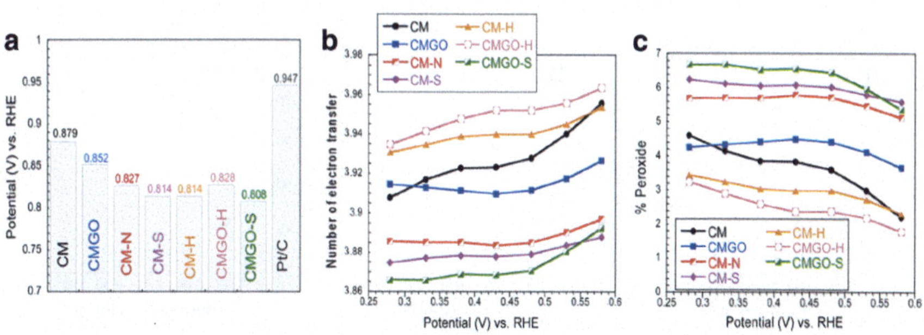

Fig. 6 Comparions of the onset potentials (**a**), number of electron transfer (**b**) and the percentage of oxygen reduced to peroxide (**c**). Reprinted with permission from Ref. [27]. Copyright 2017, American Chemical Society.

the ORR electrocatalysts [29]. The carbons were obtained from sucrose in the SBA-15 silica template. To introduce nitrogen to the surface, the initial carbon sample was oxidized with HNO_3 and impregnated with urea and heat -treated at 600, 800 and 950 °C. The oxidized sample was also heated at 950 °C. The treatment applied markedly affected the porosity, although the ordered structure was still preserved. The non-nitrogen treated samples had traces of nitrogen and after the treatment with urea 1.4 at.% of N was on the surface. While the 600- and 800 °C -treated samples had pyridines an pyrroles introduced to the carbon matrix, on the 950 °C -treated sample quaternary nitrogen was in majority. Even though the introduction of nitrogen had a very positive effect on the ORR efficiency and for the sample with the majority of quaternary nitrogen the number of electron transfer was 3.95, the oxidized and then reduced sample without the urea treatment showed also very good results with n comparable to the best N-treat sample. Interestingly, the most positive onset potentials were found for the initial sample and that oxidized one. The introduction of nitrogen resulted in the less positive potentials. To explain the trend in the obtained data, the authors analyzed the dependence of the number of electron transfer on the volume of ultramicropores and on the predominant size of those ultramicropores (peak maximum) (Fig. 7a, b). The former one showed two opposite trends distinguishing N-modified carbons from those non-modified. While for the urea treated sample, n increased linearly with an increase in the volume of ultramicropores, for the latter group a decrease was found. When the size of the predominant ultramicropores was taken into consideration, for the N-modified carbon sample a clear increase in n with a decrease in the predominate size of ultramicropores was found. No trend for another group of catalysts was detected. These analyses, although did not bring sufficient information to clearly explain the excellent behavior of the non-N-treated samples, they indicated that ORR might take place following two distinct mechanisms. Since the system in which the reduction of oxygen takes place is rather complex and oxygen is dissolved in an electrolyte and thus needs to be transported to the catalytic centers, the dependence of ECSA on the oxygen density on the surface was analyzed (Fig. 7c). Those oxygen groups, which in fact could exist only in mesopores or at the entrances of micropores were indicated as helping with the mass

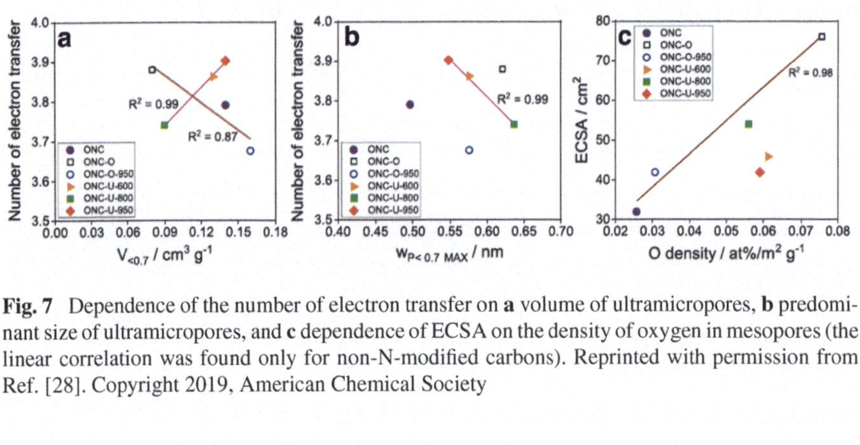

Fig. 7 Dependence of the number of electron transfer on **a** volume of ultramicropores, **b** predominant size of ultramicropores, and **c** dependence of ECSA on the density of oxygen in mesopores (the linear correlation was found only for non-N-modified carbons). Reprinted with permission from Ref. [28]. Copyright 2019, American Chemical Society

transfer of the electrolyte to the small hydrophobic pores. The linear dependence of ECSA on the density of oxygen was found for the non-N-treated samples implying the importance of oxygen accessibility to small pores. The results also suggested that the samples with the smaller ultramicropores (distribution started at the smallest pores) had the most positive onset potential. It was suggested that even though the micropores are generally considered as negatively affecting the mass transfer process, with the sufficient wetting of the surface, their entrances can be easily reached by oxygen and inside these pores O_2 is reduced via the 4 e transfer mechanism. The positive influence of defects and quinine groups in ORR was also pointed out.

Mesoporous carbon spheres with improved conductivity have been investigated as the ORR catalysts by Perez-Cadenas et al. [100]. Their carbons were prepared by a sol–gel synthesis from resinol and formaldehyde and were additionally modified by an addition of nonionic and ionic surfactants during the synthesis. The surface areas were between 500 and 600 $m^2\ g^{-1}$ and the samples had a micro/mesoporous nature. When tested as the ORR catalysts, the samples revealed current density about 5 mA cm^{-2} and n between 2.9 and 3.5. Even though the effect of the porosity was not discussed, the best performing samples had the highest volume of micropores determined from CO_2 adsorption and the DR approach. The authors indicated that the presence of mesopores is important for ORR.

Interesting microporous carbons were investigated by Pereira et al. [101]. Their catalysts were obtained from glucose and the authors managed to tailor their properties to some extent through an activation process. Even though one of their main objectives was an increase in the ORR activity by the nitrogen doping and melamine was used for this purpose, they also investigated in details a series of non-doped carbons. On the non-doped samples, the current density was between 2 and 4 mA cm^{-2} and it increased with an increase in the extent of activation. The treatment with nitrogen not always resulted in an increase in the current density. The onset potential on some non-modified samples was equal to that on the nitrogen modified ones and was higher on the porous/activated samples than on those nonactivated. The number of electron transfer on the latter samples was only 2.3. Nevertheless, when the dependence of the current density of all samples on the BET surface area was analyzed, two different linear trends were found distinguishing the undoped and N-doped samples (Fig. 8). The authors indicated that there is some effect of microporosity on the ORR performance but that effect could not be compared between doped and undoped carbons. This finding supports two different mechanisms hypothesized by Bandosz et al. [27, 28]. Interestingly, the slope was higher for the undoped samples, which might suggest the stronger influence of microporosity for this group of catalysts. The authors also indicated that for their samples the combination of right N -based chemistry (pyridinic and quaternary nitrogen) and microporosity led to the very efficient catalysts with onset potential of 0.82 V, n close to 4, and H_2O_2 production less than 7%.

Following the line of research on the importance of microporosity for ORR, Bandosz et al. investigated ultramicropores-enhanced ORR process on the series of porous carbons, doped with N and undoped [102]. Their catalysts were obtained from sucrose in KIT-6 and ammonia was a source of nitrogen. As in the previous

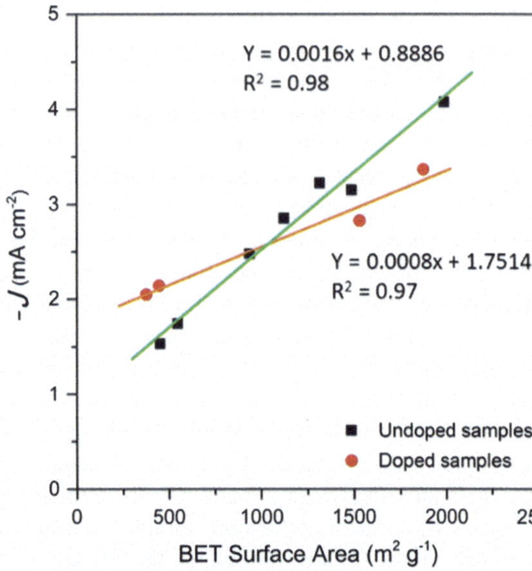

Fig. 8 Relationship between BET surface area and limiting current density of undoped and doped samples. Reprinted from Ref. [101]. License (https://creativecommons.org/licenses/by/4.0)

report, to modify the surface features the samples were heated at 600, 800 and 950 °C. However, after oxidation and further treatments carbons lost their ordered mesoporous structure. While the surface areas of undoped samples ranged between 780 and 1130 $m^2\ g^{-1}$, all N-doped carbons have similar porosity with surface areas ~1100 $m^2\ g^{-1}$. There was only a slight difference in the content of nitrogen on the surface of the latter samples (between 1.4 and 2.1 at.%), however, the distributions of nitrogen species were similar. Only pyridines and pyrroles/amines were detected on the surface and the former ones were in majority. On the samples obtained the kinetic current density reached 5 mA cm^{-2} and only small differences in the onset potentials were found (between 0.806 and 0.816 V vs. RHE) (Fig. 9). The number of electron transfer was 3.96 on the best performing samples with H_2O_2 production less than 3%. Tafel slopes were comparable to that on Pt/C. In this series of samples, the undoped ones were the best performing, and interestingly, either oxidized or reduced ones. While the oxidized sample had the oxygen content 13.4%, the reduced one-7.3%, indicating the higher hydrophobicity of the later sample. The latter one had also slightly larger surface area (892 vs. 784 $m^2\ g^{-1}$ and) and the 50% higher volume of ultramicropores. The authors indicated that owing to the complexity of the ORR process various factors should be analyzed including the apparently contradictory ones such as the presence of the high volume of small pores and the easy access of an electrolyte to these pores. To account for this, a new factor addressing the effect of the small pores' high adsorption potential along with the oxygen accessibility to those pores was proposed. It is referred to as Pore Influence Factor (PIF) and it is the product of the volume of ultramicropores and the ratio of the number of dissociating groups and ECSA. The latter (ratio) is related to the accessibility of oxygen to the pores/catalytic centers and the former represents the effect of oxygen adsorption. A

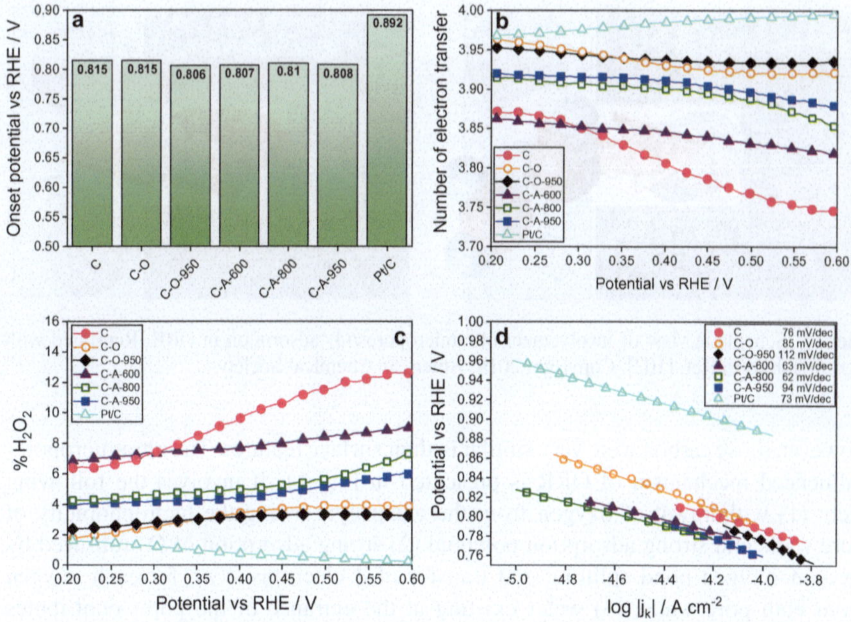

Fig. 9 **a** Onset potentials, **b** number of electron transfer, **c** The percent of H_2O_2 formed and **d** Tafel plots with linear fits (R^2 between 0.992 and 0.999) constructed from the data at 2000 rpm (slopes are listed in the legend). Reprinted with permission from Ref. [102]. Copyright 2019, American Chemical Society

linear dependence between n and PIF was found for the undoped carbons (Fig. 10). Moreover, following the approach of Pereira and coworkers also a direct dependence was found between the kinetic current density and V_{mic} or $V_{<0.7\ nm}$. The lack of any dependence for the N-doped carbon samples might suggest different mechanisms,

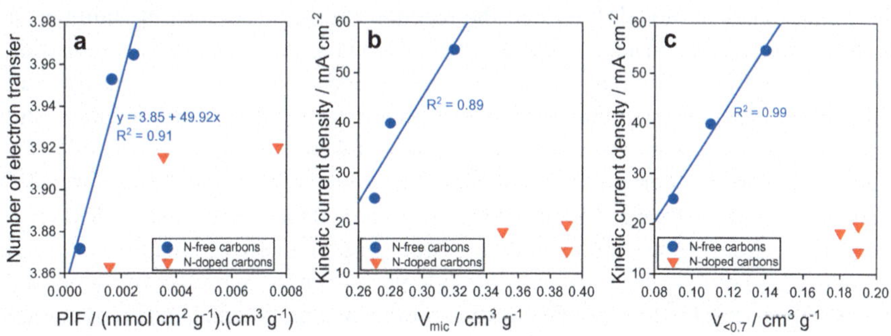

Fig. 10 **a** Dependence of the number of electron transfer on PIF; dependence of kinetic current density on **b** volume of micropores and **c** volume of ultramicropores. Reprinted with permission from Ref. [102]. Copyright 2019, American Chemical Society

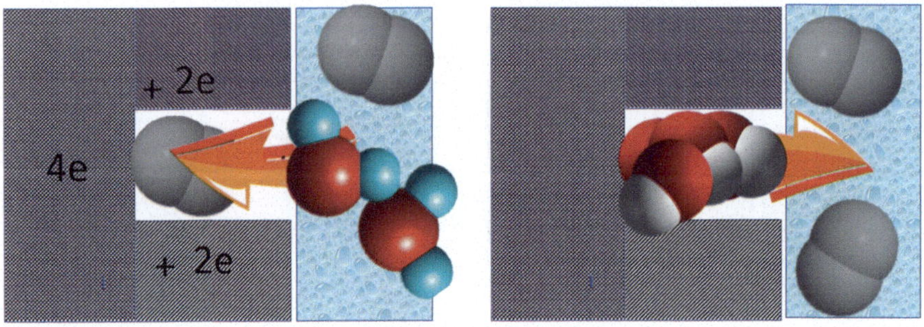

Fig. 11 Schematic view of involvement ultramicropores/O_2 adsorption in ORR. Reprinted with permission from Ref. [102]. Copyright 2019, American Chemical Society

however, those carbon were very similar in their surface features. The ultramicropore-influenced mechanism of ORR is presented in Fig. 11. It involves the following step: (1) withdrawal of oxygen from the electrolyte owing the hydrophobicity of pore walls and strong adsorption potential (2) strong adsorption of O_2 followed by oxygen–oxygen bond splitting and transfer of 4 electrons, (two for each oxygen from both pore walls; (3) water existing at the entrance of the pores contributes to the reaction; (4) formed OH- ions have a high affinity to water and they leave the pore/catalytic centers making them available for adsorption of other oxygen molecules.

A significant contribution to understand the effect of microporosity came from the group of Cazorla-Amoros [103]. They modeled ORR in microporous carbons and indicated a crucial effect of microporosity for this process. The results were compared to those obtained on nonporous carbon black. A two-wave dependence of the current on the potential was analyzed. It was proposed that it represents two stages of oxygen reduction. The first wave at the intermediate potentials represents reduction of O_2 to peroxide, a second, at lower potentials-reduction of peroxide to hydroxide. The authors also proposed that the onset potentials are governed by the microporosity. Their mathematical model described the reaction rate and provided the number of electrons transfer. It considered the O_2 and H_2O_2 mass transfer rates, two reduction reactions and different activity of ultramicropores and supermicropores (narrow and wide micropores). Their findings, also indicated that the adsorption potentials and mass transfer between two type of pores might be linked to the high activity of H_2O_2 reduction in narrow micropores. No direct relationship between the electron transfer coefficient and porosity was found. The model assumed that narrow and wide micropores are randomly distributed and their entrances are equally accessible to oxygen (oxygen does not have to transfer from wide pores to narrow ones). The proposed scheme is presented in Fig. 12. That recent combined experimental and computation approach of Cazorla-Amoros and his group is a marked step to understand the role of carbon small pores in ORR and to develop metal-free inexpensive catalysts.

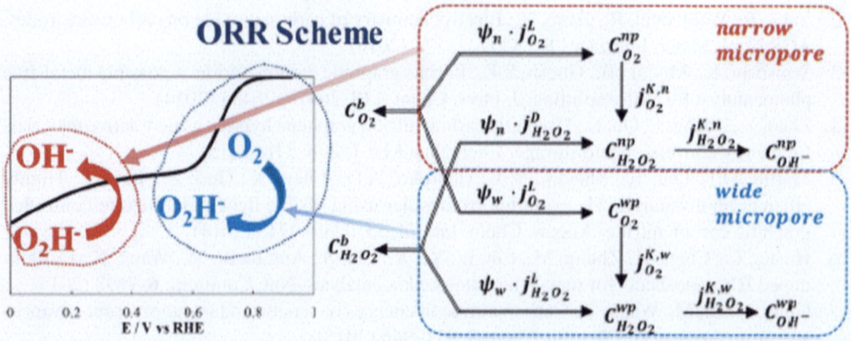

Fig. 12 The LSV curve and model proposed for the mathematical description of the ORR mechanisms on porous carbon materials with heterogenous micropores. Reprinted with permission from Ref. [103]. Copyright 2019, Elsevier

6 Concluding Remarks

The application of porous carbons in ORR is still in its infancy, although the 3-D engineered structures have been addressed quite extensively. After the first announcement of pristine graphene and the comprehension of the necessity for its extensive modifications needed for numerous catalytic or sensing applications, nanoporous carbons, owing to their pseudo-graphene nature, combined with a unique porosity and a vast range of possible surface chemistry modifications, seem to be a very attractive target of research reaching *beyond adsorption and towards energy*. Although it is without any doubts that both heteroatom-based catalytic centers and other defects in carbons certainly enhance ORR, the recent trend and the extensive efforts on the development of various exotic 3-D structures of carbons, testify on the important role of porosity for this process. The effect of oxygen adsorption and thus adsorption-driven ORR mechanisms have not been yet explored in details. Considering the complex mechanism of ORR even on "flat" carbons and rather impossible scenario to build defect-free nanoporous carbons, a direct support of this mechanism is foreseen as difficult, although not impossible. One finding seems to be true: pores in carbons, not only heteroatoms and structural defects, enhance the ORR process.... And those pores are a unique asset of this kind of materials.

Acknowledgements The author would like to highly acknowledge the extensive contribution of Prof. Francisco Rodriguez-Reinoso works to her development as a carbon scientist.

References

1. Liu, X., Dai, L.: Carbon-based metal-free catalysts. Nat. Rev. Mater. **1**, 16064 (2016)

2. Yang, S., Waldvogel, R., Jiang, X.: Electrochemistry of carbon dioxide on carbon electrodes. ACS Appl. Mater. Interfaces **8**, 28357–28371 (2016)
3. Srinsivau, K., Modak, B., Ghosh, S.K.: Porous graphitic carbon nitride: a possible metal-free photocatalyst for water splitting. J. Phys. Chem. **118**, 26479–26484 (2014)
4. Zhao, Y., Zhang, J., Qu, L.: Graphitic carbon nitride/graphene hybrids as new active materials for energy conversion and storage. ChemNanoMat **1**, 298–318 (2015)
5. Martin, D.J., Qiu, K., Shevlin, S.A., Hanfoko, A.D., Chen, X., Guo, Z., Tang, J.: Highly efficient photocatalytic H_2 evolution from water using visible light and structure-controlled graphitic carbon nitride. Angew. Chem. Int. Ed. **53**, 9240–9245 (2014)
6. Huang, C., Chen, C., Zhang, M., Lin, L., Ye, X., Lin, S., Antonietti, M., Wang, X.: Carbon-doped BN nanosheets for metal-free photoredox catalysis. Nat. Commun. **6**, 7698 (2015)
7. Gong, Y., Li, M., Wang, Y.: Carbon nitride in energy conversion and storage: recent advances and future prospects. ChemSusChem **8**, 931–946 (2015)
8. Largeot, C., Portet, C., Chmiola, J., Taberna, P.L., Gogotsi, Y., Simon, P.: Relation between the ion size and pore size for an electric double-layer capacitor. J. Am. Chem. Soc. **130**, 2730–2731 (2008)
9. Raymundo-Pinero, E., Kierzak, K., Machnikowski, J., Beguin, F.: Relationship between the nanoporous texture of activated carbons and their capacitance properties in different electrolytes. Carbon **44**, 2498–2507 (2006)
10. Eliad, J., Salitra, G., Soffer, A., Aurbach, D.: Ion sieving effects in the electrical double layer of porous carbon electrodes: estimating effective ion size in electrolytic solutions. J. Phys. Chem. B **105**, 6880–6887 (2001)
11. Chmiola, J., Yushin, G., Gogotsi, Y., Portet, C., Simon, P., Taberna, P.L.: Anomalous increase in carbon capacitance at pore sizes less than 1 nanometer. Science **313**, 1760–1763 (2006)
12. Kiyohara, K., Sugino, T., Asaka, K.: Phase transition in porous electrodes. J. Chem. Phys. **131**, 154710 (2011)
13. Kondrat, S., Georgi, N., Fedorov, M.V., Kornyshev, A.A.: A superionic state in nano-porous double-layer capacitors: insights from Monte Carlo simulations. Phys. Chem. Chem. Phys. **13**, 11359–11366 (2011)
14. Bandosz, T.J.: Nanoporous carbons: Looking beyond their perception as adsorbents, catalyst supports and supercapacitors. Chem. Rec. **16**, 205–218 (2016)
15. Ania, C.O., Armstrong, P.A., Bandosz, T.J., Beguin, F., Carvalho, A.P., Celzard, A., Frackowiak, E., Gilarranz, M.A., Laszlo, K., Matos, J., Pereira, M.F.R.: Engaging nanoporous carbons in "beyond adsorption" applications: characterization, challenges and performance. Carbon **164**, 69–84 (2020)
16. Guo, J., Morris, J.R., Ihm, Y., Contescu, C.I., Gallego, N.C., Duscher, G., Pennycook, S.J., Chisholm, M.F.: Topological defects: origin of nanopores and enhanced adsorption performance in nanoporous carbon. Small **8**, 3283–3288 (2012)
17. Bandosz, T.J., Ania, C.O.: Surface chemistry of activated carbon and its characterization. In: Bandosz, T.J. (ed.) Activated Carbon Surfaces in Environmental Remediation, pp. 159–230. Elsevier, Oxford (2006)
18. Yang, Z., Nie, H., Chen, X., Chen, X., Huang, S.: Recent progress in doped carbon nanomaterials as effective cathode catalysts for fuel cell oxygen reduction reaction. J. Power Sources **236**, 238–249 (2013)
19. Zhang, J., Xia, Z., Dai, L.: Carbon-based electrocatalysts for advanced energy conversion and storage. Sci. Adv. **1**, e150056 (2015)
20. Yu, L., Pan, X., Cao, X., Hu, P., Bao, X.: Oxygen reduction reaction mechanism on nitrogen-doped grapheme. A density functional theory study. J. Catal. **282**, 1183–1190 (2011)
21. Shao, M., Chang, Q., Dodelet, J.P., Chenitz, R.: Recent advances in electrocatalysts for oxygen reduction reaction. Chem. Rev. **116**, 3594–3657 (2016)
22. Ma, R., Lin, G., Zhou, Y., Liu, Q., Zhang, T., Shan, G., Yang, M., Wang, J.: A review of oxygen reduction mechanisms for metal-free carbon-based electrocatalysts. Comput. Mater. **5**, 78 (2019)

23. Chen, L., Xu, X., Yang, W., Jia, J.: Recent advances in carbon-based electrocatalysts for oxygen reduction reaction. Chin. Chem. Lett. **31**, 626–634 (2020)
24. Jorge, A.A., Jervis, R., Teriasamy, A.P., Qiao, M., Feng, J., Tran, L.N., Titirici, M.M.: 3D-carbon materials for efficient oxygen and hydrogen electrocatalysis. Adv. Energy. Mater. **10**, 1902494 (2019).
25. Yan, X., Jia, Y., Yao, X.: Defects on carbon for electrocatalytic oxygen reduction. Chem. Soc. Rev. **47**, 7628–7658 (2018)
26. Seredych, M., Szcurek, A., Fierro, V., Celzard, A., Bandosz, J.: Electrochemical reduction of oxygen on hydrophobic ultramicroporous PolyHIPE carbon. ACS Catal. **6**, 5618–5628 (2016)
27. Encalada, J., Savaram, K., Travlou, N.A., Li, W., Li, Q., Delgado-Sanchez, C., Fierro, V., Celzard, A., He, H., Bandosz, T.J.: Combined effect of porosity and surface chemistry on the electrochemical reduction of oxygen on cellular vitreous carbon foam catalysts. ACS Catal. **7**, 7466–7478 (2017)
28. Florent, M., Wallace, R., Bandosz, T.J.: Oxygen electroreduction on nanoporous carbons: textural features vs nitrogen and boron catalytic centers. ChemCatChem **11**, 851–860 (2019)
29. Barrera, D., Florent, N., Sapag, K., Bandosz, T.J.: Insight into the mechanism of oxygen reduction reaction on micro/mesoporous carbons: ultramicropores versus nitrogen-containing catalytic centers in ordered pore structure. ACS Appl. Energy Mater. **2**, 7412–7424 (2019)
30. Winter, M., Brodd, V.: What are batteries, fuel cells, and supercapacitors? Chem. Rev. **104**, 4245–4270 (2004)
31. Masa, J., Xia, W., Muhler, M., Schuhmann, W.: On the role of metals in nitrogen-doped carbon electrocatalysts for oxygen reduction. Angew. Chem. Int. Ed. **54**, 10102–10120 (2015)
32. Tiwari, J.N., Tiwari, R.N., Singh, G., Kim, K.S.: Recent progress in the development of anode and cathode catalysts for direct methanol fuel cells. Nano Energy **2**, 553–578 (2013)
33. Wu, G., Zelenay, P.: Nanostructured nonprecious metal catalysts for oxygen reduction reaction. Acc. Chem. Res. **46**, 1878–1889 (2013)
34. Stamenkovic, V.R., Mun, B.S., Arenz, M., Mayrhofer, K.J.J., Lucas, C.A., Wang, G., Ross, P.N., Markovic, N.M.: Trends in electrocatalysis on extended and nanoscale Pt-bimetallic alloy surfaces. Nat. Mater. **6**, 241–247 (2007)
35. Jiang, S., Ma, Y., Jian, G., Tao, H., Wang, X., Fan, Y., Lu, Y., Hu, Z., Chen, Y.: Electrocatalysts: Facile construction of Pt-Co/CN_x nanotube electrocatalysts and their application to the oxygen reduction reaction. Adv. Mater. **21**, 4953–4956 (2009)
36. Tian, N., Zhou, Z.-Y., Sun, S.-G., Ding, Y., Wang, Z.L.: Synthesis of tetrahexahedral platinum nanocrystals with high-index facets and high electro-oxidation activity. Science **31**, 732–735 (2007)
37. Wang, C., Daimon, H., Onodera, T., Koda, T., Sun, S.: A general approach to the size- and shape-controlled synthesis of platinum nanoparticles and their catalytic reduction of oxygen. Angew. Chem. **120**, 3644–3647 (2008)
38. Dai, L., Xue, Y., Qu, L., Choi, H.J., Baek, J.B.: Metal-free catalysts for oxygen reduction reaction. Chem. Rev. **115**, 4823–4892 (2015)
39. Chen, J., Zhang, H., Liu, P., Li, Y., Li, G., An, T., Zhao, H.: Thiourea sole doping reagent approach for controllable N, S co-doping of pre-synthesized large-sized carbon nanospheres as electrocatalyst for oxygen reduction reaction. Carbon **92**, 339–347 (2015)
40. Zhang, J., Dai, L.: Heteroatom-doped graphitic carbon catalysts for efficient electrocatalysis of oxygen reduction reaction. ACS Catal. **5**, 7244–7253 (2015)
41. Wu, J., Yang, Z., Li, X., Sun, Q., Jin, C., Strasser, P., Yang, R.: Phosphorus-doped porous carbons as efficient electrocatalysts for oxygen reduction. J. Mater. Chem. A **1**, 9889–9896 (2013)
42. Wu, G., More, K.L., Johnston, C.M., Zelenay, P.: High-performance electrocatalysts for oxygen reduction derived from polyaniline, iron, and cobalt. Science **332**, 443–447 (2011)
43. Watson, V.J., Delgado, C.N., Logan, B.E.: Influence of chemical and physical properties of activated carbon powders on oxygen reduction and microbial fuel cell performance. Environ. Sci. Technol. **47**, 6704–6710 (2013)

44. Choi, C.H., Park, S.H., Woo, S.: Binary and ternary doping of nitrogen, boron, and phosphorus into carbon for enhancing electrochemical oxygen reduction activity. ACS Nano **6**, 7084–7091 (2012)
45. Deak, D., Biddinger, E.J., Luthman, K., Ozkan, U.S.: The effect of phosphorus in nitrogen-containing carbon nanostructures on oxygen reduction in PEM fuel cells. Carbon **48**, 3637–3659 (2010)
46. Zhong, M., Kim, E.K., McGann, J.P., Chun, S.E., Whitacre, J.F., Jaroniec, M., Matyjaszewski, K., Kowalewski, T.: Electrochemically active nitrogen-enriched nanocarbons with well-defined morphology synthesized by pyrolysis of self-assembled block copolymer. J. Am. Chem. Soc. **134**, 14846–14857 (2012)
47. Wang, S., Zhang, L., Xia, Z., Roy, A., Chang, D.W., Baek, J.-B., Dai, L.: BCN graphene as efficient metal-free electrocatalyst for the oxygen reduction reaction. Angew. Chem. Int. Ed. **51**, 4209–4212 (2012)
48. Li, H., Kang, W., Wang, L., Yue, Q., Xu, S., Wang, H., Liu, J.: Synthesis of three-dimensional flowerlike nitrogen-doped carbons by a copyrolysis route and the effect of nitrogen species on the electrocatalytic activity in oxygen reduction reaction. Carbon **54**, 249–257 (2013)
49. Liang, J., Zheng, Y., Chen, J., Liu, J., Hulicova-Jurcakova, D., Jaroniec, M., Qiao, S.Z.: Facile oxygen reduction on a three-dimensionally ordered macroporous graphitic C_3N_4/carbon composite electrocatalyst. Angew. Chem. Int. Ed. **51**, 3892–38926 (2012)
50. Liang, J., Jiao, Y., Jaroniec, M., Qiao, S.Z.: Sulfur and nitrogen dual-doped mesoporous graphene electrocatalyst for oxygen reduction with synergistically enhanced performance. Angew. Chem. Int. Ed 51:1–6 (2012). Angew. Chem. 124, 11664–11668 (2012)
51. Liu, M., Zhang, R., Chen, W.: Graphene-supported nanoelectrocatalysts for fuel cells: synthesis, properties, and applications. Chem. Rev. **114**, 5117–5160 (2014)
52. Yang, Z., Yao, Z., Li, G., Fang, G., Nie, H., Liu, Z., Zhou, X., Chen, X., Huang, S.: Sulfur-doped graphene as an efficient metal-free cathode catalyst for oxygen reduction. ACS Nano **6**, 205–211 (2012)
53. Yang, S., Zhi, L., Tang, K., Feng, X., Maier, J., Mullen, K.: Efficient synthesis of heteroatom (N or S)-doped graphene based on ultrathin graphene oxide-porous silica sheets for oxygen reduction reactions. Adv. Funct. Mater. **22**, 3634–3640 (2012)
54. Zhang, C., Hao, R., Lian, H., Hou, Y.: Synthesis of amino-functionalized graphene as metal-free catalyst and exploration of the roles of various nitrogen states in oxygen reduction reaction. Nano Energy **2**, 88–97 (2013)
55. Gong, K.P., Du, F., Xia, Z.H., Durstock, M., Dai, L.M.: Nitrogen-doped carbon nanotube arrays with high electrocatalytic activity for oxygen reduction. Science **323**, 760–764 (2009)
56. Silva, R., Voiry, D., Chhowalla, M., Asefa, T.: Efficient metal-free electrocatalysts for oxygen reduction: polyaniline-derived N- and O-doped mesoporous carbons. J. Am. Chem. Soc. **135**, 7823–7826 (2013)
57. Tao, G., Zhang, L., Chen, L., Cui, X., Hua, Z., Wang, M., Wang, J., Chen, Y., Shi, J.: N-doped hierarchically macro/mesoporous carbon with excellent electrocatalytic activity and durability for oxygen reduction reaction. Carbon **86**, 108–117 (2015)
58. Chen, J., Zhang, H., Liu, P., Li, Y., Li, G., An, T., Zhao, H.: Nitrogen doped carbon nanotubes and their impact on the oxygen reduction reaction in fuel cells. Carbon **92**, 339–347 (2015)
59. Seredych, M., László, K., Rodríguez-Castellón, E., Bandosz, T.J.: S-doped carbon aerogels/GO composites as oxygen reduction catalysts. J. Energy Chem. **25**, 236–245 (2016)
60. Seredych, M., László, K., Bandosz, T.J.: Sulfur-doped carbon aerogel as a metal-free oxygen reduction catalyst. ChemCatChem **7**, 2924–2931 (2015)
61. Seredych, M., Biggs, M., Bandosz, T.J.: Oxygen reduction on chemically heterogeneous iron-containing nanoporous carbon: the effects of specific surface functionalities. Micro. Meso. Mater **221**, 137–149 (2016)
62. Wang, L., Ambrosi, A., Pumera, M.: "Metal-free" catalytic oxygen reduction reaction on heteroatom-doped graphene is caused by trace metal impurities. Angew. Chem. **125**, 14063–14066 (2013)

63. Ozaki, J.I., Anahara, T., Kimura, N., Oya, A.: Simultaneous doping of boron and nitrogen into a carbon to enhance its oxygen reduction activity in proton exchange membrane fuel cells. Carbon **44**, 3358–3361 (2006)
64. Ozaki, J.I., Kimura, N., Anahara, T., Oya, A.: Preparation and oxygen reduction activity of BN-doped carbons. Carbon **45**, 1847–1853 (2007)
65. Huang, S.F., Terakura, K., Ozaki, T., Ikeda, T., Boero, M., Oshima, M., Ozaki, J.I., Miyata, S.: First-principles calculation of the electronic properties of graphene clusters doped with nitrogen and boron: analysis of catalytic activity for the oxygen reduction reaction. Phys. Rev. B Condens. Matter Mater. Phys. **80**, 235410 (2009)
66. Chisaka, M., Iijina, T., Ishigara, Y., Suzuki, Y., Inadar, R., Sakurai, Y.: Carbon catalyst codoped with boron and nitrogen for oxygen reduction reaction in acid media. Electrochim. Acta **85**, 399–410 (2012)
67. Wang, S., Iyyamperumal, E., Roy, A., Xue, Y., Yu, D., Dai, L.: Vertically aligned BCN nanotubes as efficient metal-free electrocatalysts for the oxygen reduction reaction: a synergetic effect by co-doping with boron and nitrogen. Angew. Chem. Int. Ed. **50**, 11756–11760 (2011)
68. Zhao, Y., Yang, L., Chen, S., Wang, X., Ma, Y., Wu, Q., Jiang, Y., Qian, W., Hu, Z.: Can boron and nitrogen co-doping improve oxygen reduction reaction activity of carbon nanotubes? J. Am. Chem. Soc. **135**, 1201–1204 (2013)
69. Choi, C.H., Chung, M.W., Kwon, H.C., Park, S.H., Woo, S.I.: B, N- and P, N-doped graphene as highly active catalysts for oxygen reduction reactions in acidic media. J. Mater. Chem. A **1**, 3694–3699 (2013)
70. Baik, S., Suh, B.L., Byeon, A., Kim, J., Lee, J.W.: In-situ boron and nitrogen doping in flue gas derived carbon materials for enhanced oxygen reduction reaction. J. CO_2 Utiliz. **20**, 73–80 (2017)
71. Paraknowitsch, J.P., Thomas, A.: Doping carbons beyond nitrogen: an overview of advanced heteroatom doped carbons with boron, sulfur and phosphorus for energy applications. Energy Environ. Sci. **6**, 2839–2854 (2013)
72. Zhao, Z., Xia, Z.: Design principles for dual-element-doped carbon nanomaterials as efficient bifunctional catalyst for oxygen reduction and evolution reactions. ACS Catal. **6**, 1553–1558 (2016)
73. Jiang, Y., Yang, L., Sun, T., Zhao, J., Lyu, Z., Zhuo, O., Wang, X., Wu, Q., Ma, J., Hu, Z.: Significant contribution of intrinsic carbon defects to oxygen reduction activity. ACS Catal. **5**, 6707–6712 (2015)
74. Xue, L., Li, Y., Liu, X., Liu, Q., Shang, J., Duan, H., Dai, L., Shui, J.: Zigzag carbon as efficient and stable oxygen reduction electrocatalyst for proton exchange memrane fuel cells. Nat. Comm. **9**, 3819 (2018)
75. Li, Y., Zhou, W., Wang, H., Xie, L., Liang, Y., Wei, F., Idrobo, J.-C., Pennycook, S.J., Dai, H.: An oxygen reduction electrocatalyst based on carbon nanotube-graphene complexes. Nat. Nanotechnol. **7**, 394–400 (2012)
76. Wang, X., Lee, J.S., Zhu, Q., Liu, J., Wang, Y., Dai, S.: Ammonia -treated ordered mesoporous carbons as catalytic materials for oxygen reduction reactions. Chem. Mater. **22**, 2178–2180 (2010)
77. Gavrilov, N., Pasti, I.A., Mitric, M., Trava-Sejdic, J., Ciric-Marjanonic, G., Mantus, S.V.: Electrocatalysis of oxygen reduction reaction on polyaniline-derived nitrogen-doped carbon nanoparticle surfaces in alkaline media. J. Power Sources **220**, 306–316 (2012)
78. Zhao, X., Zhao, H., Zhang, T., Yan, X., Yuan, Y., Zhang, H., Zhao, H., Zhang, D., Zhu, G., Yao, X.: One-step synthesis of nitrogen doped microporous carbon materials as metal free electrocatalysts for oxygen reduction reaction. J. Mater. Chem. A **2**, 11666–11671 (2014)
79. Liu, Y.-L., Shi, C.-S., Xu, X.-Y., Sun, P.-C., Chen, T.-H.: Nitrogen doped hierarchically porous carbon spheres as efficient metal-free electrocatalyst for an oxygen reduction reaction. J. Power. Sources **282**, 389–396 (2015)
80. He, Y., Han, X., Du, Y., Song, B., Xu, P., Zhang, B.: Bifunctional nitrogen-doped microporous carbon microspheres derived from poly(o-methylamine) for oxygen reduction and supercapacitors. ACS Appl. Mater. Interfaces **8**, 3601–3608 (2016)

81. Ferrero, G.A., Preuss, K., Fuertes, A.A., Sevilla, M., Titirici, M.-M.: The influence of pore size distribution on the oxygen reduction reaction performance in nitrogen doped carbon microspheres. J. Mater. Chem. **4**, 2581–2589 (2016)
82. Eisneberger, D., Prinsen, P., Geels, N.J., Stroek, W., Yan, N., Hua, B., Kuo, J.-L., Rothenberg, G.: The evolution of chierarchical porosity in self templated nitrogen-doped carbon and its effects on oxygen reduction electrocatalysis. RCS Adv. **6**, 80398–80407 (2016)
83. Quilez-Bermejo, J., Gonzalez-Gaitan, C., Morallon, E., Cazorla-Amoros, D.: Effect of carbonization conditions of polyaniline on its catalytic activity towards ORR. Some insights about the nature of active sites. Carbon **119**, 62–71 (2017)
84. Quilez-Bermejo, J., Morallon, E., Cazorla-Amoros, D.: Oxygen-reduction catalysis of N-doped carbons prepared via heat treatment of polyaniline at over 1100 °C. Chem. Comm. **54**, 441–4444 (2018)
85. Zhang, H., Lv, K., Fang, B., Forster, M.C., Derivighu, R., Andreas, L.B., Zhang, K., Chen, S.: Crucial role for oxygen functional groups in the oxygen reduction reaction electrocatalytic activity of nitrogen-doped carbons. Electrochem. Acta **292**, 942–950 (2018)
86. Li, C., Li, X., Sun, X., Zhang, X., Duan, L., Yang, X., Wang, L., Lu, W.: Porous carbón networks derived from graphitic carbon nitride for efficient oxygen reduction reaction. Nanoscale Res. Lett. **16**, 249–2457 (2019)
87. Mostazo-Lopez, M.J., Salina-Torres, D., Ruiz-Rosas, R., Morallon, E., Cazorla-Amoros, D.: Nitrogen-doped superporous activated carbons as electrocatalysts for the oxygen reduction reaction. Materials **12**, 1346–1362 (2019)
88. Xie, B., Zhang, Y., Zhang, R.: Pure nitrogen-doped graphene aerogel with rich micropores yield high ORR performance. Mat. Sci. Eng. **242**, 1–5 (2019)
89. Li, D., Jia, Y., Chang, G., Chen, J., Liu, H., Wang, J., Hu, Y., Xia, Y., Yang, D., Yao, X.: A defect-driven metal free electrocatalyst for oxygen reduction in acidic electrolyte. Chem **4**, 2345–2356 (2018)
90. Wang, H., Bo, X., Zhang, Y., Guo, L.: Sulfur-doped ordered mesoporous carbon with high electrocatalytic activity for oxygen reduction. Electrochim. Acta **108**, 404–411 (2013)
91. Seredych, M., Idrobo, J.C., Bandosz, T.J.: Effect of confined space reduction of graphite oxide followed by sulfur doping on oxygen reduction reaction in neutral electrolyte. J. Mater. Chem. A **1**, 7059–7067 (2013)
92. Seredych, M., Bandosz, T.J.: Confined space reduced graphite oxide doped with sulfur as metal-free oxygen reduction catalyst. Carbon **66**, 227–233 (2014)
93. Baik, S., Lee, J.W.: Effect of boron-nitrogen bonding on oxygen reduction reaction activity of BN Co-doped activated porous carbons. RSC Adv. **5**, 24661–24669 (2015)
94. Ishii, T., Maie, T., Kimura, N., Kobori, Y., Imashiro, Y., Ozaki, J.I.: Enhanced catalytic activity of nanoshell carbon co-doped with boron and nitrogen in the oxygen reduction reaction. J. Hydrogen Energy **42**, 15489–15496 (2017)
95. Zhang, Y., Huang, N., Zhang, J., Qiao, S., Sun, Y., Gao, Z.: P-doped three dimentional porous carbon networks as efficient metal-free electrocatalysts for ORR. ECS J. Solid State. Sci. Technol. **7**, M123–M127 (2018)
96. Sun, Y.N., Zhang, N.L., Zhao, L., Sui, Z.-Y., Sun, Z.-Y., Han, B.-H.: A N, P dual-doped carbon with high porosity as an advanced metal-free oxygen reduction catalyst. Adv. Mater. Interf. **6**, 1900592 (2019)
97. Tang, C., Zhang, Q.: Nanocarbon for oxygen reduction electrocatalysis: dopants, edges, and defects. Adv. Mater. **28**, 1604103 (2017)
98. Kruusenberg, I., Keis, J., Aulepp, M., Tammeveski, K.: Oxygen reduction on carbon nanomaterials-modified glassy carbon electrodes in alkaline solution. J. Solid State Electrochem. **14**, 1269–1277 (2010)
99. Wei, W., Tao, Y., Lv, W., Su, F.-Y., Ke, L., Li, J., Wang, D.-W., Li, B., Kang, F., Yang, Q.-H.: Unusual high oxygen reduction performance in all-carbon electrocatalysts. Sci. Rep. **4**, 6289 (2014)
100. Castelo-Quiben, J., Bailon-Garcia, E., Perez-Fernandez, F.J., Carrasco-Marin, F., Perez-Cadenas, A.F.: Mesoporous carbon nanospheres with improved conductivity for electrocatalytic reduction of O_2 and CO_2. Carbon **155**, 88–99 (2019)

101. Gomes-Morai, R., Rey-Raap, N., Figueiredo, J.L., Ribeiro Pereira, M.F.: Glucose-derived carbon materials with tailored properties as electrocatalyst for the oxygen reduction reaction. Beilstein J. Nanotechnol. **10**, 1089–1102 (2019)
102. Barrera, D., Florent, M., Kulko, M., Bandosz, T.J.: Ultramicropores-influenced mechanism of oxygen electroreduction on metal-free carbon catalysts. J. Mater. Chem. A **7**, 27110–27123 (2019)
103. Gabe, A., Ruiz-Rosa, R., Gonzalez-Gaitan, C., Morallon, E., Cazorla-Amoros, D.: Modeling of oxygen reduction reaction in porous carbon materials in alkaline medium. Effect of microporosity. J. Power Sources **412**, 451–464 (2019)

Computer Simulation and Experimental Studies of Various Environmental Gases (NH_3, CH_2O, SO_2, H_2S, Benzene, Water) on Carbon Materials

Lumeng Liu, Shiliang (Johnathan) Tan, and D. D. Do

Abstract Adsorption by carbonaceous materials is an indispensable technique for gas purification. Its process is fundamentally driven by the interplay between the adsorbate-adsorbate and adsorbate-adsorbent interactions. The extent of these interactions depends on the temperature, the properties of the adsorbate (the saturation vapour pressure, the polarizability) and the physical and chemical properties of the adsorbent (the surface area, the pore size distribution and the functional groups). Conventional modeling and correlations are practical, but they are not informative in elucidating adsorption mechanism, which is critically important for rational design of high-performance adsorbents and proper management of adsorption processes. To this end, this chapter presents molecular simulation of adsorption for five common gaseous pollutants, benzene, formaldehyde, sulfur dioxide, hydrogen sulfide, and ammonia, as well as water on graphite, functional groups, graphitic pores and functionalized pores. These adsorption systems were classified based on the parameters Ω, which is the ratio of the isosteric heat of adsorption to the heat of condensation of the bulk), and they are analyzed with the Henry Constant, the isotherm, the isosteric heat and the density distribution of adsorbate within the system. Finally, the competitive and cooperative phenomena observed for multi-component adsorption were briefly presented and discussed with some preliminary molecular simulation results.

L. Liu
School of Civil Engineering, Chongqing University, Chongqing 400045, China

S. Tan
School of Chemical and Biomedical Engineering, Nanyang Technological University, Singapore 637459, Singapore

D. D. Do (✉)
School of Chemical Engineering, University of Queensland, St Lucia 4072, Australia
e-mail: d.d.do@uq.edu.au

J. C. Moreno-Piraján et al. (eds.), *Porous Materials*, Engineering Materials,
https://doi.org/10.1007/978-3-030-65991-2_3

1 Introduction

The hazardous gases and vapors from various industrial processes (e.g., chemical, pharmaceutical, etc.), combustion exhaust and in indoor/outdoor environments (e.g., buildings, automobiles, aircrafts, etc.) must be tackled properly to meet stringent environmental regulations for human health. Table 1 presents the properties, the impact on health and the immediately dangerous to life or health (IDLH) of selected gaseous pollutants.

There are various air cleaning technologies, such as adsorption, catalytic oxidation, photocatalysis, etc., among which adsorption by carbonaceous adsorbents is widely used because it is reliable and economically favorable. There are a number of factors that govern the performance of an adsorption process: the boiling point, the vapor pressure, and the polarizability of the adsorbate; the surface area, the pore size distribution and functionalities of the adsorbent; and the temperature at which the adsorption occurs [3–6]. Numerous studies have put efforts into correlating those factors to derive useful information for practical applications, for example correlations involving surface functional groups (SFG), micropore volume of solids, polarizability of gases and pre-adsorbed water (multi-component adsorption) are given in Fig. 1.

As practical as they can be, the correlations per se are not informative in elucidating adsorption mechanism, which is critically important for rational design of high-performance adsorbents and proper management of adsorption processes. Adsorption is fundamentally driven by the forces of interaction between all entities in a given system, including the attractive dispersion forces, the short-range repulsive forces, and electrostatic forces, balanced by the interplay between the energy and the entropy to minimize the free energy. Therefore, if the interactions between adsorbate and adsorbent are correctly described, the critical details obscured by the interplay of those factors can be unraveled by way of molecular simulation. This is the path taken by many researchers in the past many decades, and significant progress has been made on the development of theory and application of molecular simulation of adsorption together with high-resolution experimental data [11–17].

Table 1 Properties, health effects and IDLH of selected gaseous pollutants [1, 2]

Gas	Boiling point (K)	Dipole moment (D)	Healthy effect	IDLH (ppm)
Formaldehyde	253.9	2.33168	Carcinogen (nasopharyngeal)	30
Benzene	351	0	Carcinogen (leukemia)	3000
Ammonia	239.8	1.4718	Irrigating eyes, skin and respiratory	300
Sulfur dioxide	263	1.62	Respiratory disease	100
Hydrogen sulfide	212.84	0.97	Respiratory disease	100
Water	373.15	1.85	–	

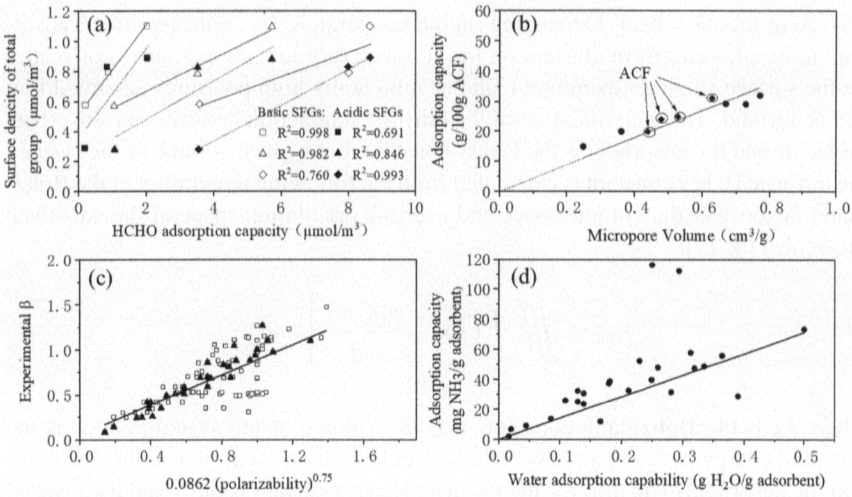

Fig. 1 Correlations between **a** formaldehyde adsorption capacity and density of functional groups [7], **b** benzene adsorption and micropore volume [8], **c** affinity coefficient of Dubinin model and polarizability [9], and **d** ammonia adsorption and amounts of pre-adsorbed water [10]

In this chapter, we present molecular simulation of adsorption for five representative gaseous pollutants, benzene (representing volatile organic compounds), formaldehyde (indoor carcinogen), sulfur dioxide (flue gas impurity), hydrogen sulfide (foul odor), and ammonia (corrosive gas), as well as water (because it impedes the removal of the pollutants) on graphite, functional groups, graphitic pores and functionalized pores. The goal is to delineate the role of various components of a carbonaceous adsorbent on their interactions with gaseous pollutants, from which we can derive the underlying mechanism of adsorption.

2 Adsorption on Surface

Practical carbonaceous adsorbents are porous and adsorption in pores at a given temperature is affected by many factors: pore size, shape and connectivity, surface affinity and functionalities. Among these factors, the surface affinity is the most critical one as it is the measure of interaction between adsorbate and adsorbent. Therefore, we isolate other factors and study this interaction and quantify it with the Henry constant and the isosteric heat at zero loading. Once this has been understood, we carry out molecular simulation of adsorption to study the interplay between the adsorbate–adsorbate interaction and the adsorbate-adsorbent interaction.

When the adsorbate–adsorbate interaction is less than the adsorbate-adsorbent interaction, adsorption is not favorable. On the other hand, when the interaction between adsorbate and adsorbent is greater, adsorbate molecules will adsorb on the surface of the adsorbent. Depending on the temperature (i.e. entropy effect) adsorbate molecules can form clusters on the surface (very low temperatures) or scatter on the surface (high temperatures), and at sufficiently high pressures adsorbed film can be formed. The quantification of the intrinsic interaction between one adsorbate molecule and the adsorbent is the Henry constant. For a given volume of the system, the intrinsic Henry constant is calculated from the following integration of the Boltzmann factor over the volume space and over the orientation space of the adsorbate molecule [18, 19]:

$$K_{\text{int}} = \iiint_V \int_\omega \exp\left[-\frac{\varphi(\underline{r}, \omega)}{k_B T}\right] d\omega dV \tag{1}$$

where k_B is the Boltzmann constant, V is the volume of the system, and φ is the potential energy between an adsorbate molecule with orientation ω at the position $\underline{r}$ and the adsorbent. The unit for the intrinsic Henry constant is nm^3, and its physical meaning is that it is the volume occupied by the molecules in the system if they are allowed to expand to the same density as the bulk gas [20]. For strong adsorbents this intrinsic Henry constant is much greater than the volume of the system. If the adsorbed density for surface adsorption is expressed as molecules per unit area.

$$\Gamma = K \rho_G \tag{2}$$

where K is the surface Henry constant, $K = K_{\text{int}}/S$, having a unit of length.

The isosteric heat at zero loading can also be computed from [20]:

$$q_{st}^{(0)} = k_B T \frac{\iiint_V \frac{\varphi(\underline{r},\omega)}{k_B T} \exp\left[-\frac{\varphi(\underline{r},\omega)}{k_B T}\right] d\underline{r} d\omega}{K_{\text{int}}} \tag{3}$$

shows the surface Henry constant for the adsorbates studied in this chapter interacting with graphite as a function of temperature. It is seen that graphite has the highest affinity towards benzene because there are six carbon atoms and six hydrogen atoms, resulting in a very strong dispersive interaction with graphite. The order of the affinity for a given temperature is.

$$\text{Benzene} > SO_2 > H_2S \sim \text{Formaldehyde} > NH_3 > \text{water}$$

The Henry constant decreases with temperature, and the extent of this decrease is greater for adsorbates having high energy of interaction with graphite, for example the Henry constant for benzene decreases by a factor of 100 when the temperature increases from 250 to 350 K while the Henry constant for water decreases by a factor of approximately 2 over the same temperature range (Fig. 2).

The variation of the Henry constant with temperature is reflected in the isosteric heat at zero loading, shown as solid lines in Fig. 3 as a function of temperature. Also

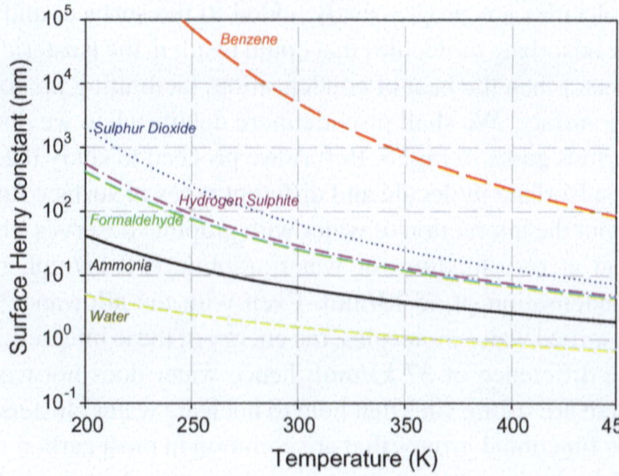

Fig. 2 Surface Henry constant (Eq. 2) as a function of temperature for various adsorbates on graphite without functional groups

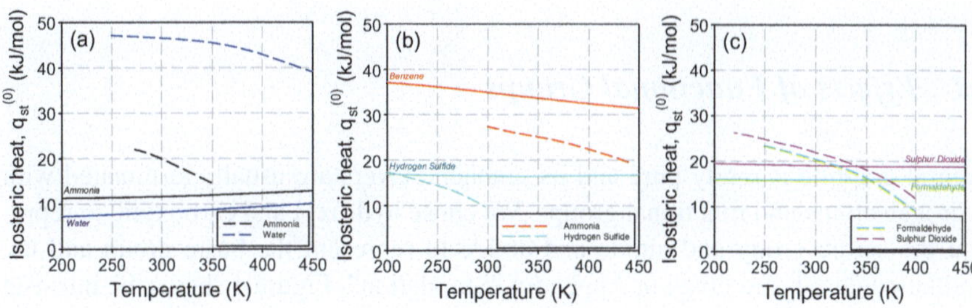

Fig. 3 Isosteric heat at zero loading (solid lines) for adsorption of various pollutants on pure graphite and the heat of condensation (dash line) as a function of temperature

plotted in the same figure is the heat of condensation (dashed lines) so that we can evaluate the adhesiveness between adsorbate and adsorbent versus the cohesiveness between adsorbate molecules. There are two extreme cases: (1) isosteric heat at zero loading $q_{st}^{(0)}$ is always greater than the bulk heat of condensation λ for the range of temperatures investigated in this study, as seen with benzene and hydrogen sulfide, and (2) $q_{st}^{(0)} < \lambda$ for the case of ammonia and water. Falling between these two extremes is observed with sulfur dioxide and formaldehyde, i.e. $q_{st}^{(0)} < \lambda$ for low temperatures and $q_{st}^{(0)} > \lambda$ for high temperatures. The implication of this relative difference is the behavior of adsorption isotherms at different temperatures for a given reduced pressure.

For case (1) benzene always adsorb favorably on a carbon surface, while for case (2) water does not adsorb on pure graphite. The interesting cases are formaldehyde and sulfur dioxide because the isosteric heat at zero loading is lower than the heat of condensation at low temperatures and becomes greater at higher temperatures. This does not mean that adsorption does not occur for ammonia, formaldehyde and sulfur dioxide at low temperatures because Fig. 3 only provides information of one adsorbate molecule interacting with graphite. As the pressure is increased, more molecules are progressively added to the surface, and it is the interaction between the adsorbate molecules that could result in the isosteric heat of adsorption to become greater than the heat of condensation, facilitating the formation of adsorbed layer on the surface. We shall provide more details when we discuss adsorption isotherms of various gases in Fig. 6. Before we proceed to study the intrinsic interaction between an adsorbate molecule and different types of surface functional groups, an argument about the interaction of water with graphite deserves a brief discussion. The isosteric heat at zero loading for water/graphite is 8 kJ/mol, much lower than the heat of condensation of 45 kJ/mol. Even with the allowance for the interactions between adsorbed water molecules, the energy of these interactions is not enough to overcome the difference of 37 kJ/mol; hence water does not wet the graphite surface unless there are strong sites that help to nucleate water clusters; examples of the strong sites are functional groups that are common in most carbon materials and very fine cracks of dimensions commensurate to the molecular size of water.

2.1 Effects of Functional Groups

Natural graphite is rarely pure and its graphene layers are usually terminated with hydrogen atoms and functional groups. We chose hydroxyl and carboxylic to represent the acidic group and amine and amide to represent the basic group and the simulation details are given in Appendix "Simulation". Figure 4 shows the intrinsic Henry constants (calculated with Eq. 1) for the various pollutants interacting with graphite grafted with one functional group mounted in the center of the top surface of graphite, and we classify them into three groups; Group 1: water and ammonia, Group 2: formaldehyde and sulfur dioxide, and Group 3: benzene and hydrogen sulfide.

2.1.1 Group 1: Water and Ammonia

The Henry constants for water and ammonia on graphite grafted with functional groups are not only greater than the corresponding Henry constants on a pure graphite surface (black line), but also very sensitive to the type of the functional group. This is due to the strong hydrogen bonding between water (ammonia) molecule and the functional group, by virtue of the small hydrogen atom that facilitates the closer approach between a water (ammonia) molecule and the functional group in order to maximize the electrostatic interactions. These interactions are especially stronger for the functional groups that are extruding further from the graphite surface, for example carboxylic and amide, allowing water (ammonia) molecule to achieve the optimal orientation to maximize the electrostatic interactions, compared to the shorter hydroxyl and amine.

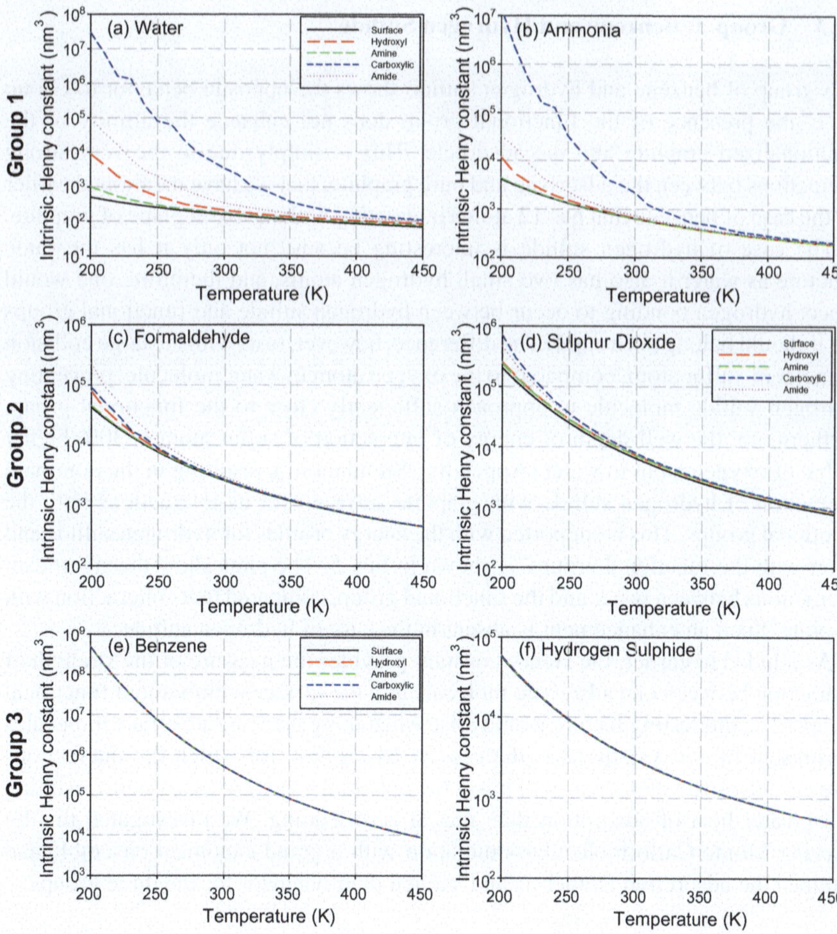

Fig. 4 Volumetric Henry constant (Eq. 1) as a function of temperature for various pollutants on graphite grafted with one functional group (hydroxyl, amine, carboxylic and amide) at the centre of the surface

2.1.2 Group 2: Formaldehyde and Sulfur Dioxide

The presence and the type of functional group only moderately enhance the Henry constant. Although formaldehyde and sulfur dioxide have high dipole moments, these molecules can't approach the functional group close enough because of their large collision diameters, resulting in only moderate electrostatic interactions.

2.1.3 Group 3: Benzene and Hydrogen Sulfide

This group of benzene and hydrogen sulfide shows the opposite behavior to Group 1, i.e. the presence of the functional group does not enhance the affinity of the functionalized graphite and the adsorbate. This is simply due to the very strong interactions between the adsorbate and pure graphite, that we have mentioned earlier for the case of benzene that has 12 atoms interacting with the basal plane of graphite.

The case of hydrogen sulfide is interesting because not only it has the same structure as water it also has two small hydrogen atoms, and therefore, one would expect hydrogen bonding to occur between hydrogen sulfide and functional groups and it should belong to Group 1. The difference, however, rests with the large collision diameter of sulfur atom, compared to the oxygen atom in water molecule, preventing hydrogen sulfide molecule to approach sufficiently close to the functional group. Furthermore, the well-depth of energy of interaction of sulfur atom is 30% higher to that of oxygen atom in water (Appendix "Simulation"), resulting in the dominant interaction of hydrogen sulfide with graphite surface over its interactions with the functional groups. This is supported with the energy profiles for hydrogen sulfide and water with the functional groups as shown in Fig. 5. The plots show the significant interactions between water and the functional group, compared to its interaction with graphite. Such an enhancement is absent in the case of hydrogen sulfide.

As alluded to earlier, the Henry constant provides the measure of the strength of interaction between one adsorbate molecule and the surface with/without functional groups. The interesting feature would arise when more than one adsorbate molecules are present in the system, as is the case in adsorption, for which the interactions between adsorbate molecules will lower the potential energy of the system, resulting in a greater heat of adsorption than that at zero loading. We investigated this by carrying Monte Carlo molecular simulation with a grand canonical ensemble and obtained the adsorption isotherms and studied their behavior for the three groups.

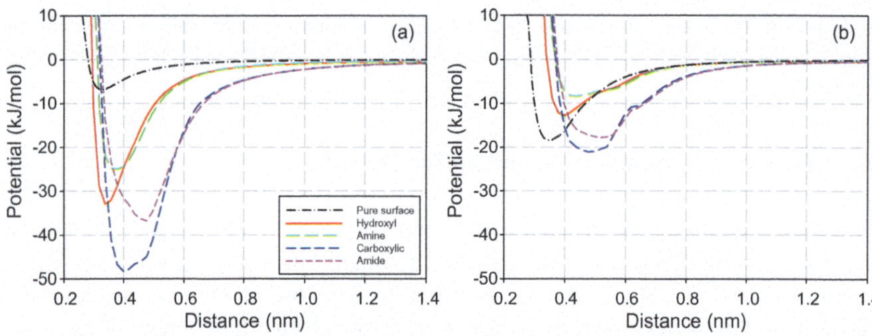

Fig. 5 Potential energies of interaction between adsorbate and one functional group mounted on a hard wall versus their separation distance for **a** water and **b** hydrogen sulfite. The separation distance is the distance between the center of geometry of and adsorbate and the mounting point of the functional group on the hard wall. The corresponding energies between adsorbate and pure graphite are also shown (black dot-dashed line) for comparison

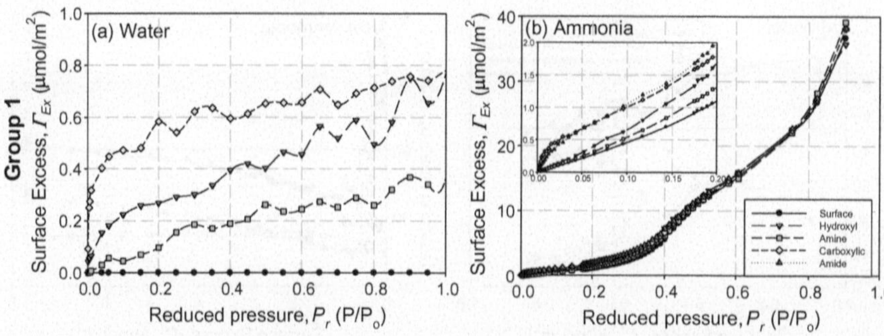

Fig. 6 Isotherms for **a** water, **b** ammonia adsorption at 298 K on graphite with 25 functional groups grafted at the centre of the top surface and their interspacing is 0.5 nm as shown in Fig. 23

Figure 6 shows the isotherms of water and ammonia adsorption at 298 K on graphite grafted with various types of functional groups (simulation details are given in Appendix "Simulation"). Over the full range of pressure, these isotherms apparently exhibit Type III, according to the IUPAC classification, in that the adsorbed density is very low for low pressures and it increases sharply at moderate and high pressures. However, we take ammonia as an example to show the fine features of the isotherm. Careful analysis of the isotherms at low loadings (Henry law region) shows that the adsorbed density for ammonia is greater for graphite grafted with carboxylic and amide functional groups compared to that for amine and hydroxyl groups, in agreement with the Henry constants that we reported in Fig. 4. Another interesting feature is that the isotherms for ammonia adsorption on graphite with carboxylic and amide groups at low loadings follow Langmuirian type. This is a result of the large Henry constant and the progressive coverage of the functional groups until they are fully covered with ammonia molecules. Over the moderate loading region, clusters of ammonia molecules begin to span over the basal plane of graphite, resulting in the concave shape of the isotherm because of the greater interactions (hydrogen bonding) between the adsorbed molecules that make the isosteric heat of adsorption greater than the heat of condensation as shown in Fig. 7a.

The isosteric heat of adsorption is decomposed into the contributions from the adsorbate–adsorbate interactions and the adsorbate-adsorbent interactions to understand the roles of these interactions. Figure 7a shows that the decrease of the isosteric heat for water is mainly due to the interaction between the adsorbate-adsorbent as the functional groups are being progressively occupied. Eventually, the isosteric heat falls below the heat of condensation once all functional groups have been saturated. The isosteric heat of adsorption for ammonia also has a similar decrease, due to the progressive occupancy of the strong sites; however, unlike water, the isosteric heat increases with loading after the functional groups have been saturated, and it is due to the combined contributions from the adsorbate–adsorbate interactions and the interactions between ammonia and graphite. The isosteric heat eventually becomes greater than the heat of vaporization, resulting favorable adsorption as seen in the isotherm in Fig. 6b.

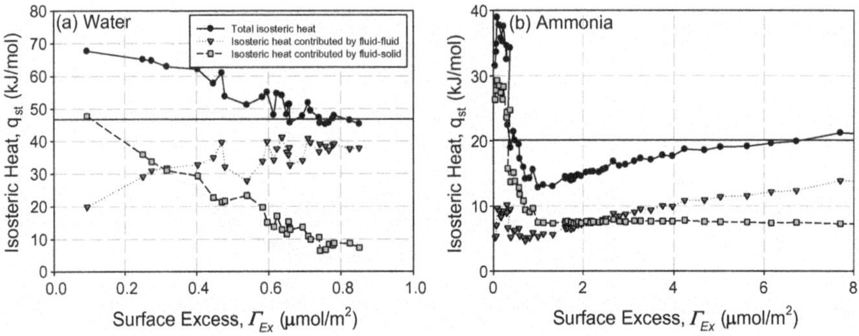

Fig. 7 Isosteric heats versus loading for **a** water, **b** ammonia adsorption at 298 K on graphite grafted with 25 carboxylic functional groups, separated by 0.5 nm, at the centre on the surface. The configuration of functional groups is shown in Fig. 23 (Appendix "Simulation")

The isotherms at 298 K for Group 2 with formaldehyde and sulfur dioxide are shown in Fig. 8 and they are of Type III at low loadings and of Type II at moderate and high loadings. The progress of adsorption is due to the formation of molecular

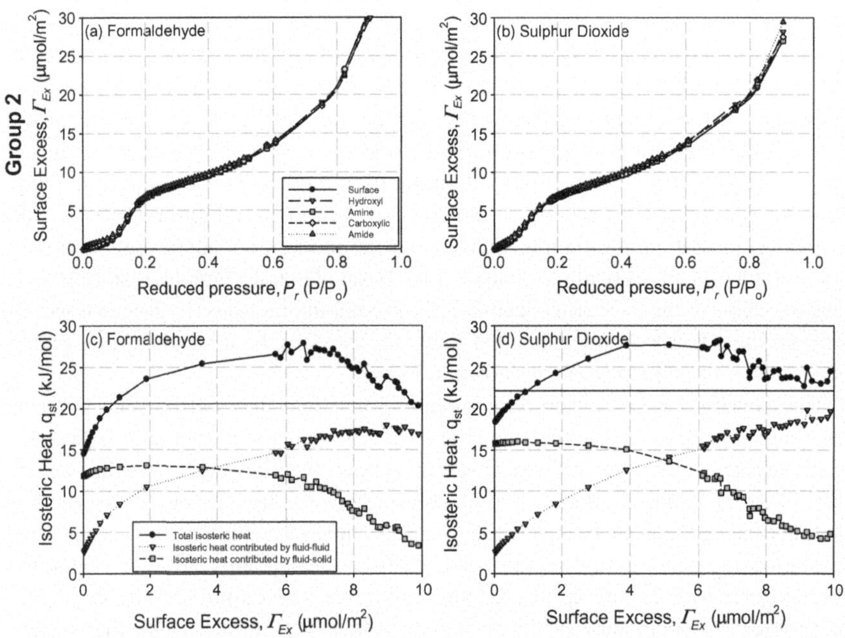

Fig. 8 Isotherms **a** formaldehyde and **b** sulfur dioxide adsorption at 298 K on graphite grafted with 25 functional groups (separated by 0.5 nm) at the centre of the surface. Their corresponding isosteric heats and their contributions from the adsorbate–adsorbate interaction and the adsorbate-adsorbent interaction for the case of carboxylic functional group are shown in (**c**) and (**d**)

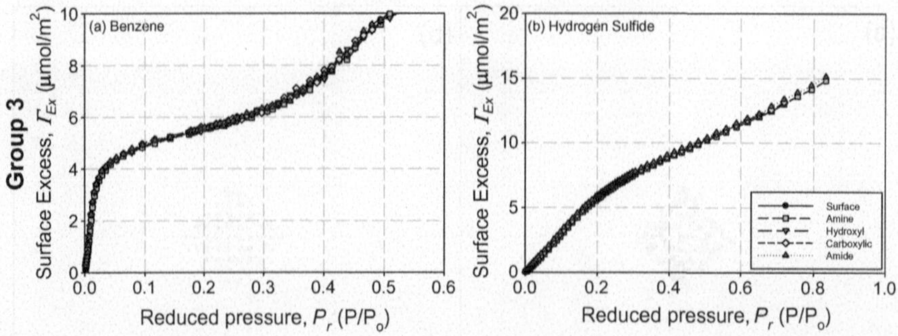

Fig. 9 Adsorption isotherms for **a** benzene and **b** hydrogen sulfide adsorption at 298 K on graphite grafted with 25 functional groups, separated by 0.5 nm, at the centre of the surface (see Fig. 23)

layers on the surface, which is attributed to the interactions between the adsorbed molecules that readily make up the relatively small difference between the isosteric heat of adsorption at zero loading and the heat of condensation as shown in Fig. 8c, d, where we see that the isosteric heat is greater than the heat of condensation at a loading of 1 μmol/m^2, which is well below the monolayer density.

The isotherms for Group 3 with benzene and hydrogen sulfide exhibit Type I (Fig. 9), which is due to their strong interactions with the basal plane of graphite (see Fig. 2). The isosteric heat at zero loading for this group is greater than the heat of condensation, which gives rise to the convex shape of the isotherm. For benzene, the isotherm on the surface is very strong due to the high dispersive interaction, therefore, the addition of functional groups hardly enhances the adsorption. The same is observed for hydrogen sulfide as well and although the affinity of hydrogen sulfide on the surface is not as strong as benzene, the sulfur atom is too large for H2S molecules to approach close enough to the functional groups to exert high electrostatic interactions. Therefore, the presence of the functional groups also hardly enhances the adsorption.

In summary, these three groups show different adsorption mechanisms.

1. Group 1: stronger interaction with the functional group, compared to the basal plane of graphite, which induces a clustering mechanism.
2. Group 2: moderate interaction with both functional group and surface which induce molecular layering with initial adsorption at the functional group.
3. Group 3: strong interaction with the surface which induces molecular layering

These mechanisms are illustrated in Fig. 10 with snapshots of the molecular configuration of adsorbed molecules at low loadings. The adsorption process for Group 1 follows a clustering mechanism where adsorption occurs around the functional groups. Ammonia have a moderate interaction with the surface and therefore a spill-over mechanism is observed [21] at moderate loadings. For Group 2, adsorption occurs on both the functional group and the surface. And finally, due to the strong interaction between the adsorbate and the adsorbents of Group 3, a large proportion of adsorbate molecules prefers to reside on the basal plane.

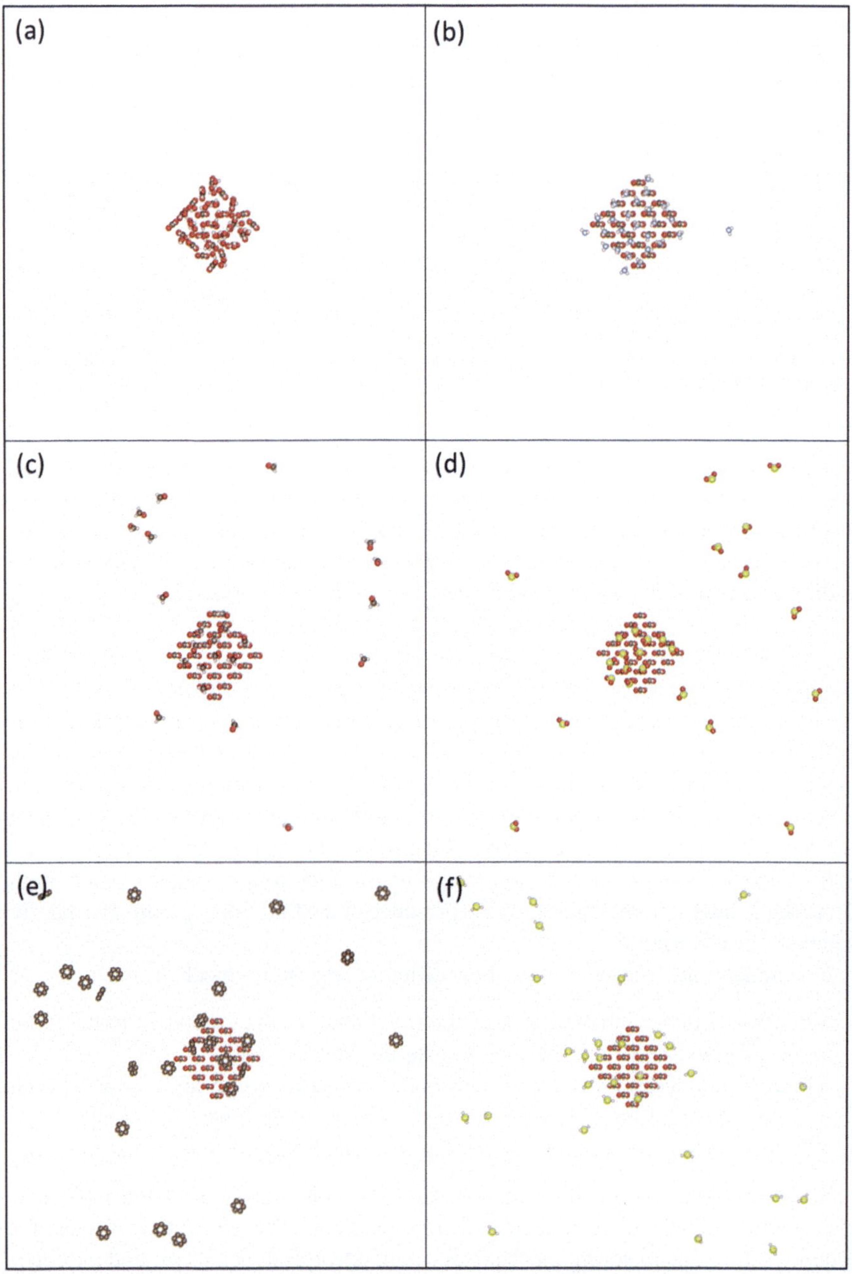

Fig. 10 Snapshots of the molecular configuration at a loading of 0.5 μmol/m^2 for **a** water, **b** ammonia, **c** formaldehyde, **d** sulfur dioxide, **e** benzene and **f** hydrogen sulfite adsorption on graphite grafted with 25 functional groups, separated by 0.5 nm, on the surface

3 Adsorption in Pores

3.1 Pore Models

Activated carbon (AC) has very complex pore structures and functionalities, and a complete atomistic modeling would lead to intractable results. Simplification that retains key features of real solids is often adopted and is done in conjunction with the advanced characterizations, such as X-ray scattering and diffraction for structural information; X-ray photoelectron spectroscopy (XPS) and Fourier transform infrared spectroscopy (FTIR) for heteroatoms and functional groups. For example, the X-ray studies by Franklin showed that AC consists of small graphitic crystallites, which are comprised of stacks of graphene sheets [22]. The pore structure, i.e., shape and size, is determined by the spatial arrangement of these crystallites. Slit pore is simplest and most popular geometry whose pore size is defined by the distance between the two parallel micro-crystallites, the surfaces of which constitute the pore walls. A deviation from this slit model is the wedge pore model for which its behavior is similar to the slit model [23–26] and for simplicity we will present results from the slit pore to show the effects of structural parameters, the functional groups on the adsorption behavior of different pollutants in the confined space. A brief summary of the molecular model for carbonaceous materials can be found in Bock et al. [27].

The dispersive-repulsive interaction between a gas molecule and a graphene sheet can be obtained by either a direct summation of pairwise potential energies between the adsorbate molecule and all the solid atoms or integration of the pairwise interaction energy over the solid volume (coarse-grained approach). The latter has been the method of choice because the computational time can be reduced substantially, provided that the distance between the adsorbate molecule and the surface is greater than the interatomic distance between neighboring solid atoms or the temperature is not too low. For adsorption of pollutant adsorbate on graphite the closest distance between an adsorbate molecule and graphite is about 0.4 nm and the separation distance between carbon atoms in a graphene layer is 0.142 nm, justifying the use of the coarse-grained approach. Using this approximation method, a number of solid–fluid potential energy equations have been reported in the literature, for example: the Crowell 10–4 equation for a single flat layer of infinite extent in the directions parallel to the surface [28], the 10–4–3 Steele equation which is an excellent approximation for a layered substrate, such as graphite, composed of many equally spaced graphene layers [29], the Bojan-Steele equation for a planar substrate of finite extent in one direction and infinite in extent in the other direction [30], or the Everett-Powl [31] equation for a layered cylinder of infinite extent in the axial direction. Liu et al. reported comprehensive derivations of solid–fluid potentials for a wide range of solids: layered substrates with constant surface atom density or solid substrates with constant volumetric atom density, and various curvatures and dimensions [32].

Other than the pore structure, the functional groups, discussed in Sect. 2.1, are incorporated in the pore model to simulate the interactions between the adsorbate molecule and impurities on the carbon surface. The hydroxyl group is the most abundant functional group in carbonaceous materials [33] and was chosen as the representative in molecular simulation studies [17, 34–36]. While early studies assumed the locations of functional groups along the edges of graphene layers [4, 37], recent solid-state nuclear magnetic resonance (NMR) [38], ultra-high-resolution transmission electron microscopy (TEM) [33] and near-edge X-ray absorption fine-structure [39] studies suggest that the hydroxyl groups tend to attach to the defective sites on the basal plane of the graphene layers rather than at their edges.

3.2 Adsorption in Graphitic Slit Pore

The behavior of the Henry constant for adsorption of various pollutants in graphitic pores with respect to temperature is the same as that observed with adsorption on graphite (Fig. 3), with the only difference being the greater Henry constant for pores because the enhancement in the interactions between the adsorbate and the pore. The enhancement is greater for smaller graphitic pores as shown in Fig. 11 for 0.7 and 1.6 nm pore width, which is defined as the distance between the plane passing through the centres of carbon atoms of the outermost layer of one wall to the corresponding plane of the opposite wall. These two pores represent ultra-micropore and micropore, according to the IUPAC classification.

As discussed in Fig. 3, the relative difference between the isosteric heat at zero loading and the heat of condensation does play a role on how adsorption on graphite would behave. This is also the case for graphitic pores for which the enhancement in the potential energy increases the isosteric heat at zero loading, compared to that for graphite. To quantify this, we introduced an energetic parameter Ω, defined as the ratio of the isosteric heat of adsorption to the heat of condensation. Figure 12 shows Ω at zero loading [40] as a function of pore width for various pollutants adsorption in

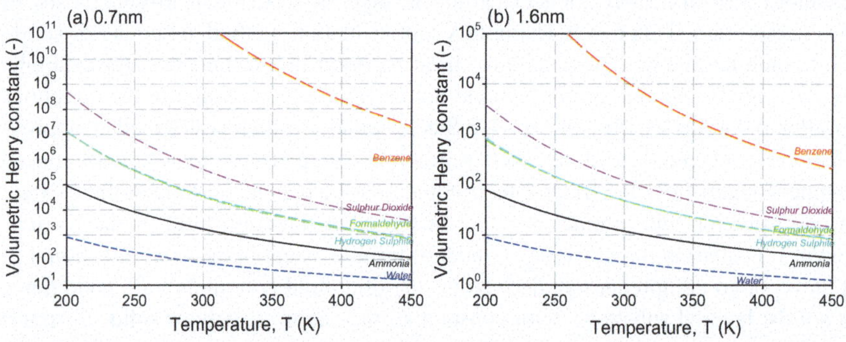

Fig. 11 Volumetric Henry Constant, defined as the ratio of the intrinsic constant to the pore accessible volume, as a function of temperature for various species adsorbing in **a** 0.7 nm and **b** 1.6 nm graphitic pores

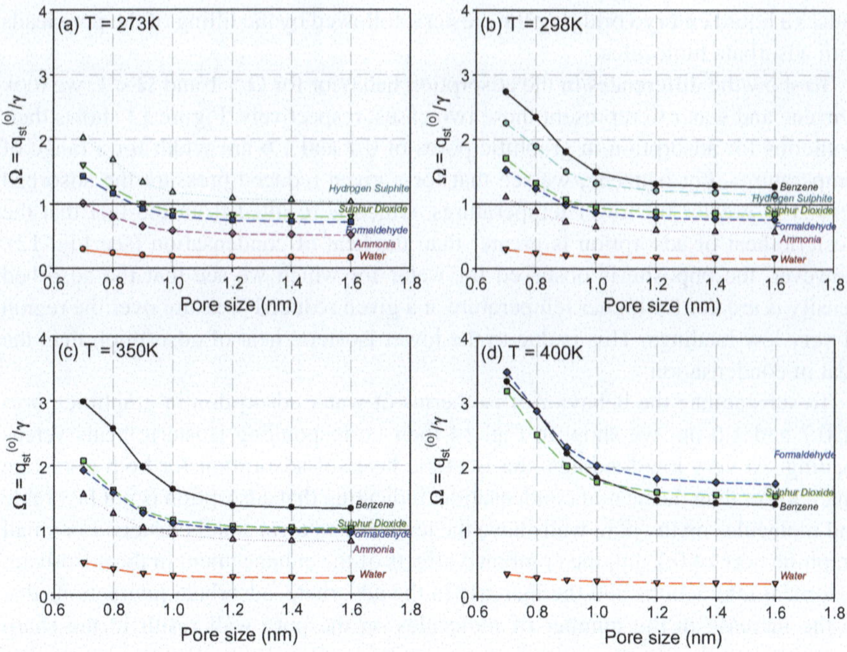

Fig. 12 Plots of Ω as a function of pore size for various adsorbates at **a** 273 K, **b** 298 K, **c** 350 K and **d** 400 K

graphitic pores at 273, 298, 350 and 400 K. This energetic parameter Ω decreases with pore width and approaches the value for graphite for widths greater than 1.6 nm for which there is no enhancement in the potential energy. When the isosteric heat at zero loading is greater than the heat of condensation ($\Omega > 1$), adsorbate molecules interact strongly with the adsorbent and therefore, they tend to form adsorbed films on the pore walls at low loadings. On the other hand, when the isosteric heat at zero loading is less than the heat of condensation ($\Omega < 1$), how the system would behave depends on how small the value of Ω is. For $0.5 < \Omega < 1$, the interactions between adsorbate molecules are comparable to those between adsorbate molecules and the solid atoms. At low loadings adsorption on the pore walls is by way of clustering because of the lower isosteric heat compared to the heat of condensation, but as the loading is increased the interactions between adsorbate molecules become greater because of the higher number of neighboring molecules, resulting in a greater isosteric heat than the heat of condensation. This leads to a transition from a clustering mechanism to molecular layering, and for small pores this would make the transition directly to complete filling of the pore. Finally, when the isosteric heat at zero loading is much less than the heat of condensation ($\Omega < 0.5$), the surface forces are not strong enough to spread adsorbate molecules over the surface, but rather they prefer to form clusters on the pore walls. As the loading is increased, in the case of small pores, clusters on one pore wall interact with those on the opposite wall and it is this interaction that causes a liquid embryo bridging the clusters, followed by the filling of the pore leads with adsorbate molecules.

To show the differences in the adsorption behavior for $\Omega > 1$ and $\Omega < 1$, we took benzene and water to represent these two cases, respectively. Figure 13 shows their isotherms for adsorption in graphitic pores of 0.7 and 1.6 nm width for a range of temperatures. For benzene, we see that for a given reduced pressure the adsorbed density is greater for lower temperatures, which is resulted from the fact that the isosteric heat of adsorption is greater than the heat of condensation (see Fig. 12). However, the opposite is observed for water for which we see that the adsorbed density is greater for higher temperature at a given reduced pressure over the region of very low loadings. This is due to the lower isosteric heat of adsorption than the heat of condensation.

To substantiate the behavior of isotherms of water adsorption in graphitic pores of 0.7 and 1.6 nm we show in Fig. 14 their corresponding isosteric heats versus loading. At very low loadings, the isosteric heats of adsorption for both pores are much lower than the heat of condensation, indicating that adsorption is not favorable and molecules on the pore walls have the tendency to form small clusters. For small graphitic pore of 0.7 nm, the combined effects of the enhancement in the adsorbate-adsorbent interactions and the increase in the adsorbate–adsorbate interactions due to the increase in the number of molecules on the pore wall result in the sharp

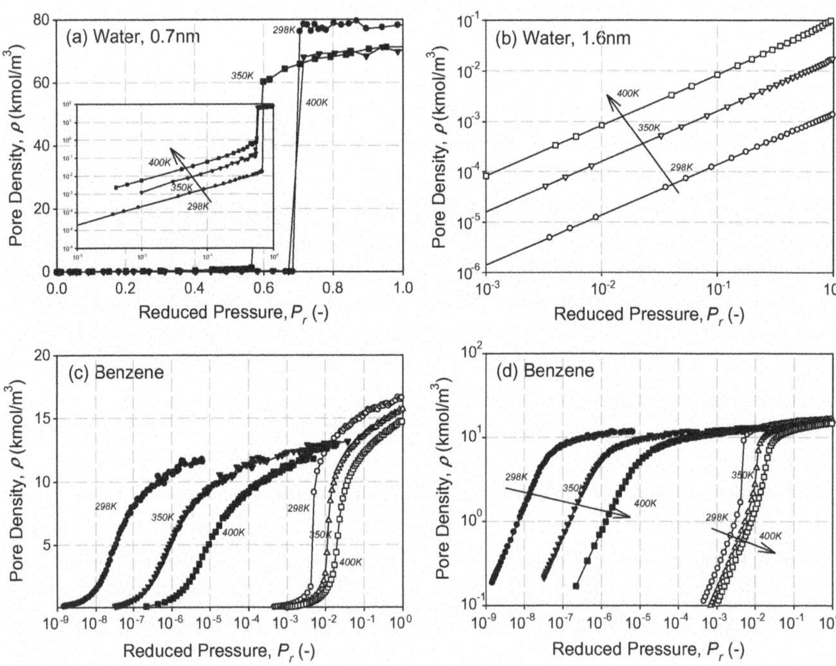

Fig. 13 Isotherms for **a**, **b** water and **c**, **d** benzene adsorption at 298, 350 and 400 K in 0.7 nm (solid symbols) and 1.6 nm (open symbols) graphitic pores

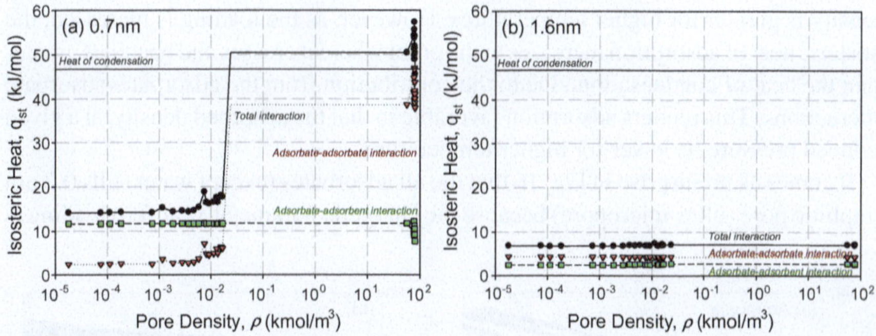

Fig. 14 Isosteric heat as a function of loading for water adsorption at 298 K in 0.7 nm (**a**) and 1.6 nm (**b**) graphitic pores

increase in the isosteric heat for it to become comparable to the heat of condensation as seen in Fig. 14a. This is the reason for the filling of 0.7 nm graphitic pore with water molecules at reduced pressures around 0.6. On the other hand, in the larger graphitic pore of 1.6 nm where there is no enhancement in the adsorbate-adsorbent interactions, and as a result the number of molecules on the pore walls increases slowly and as the saturation vapor pressure has been reached the contribution from the adsorbate–adsorbate interactions is not sufficient to make up for the difference between the heat of condensation and the isosteric heat at zero loading. As a result, water does not fill the 1.6 nm-graphitic pore (Fig. 14b).

Figure 15 shows the isotherms for formaldehyde adsorption in 1.6 nm graphitic pore at 298, 350 and 400 K. This is an interesting case because the energetic parameter Ω is less than 1 for 298, 350 K while it is greater than 1 for 400 K (see Fig. 12). This suggests that adsorption is favorable for all loadings at 400 K. For 298 and 350 K the isosteric heat at zero loading is less than the heat of condensation and therefore adsorption is unfavorable in that at a given reduced pressure the adsorbed

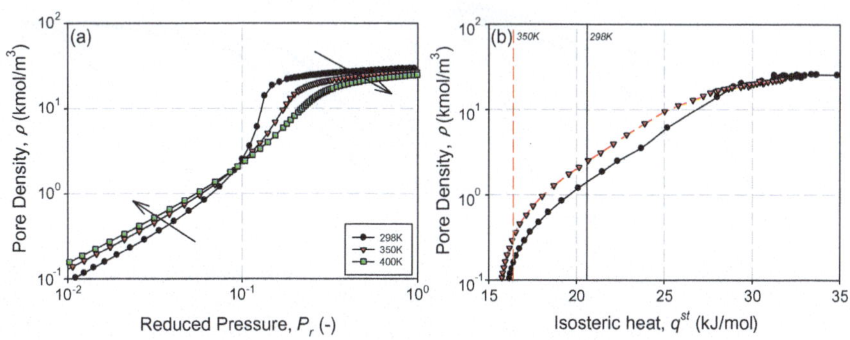

Fig. 15 **a** Isotherms for formaldehyde adsorption at 298, 350 and 400 K in 1.6 nm graphitic pore; and **b** Isosteric heat versus loading for the same systems with the dashed and vertical lines mark the heats of condensation at 350 and 298 K

density is greater for higher temperatures. However, as the loading is increased, the isosteric heat of adsorption increases to the point where it crosses and becomes greater than the heat of condensation, due to the contribution from the adsorbate–adsorbate interactions. This renders adsorption favorable in that the adsorbed density, at a given reduced pressure, is lower for higher temperatures.

It comes as no surprise in Fig. 16 that the all adsorbates (except water) fills 0.7 nm graphitic pore (ultra-micropore) because the energetic parameter Ω at zero loading is

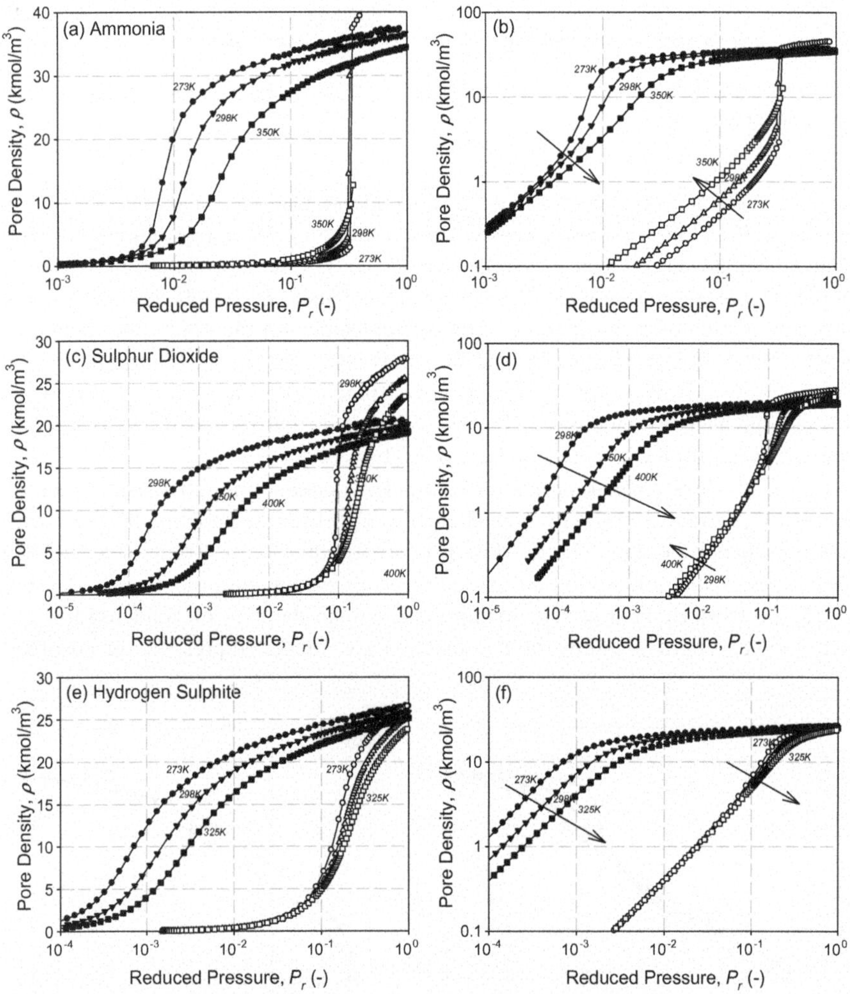

Fig. 16 Isotherms for **a**, **b** ammonia, **c**, **d** sulfur dioxide and **e**, **f** hydrogen sulfite adsorption at 298, 350 and 400 K in 0.7 nm (solid symbols) and 1.6 nm (open symbols) graphitic pores. Plots in **a**, **c**, **e** are in semi-log scales and those in **b**, **d**, **f** are in log–log scales

greater than unity. However, for larger graphitic pore of 1.6 nm width (large micropore), Ω falls below unity for ammonia and sulfur dioxide at low temperatures, indicating that adsorption is not favorable at low loadings with molecules on the surface forming clusters because of the stronger adsorbate–adsorbate interaction than the adsorbate-adsorbent interaction. As the loading is increased, clusters on the opposite pore walls coalesce to form condensate bridges for which the adsorbate–adsorbate interaction is stronger because of the greater number of neighboring molecules. This eventually leads to pore filling as the pressure is further increased. For hydrogen sulfide the isotherm behavior is similar to benzene for 0.7 nm pore because the isosteric heat of adsorption is greater than the heat of condensation over the temperature range investigated; however for 1.6 nm pore the adsorbed density is insensitive to temperature for a given reduced pressure over low loadings because the isosteric heat is almost the same as the heat of condensation. Beyond the Henry law region, the adsorbed density is greater for lower temperatures at a given reduced pressure, due to the contribution from the adsorbate–adsorbate interactions that make the isosteric heat greater than the heat of condensation.

3.3 Adsorption in Functionalized Slit Pore

The effects of the interplay between the interactions between adsorbate molecules and the interactions between the adsorbate and the adsorbent on the adsorption behavior of different pollutants in homogeneous graphitic micropores have been discussed in the previous section to study the role of graphite on adsorption. However, practical carbonaceous solids are heterogeneous, with adsorbing sites such as heteroatoms, defects, and functional groups, and they play a critical role in attracting highly associating adsorbates, such as water and ammonia. The combining effects of functional groups (by way of electrostatic interactions) and pore confinement (by way of enhancement in the dispersive interactions) on the capture of water [17, 34], formaldehyde [34, 35], ammonia [34, 35], and benzene [36] have been explored in a functionalized slit pore. The details of a functionalized pore model were given in Sect. 3.1 and its schematic diagram is shown in Fig. 17 for a slit pore with two ends open to the bulk gas surroundings. Each pore wall consists of three graphene layers that are finite in the x-direction and infinite in the y-direction (normal to the page), the boundaries for the latter direction are modelled with the periodic boundary conditions. Several rows of hydroxyl groups are grafted on the pore walls along the y-direction near the two open ends, with the interspacing between the two rows at each end being 0.426 nm (3 times the length of C–C bond, i.e., 0.142 nm).

Understanding how the properties of an adsorbate affect adsorption is important not only from the scientific point of view, but also implication in practical application because it is impracticable to develop a specific adsorbent for each pollutant. In doing so, the simulated isotherms, scaled against the maximum density at the saturation vapor pressure, for water, ammonia, formaldehyde, and benzene in the 1.55 nm slit carbon pore with functional groups at 298 K are shown in Fig. 18.

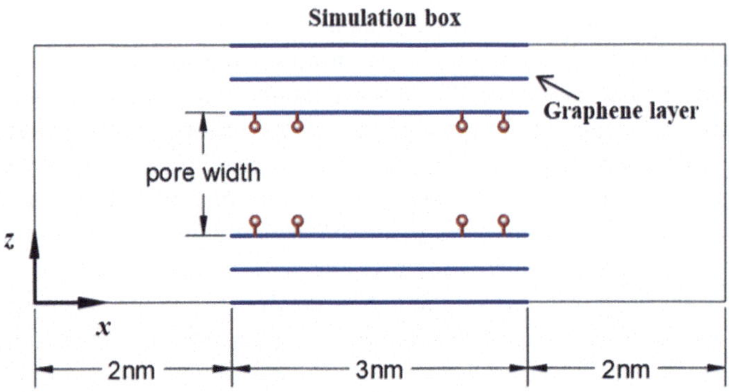

Fig. 17 Schematic diagram of the functionalized slit pore

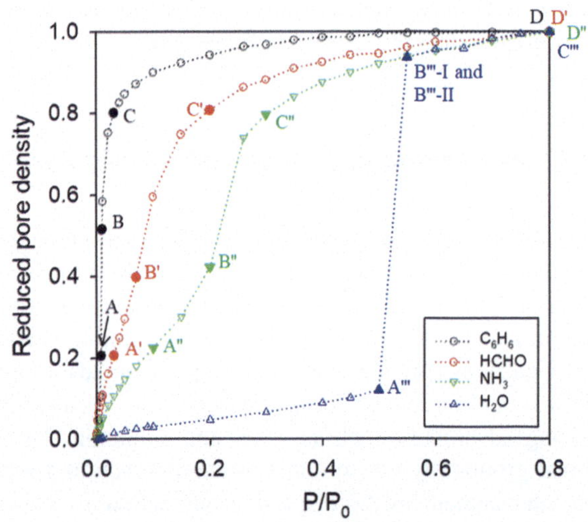

Fig. 18 Simulated isotherms for water [17], ammonia [35], formaldehyde [35] and benzene [36] adsorption in 1.55 nm graphitic pores with functional groups

A visualized demonstration of how these adsorbates behave microscopically as a function of loading is valuable to understand the fundamental differences between those adsorbates. Figure 19 shows the 2D-density distributions at various points as marked in Fig. 18. Benzene, a non-polar adsorbate, adsorbs in the interior region of the pore because its adsorption is dominated by the strong interactions between benzene and the graphitic basal planes, as also reflected in its isotherm of Type I (Point A). As the pressure is increased, benzene molecules begin to fill the pore, including the regions near the pore mouths where the functional groups reside (Points B and C) until the pore is completely filled (Point D). Formaldehyde, ammonia, and water, being polar adsorbate, exhibit different microscopic behavior compared to benzene. At low loadings (Point A′), formaldehyde adsorbs and forms clusters around the functional groups at both ends of the pore. As the loading is increased to Point B′, the clusters grow laterally along the pore wall, because of the adsorbate-adsorbent

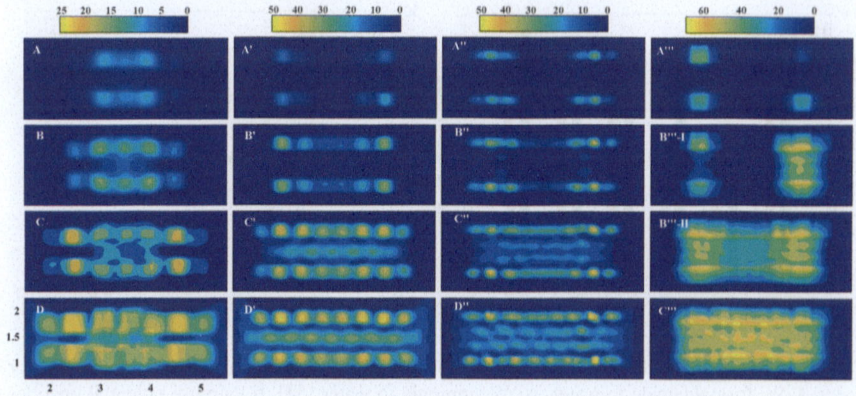

Fig. 19 2D-density distributions as a function of the vertical distance from the pore wall and the axial distance along the pore, at the selected points marked in Fig. 18. Columns from left to right: benzene, formaldehyde, ammonia, and water. The axes are in units of nm, and the scale bars are in units of kmol/m^3

interactions, and finally merge to form a monolayer on each of the pore walls. When the pressure is further increased to Point C′ and D′, the pore is filled with an additional layer at the center of the pore. While the process for water adsorption shares features similar to that for formaldehyde, one major difference stands out in that the direction of the growth of the clusters. Water clusters grow normally from the surfaces of the opposite pore walls, and coalesce forming bridges spanning across the pore (Point B″-I) at the filling pressure, beyond which pore filling occurs, induced by the growth of the two bridges of adsorbate in the axial direction (Points B‴-II to C″). Ammonia is intermediate between formaldehyde and water, and clusters grow in both directions as shown for Points A″-D″ in Fig. 19.

Figure 19 shows two further important observations; provided that the filling has not occurred, i.e. before Points B‴-II, (1) water molecules only occupy a small fraction of the total pore volume and they reside around the functional groups near the two ends of the pore, while benzene molecules occupy the interior of the pore where the dispersive interaction is strong, and (2) polar adsorbates, formaldehyde and ammonia, like water, initially form clusters around the functional groups, and therefore in the presence of water there will be competition between these species for the formation of clusters around functional groups. For non-polar adsorbates, the interference of water can be minimized by operating at pressures below the filling pressure for water. We have demonstrated in an extensive review [41] and a molecular simulation study [17] that water adsorption in carbon nanopores is extremely sensitive to both pore size and the functional groups because of the clustering mechanism of water around the functional groups and the possible coalescence of water clusters across the pore walls in small pores. Experimentally, increasing the hydrophobicity of adsorbents can enhance the capacity for the non-polar benzene under humid condition [42, 43], but would also hamper the capture of polar pollutants [35, 44]. For polar pollutants, like ammonia and formaldehyde, functional groups are beneficial in attracting their

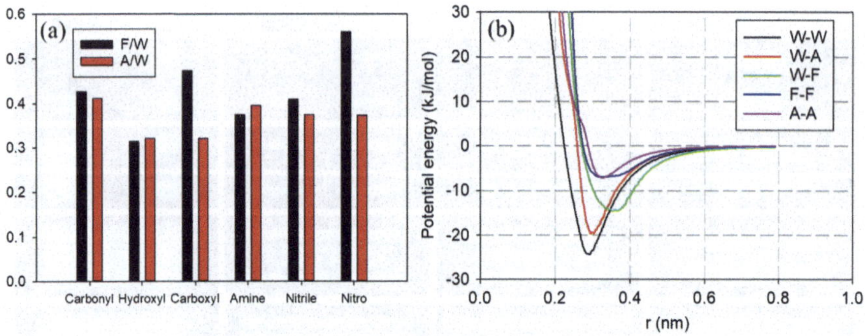

Fig. 20 **a** The isosteric heats at zero loading for formaldehyde and ammonia relative to that for water; F/W: formaldehyde to water, A/W: ammonia to water [34]. **b** Pairwise potential energies of interaction between water–water (W–W), water–ammonia (W–A), water–formaldehyde (W–F), formaldehyde–formaldehyde (FF), and ammonia–ammonia (A-A) [34]

molecules to nucleate clusters, but the concentration of functional groups must be carefully tailored because they have higher affinity toward water (Fig. 20a) [34]. Nonetheless, water is not necessarily detrimental to adsorption of polar pollutants in all circumstances because the polarity of water allows it to act as anchors for other polar gases by way of electrostatic interactions such that polar adsorbates can form cluster around the water clusters, and there is possible "absorption" of polar pollutant inside the water clusters. This is indeed shown in Fig. 20b that the energies of interactions between water and formaldehyde and those between water and ammonia are stronger than the intermolecular interactions of formaldehyde and the intermolecular interactions of ammonia [34]. The cooperative adsorption between water and polar gases are indeed observed experimentally and will be discussed in detail in the next section. Regardless of the properties of the pollutants, the filling of the pore with water molecules must be avoided so that the desired removal of the polar pollutants is not severely affected.

4 Multi-components Adsorption: Impacts of Water

Ambient moist can have considerable effects on adsorption of gaseous pollutants. The nature and the extent of these effects depend on the pollutant properties and the relative humidity (RH). For non- or very weakly polar gases, water impedes the adsorption, and more markedly at elevated RH [45–48]; for example, Cal et al. found that the adsorptive capacity of benzene adsorption in an activated carbon cloth is not noticeably reduced until 65% RH or above has been reached [45]. This is due to the fact that the onset of water filling in activated carbon occurs around 50% RH [41], below which water molecules occupy a small fraction of the pore volume as discussed in the last section (Fig. 19). The role that water plays in adsorption of polar adsorbates is more complicated. In the same study by Cal et al., the adsorptive capacity of acetone in activated carbon was not noticeably affected even at water humidity as high as 90% RH [45]. Qi and LeVan found that at low humidity water behaves cooperatively in adsorption of methyl ethyl ketone and methyl isobutyl ketone but competitively at high humidity [49]. Such complicated effects of water have also been observed for formaldehyde [50–52] and ammonia [10, 53–55]. Besides the RH and the properties of adsorbate, solid characteristics also play a role, e.g., the cooperative effect of water on ammonia adsorption is more pronounced for the activated carbon substrates with lower density of oxygen groups [54].

To have a better exposure on the role that water plays in mixture adsorption, we show in Fig. 21 the co-adsorption of water with n-hexane (a and b) and four alcohols (c and d) from the experiments by LeVan and co-workers [56, 57]. In their experiments, certain doses of hexane and the alcohols were pre-adsorbed on the

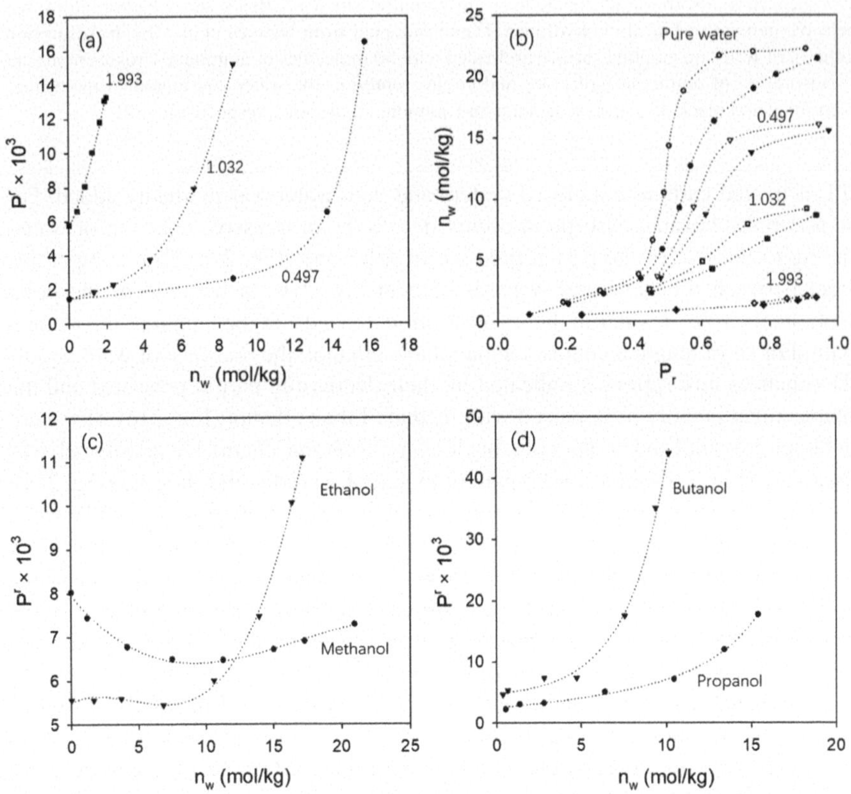

Fig. 21 **a** Effect of water loadings (n_W) on the partial pressure of hexane with different loadings (0.497, 1.032 and 1.993 mol/kg) at 298 K [56]; **b** effect of hexane loadings on water isotherms at 298 K [56], with solid symbols for adsorption and open symbols for desorption; Effect of water loadings (n_W) on the partial pressure of **c** methanol and ethanol and **d** propanol and butanol at 298 K [57]

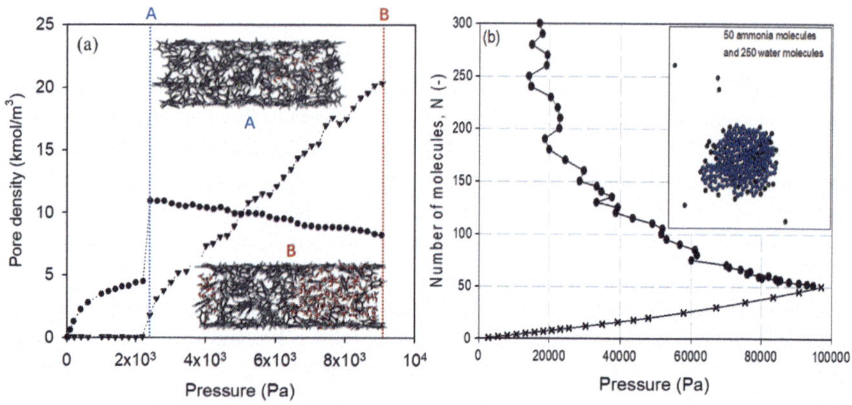

Fig. 22 **a** Isotherms and the snapshots of molecular configuration for benzene—water mixture (50% benzene) adsorption at 303 K in 2 nm graphitic slit pore. Black stick: carbon atom, red stick: oxygen, and white stick: hydrogen. Data is adopted from Nguyen et al. [58]; **b** Adsorption isotherm of water on graphite surface preloaded with 50 molecules of ammonia. Crosses represent the adsorption of ammonia molecules and circles represent the water and ammonia molecules, insert: blue and black dots represent water and ammonia molecules, respectively [59]

BPL activated carbon in a closed system and then water was gradually added. For the non-polar hexane, desorption occurred with the progressive injection of water, leading to the rise of the partial pressure of n-hexane (Fig. 21a). The competitive effect between n-hexane and water is seen in Fig. 21b. In the case of the polar alcohols, the role of water in the co-adsorption depends on the length of the alcohol chain. For short length alcohols, methanol and ethanol, the presence of water at low RH enhances adsorption as indicated by their decreasing partial pressures and the enhancement is more pronounced with methanol than ethanol (Fig. 21c). However, for larger propanol and butanol (having long hydrophobic chain), water only behaves negatively (Fig. 21d), in a similar pattern to that observed with n-hexane (Fig. 21a).

Figure 22a shows the GCMC simulation of co-adsorption of water and benzene in a 2 nm graphitic pore at 303 K [58]. Water displaced some benzene molecules by forming clusters in the pore at high water loadings, as seen in the snapshots (inset). This is due to the interactions between water molecules are greater than the intermolecular interactions of benzene. Figure 22b shows the Monte Carlo simulation of adsorption of a mixture of water and ammonia on a graphite surface [59]. The simulation was conducted in the canonical ensemble and initiated with introducing ammonia molecules into the simulation box as seen in the increase in pressure with the progressive adding ammonia molecules up to 50 molecules, and once the ammonia cluster has been formed around the functional groups water molecules are introduced into the system. The pressure of the system decreased with the increase of number of water molecules, indicative of the cooperative adsorption between water and ammonia around the functional groups, that results in lesser molecules in the gas phase. The net result is a complex cluster of water-ammonia mixture which could be viewed as ammonia molecules "absorbed" within a water cluster. Such a complex interaction is under progress and results will be reported in a future correspondence.

Appendix

Simulation

The Henry constant and the isosteric heat at zero loading are determined by Monte Carlo integration of the Boltzmann factor [19]. For a given temperature, 10^8 trial insertions of a molecule were carried out for the integration. The Henry constant in Fig. 4 is computed based from a single functional group mounted on the center of a 10 nm × 10 nm graphite, and the adsorption isotherms presented in Figs. 6, 7, 8, 9 and 10 are determined with 25 functional groups, separated by 0.5 nm, mounted in the center of the graphite surface as illustrated in Fig. 23.

The grand canonical ensemble was employed in Monte Carlo simulation to obtain the isotherm with the chemical potential as the input to a simulation. In a simulation,

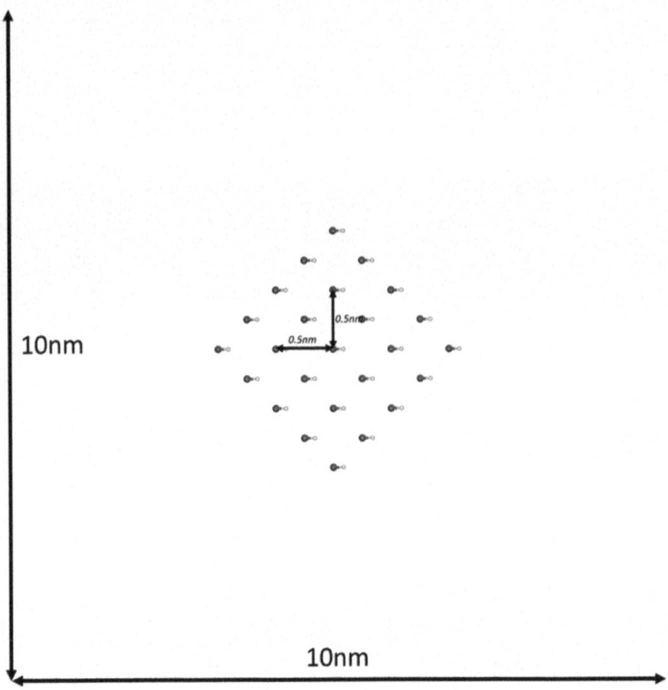

Fig. 23 Configuration of 25 functional groups grafted on top of the surface of graphite, separated by 0.5 nm

we used at least 200 million configurations to obtain simulation results with high accuracy. For adsorption on a surface, a 10 nm × 10 nm graphite was placed at the bottom of the simulation box of 4 nm height, and the potential energy between an LJ-site with graphite was calculated with the 10–4–3 equation [29] with the interlayer spacing of 0.3354 nm. The graphite is infinite in extent in the directions parallel to the surface and the infinite extent is modelled with periodic boundary conditions imposed at the boundaries. For adsorption in pores of finite length, the pore walls consist of three homogeneous graphene layers with interlayer spacing of 0.3354 nm, and the potential energy between a LJ-site and a layer is calculated with the Bojan-Steele equation [60]. The atomic carbon density of graphene layer is 38.2 nm^{-2}, and the molecular parameters for a carbon atom in graphene are collision diameter (σ) = 0.34 nm and the well-depth of interaction energy (ε/k_B) = 28 K. The cut-off radius of five times the collision diameter of the atom (marked * in Table 2) for each adsorbate was used in the calculation of the interaction energy. The molecular parameters for

Table 2 Molecular parameters of adsorbates and functional groups

Fluid	Parameter	Symbols	Unit	Value
Adsorbates				
Sulfur Dioxide [61]	S	σ	nm	0.339
		ε/k_B	K	73.8
		q	e	0.59
	O*	σ	nm	0.305
		ε/k_B	K	79
		q	e	−0.295
	Angle	O = S = O	degree	119.3
	Atomic distance	S = O	nm	0.1432
Benzene [62]	CH*	σ	nm	0.374
		ε/k_B	K	48
	Center site	q	e	2.42
	Pi site	q	e	−1.21
	Atomic distance	Center-Pi site	nm	0.0785
		CH–CH	nm	0.14
Hydrogen Sulfide [63]	S*	σ	nm	0.36
		ε/k_B	K	122
	X	q	e	−0.42
	H	σ	nm	0.25
		ε/k_B	K	50
		q	e	0.21
	Atomic distance	S–H	nm	0.134
		S–X	nm	0.03
	Angle	H–S–H	degree	92

(continued)

Table 2 (continued)

Fluid	Parameter	Symbols	Unit	Value
Formaldehyde [64]	C*	σ	nm	0.375
		ε/k_B	K	52.9
		q	e	0.45
	O	σ	nm	0.296
		ε/k_B	K	105.8
		q	e	−0.45
	H	σ	nm	0.242
		ε/k_B	K	7.6
	Atomic distance	C–O	nm	0.1203
		C–H	nm	0.1101
	Angle	H–C–H	degree	116.4
Water, SPC/E [65]	O*	σ	nm	0.3166
		ε/k_B	K	78.2
		q	e	−0.8476
	H	q	e	0.4238
	Atomic distance	O–H	nm	0.1
	Angle	H–O–H	degree	109.47
Ammonia [66]	N*	σ	nm	0.3385
		ε/k_B	K	170
		q	e	−1.035
	H	q	e	0.345
	Atomic distance	O–H	nm	0.1024
	Angle	H–O–H	degree	106.68
Functional group				
Hydroxyl [67]	O	σ	nm	0.307
		ε/k_B	K	78.2
		q	e	−0.64
	H	q	e	0.44
	C (atom is in the carbon surface)	q	e	0.2
	Atomic distance	C–O	nm	0.1364
		O–H	nm	0.096
	Angle	C–O–H	degree	110.5
Carboxyl [67]	C (COOH)	σ	nm	0.375
		ε/k_B	K	52.8
		q	e	0.55
	=O	σ	nm	0.296
		ε/k_B	K	105.7

(continued)

Table 2 (continued)

Fluid	Parameter	Symbols	Unit	Value
		q	e	−0.5
	– O	σ	nm	0.3
		ε/k_B	K	85.6
		q	e	−0.58
	H	q	e	0.45
	C (atom is in the carbon surface)	q	e	0.08
	Atomic distance	C–C	nm	0.152
		C=O	nm	0.1214
		C–O	nm	0.1364
		O–H	nm	0.097
	Angle	C–C–O	degree	111
		O=C–O	degree	123
		C–O–H	degree	107
Amine [68]	N	σ	nm	0.334
		ε/k_B	K	111
		q	e	−0.892
	H	q	e	0.356
	C (atom is in the carbon surface)	q	e	0.18
	Atomic distance	C–O	nm	0.1448
		N–H	nm	0.101
	Angle	C–N–H	degree	112.9
		H–N–H	degree	106.4
Amide [68]	C (CON)	σ	nm	0.372
		ε/k_B	K	34
		q	e	0.424
	=O	σ	nm	0.305
		ε/k_B	K	79
		q	e	−0.424
	– N	σ	nm	0.334
		ε/k_B	K	111
		q	e	−0.8
	H	q	e	0.4
	Atomic distance	C–C	nm	0.152
		C=O	nm	0.1229
		C–N	nm	0.1448
		N–H	nm	0.101
	Angle	C–C–N	degree	115.7
		C–C=O	degree	121.4
		H–N–H	degree	106.4

the adsorbates and the functional groups, dealt with in this chapter, are given in Table 2.

Isosteric Heat in Pore

See Fig. 24.

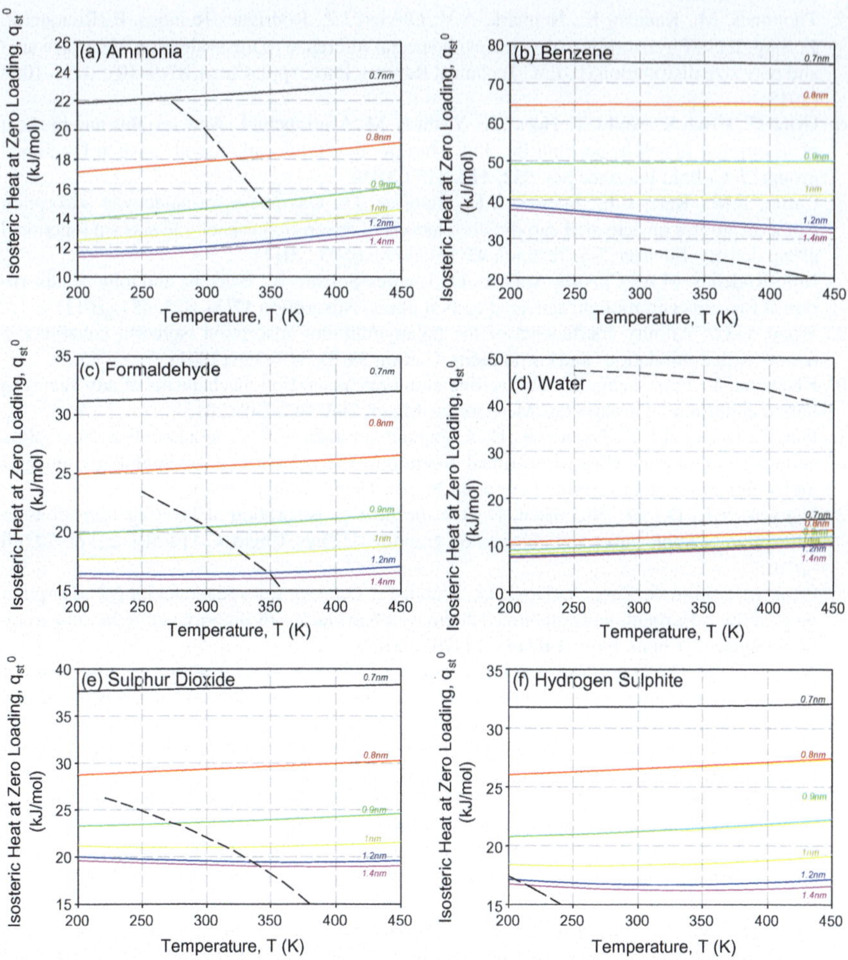

Fig. 24 Isosteric heat at zero loading (solid lines) and the heat of condensation (dashed lines) as a function of temperature for various adsorbates in graphitic pores of various widths

References

1. Barsan, M.E.: NIOSH Pocket Guide to Chemical Hazards. National Institute for Occupational Safety and Health, Cincinnati, Ohio (2007)
2. WHO: Selected Pollutants: WHO Guideline for Indoor Air Quality. Copenhagen, Denmark (2010)
3. Rodriguez-Reinoso, F., Molina-Sabio, M., Munecas, M.: Effect of microporosity and oxygen surface groups of activated carbon in the adsorption of molecules of different polarity. J. Phys. Chem. **96**(6), 2707–2713 (1992)
4. Boehm, H.: Surface oxides on carbon and their analysis: a critical assessment. Carbon **40**(2), 145–149 (2002)
5. Thommes, M., Kaneko, K., Neimark, A.V., Olivier, J.P., Rodriguez-Reinoso, F., Rouquerol, J., Sing, K.S.: Physisorption of gases, with special reference to the evaluation of surface area and pore size distribution (IUPAC Technical Report). Pure Appl. Chem. **87**(9–10), 1051–1069 (2015)
6. Urita, C., Urita, K., Araki, T., Horio, K., Yoshida, M., Moriguchi, I.: New insights into the heat of adsorption of water, acetonitrile, and n-hexane in porous carbon with oxygen functional groups. J. Colloid Interface Sci. **552**, 412–417 (2019)
7. Carter, E.M., Katz, L.E., Speitel, G.E., Ramirez, D.: Gas-phase formaldehyde adsorption isotherm studies on activated carbon: correlations of adsorption capacity to surface functional group density. Environ. Sci. Technol. **45**(15), 6498–6503 (2011)
8. Lillo-Ródenas, M.A., Cazorla-Amorós, D., Linares-Solano, A.: Benzene and toluene adsorption at low concentration on activated carbon fibres. Adsorption **17**(3), 473–481 (2011)
9. Wood, G.O.: Affinity coefficients of the polanyi/dubinin adsorption isotherm equations: a review with compilations and correlations. Carbon **39**(3), 343–356 (2001)
10. Khabzina, Y., Farrusseng, D.: Unravelling ammonia adsorption mechanisms of adsorbents in humid conditions. Microporous Mesoporous Mater. **265**, 143–148 (2018)
11. Fan, C., Do, D., Li, Z., Nicholson, D.: Computer simulation of argon adsorption on graphite surface from subcritical to supercritical conditions: the behavior of differential and integral molar enthalpies of adsorption. Langmuir **26**(20), 15852–15864 (2010)
12. Nguyen, V.T., Do, D., Nicholson, D.: On the heat of adsorption at layering transitions in adsorption of noble gases and nitrogen on graphite. J. Phys. Chem. C **114**(50), 22171–22180 (2010)
13. Prasetyo, L., Tan, S., Zeng, Y., Do, D.D., Nicholson, D.: An improved model for N2 adsorption on graphitic adsorbents and graphitized thermal carbon black—the importance of the anisotropy of graphene. J. Chem. Phys. **146**(18), 184702 (2017)
14. Nguyen, V.T., Do, D.D., Nicholson, D., Jagiello, J.: Effects of temperature on adsorption of methanol on graphitized thermal carbon black: a computer simulation and experimental study. J. Phys. Chem. C **115**(32), 16142–16149 (2011)
15. Nguyen, V.T., Horikawa, T., Do, D.D., Nicholson, D.: Water as a potential molecular probe for functional groups on carbon surfaces. Carbon **67**, 72–78 (2014)
16. Loi, Q.K., Prasetyo, L., Tan, S., Do, D.D., Nicholson, D.: Nonwetting/prewetting/wetting transition of ammonia on graphite. Langmuir **35**(3), 641–652 (2019)
17. Liu, L., Zeng, Y., Tan, S.J., Xu, H., Do, D.D., Nicholson, D., Liu, J.: On the mechanism of water adsorption in carbon micropores—a molecular simulation study. Chem. Eng. J. **357**, 358–366 (2019)
18. Do, D.D., Do, H.D., Wongkoblap, A., Nicholson, D.: Henry constant and isosteric heat at zero-loading for gas adsorption in carbon nanotubes. Phys. Chem. Chem. Phys. **10**, 7293–7303 (2008)
19. Zeng, Y., Xu, H., Horikawa, T., Do, D.D., Nicholson, D.: Henry constant of water adsorption on functionalized graphite: importance of the potential models of water and functional group. J. Phys. Chem. C **122**(42), 24171–24181 (2018)
20. Do, D.D., Nicholson, D., Do, H.D.: On the Henry constant and isosteric heat at zero loading in gas phase adsorption. J. Colloid Interface Sci. **324**(1–2), 15–24 (2008)

21. Do, D.D., Johnathan Tan, S.L., Zeng, Y., Fan, C., Nguyen, V.T., Horikawa, T., Nicholson, D.: The interplay between molecular layering and clustering in adsorption of gases on graphitized thermal carbon black—spill-over phenomenon and the important role of strong sites. J. Colloid Interface Sci. **446**, 98–113 (2015)
22. Franklin, R.E.: Crystallite growth in graphitizing and non-graphitizing carbons. Proc. Royal Soc. London. Series A. Math. Phys. Sci. **209**(1097), 196–218 (1951)
23. Yu, D., Ghosh, P., Snurr, R.Q.: Hierarchical modeling of ammonia adsorption in functionalized metal-organic frameworks. Dalton Trans. **41**(14), 3962–3973 (2012)
24. Zeng, Y., Phadungbut, P., Do, D., Nicholson, D.: Wedge pore model as an alternative to the uniform slit pore model for the determination of pore size distribution in activated carbon. J. Phys. Chem. C **119**(46), 25853–25859 (2015)
25. Klomkliang, N., Do, D., Nicholson, D.: Hysteresis loop and scanning curves of argon adsorption in closed-end wedge pores. Langmuir **30**(43), 12879–12887 (2014)
26. Loi, Q.K., Prasetyo, L., Tan, S., Do, D.D., Nicholson, D.: Order-disorder transition of an argon adsorbate in graphitic wedge pores. Chem. Eng. J. **384**, 123286 (2020)
27. Bock, H., Pikunic, J., Gubbins, K.E.: Models of porous carbons. In: Adsorption by Carbons, Elsevier (2008)
28. Crowell, A., Steele, R.: Interaction potentials of simple nonpolar molecules with graphite. J. Chem. Phys. **34**(4), 1347–1349 (1961)
29. Steele, W.A.: The physical interaction of gases with crystalline solids: I. Gas-solid energies and properties of isolated adsorbed atoms. Surf. Sci. 36(1), 317–352 (1973)
30. Bojan, M.J., Steele, W.A.: Computer simulation of physisorption on a heterogeneous surface. Surf. Sci. **199**(3), 395–402 (1988)
31. Everett, D.H., Powl, J.C.: Adsorption in slit-like and cylindrical micropores in the henry's law region. A model for the microporosity of carbons. J. Chem. Soc. Faraday Trans. **72**, 619–636 (1976)
32. Liu, L., Zeng, Y., Do, D.D., Nicholson, D., Liu, J.: Development of averaged solid–fluid potential energies for layers and solids of various geometries and dimensionality. Adsorption **24**(1), 1–9 (2018)
33. Erickson, K., Erni, R., Lee, Z., Alem, N., Gannett, W., Zettl, A.: Determination of the local chemical structure of graphene oxide and reduced graphene oxide. Adv. Mater. **22**(40), 4467–4472 (2010)
34. Zhang, D., Liu, J., Liu, L.: On the capture of polar indoor air pollutants at sub-ppm level—a molecular simulation study. Build. Simul. (2020)
35. Liu, L., Liu, J., Zeng, Y., Tan, S.J., Do, D.D., Nicholson, D.: Formaldehyde adsorption in carbon nanopores—new insights from molecular simulation. Chem. Eng. J. **370**, 866–874 (2019)
36. Zhang, D., Liu, J., Liu, M., Liu, L., Do, D.D.: On the capture of ultralow-level benzene in indoor environments: experiments, modeling and molecular simulation. Sep. Purif. Technol. **251**, 117306 (2020)
37. Boehm, H.: Some aspects of the surface chemistry of carbon blacks and other carbons. Carbon **32**(5), 759–769 (1994)
38. Cai, W., Piner, R.D., Stadermann, F.J., Park, S., Shaibat, M.A., Ishii, Y., Yang, D., Velamakanni, A., An, S.J., Stoller, M., An, J., Chen, D., Ruoff, R.S.: Synthesis and solid-state NMR structural characterization of ¹³C-labeled graphite oxide. Science **321**(5897), 1815–1817 (2008)
39. Pacilé, D., Meyer, J.C., Fraile Rodríguez, A., Papagno, M., Gómez-Navarro, C., Sundaram, R.S., Burghard, M., Kern, K., Carbone, C., Kaiser, U.: Electronic properties and atomic structure of graphene oxide membranes. Carbon **49**(3), 966–972 (2011)
40. Prasetyo, L., Loi, Q.K., Tan, S.J., Do, D.D., Nicholson, D.: Effects of temperature on the transition from clustering to layering for argon adsorption on substrates of different strength—parametric map of wetting, pre-wetting and non-wetting. Microporous Mesoporous Mater. **304**, 109239 (2020)
41. Liu, L., Tan, S., Horikawa, T., Do, D.D., Nicholson, D., Liu, J.: Water adsorption on carbon—a review. Adv. Coll. Interface. Sci. **250**, 64–78 (2017)

42. Liu, H.-B., Yang, B., Xue, N.-D.: Enhanced adsorption of benzene vapor on granular activated carbon under humid conditions due to shifts in hydrophobicity and total micropore volume. J. Hazard. Mater. **318**, 425–432 (2016)
43. Li, X., Zhang, L., Yang, Z., He, Z., Wang, P., Yan, Y., Ran, J.: Hydrophobic modified activated carbon using PDMS for the adsorption of VOCs in humid condition. Sep. Purif. Technol. **239**, 116517 (2020)
44. Wang, L., Liang, X.-Y., Chang, Z.-Y., Ding, L.-S., Zhang, S., Li, B.-J.: Effective formaldehyde capture by green cyclodextrin-based metal-organic framework. ACS Appl. Mater. Interfaces. **10**(1), 42–46 (2018)
45. Cal, M.P., Rood, M.J., Larson, S.M.: Removal of VOCs from humidified gas streams using activated carbon cloth. Gas Sep. Purif. **10**(2), 117–121 (1996)
46. Russell, B.P., LeVan, M.D.: Coadsorption of organic compounds and water vapor on BPL activated carbon. 3. Ethane, propane, and mixing rules. Ind. Eng. Chem. Res. **36**(6), 2380–2389 (1997)
47. Marbán, G., Fuertes, A.B.: Co-adsorption of n-butane/water vapour mixtures on activated carbon fibre-based monoliths. Carbon **42**(1), 71–81 (2004)
48. Biron, E., Evans, M.J.B.: Dynamic adsorption of water-soluble and insoluble vapours on activated carbon. Carbon **36**(7), 1191–1197 (1998)
49. Qi, N., LeVan, M.D.: Coadsorption of organic compounds and water vapor on BPL activated carbon. 5. Methyl ethyl ketone, methyl isobutyl ketone, toluene, and modeling. Ind. Eng. Chem. Res. **44**(10), 3733–3741 (2005)
50. Kim, D.I., Park, J.H., Kim, S.D., Lee, J.-Y., Yim, J.-H., Jeon, J.-K., Park, S.H., Park, Y.-K.: Comparison of removal ability of indoor formaldehyde over different materials functionalized with various amine groups. J. Ind. Eng. Chem. **17**(1), 1–5 (2011)
51. Thevenet, F., Debono, O., Rizk, M., Caron, F., Verriele, M., Locoge, N.: VOC uptakes on gypsum boards: sorption performances and impact on indoor air quality. Build. Environ. **137**, 138–146 (2018)
52. Pei, J., Zhang, J.S.: On the performance and mechanisms of formaldehyde removal by chemisorbents. Chem. Eng. J. **167**(1), 59–66 (2011)
53. Bandosz, T.J., Petit, C.: On the reactive adsorption of ammonia on activated carbons modified by impregnation with inorganic compounds. J. Colloid Interface Sci. **338**(2), 329–345 (2009)
54. Gonçalves, M., Sánchez-García, L., Oliveira Jardim, E.D., Silvestre-Albero, J., Rodríguez-Reinoso, F.: Ammonia removal using activated carbons: effect of the surface chemistry in dry and moist conditions. Environ. Sci. Technol. **45**(24), 10605–10610 (2011)
55. Kang, S., Chun, J., Park, N., Lee, S.M., Kim, H.J., Son, S.U.: Hydrophobic zeolites coated with microporous organic polymers: adsorption behavior of ammonia under humid conditions. Chem. Commun. **51**(59), 11814–11817 (2015)
56. Rudisill, E.N., Hacskaylo, J.J., LeVan, M.D.: Coadsorption of hydrocarbons and water on BPL activated carbon. Ind. Eng. Chem. Res. **31**(4), 1122–1130 (1992)
57. Taqvi, S.M., Appel, W.S., LeVan, M.D.: Coadsorption of organic compounds and water vapor on BPL activated carbon. 4. Methanol, ethanol, propanol, butanol, and modeling. Ind. Eng. Chem. Res. **38**(1), 240–250 (1999)
58. Nguyen, P.T., Do, D., Nicholson, D.: Computer simulation of benzene-water mixture adsorption in graphitic slit pores. J. Phys. Chem. C **116**(26), 13954–13963 (2012)
59. Tan, S., Do, D.D., Nicholson, D.: Consistency of NVT, NPT, μVT and Gibbs (NV2T and NPT) with kinetic Monte Carlo schemes. Chem. Eng. J. **401**, 126056 (2020)
60. Bojan, M.J., Steele, W.A.: Computer simulation of physisorption on a heterogeneous surface. Surf. Sci. **199**(3), L395–L402 (1988)
61. Ketko, M.H., Kamath, G., Potoff, J.J.: Development of an optimized intermolecular potential for sulfur dioxide. J. Phys. Chem. B **115**(17), 4949–4954 (2011)
62. Wick, C.D., Siepman, J.I., Klotz, W.L., Schure, M.R.: Temperature effects on the retention of n-alkanes and arenes in helium-squalane gas-liquid chromatography. Experiment and molecular simulation. J. Chromatogr. A **954**(1–2), 181–190 (2002)

63. Shah, M.S., Tsapatsis, M., Siepmann, J.I.: Development of the transferable potentials for phase equilibria model for hydrogen sulfide. J. Phys. Chem. B **119**(23), 7041–7052 (2015)
64. Jorgensen, W.L., Maxwell, D.S., Tirado-Rives, J.: Development and testing of the OPLS all-atom force field on conformational energetics and properties of organic liquids. J. Am. Chem. Soc. **118**(45), 11225–11236 (1996)
65. Berendsen, H.J.C., Grigera, J.R., Straatsma, T.P.: The missing term in effective pair potentials. J. Phys. Chem. B **91**(24), 6269–6271 (1987)
66. Kristof, T., Vorholz, J., Liszi, J., Rumpf, B., Maurer, G.: A simple effective pair potential for the molecular simulation of the thermodynamic properties of ammonia. Mol. Phys. **97**(10), 1129–1137 (1999)
67. Jorge, M., Schumacher, C., Seaton, N.A.: Simulation study of the effect of the chemical heterogeneity of activated carbon on water adsorption. Langmuir **18**(24), 9296–9306 (2002)
68. Wick, C.D., Stubbs, J.M., Rai, N., Siepmann, J.I.: Transferable potentials for phase equilibria. 7. Primary, secondary, and tertiary amines, nitroalkanes and nitrobenzene, nitriles, amides, pyridine, and pyrimidine. J. Phys. Chem. B **109**(40), 18974–18982 (2005)

Recent Developments in the Electrophoretic Deposition of Carbon Nanomaterials

Artur P. Terzyk, Monika Zięba, Stanisław Koter, Emil Korczeniewski, Wojciech Zięba, Piotr Kowalczyk, and Joanna Kujawa

Abstract In this chapter we describe the basics of electrophoretic deposition (EPD) process in relation to carbon nanomaterials. We start from the general description of the process and the parameters influencing it. The screening—factors Taguchi optimization method, often applied for the optimization of the EPD parameters is described. Next we discuss recent developments in the deposition of carbon nanomaterials, mainly graphene, nanotubes, nanohorns, fullerenes, and nanodiamonds. The DLVO theory is shortly discussed in application to the description of electrophoretic deposition of carbon nanomaterials. Finally, we present new results on the use of EPD for the creation of carbon nanohorn layers on hydrophobic polymers using Taguchi approach. Created layers are good starting surfaces for subsequent studies.

1 EPD Fundamentals

The EPD process was developed by Russian scientist Ruess in 1808, and the basics were given in the 1940 by Hamaker et al. [1, 2]. This method is now one of the most popular ways allowing the creation of carbon nanomaterial—containing surfaces and films [3, 4]. Diba et al. [5] while discussing the main advantages of the EPD process mentioned, among others: low costs, simple equipment, and size-scalability. On the other hand, among the limitations one should remember the limited thickness of created layers, the possibility of side reactions, not so good adhesion of layers as obtained using other methods (this adhesion can be measured via the procedure

A. P. Terzyk (✉) · M. Zięba · E. Korczeniewski · W. Zięba
Faculty of Chemistry, Physicochemistry of Carbon Materials Research Group, Nicolaus Copernicus University in Toruń, Gagarin Street 7, 87-100, Toruń, Poland
e-mail: aterzyk@chem.umk.pl

S. Koter · J. Kujawa
Faculty of Chemistry, Chair of Physical Chemistry and Physiochemistry of Polymers, Nicolaus Copernicus University in Toruń, Gagarin Street 7, 87-100 Toruń, Poland

P. Kowalczyk
College of Science, Health, Engineering and Education, Murdoch University, Perth, WA 6150, Australia

J. C. Moreno-Piraján et al. (eds.), *Porous Materials*, Engineering Materials,
https://doi.org/10.1007/978-3-030-65991-2_4

described in [6]), for example the CVD process [7], and the problems with the application of unstable suspensions. Also the range of the dielectric constant values of solvents applied in EPD is very narrow, therefore the most commonly applied are water, methanol, ethanol, and acetone. The difficulties in adhesion and poor stability of deposited nanocarbon layers can be solved by using for example, sodium cholate [8]. This compound is not only advanced charged dispersant, but the coating-forming agent as well. EPD is not applicable for uncharged, hardly dispersable particles like for example, a graphene sheet. In this case, charged additives are usually applied. For example, magnesium nitrate was used to enable the EPD process of graphene from isopropanol on polished silicon wafer [9]. Also charging of graphene with acid has been used to solve this problem [10].

Fundamentals of the EPD application and the basic nomenclature were provided in [11]. The authors [11] described the Hamaker's model in detail, discussing the parameters characterizing the EPD process. These factors can be divided into two groups, i.e. the parameters defining the suspension (particle size, dielectric constant of liquid, stability conductivity and viscosity of suspension, zeta potential, etc.) and the parameters influencing deposition process (deposition time, voltage, concentration of solid, conductivity of substrate, etc.). Not only aqueous but non aqueous EPD process was described, together with multicomponent deposition. Excellent review considering the forces occurring during a process of nanoparticles film formations can be found in [12]. The author discussed different methods of nanoparticles surfaces creation, including EPD. The mechanism of the process was additionally discussed [12]. The author mentioned the differences between EPD and electrolytic deposition (i.e. ELD) where ions are deposited [13], and described different applications of EPD.

Generally, EPD process can be characterized by five steps: convection of nanoparticles surrounded by charged species toward the electrode, diffusion through the hydrodynamic boundary layer and next, through a concentration boundary layer. Finally, adsorption of nanoparticles on an electrode. Recently Obregon et al. [14] pointed out that the EPD process mechanism is not simple since the process is affected by several parameters which are commonly related. It was also proved [15] by the impedance measurements during EPD that more homogeneous films are created from high conductivity suspensions and in contrast, if the suspension has low conductivity, a surface is more heterogeneous.

The most important equation describing the mass of a deposit $w(t)$ in the EPD process is the Hamaker's law:

$$w(t) = \int_{t_o}^{t} f \mu E A C dt \tag{1}$$

where f is the efficiency of deposition process ($f \leq 1$), μ is the electrophoretic mobility of particles, E is the electrical field strength, A is the area of the electrode, C is the concentration of particles in a solution, and t is the deposition time.

Hamaker's law applicability was recently studied in [16] for the deposition of graphene oxide onto indium tin oxide (ITO) covered glass. The authors stated that the process is not a typical reduction, but rather CO_2 elimination from the system and the product is reduced graphene. The authors confirmed the validity of Eq. 1. The comparison of the kinetics of graphene oxide EPD from aqueous and organic solutions was collected in [17]. At low voltages (EPD from water) good applicability of Hamaker's model was reported. The deviations from this law, and the future applications of EPD in pulsed and alternating fields, were discussed in [18] and in [19]. Kinetics of the EPD process was described in [20]. Carbon nanotubes deposition on stainless steel was studied from different polyethyleneimine containing alcohols, and it was shown that the deposition is faster for larger values of the ζ potential. Kinetics of the EPD process was also mathematically described in [21]. The authors take into account the potential change during the process, and discuss the modification of Hamaker's model. Some facts and myths considering the EPD were discussed in [22].

Among the mentioned above parameters characterizing the process, the temperature has a very strong influence on the morphology of the surfaces obtained by the application of EPD [23]. The surfaces produced at lower temperatures are smoother and not so porous, as obtained at larger temperatures [23]. However, some authors postulate that the EPD at higher temperatures leads to more homogeneous coverage [24]. Also the mass of deposits and their specific surface areas increase with the rise in EPD temperature. Some additional, more or less exotic factors can also influence the EPD process. Thus, for example, it was shown that water magnetization [25] by increasing conductivity leads to a reduction in the EPD time and to better control of water electrolysis. It has also influence on the morphology of deposited layers. Ultrasonic treatment changes the rate of EPD and the structure of the surface layer [26] and if applied for deposited nanotubes, it can change their arrangement after the EPD process [27]. Moreover, during the sonically assisted EPD [28] the aggregation of colloids at the bottom of the deposition bath is reduced, as well as the bubbles from water electrolysis are easily removed, improving the deposition rate.

Sometimes the EPD process, if applied for porous solids, can lead to the blocking of pores located on the external surface. In this way, the internal porous space becomes unavailable. To avoid this, a new EPD method (called periodic EPD—PEPD) was proposed. The procedure bases on the series of EPD/ critical point drying procedures [29]. Studying the literature one can conclude that the EPD on semi and nonconducting surfaces is still a challenge [30, 31]. To overcome this, the modification of the EPD cell was proposed by using a membrane. However, in this case, the deposition time is longer. Recently also continuous EPD process of nanotubes was reported [32].

Some mixed techniques have been proposed thus, for example electrospray—assisted EPD, eliminating the functionalization of carbon material before deposition [33, 34]. Also the electrophoretically—guided micro additive manufacturing process [35] seems to be a very interesting method of particles deposition from a colloid solutions. In this method, the modulated electric field is applied for manipulation with particles, and the EPD is combined with dielectrophoresis. Also a very interesting procedure called light—directed EPD [36] was proposed. In this method photocond-cting electrodes were applied to form a pattern on a surface. It was also proved that EPD under modulated electric field [37] applying pulsed direct current offers more homogeneous layers that the typical EPD.

2 Application of Taguchi Method in the EPD

Taguchi method is widely applied for the improvement of the quality of the final product, and it is based on the screening of factors and the identification which are relevant for explaining a process variation [38]. Thus, during EPD process it can be applied for the optimization of different properties like for example, the optical features of deposited nanotubes [38], to maximize the ionic conductivity of solid electrolytes [39], and/or in preparation of the optimum electrodes [40]. Wang et al. [41] showed the application of the Taguchi method in the process of carbon nanofibers EPD on carbon fibers.

In Taguchi's approach, the three possible variants, depending on the goal, can be chosen considering η values: smaller is better, larger is better, and nominal is the best. For the "larger is better" method the signal to noise ratio ($\eta = \text{S/N}$) is calculated using [42]:

$$\eta_{ij} = -10\log\left(\frac{1}{l}\sum_{k=1}^{l} 1/y_{ijk}^2\right), \qquad 0 \le y_{ijk} \le \infty \tag{2}$$

where y_{ijk} is the observed data for the j-th response in the i-th trial and the k-th repetition.

The overall value of S/N is computed by:

$$\eta = \sum_{\mathrm{ij}} \eta_{\mathrm{ij}} \tag{3}$$

Wang et al. [36] used the Taguchi optimization to get the set of the EPD process parameters leading to the maximum electrical conductivity of a layer obtained from the deposition of carbon nanotubes on fabric. In the current study, this procedure will be applied for the optimization of the water contact angle (WCA) on carbon nanohorn containing surfaces obtained using the EPD.

3 EPD of Graphene, Carbon Nanotubes and Related Materials

Generally EPD covering of a surface by carbon nanomaterials is applied to improve mechanical strength [43–57] tensile properties [58, 59], colour [60], ablation resistance [61] electrical conductivity [62], and interfacial adhesion [63]. It has found application during the preparation of carbon nanotube films [64–66], composites [67–70], coatings [71–74], including anticorrosion [75–80], and lubricating coatings [81, 82], buckypapers [83], hybrid composites [84], implants [85–87], and fibers [88], electrodes [89] including transparent electrodes [90–92], applied for example in solar cells [93], electrodes for in vivo electrochemical measurements [94–98], sensors [99], field emitters, [100–104], transistors [105], biomaterials [106] and antibacterial layers [107, 108], capacitors [109] supercapacitors [110–114], and heat transfer blockers [115].

The review on the application of different nanomaterials for the strengthening of carbon fiber reinforced polymer composites has been recently published in [116]. The mini review on the possible application of EPD created surfaces made of carbon nanotubes was presented by Korczeniewski et al. [117]. Diba et al. [5] reviewed recent developments in the EPD of graphene related materials (GRM). Especially they discussed few layered graphene, graphene oxide, and reduced graphene oxide. Hamaker's law is usually cited as applicable to discuss the kinetics of the EPD process however, the deviations from this law are possible. The authors also pointed out that the mechanisms of EPD processes are rarely studied in the literature. Usually, GRM charge neutralization leads to the formation of deposits. Water electrolysis can lead to local protonation of surface groups and the protonation of surface acids. In this way the repulsive forces, stabilizing the colloids, can vanish and this process accelerates the coagulation. It is also possible that metal ions from an electrode participate in this process. Generally, the reduction (or more precisely deoxygenation) of GRM can also occur during the EPD process [118]. The authors [5] also discussed the influence of EPD conditions on the properties of the deposited film, i.e. thickness, porosity, density and roughness. A similar study considering the influence of EPD conditions on the properties of suspended graphene—based layers, was published in [119]. Recent studies in this field show the important role of protons in the electrochemical reduction of graphene oxide [120]. Gao and Duan [121] described recently the application of EPD for the creation of 2D and 3D surfaces of GRM. Also Ma et al. [122] discussed the same subject, referring to the papers published during the last 5 years. The authors concluded that the mechanism of the process needs further detailed study. To do this Politano et al. [123] studied the properties of electrophoretically deposited graphene using electrical impedance, Raman, SEM, and X-ray reflectometry.

Applications of graphene EPD was reviewed by [124]. The authors stipulate that the alternating current (AC) EPD can found future applications in the area of graphene deposition. This method is postulated in water to avoid electrolysis however, the experimental setup and the parameters of the process need additional optimization. The progress in carbon fibers surface modifications by graphene was recently reviewed in [125]. The authors, considering future outlook, pointed out that the mechanism of fibers strengthening by graphene is still unknown. The purity of graphene, the material characterization methods and the control of the EPD process are still the challenge.

EPD process applied for the creation of surfaces from carbon nanotubes was reviewed in [126]. Boccaccini et al. [127] reviewed also the development in carbon nanotube—ceramic (SiO_2, TiO_2, MnO_2, Fe_3O_4, hydroxyapatite, and bioactive glass) composites, and the deposition of carbon nanotubes on fibers has been recently reviewed in [128]. The authors also discussed differences between different methods of deposition, i.e. dipping deposition, the EPD, and CVD in situ growth. The equipment applied during the process was discussed in details, together with the details of a setup developed in the author's group. Special attention was paid to the nanotube modified: glass fibers, carbon fibers and aramid fibers. Benko et al. [129] discussed the influence of diluent on the morphology of carbon nanotubes deposited on a titanium plate. Water and the mixture of organic solvents were applied. It was concluded that for water (more conductive and in this way containing more charge carriers) the density of current is higher, and in this way, the process is faster and deposited layer thicker. In organic solvents, the mobility of tubes is lower, and the process is slower. Thinner layers are observed however, possessing higher density. Recent developments in the EPD of carbon nanotubes were given in the review by Ata et al. [130]. In this study, the authors paid special attention to the application of cationic and anionic dispersants during the EPD process. The attention was paid to so called bile acids (BAS) and dyes. It was concluded that BAS and dyes possess film-forming properties, and this opens new possibilities in the fabrication of nanotube surfaces by EPD process. Also other dispersants and coating forming agents can be applied [131], allowing the formation of Teflon, diamond and carbon dots layers on stainless steel. EPD of functionalized CNTs have been recently applied to control the thickness of tubes on glass [132]. The process was fully controlled, including the homogeneity of the coverage [133, 134]. Recently Korczeniewski et al. [117] used unsheathed carbon nanotubes for the creation of surfaces with controlled wettability. Because during unsheathing the number of nanotube walls decreases, at the same time decreases the number of atoms under the bottom of the droplet. Since the surface tension of the droplet/the work of droplet adhesion ratio changes the WCA increases.

EPD is also applied for the modification of nanotubes with metals [135] to improve lighting strike protection properties in the aircraft industry. Metal—modified nanotubes can be deposited and, by the control of the process conditions, one can change the contact angle and tubes orientation [136]. EPD is also applied for deposition on the surfaces of carbon nanotubes. For example, gold [137] nanoparticles were deposited on carbon nanotubes, and it was shown that the responsibility of created in such a way NO_2 sensor depends on the surface concentration of deposited gold.

4 EPD of Carbon Nanohorns

Oakes et al. [138] reported the application of EPD for the deposition of carbon nanohorns on different surfaces: stainless steel, nickel foam, Al_2O_3 and PTFE. The authors compared the morphology of surfaces obtained using different solvents during the process: THF, acetone and water, and performing the deposition with and without a surfactant. Generally, the presence of surfactant only changed the mass of deposits, and no remarkable influence on the morphology of samples was observed. During the deposition without surfactant, deposition form THF leads to more homogeneous surfaces than obtained from acetone. The authors also discussed the mechanism of EPD process and obtained results suggest that the process is driven not by the electroosmotic flow, but by the aggregation and flocculation at the electrode—solvent interface. Moreover, the same authors [138] suggest that a similar mechanism is applicable for other carbon nanomaterials. Recently [139] carbon nanohorns were electrophoretically deposited on stainless steel. The purpose was to obtain the friction reduction due to a solid nanohorn—surface lubricant. As it can be seen, there is still a little number of papers where carbon nanohorn surfaces are created and studied. This is why we decided in this study to perform the deposition of carbon nanohorns on a polymer surface, and obtained material is a good starting point for the creation of more complex surfaces based on carbon nanohorns.

5 EPD of Other Carbon Nanomaterials and Codeposition

Zhitomirsky [140] described the results obtained during the EPD of diamond particles on different supports (metals, carbon fiber, steel, silica etc.). The field emission properties of nanodiamonds electrophoretically deposited on titanium [141] were strongly temperature dependent. Also Goto et al. [142] presented the results of diamond deposition on glass coated with gold. Thermal annealing increased the stability of obtained layers, and they were more stable than obtained using a spin coating technique. Stable SiC/diamonds composites were fabricated on graphite [143] using combined CVD and EPD methods. Detonation nanodiamonds deposition on mica was studied in [144]. Monolayers were obtained after centrifugation of a nanomaterial. The kinetics of the process was described. The optimal parameters for EPD of nanodiamonds on nickel foam were described in [145] and obtained layers possessed very good mechanical stability. In [146] the process of nanodiamonds fluorination was described. Next the fluorinated materials were electrophoretically deposited on a surface of stainless steel. Affoune et al. [147] deposited nanodiamonds on HOPG surface. It is also possible to obtain homogeneous coverings form from nanodiamonds in the pores of silicon [148]. In [149] it was shown that using the EPD process from acetonitrile, water and DMSO, it is possible to obtain vertically standing nanodiamonds arrays. The process based on the application of membranes as templates and the thermal conductivity of arrays depend on the type of applied solvent. The authors explain differences between arrays taking into account the differences in the mechanism of EPD process. Similar arrays can be obtained by combining EPD in porous aluminum oxide with a chemical etching of alumina [150]. The application of different methods (including the EPD) in the template-based synthesis of arrays has been reviewed by Cao and Liu [151]. Ohtani et al. [152] presented the study on co-deposition of nanodiamonds with porphyrin to form surfaces applied for photo-energy conversion.

Amorphous carbon nanopowder was deposited on copper [153] to increase the melting point and to improve functionality in fuel cells and batteries. Carbon spheres obtained by hydrothermal treatment of glucose were deposited on fluorine tin oxide coated glass [154]. It was shown that the hydrophobicity of a surface can be changed by the annealing temperature. Activated carbon powder was deposited on a metal electrode to obtain supercapacitors [155]. To assure good adhesion of carbon to electrode, the authors used a binder—ethyl cellulose. Carbon black powder and carbon fibers deposition was used for the creation of superhydrophobic [156] surfaces. Carbon nanofibers deposition was provided for the first time by Yang et al. in 2007 [157]. The material was successfully deposited on ITO from three different solvents.

Fullerenes C_{60} were co-deposited [158] to obtain the donor—acceptor nanocomposite—a promising nanomaterial for application in photocells. The same fullerene was deposited [159] on ITO—coated glass electrode. In this case, the suspension of C_{60} in water was prepared using a laser ablation method, leading to a very stable colloid solution. Chugh et al. [160] reported the EPD of C_{60} on a graphene. The material can find applications as electrodes and/or for the construction of photovoltaic devices. Also C_{60} deposition on a graphene membrane was described [161] to obtain hybrid structures showing the potential application in optoelectronics. The mixture of C_{60} and nanodiamonds was deposited by Umeyama et al. [162] for the creation of hierarchical surfaces deposited on SnO_2 to obtain the materials used in optoelectronics. C_{70} was deposited on alumina membrane [163] to form nanowire arrays. Onion—like carbon spherical nanoparticles [164] were deposited on stainless steel to produce lubricant surfaces. Graphene quantum dots were deposited on ITO coated glass [165] to construct an electrochemical immunosensor. Tjandra et al. [166] deposited graphene quantum dots on carbon cloth to enhance the capacitance and to obtain the electrodes for supercapacitors. The same material was deposited on ITO glass [167]. The application of EPD for deposition of other quantum dots can be found in [168].

Carbon black and polymer codeposition [169] was studied on copper, and it was proved that the properties of the product depend on time and suspension viscosity. The mixture of carbon nanotubes and graphene was applied for the codeposition [170] on Ti film to form composites for electron field emission applications. Also graphene oxide/nanotubes mixture was deposited on carbon fabrics [171]. Yan et al. [172] described the procedure of codeposition of carbon nanotubes and copper on the surfaces of carbon—fiber reinforced polymers to improve thermal properties.

6 The DLVO Theory

In the EPD, the suspensions of charged particles are used. Such a suspension should be stable however, not too stable; otherwise the deposition would be poor or none [173]. The stability of charged particles is explained by the DLVO theory (Deryaguin–Landau–Verwey–Overbeek). According to this theory, the total interaction potential energy, V_T, is the sum of attractive dispersion energy, V_A, and the electrostatic repulsive energy, V_R [174]:

$$V_t = V_A + V_R \tag{4}$$

The expressions for V_A and V_R are derived for simple geometries of charged surfaces.

For example, V_A for two spheres from the same material and of the same radius *a* in a dispersion medium is the following function of the ratio of separation distance to the radius, $k = H/a$ (unretarded case [174]):

$$V_A = -\frac{A_H}{6}\left(2\left(\frac{1}{(2+k)^2} + \frac{1}{k(4+k)}\right) + ln\left(\frac{k(4+k)}{(2+k)^2}\right)\right) \tag{5}$$

where A_H is the Hamaker constant is given by the approximated formula:

$$A_H \approx \left(\sqrt{A_{11}} - \sqrt{A_{22}}\right)^2 \tag{6}$$

and A_{11}, A_{22} are the Hamaker constants describing the properties of the particle and dispersion medium material, respectively.

Regarding V_R the situation is more complicated because the analytical solutions exist for special cases only, for example for spheres of the same radius and the Stern potential, ψ_σ, when the ratio of a to the Debye length is lower than 5, V_R is given by (the constant potential condition) [174]:

$$V_R = 2\pi\varepsilon_0\varepsilon_r a\psi_\sigma^2 \exp(-\chi H) \tag{7}$$

Formulae for other cases can be found in [174, 175] and other sources.

In Fig. 1, the total potential energy is shown for the graphite particles in water for various values of ionic strength. The radius of particles was assumed to be 115 nm, the zeta potential ζ = -30 mV [176], it was assumed that $\psi_\sigma \approx \zeta$. The Hamaker constant for graphite ($A_{11} = 46.9 \times 10^{-20}$ J) was taken from [177], for water ($A_{22} = 4.35 \times 10^{-20}$ J) from [178].

It is seen that at low ionic strength (<0.2 mM) the energetic barrier, $V_{T,\max}$, exceeds 25 times the thermal energy, k_BT, which ensures the colloid stability. By increasing the ionic strength $V_{T,\max}$ decreases which leads to the colloid destabilization. Thus,

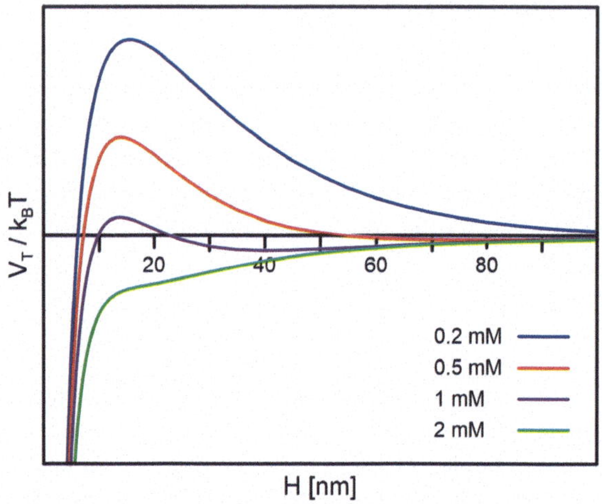

Fig. 1 Total potential energy, V_T/k_BT, vs. the distance, H, between the graphite particles of radius 135 nm in water for various ionic strengths, $\zeta = -30$ mV. The primary minimum of energy is not visible because the repulsion energy of short range is not taken into account

the ionic strength should be properly adjusted to give proper colloid stability (not too high, not too low).

7 Experimental

To illustrate the process of EPD and the Taguchi procedure (Eqs. 2 and 3) the deposition of carbon nanohorns on a polymer membrane is described. Since we are looking for the optimal hydrophobic surface, we use hydrophobic support. Created layers are good starting surfaces for subsequent studies, and the results will be presented in the future.

7.1 Single-Walled Carbon Nanohorns (SWCNH) Modification and Characterisation

The SWCNH samples (NEC, Tokyo, Japan, ~200 mg each) were placed in open quartz tube and heated in air atmosphere at 450, 500 and 550 °C (the samples were labeled as SWCNH-x °C, where x is the heating temperature), with heating rate 2 °C/min and stored in the indicated temperature for 10 min, next slowly cooled down up to the room temperature.

Obtained materials were characterized by using the DLS and ζ-potential measurements (25 °C, Micromeritics, Particulate Systems, Nano Plus HD, Norcross, GA, USA) in concentration 0.01 mg/mL in acetone (30 mL), sonicated for 30 s at 50 W (BANDELIN, Sonoplus HD4100, Berlin, Germany). The results are summarized in Table 1.

Table 1 The results of the oxidised carbon nanohorn characterization

Sample	DLS [nm]	ς -potential [mV]
SWCNH-450 °C	202.9 ± 8.4	−81.7 ± 7.1
SWCNH-500 °C	515 ± 41	−83.6 ± 4.6
SWCNH-550 °C	580 ± 87	−75.0 ± 6.4

7.2 Polymer Treatment in Plasma

As a substrate for the deposition a polyvinylidene fluoride (PVDF) hydrophobic membrane (Durapore Membrane Filter, Merck) with ~0.22 μm pore diameter was chosen. The plasma treatment was carried out for 10 min in ammonia (99.999%) and air atmosphere, using HPT 100 reactor (EDF Electronics, UK) with 50% power, gas flow 10 sccm, and working pressure 0.4 mbar. The samples were used for deposition just after the modification.

7.3 EPD Process

EPD was carried out from solutions of oxidized SWCNH in acetone with a concentration 0.01 mg/mL. The Al electrodes were used, placed at a distance of 10 mm, and immersed in 10 mL of solution. The positively charged one was covered with PVDF membrane. Each portion was sonicated just before the deposition process for 30 s at 50 W.

Following the Taguchi procedure (Eqs. 2, 3), four different parameters were studied in three variants—see Table 2.

In Table 3 the designed parameters for each experiment there are presented, together with the WCA (T = 25 °C) values for reference substrate materials and obtained samples. The WCA values were measured using the goniometer described in details in [179]. The changes in WCA during the procedure are also presented on Fig. 2 (Tables 4 and 5).

In this way, we obtain the optimal set of parameters, and the surfaces created will be applied in our future studies.

Table 2 The parameters for the Taguchi's method

parameter	level 1	level 2	level 3
Substrate	Pristine (PVDF)	Air plasma treated (PVDF-AIR)	Ammonia plasma treated (PVDF_NH_3)
Nanomaterial	SWCNH-450 °C	SWCNH-500 °C	SWCNH-550 °C
deposition voltage	140 V	160 V	200 V
Deposition time	10 min	20 min	30 min

Table 3 The parameters of designed experiments for Taguchi procedure and WCA results for reference and experimental samples

	Time (min)	Voltage (V)	SWCNH oxidation temp. (°C)	Substrate	WCA [°]
				PVDF	129.7 ± 1.1
				PVDF-AIR	84.7 ± 1.9
				PVDF_NH_3	62.2 ± 2.6
1	10	140	450	PVDF	97.1 ± 5.4
2	10	160	500	PVDF-AIR	63.7 ± 8.5
3	10	200	550	PVDF_NH_3	20.3 ± 1.7
4	20	140	500	PVDF-AIR	117.4 ± 3.8
5	20	160	450	PVDF_NH_3	25.2 ± 2.5
6	20	200	550	PVDF	62.7 ± 2.5
7	30	140	550	PVDF_NH_3	42.7 ± 8.0
8	30	160	500	PVDF	114.5 ± 4.1
9	30	200	450	PVDF-AIR	80.3 ± 8.8

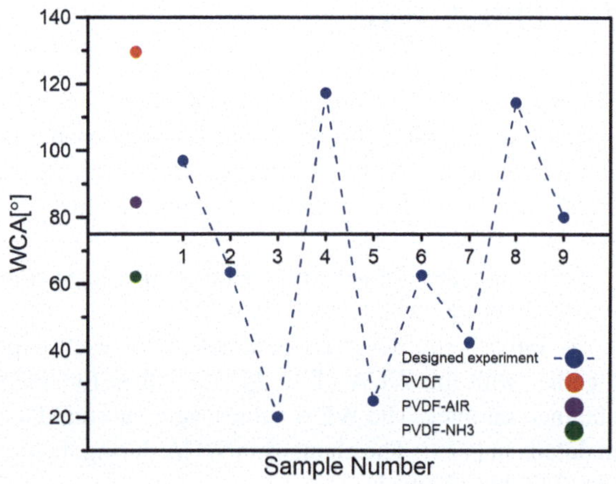

Fig. 2 Results of WCA measurements for designed experiments

Table 4 The analysis of the means for design factors

Factor	Level			Optimum level
	1	2	3	
Time	60.37	68.43	79.17	30 min
Voltage	85.73	50.85	54.43	140 V
SWCNH oxidation temp	67.53	98.53	41.90	500 °C
Substrate	91.43	87.13	29.40	PVDF

Table 5 The S/N ratios of design factors for case: larger the better

	Time	Voltage	SWCNH oxidation temp	Substrate
Level 1	30.33	36.16	32.13	38.37
Level 2	31.98	31.99	38.80	38.01
Level 3	35.85	30.24	29.68	28.19
Range (delta)	5.52	5.91	9.12	10.18
Rank	4	3	2	1

The results collected in Fig. 2 show that plasma treatment of the PVDF membrane leads to its hydrophilization. However, the incorporation of carbon nanohorns, depending on the procedure of modification and EPD conditions (Table 3) can lead to further hydrophilization or hydrophobization of PVDF surface. Thus by the EPD application, one can easily change the properties and morphology of this membrane (Fig. 3). Since carbon introduction will improve mechanical and porous properties, one can easily expect the improvement of PVDF separation abilities in selected processes. This will be the subject of our future studies.

8 Summary

In this chapter, the basic literature on the EPD of carbon nanomaterials was reviewed. We show the cons and profs of this method, showing possible solving of existing problems, as well as future directions. In our opinion, this method will be developed as the most promising and perspective tool for carbon nanomaterials surfaces creations. We provide the basics of DLVO theory. Since our study shows that little is known about the EPD of carbon nanohorns we provide the results of preliminary studies on the deposition of oxidized nanohorns on a hydrophobic membrane. Taguchi procedure was applied to optimize the deposition parameters. Up to our knowledge, the SWCNH were deposited on a hydrophobic polymer membrane for the first time. The properties of this new system and its application for the creation of more complicated surfaces will be reported in the future.

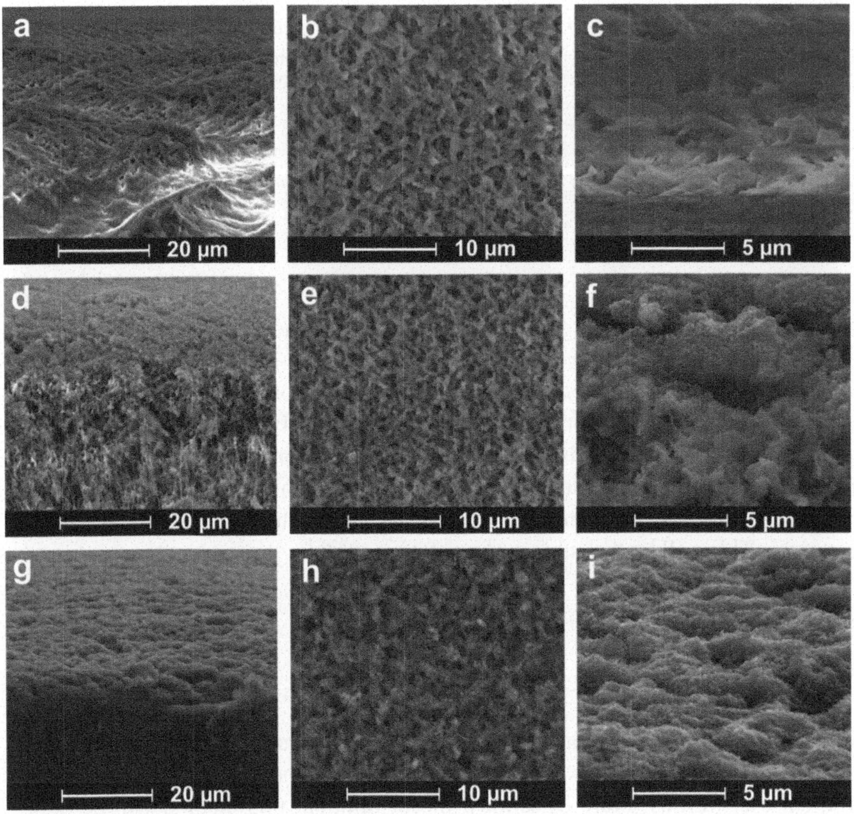

Fig. 3 SEM images for PVDF membranes (**a**,**b**,**c**) and the samples with maximum WCA values: Sample 4 (**d**,**e**,**f**) and Sample 8 (**g**,**h**,**i**)

Acknowledgements This chapter was prepared to honor the great Spanish scientist Francisco Rodriguez-Reinoso from the University of Alicante (Universidad de Alicante), Spain. A.P.T. thanks to the editors: Prof. Liliana Giraldo and Prof. Fernando Gómez, and Juan Carlos Moreno Pirajan for invitation. This research was funded by Polish NCN, grant number OPUS 13 UMO—2017/25/B/ST5/00975.

References

1. Hamaker, H.C.: The influence of particle size on the physical behaviour of colloidal systems. Trans. Faraday Soc. **35**, 186 (1940). https://doi.org/10.1039/tf9403500186
2. Hamaker, H.C., Verwey, E.J.W.: Part II.—(C) Colloid stability. The role of the forces between the particles in electrodeposition and other phenomena. Trans. Faraday Soc. **35**, 180–185 (1940). https://doi.org/10.1039/TF9403500180

3. Tao, Y., Endo, M., Inagaki, M., Kaneko, K.: Recent progress in the synthesis and applications of nanoporous carbon films. J. Mater. Chem. **21**, 313–323 (2011). https://doi.org/10.1039/C0JM01830A
4. Yang, S.-M., Jang, S.G., Choi, D.-G., Kim, S., Yu, H.K.: Nanomachining by Colloidal Lithography. Small **2**, 458–475 (2006). https://doi.org/10.1002/smll.200500390
5. Diba, M., Fam, D.W.H., Boccaccini, A.R., Shaffer, M.S.P.: Electrophoretic deposition of graphene-related materials: A review of the fundamentals. Prog. Mater Sci. **82**, 83–117 (2016). https://doi.org/10.1016/j.pmatsci.2016.03.002
6. Russ, B.E., Talbot, J.B.: A method for measuring the adhesion strength of powder coatings. J. Adhesion **68**, 257–268 (1998). https://doi.org/10.1080/00218469808029257
7. Raza, M.A., Ali, A., Ghauri, F.A., Aslam, A., Yaqoob, K., Wasay, A., Raffi, M.: Electrochemical behavior of graphene coatings deposited on copper metal by electrophoretic deposition and chemical vapor deposition. Surf. Coat. Technol. **332**, 112–119 (2017). https://doi.org/10.1016/j.surfcoat.2017.06.083
8. Liu, X., Zhao, Q., Veldhuis, S., Zhitomirsky, I.: Cholic acid is a versatile coating-forming dispersant for electrophoretic deposition of diamond, graphene, carbon dots and polytetrafluoroethylene. Surf. Coat. Technol. **384**, 125304 (2020). https://doi.org/10.1016/j.surfcoat.2019.125304
9. Shen, B., Hong, H., Chen, S., Chen, X., Zhang, Z.: Cathodic electrophoretic deposition of magnesium nitrate modified graphene coating as a macro-scale solid lubricant. Carbon **145**, 297–310 (2019). https://doi.org/10.1016/j.carbon.2019.01.046
10. Lee, S., Cho, M.S., Lee, H., Pu, L.S., Lee, Y.: Electrodeposition of graphene layers doped with Bronsted acids. J. Mater. Sci. **48**, 6891–6896 (2013). https://doi.org/10.1007/s10853-013-7493-4
11. Besra, L., Liu, M.: A review on fundamentals and applications of electrophoretic deposition (EPD). Prog. Mater Sci. **52**, 1–61 (2007). https://doi.org/10.1016/j.pmatsci.2006.07.001
12. Bensebaa, F.: Nanoparticle assembling and system integration. In: Interface Science and Technology. Elsevier, pp. 185–277 (2013)
13. Mersagh Dezfuli, S., Sabzi, M.: Deposition of ceramic nanocomposite coatings by electroplating process: A review of layer-deposition mechanisms and effective parameters on the formation of the coating. Ceram. Int. **45**, 21835–21842 (2019). https://doi.org/10.1016/j.ceramint.2019.07.190
14. Obregón, S., Amor, G., Vázquez, A.: Electrophoretic deposition of photocatalytic materials. Adv. Coll. Interface. Sci. **269**, 236–255 (2019). https://doi.org/10.1016/j.cis.2019.05.003
15. Stappers, L., Zhang, L., Van der Biest, O., Fransaer, J.: Study of the deposit resistance during electrophoretic deposition. KEM **412**, 9–14 (2009). https://doi.org/10.4028/www.scientific.net/KEM.412.9
16. Diba, M., García-Gallastegui, A., Klupp Taylor, R.N., Pishbin, F., Ryan, M.P., Shaffer, M.S.P., Boccaccini, A.R.: Quantitative evaluation of electrophoretic deposition kinetics of graphene oxide. Carbon **67**, 656–661 (2014). https://doi.org/10.1016/j.carbon.2013.10.041
17. Hajizadeh, A., Aliofkhazraei, M., Hasanpoor, M., Mohammadi, E.: Comparison of electrophoretic deposition kinetics of graphene oxide nanosheets in organic and aqueous solutions. Ceram. Int. **44**, 10951–10960 (2018). https://doi.org/10.1016/j.ceramint.2018.03.168
18. Neirinck, B., Van der Biest, O., Vleugels, J.: A current opinion on electrophoretic deposition in pulsed and alternating fields. J. Phys. Chem. B **117**, 1516–1526 (2013). https://doi.org/10.1021/jp306777q
19. Chávez-Valdez, A., Boccaccini, A.R.: Innovations in electrophoretic deposition: Alternating current and pulsed direct current methods. Electrochim. Acta. **65**, 70–89 (2012). https://doi.org/10.1016/j.electacta.2012.01.015
20. Farrokhi-Rad, M., Mohammadalipour, M., Shahrabi, T.: Electrophoretically deposited halloysite nanotubes coating as the adsorbent for the removal of methylene blue from aqueous solution. J. Eur. Ceram. Soc. **38**, 3650–3659 (2018). https://doi.org/10.1016/j.jeurceramsoc.2018.03.048

21. Ciou, S.-J., Fung, K.-Z., Chiang, K.-W.: The mathematical expression for kinetics of electrophoretic deposition and the effects of applied voltage. J. Power Sources **172**, 358–362 (2007). https://doi.org/10.1016/j.jpowsour.2007.07.046
22. Fukada, Y., Nagarajan, N., Mekky, W., Bao, Y., Kim, H.-S., Nicholson, P.S.: Electrophoretic deposition—mechanisms, myths and materials. J. Mater. Sci. **39**, 787–801 (2004). https://doi.org/10.1023/B:JMSC.0000012906.70457.df
23. Moore, J.J.E., Kang, J.H., Jayaram, S.H., Wen, J.Z.: Performance of nanotube-based electrodes from temperature-controlled electrophoretic deposition. J. Appl. Electrochem. **42**, 501–508 (2012). https://doi.org/10.1007/s10800-012-0428-y
24. Can-Ortiz, A., Oliva-Avilés, A.I., Gamboa, F., May-Pat, A., Velasco-Santos, C., Avilés, F.: Electrophoretic deposition of carbon nanotubes onto glass fibers for self-sensing relaxation-induced piezoresistivity of monofilament composites. J. Mater. Sci. **54**, 2205–2221 (2019). https://doi.org/10.1007/s10853-018-2965-1
25. Bazubandi, B., Moaseri, E., Baniadam, M., Maghrebi, M., Gholizadeh, M.: Fabrication of multi-walled carbon nanotube thin films via electrophoretic deposition process: effect of water magnetization on deposition efficiency. Appl. Phys. A **120**, 495–502 (2015). https://doi.org/10.1007/s00339-015-9276-z
26. Jiang, J., Xu, C., Su, Y., Guo, Q., Liu, F., Deng, C., Yao, X., Zhou, L.: Influence of Carbon Nanotube Coatings on Carbon Fiber by Ultrasonically Assisted Electrophoretic Deposition on Its Composite Interfacial Property. Polymers **8**, 302 (2016). https://doi.org/10.3390/polym8080302
27. Kim, S.-K., Lee, H., Tanaka, H., Weiss, P.S.: Vertical alignment of single-walled carbon nanotube films formed by electrophoretic deposition. Langmuir **24**, 12936–12942 (2008). https://doi.org/10.1021/la802266y
28. Zhou, G., Byun, J.-H., Wang, Y.-Q., Cha, H.-J., Lee, J.-U., Jung, B.-M., Song, J.-I., Kim, B.-S., Chou, T.-W.: Mechanism of sonication-assisted electrophoretic deposition of carbon nano-fiber on carbon fabrics. Compos. Sci. Technol. **107**, 29–35 (2015). https://doi.org/10.1016/j.compscitech.2014.11.011
29. Taale, M., Krüger, D., Ossei-Wusu, E., Schütt, F., Rehman, M.A.U., Mishra, Y.K., Marx, J., Stock, N., Fiedler, B., Boccaccini, A.R., Willumeit-Römer, R., Adelung, R., Selhuber-Unkel, C.: Systematically designed periodic electrophoretic deposition for decorating 3D Carbon-based scaffolds with bioactive nanoparticles. ACS Biomater. Sci. Eng. **5**, 4393–4404 (2019). https://doi.org/10.1021/acsbiomaterials.9b00102
30. Zhao, P., LeSergent, L.J., Farnese, J., Wen, J.Z., Ren, C.L.: Electrophoretic deposition of carbon nanotubes on semi-conducting and non-conducting substrates. Electrochem. Commun. **108**, 106558 (2019). https://doi.org/10.1016/j.elecom.2019.106558
31. Vilarinho, P.M., Fu, Z., Wu, A., Axelsson, A., Kingon, A.I.: Electrophoretic deposition on nonconducting substrates: A demonstration of the application to microwave devices. Langmuir **31**, 2127–2135 (2015). https://doi.org/10.1021/la504184k
32. Gong, G., Nyström, B., Sandlund, E., Eklund, D., Noël, M., Westerlund, R., Stenberg, S., Pupure, L., Pupurs, A., Joffe, R.: Development of electrophoretic deposition prototype for continuous production of carbon nanotube-modified carbon fiber fabrics used in high-performance multifunctional composites. Fibers **6**, 71 (2018). https://doi.org/10.3390/fib6040071
33. Kanakamedala, K., DeSoto, J., Sarkar, A., Race, T.D.: Study of electrospray assisted electrophoretic deposition of carbon nanotubes on insulator substrates. Electron. Mater. Lett. **11**, 949–956 (2015). https://doi.org/10.1007/s13391-015-5108-8
34. Maulik, S., Sarkar, A., Basu, S., Daniels-Race, T.: Voltage-controlled spray deposition of multiwalled carbon nanotubes on semiconducting and insulating substrates. J. Elec. Mater. **47**, 4604–4609 (2018). https://doi.org/10.1007/s11664-018-6316-4
35. Pritchet, D., Ehmann, K., Cao, J., Huang, J.: Manipulation and localized deposition of particle groups with modulated electric fields. Micromachines **11**, 226 (2020). https://doi.org/10.3390/mi11020226

36. Pascall, A.J., Qian, F., Wang, G., Worsley, M.A., Li, Y., Kuntz, J.D.: Light-directed electrophoretic deposition: a new additive manufacturing technique for arbitrarily patterned 3D composites. Adv. Mater. **26**, 2252–2256 (2014). https://doi.org/10.1002/adma.201304953
37. Ammam, M.: Electrophoretic deposition under modulated electric fields: A review. RSC Adv. **2**, 7633 (2012). https://doi.org/10.1039/c2ra01342h
38. Pander, A., Ishimoto, K., Hatta, A., Furuta, H.: Significant decrease in the reflectance of thin CNT forest films tuned by the Taguchi method. Vacuum **154**, 285–295 (2018). https://doi.org/10.1016/j.vacuum.2018.05.020
39. Wei, X., Xia, Y., Liu, X., Yang, H., Shen, X.: Preparation of sodium beta″-alumina electrolyte thin film by electrophoretic deposition using Taguchi experimental design approach. Electrochim. Acta **136**, 250–256 (2014). https://doi.org/10.1016/j.electacta.2014.05.096
40. Ayoubi-Feiz, B., Soleimani, D., Sheydaei, M.: Taguchi method for optimization of immobilized Dy_2O_3/graphite/TiO_2/Ti nanocomposite preparation and application in visible light photoelectrocatalysis process. J. Electroanal. Chem. **849**, 113377 (2019). https://doi.org/10.1016/j.jelechem.2019.113377
41. Wang, Y.-Q., Byun, J.-H., Kim, B.-S., Song, J.-I., Chou, T.-W.: The use of Taguchi optimization in determining optimum electrophoretic conditions for the deposition of carbon nanofiber on carbon fibers for use in carbon/epoxy composites. Carbon **50**, 2853–2859 (2012). https://doi.org/10.1016/j.carbon.2012.02.052
42. Liao, H.-C.: A data envelopment analysis method for optimizing multi-response problem with censored data in the Taguchi method. Comput. Ind. Eng. **46**, 817–835 (2004). https://doi.org/10.1016/j.cie.2004.05.012
43. Huang, S.-Y., Wu, G.-P., Chen, C.-M., Yang, Y., Zhang, S.-C., Lu, C.-X.: Electrophoretic deposition and thermal annealing of a graphene oxide thin film on carbon fiber surfaces. Carbon **52**, 613–616 (2013). https://doi.org/10.1016/j.carbon.2012.09.062
44. Zhang, S., Liu, W.B., Hao, L.F., Jiao, W.C., Yang, F., Wang, R.G.: Preparation of carbon nanotube/carbon fiber hybrid fiber by combining electrophoretic deposition and sizing process for enhancing interfacial strength in carbon fiber composites. Compos. Sci. Technol. **88**, 120–125 (2013). https://doi.org/10.1016/j.compscitech.2013.08.035
45. Feng, L., Li, K., Zheng, C., Fu, Y., Zhao, Z., Song, Q.: Improving mechanical strength and thermal shock resistance of SiC/zinc aluminum silicate joint by electrophoretic deposited multi-walled carbon nanotubes. Mater. Des. **94**, 417–423 (2016). https://doi.org/10.1016/j.matdes.2016.01.061
46. Mei, H., Xia, J., Zhang, D., Li, H., Bai, Q., Cheng, L.: Mechanical properties of carbon fiber reinforced bisphenol A dicyanate ester composites modified with multiwalled carbon nanotubes: ARTICLE. J. Appl. Polym. Sci. **134**, 45100 (2017). https://doi.org/10.1002/app.45100
47. An, Q., Tamrakar, S., Gillespie, J.W., Rider, A.N., Thostenson, E.T.: Tailored glass fiber interphases via electrophoretic deposition of carbon nanotubes: Fiber and interphase characterization. Compos. Sci. Technol. **166**, 131–139 (2018). https://doi.org/10.1016/j.compscitech.2018.01.003
48. Haghbin, A., Liaghat, G.H., Arabi, A.M., Hadavinia, H., Pol, M.H.: Investigations on electrophoretic deposition of carbon nanotubes on glass textures to improve polymeric composites interface. Compos. Sci. Technol. **155**, 197–204 (2018). https://doi.org/10.1016/j.compscitech.2017.12.005
49. Yang, K., Mei, H., Han, D., Cheng, L.: Strengthening of C/SiC Composites by Electrophoretic Deposition of CNTs on a SiC Coating. J. Mater. Eng. Perform, **27**, 5762–5768 (2018). https://doi.org/10.1007/s11665-018-3693-z
50. Yang, B., Yang, K., Xuan, F.-Z., Xiang, Y., Li, D., Luo, C.: Enhanced adhesion between glass, carbon, and their hybrid fiber-bundle with epoxy at room and elevated temperatures: A comparative study between graphene and MWCNT filled interface strategies. Polym. Compos. **39**, E2370–E2380 (2018). https://doi.org/10.1002/pc.24684
51. Yuan, X., Zhu, B., Cai, X., Qiao, K., Zhao, S., Zhang, M., Yu, J.: Micro-configuration controlled interfacial adhesion by grafting graphene oxide onto carbon fibers. Compos. A Appl. Sci. Manuf. **111**, 83–93 (2018). https://doi.org/10.1016/j.compositesa.2018.05.010

52. Bhanuprakash, L., Parasuram, S., Varghese, S.: Experimental investigation on graphene oxides coated carbon fibre/epoxy hybrid composites: Mechanical and electrical properties. Compos. Sci. Technol. **179**, 134–144 (2019). https://doi.org/10.1016/j.compscitech.2019.04.034
53. Mallick, M., Arunachalam, N.: Electrophoretic deposited graphene based functional coatings for biocompatibility improvement of Nitinol. Thin Solid Films **692**, 137616 (2019). https://doi.org/10.1016/j.tsf.2019.137616
54. Mei, H., Zhao, Y., Han, D., Ji, T., Xu, L., Cheng, L., Dassios, K.G.: Depositing CNTs on the interface and surface of C/PyC/SiCs for tunable mechanical and electromagnetic properties. Ceram. Int. **45**, 23411–23417 (2019). https://doi.org/10.1016/j.ceramint.2019.08.044
55. Tang, C., Dang, A., Li, T., Zhao, T., Li, H., Jiao, S.: Mechanical properties and oxidation resistance of phenolic formaldehyde interlocking CNTs-Cf/SiC composite. Ceram. Int. **45**, 9099–9105 (2019). https://doi.org/10.1016/j.ceramint.2019.01.248
56. Sun, T., Li, M., Zhou, S., Liang, M., Chen, Y., Zou, H.: Multi-scale structure construction of carbon fiber surface by electrophoretic deposition and electropolymerization to enhance the interfacial strength of epoxy resin composites. Appl. Surf. Sci. **499**, 143929 (2020). https://doi.org/10.1016/j.apsusc.2019.143929
57. Li, L., Liu, W., Yang, F., Jiao, W., Hao, L., Wang, R.: Interfacial reinforcement of hybrid composite by electrophoretic deposition for vertically aligned carbon nanotubes on carbon fiber. Compos. Sci. Technol. **187**, 107946 (2020). https://doi.org/10.1016/j.compscitech.2019.107946
58. Zhou, J., Yuan, M., Li, Z., Meng, H., Zhang, T., Qi, L.: A great improvement of tensile properties of Cf/AZ91D composite through grafting CNTs onto the surface of the carbon fibers. Mater. Sci. Eng. A **762**, 138061 (2019). https://doi.org/10.1016/j.msea.2019.138061
59. Wang, C., Li, J., Sun, S., Li, X., Zhao, F., Jiang, B., Huang, Y.: Electrophoretic deposition of graphene oxide on continuous carbon fibers for reinforcement of both tensile and interfacial strength. Compos. Sci. Technol. **135**, 46–53 (2016). https://doi.org/10.1016/j.compscitech.2016.07.009
60. Hwang, Y.-J., Kim, B.-J., Park, J.-S.: Improvement in color properties of copper mesh electrodes via electrophoretic coating with nano-structured carbon materials. J. Appl. Phys. **121**, 095306 (2017). https://doi.org/10.1063/1.4977008
61. Qingliang, S., Hejun, L., Fengling, Z., Qiang, S., Qiangang, F.: Electrophoretic deposition of carbon nanotubes for improved ablation resistance of carbon/carbon composites. Corros. Sci. **132**, 204–213 (2018). https://doi.org/10.1016/j.corsci.2018.01.001
62. Castro, R.H.R., Hidalgo, P., Diniz, E.C.: Enhanced electrical conduction in aluminum wires coated with carbon nanotubes. Mater. Lett. **65**, 271–274 (2011). https://doi.org/10.1016/j.matlet.2010.10.011
63. Mahmood, H., Tripathi, M., Pugno, N., Pegoretti, A.: Enhancement of interfacial adhesion in glass fiber/epoxy composites by electrophoretic deposition of graphene oxide on glass fibers. Compos. Sci. Technol. **126**, 149–157 (2016). https://doi.org/10.1016/j.compscitech.2016.02.016
64. Du, C., Heldbrant, D., Pan, N.: Preparation and preliminary property study of carbon nanotubes films by electrophoretic deposition. Mater. Lett. **57**, 434–438 (2002). https://doi.org/10.1016/S0167-577X(02)00806-6
65. Du, C., Pan, N.: High power density supercapacitor electrodes of carbon nanotube films by electrophoretic deposition. Nanotechnology **17**, 5314–5318 (2006). https://doi.org/10.1088/0957-4484/17/21/005
66. Bahru, R., Mohamed, M.A.: Enhancement of thermal interface material properties using carbon nanotubes through simple electrophoretic deposition method. Int. J. Energy Res. **44**, 4944–4960 (2020). https://doi.org/10.1002/er.5221
67. Fritz, P.A., Lange, S.C., Giesbers, M., Zuilhof, H., Boom, R.M., Schroën, C.G.P.H.: Simultaneous Silicon Oxide Growth and Electrophoretic Deposition of Graphene Oxide. Langmuir **35**, 3717–3723 (2019). https://doi.org/10.1021/acs.langmuir.8b03139
68. Li, M., Liu, Q., Jia, Z., Xu, X., Cheng, Y., Zheng, Y., Xi, T., Wei, S.: Graphene oxide/hydroxyapatite composite coatings fabricated by electrophoretic nanotechnology for

biological applications. Carbon **67**, 185–197 (2014). https://doi.org/10.1016/j.carbon.2013.09.080
69. Mazurenko, I., Etienne, M., Francius, G., Vakulko, I., Walcarius, A.: Macroporous carbon nanotube-carbon composite electrodes. Carbon **109**, 106–116 (2016). https://doi.org/10.1016/j.carbon.2016.07.050
70. Liu, S., Li, H., Zhang, L., Guo, Q.: Pulsed electrodeposition of carbon nanotubes-hydroxyapatite nanocomposites for carbon/carbon composites. Ceram. Int. **42**, 15650–15657 (2016). https://doi.org/10.1016/j.ceramint.2016.07.020
71. Minnikanti, S., Skeath, P., Peixoto, N.: Electrochemical characterization of multi-walled carbon nanotube coated electrodes for biological applications. Carbon **47**, 884–893 (2009). https://doi.org/10.1016/j.carbon.2008.11.045
72. Rodríguez-Uicab, O., Avilés, F., Gonzalez-Chi, P.I., Canché-Escamilla, G., Duarte-Aranda, S., Yazdani-Pedram, M., Toro, P., Gamboa, F., Mazo, M.A., Nistal, A., Rubio, J.: Deposition of carbon nanotubes onto aramid fibers using as-received and chemically modified fibers. Appl. Surf. Sci. **385**, 379–390 (2016). https://doi.org/10.1016/j.apsusc.2016.05.037
73. Thomas, B.J.C., Boccaccini, A.R., Shaffer, M.S.P.: Multi-walled Carbon nanotube coatings using Electrophoretic Deposition (EPD). J. American Ceramic Soc. **88**, 980–982 (2005). https://doi.org/10.1111/j.1551-2916.2005.00155.x
74. Thomas, B.J.C., Shaffer, M.S.P., Freeman, S., Koopman, M., Chawla, K.K., Boccaccini, A.R.: Electrophoretic deposition of Carbon Nanotubes on metallic surfaces. KEM **314**, 141–146 (2006). https://doi.org/10.4028/www.scientific.net/KEM.314.141
75. Zhang, K., Zhang, H., Liu, P., Zhang, C., Li, W., Chen, X., Ma, F.: Electrophoretic deposition of graphene oxide on NiTi alloy for corrosion prevention. Vacuum **161**, 276–282 (2019). https://doi.org/10.1016/j.vacuum.2018.12.032
76. Singh, B.P., Nayak, S., Nanda, K.K., Jena, B.K., Bhattacharjee, S., Besra, L.: The production of a corrosion resistant graphene reinforced composite coating on copper by electrophoretic deposition. Carbon **61**, 47–56 (2013). https://doi.org/10.1016/j.carbon.2013.04.063
77. Raza, M.A., Rehman, Z.U., Ghauri, F.A., Ahmad, A., Ahmad, R., Raffi, M.: Corrosion study of electrophoretically deposited graphene oxide coatings on copper metal. Thin Solid Films **620**, 150–159 (2016). https://doi.org/10.1016/j.tsf.2016.09.036
78. Ollik, K., Rybarczyk, M., Karczewski, J., Lieder, M.: Fabrication of anti-corrosion nitrogen doped graphene oxide coatings by electrophoretic deposition. Appl. Surf. Sci. **499**, 143914 (2020). https://doi.org/10.1016/j.apsusc.2019.143914
79. Hares, E., El-Shazly, A.H., El-Kady, M.F., Hammad, A.S.: Electrophoretic Deposition of Graphene Oxide Nanosheets on Copper Pipe for Corrosion Protection. Arab J. Sci. Eng. **44**, 5559–5569 (2019). https://doi.org/10.1007/s13369-019-03872-0
80. Jena, G., Vanithakumari, S.C., Polaki, S.R., George, R.P., Philip, J., Amarendra, G.: Electrophoretically deposited graphene oxide–polymer bilayer coating on Cu-Ni alloy with enhanced corrosion resistance in simulated chloride environment. J. Coat Technol. Res. **16**, 1317–1335 (2019). https://doi.org/10.1007/s11998-019-00213-6
81. Liu, Y., Li, J., Chen, X., Luo, J.: Fluorinated Graphene: a promising macroscale solid lubricant under various environments. ACS Appl. Mater. Interf. **11**, 40470–40480 (2019). https://doi.org/10.1021/acsami.9b13060
82. Qiang, R., Hou, K., Wang, J., Yang, S.: Smooth and dense graphene quantum dots-based lubricating coatings prepared by electrophoretic deposition. Appl. Surf. Sci. **509**, 145338 (2020). https://doi.org/10.1016/j.apsusc.2020.145338
83. Rigueur, J.L., Hasan, S.A., Mahajan, S.V., Dickerson, J.H.: Buckypaper fabrication by liberation of electrophoretically deposited carbon nanotubes. Carbon **48**, 4090–4099 (2010). https://doi.org/10.1016/j.carbon.2010.07.016
84. Li, Q., Church, J.S., Naebe, M., Fox, B.L.: Interfacial characterization and reinforcing mechanism of novel carbon nanotube—Carbon fibre hybrid composites. Carbon **109**, 74–86 (2016). https://doi.org/10.1016/j.carbon.2016.07.058
85. Kaya, C., Singh, I., Boccaccini, A.R.: Multi-walled Carbon nanotube-reinforced Hydroxyapatite layers on Ti6Al4V medical implants by Electrophoretic Deposition (EPD). Adv. Eng. Mater. **10**, 131–138 (2008). https://doi.org/10.1002/adem.200700241

86. Mei, H., Han, D., Xiao, S., Ji, T., Tang, J., Cheng, L.: Improvement of the electromagnetic shielding properties of C/SiC composites by electrophoretic deposition of carbon nanotube on carbon fibers. Carbon **109**, 149–153 (2016). https://doi.org/10.1016/j.carbon.2016.07.070
87. Ordikhani, F., Ramezani Farani, M., Dehghani, M., Tamjid, E., Simchi, A.: Physicochemical and biological properties of electrodeposited graphene oxide/chitosan films with drug-eluting capacity. Carbon **84**, 91–102 (2015). https://doi.org/10.1016/j.carbon.2014.11.052
88. Deng, C., Jiang, J., Liu, F., Fang, L., Wang, J., Li, D., Wu, J.: Influence of surface properties of graphene oxide/carbon fiber hybrid fiber by oxidative treatments combined with electrophoretic deposition: Surface properties of graphene oxide/carbon fiber hybrid fiber. Surf. Interf. Anal. **48**, 212–217 (2016). https://doi.org/10.1002/sia.5944
89. Qian, W., Cao, M., Xie, F., Dong, C.: Thermo-electrochemical cells based on carbon nanotube electrodes by electrophoretic deposition. Nano-Micro. Lett. **8**, 240–246 (2016). https://doi.org/10.1007/s40820-016-0082-8
90. Kim, B.-J., Park, J.-S., Hwang, Y.-J., Park, J.-S.: Characteristics of copper meshes coated with carbon nanotubes via electrophoretic deposition. Appl. Surf. Sci. **380**, 2–7 (2016). https://doi.org/10.1016/j.apsusc.2016.02.110
91. Cherusseri, J., Sambath Kumar, K., Pandey, D., Barrios, E., Thomas, J.: Vertically Aligned Graphene-Carbon Fiber Hybrid Electrodes with Superlong Cycling Stability for Flexible Supercapacitors. Small **15**, 1902606 (2019). https://doi.org/10.1002/smll.201902606
92. Lalau, C.C., Low, C.T.J.: Electrophoretic deposition for Lithium-Ion battery electrode manufacture. Batt. Supercaps **2**, 551–559 (2019). https://doi.org/10.1002/batt.201900017
93. Zhu, G., Pan, L., Lu, T., Liu, X., Lv, T., Xu, T., Sun, Z.: Electrophoretic deposition of carbon nanotubes films as counter electrodes of dye-sensitized solar cells. Electrochim. Acta **56**, 10288–10291 (2011). https://doi.org/10.1016/j.electacta.2011.09.028
94. Xiao, T., Jiang, Y., Ji, W., Mao, L.: Controllable and reproducible sheath of carbon fibers with single-walled carbon nanotubes through electrophoretic deposition for in vivo electrochemical measurements. Anal. Chem. **90**, 4840–4846 (2018). https://doi.org/10.1021/acs.analchem.8b00303
95. Umeyama, T., Fujita, M., Tezuka, N., Kadota, N., Matano, Y., Yoshida, K., Isoda, S., Imahori, H.: Electrophoretic Deposition of Single-Walled Carbon Nanotubes Covalently Modified with Bulky Porphyrins on Nanostructured SnO_2 Electrodes for Photoelectrochemical Devices. J. Phys. Chem. C **111**, 11484–11493 (2007). https://doi.org/10.1021/jp072075w
96. Umeyama, T., Imahori, H.: Carbon nanotube-modified electrodes for solar energy conversion. Energy Environ. Sci. **1**, 120 (2008). https://doi.org/10.1039/b805419n
97. Luan, X., Chen, L., Zhang, J., Qu, G., Flake, J.C., Wang, Y.: Electrophoretic deposition of reduced graphene oxide nanosheets on TiO2 nanotube arrays for dye-sensitized solar cells. Electrochim. Acta **111**, 216–222 (2013). https://doi.org/10.1016/j.electacta.2013.08.016
98. Ghasemi, S., Hosseini, S.R., Mousavi, F.: Electrophoretic deposition of graphene nanosheets: A suitable method for fabrication of silver-graphene counter electrode for dye-sensitized solar cell. Colloids Surf. A **520**, 477–487 (2017). https://doi.org/10.1016/j.colsurfa.2017.02.004
99. Zhu, Q.-G., Sujari, A.N.A., Ab Ghani, S.: Electrophoretic deposited MWCNT composite graphite pencils and its uses to determine hyperin. J. Solid State Electrochem. **16**, 3179–3187 (2012). https://doi.org/10.1007/s10008-012-1749-9
100. Quale, S.L., Talbot, J.B.: Electrophoretic deposition of substrate-normal-oriented single-walled carbon nanotube structures. J. Electrochem. Soc. **154**, K25 (2007). https://doi.org/10.1149/1.2741116
101. Koh, A.T.T., Foong, Y.M., Pan, L., Sun, Z., Chua, D.H.C.: Effective large-area free-standing graphene field emitters by electrophoretic deposition. Appl. Phys. Lett. **101**, 183107 (2012). https://doi.org/10.1063/1.4765070
102. Maiti, U.N., Maiti, S., Das, N.S., Chattopadhyay, K.K.: Hierarchical graphene nanocones over 3D platform of carbon fabrics: A route towards fully foldable graphene based electron source. Nanoscale **3**, 4135 (2011). https://doi.org/10.1039/c1nr10383k
103. Du, J., Zhang, Y., Deng, S., Xu, N., Xiao, Z., She, J., Wu, Z., Cheng, H.: Correlation between topographic structures and local field emission characteristics of graphene-sheet films. Carbon **61**, 507–514 (2013). https://doi.org/10.1016/j.carbon.2013.05.031

104. Chen, L.F., Yu, H., Zhong, J., Wu, J., Su, W.: Using Si textured structure to build vertical graphene sharp edges for electron field emission. Mater. Res. Express **6**, 086301 (2019). https://doi.org/10.1088/2053-1591/ab19ec
105. Liu, X., Chen, C., Sharma, P., Zhang, L., Wang, Y., Hu, N., Song, C., Wei, L., Liu, Q., Zhang, Y.: Fabrication of thin-film transistor based on self-assembled single-walled carbon nanotube network. Physica E **74**, 451–456 (2015). https://doi.org/10.1016/j.physe.2015.07.036
106. Kumar, S., Nehra, M., Kedia, D., Dilbaghi, N., Tankeshwar, K., Kim, K.-H.: Nanotechnology-based biomaterials for orthopaedic applications: Recent advances and future prospects. Mater. Sci. Eng. C **106**, 110154 (2020). https://doi.org/10.1016/j.msec.2019.110154
107. Qiu, J., Geng, H., Wang, D., Qian, S., Zhu, H., Qiao, Y., Qian, W., Liu, X.: Layer-number dependent antibacterial and osteogenic behaviors of graphene oxide electrophoretic deposited on Titanium. ACS Appl. Mater. Interf. **9**, 12253–12263 (2017). https://doi.org/10.1021/acsami.7b00314
108. Luo, X., Zhong, J., Zhou, Q., Du, S., Yuan, S., Liu, Y.: Cationic Reduced Graphene Oxide as Self-Aligned Nanofiller in the Epoxy Nanocomposite Coating with Excellent Anticorrosive Performance and Its High Antibacterial Activity. ACS Appl. Mater. Interf. **10**, 18400–18415 (2018). https://doi.org/10.1021/acsami.8b01982
109. Chen, Y., Zhang, X., Yu, P., Ma, Y.: Electrophoretic deposition of graphene nanosheets on nickel foams for electrochemical capacitors. J. Power Sources **195**, 3031–3035 (2010). https://doi.org/10.1016/j.jpowsour.2009.11.057
110. Huq, M.M., Hsieh, C.-T., Ho, C.-Y.: Preparation of carbon nanotube-activated carbon hybrid electrodes by electrophoretic deposition for supercapacitor applications. Diam. Relat. Mater. **62**, 58–64 (2016). https://doi.org/10.1016/j.diamond.2015.12.014
111. Moore, J.J., Kang, J.H., Wen, J.Z.: Fabrication and characterization of single walled nanotube supercapacitor electrodes with uniform pores using electrophoretic deposition. Mater. Chem. Phys. **134**, 68–73 (2012). https://doi.org/10.1016/j.matchemphys.2012.02.030
112. Zhan, Y., Edison, E., Manalastas, W., Tan, M.R.J., Satish, R., Buffa, A., Madhavi, S., Mandler, D.: Electrochemical deposition of highly porous reduced graphene oxide electrodes for Li-ion capacitors. Electrochim. Acta **337**, 135861 (2020). https://doi.org/10.1016/j.electacta.2020.135861
113. Dong, Q., Yang, C., Lu, Y.: Electrophoretic deposition of graphene oxide on graphite fiber cloth and the electrochemical performance as electrode of supercapacitor. J. Mater. Sci. Mater. Electron. **28**, 6308–6319 (2017). https://doi.org/10.1007/s10854-016-6314-3
114. Yunhe, Xu., Li, J., Huang, W.: Porous Graphene Oxide Prepared on Nickel Foam by Electrophoretic Deposition and Thermal Reduction as High-Performance Supercapacitor Electrodes. Materials **10**, 936 (2017). https://doi.org/10.3390/ma10080936
115. Yeo, L.P., Nguyen, T.D., Ling, H., Lee, Y., Mandler, D., Magdassi, S., Tok, A.I.Y.: Electrophoretic deposition of reduced graphene oxide thin films for reduction of cross-sectional heat diffusion in glass windows. J. Sci. Adv. Mater. Dev. **4**, 252–259 (2019). https://doi.org/10.1016/j.jsamd.2019.04.002
116. Vedrtnam, A., Sharma, S.P.: Study on the performance of different nano-species used for surface modification of carbon fiber for interface strengthening. Compos. A Appl. Sci. Manuf. **125**, 105509 (2019). https://doi.org/10.1016/j.compositesa.2019.105509
117. Korczeniewski, E., Zięba, M., Zięba, W., Kolanowska, A., Bolibok, P., Kowalczyk, P., Wiertel-Pochopień, A., Zawała, J., Boncel, S., Terzyk, A.P.: Electrophoretic deposition of layer-by-layer unsheathed carbon nanotubes—a step towards steerable surface roughness and wettability. Materials **13**, 595 (2020). https://doi.org/10.3390/ma13030595
118. An, S.J., Zhu, Y., Lee, S.H., Stoller, M.D., Emilsson, T., Park, S., Velamakanni, A., An, J., Ruoff, R.S.: Thin film fabrication and simultaneous anodic reduction of deposited Graphene Oxide platelets by electrophoretic deposition. J. Phys. Chem. Lett. **1**, 1259–1263 (2010). https://doi.org/10.1021/jz100080c
119. González, Z., Pérez-Mas, A.M., Blanco, C., Granda, M., Santamaría, R.: Influence of the electrophoretic deposition parameters on the formation of suspended graphene-based films. Mater. Des. **160**, 58–64 (2018). https://doi.org/10.1016/j.matdes.2018.08.063

120. Quezada-Renteria, J.A., Ania, C.O., Chazaro-Ruiz, L.F., Rangel-Mendez, J.R.: Influence of protons on reduction degree and defect formation in electrochemically reduced graphene oxide. Carbon **149**, 722–732 (2019). https://doi.org/10.1016/j.carbon.2019.04.109
121. Gao, H., Duan, H.: 2D and 3D graphene materials: Preparation and bioelectrochemical applications. Biosens. Bioelectron. **65**, 404–419 (2015). https://doi.org/10.1016/j.bios.2014.10.067
122. Ma, Y., Han, J., Wang, M., Chen, X., Jia, S.: Electrophoretic deposition of graphene-based materials: A review of materials and their applications. J. Materiomics **4**, 108–120 (2018). https://doi.org/10.1016/j.jmat.2018.02.004
123. Politano, G.G., Versace, C., Vena, C., Castriota, M., Ciuchi, F., Fasanella, A., Desiderio, G., Cazzanelli, E.: Physical investigation of electrophoretically deposited graphene oxide and reduced graphene oxide thin films. J. Appl. Phys. **120**, 195307 (2016). https://doi.org/10.1063/1.4968000
124. Chavez-Valdez, A., Shaffer, M.S.P., Boccaccini, A.R.: Applications of Graphene Electrophoretic Deposition. A Review. J. Phys. Chem. B **117**, 1502–1515 (2013). https://doi.org/10.1021/jp3064917
125. Hung, P., Lau, K., Fox, B., Hameed, N., Lee, J.H., Hui, D.: Surface modification of carbon fibre using graphene–related materials for multifunctional composites. Compos. B Eng. **133**, 240–257 (2018). https://doi.org/10.1016/j.compositesb.2017.09.010
126. Boccaccini, A.R.: Electrophoretic deposition: fundamentals and applications in materials science. J. Mater. Sci. **41**, 8029–8030 (2006). https://doi.org/10.1007/s10853-006-0995-6
127. Boccaccini, A.R., Cho, J., Subhani, T., Kaya, C., Kaya, F.: Electrophoretic deposition of carbon nanotube–ceramic nanocomposites. J. Eur. Ceram. Soc. **30**, 1115–1129 (2010). https://doi.org/10.1016/j.jeurceramsoc.2009.03.016
128. Avilés, F., de Jesús Kú-Herrera, J., Oliva-Avilés, A.I.: Deposition of Carbon Nanotubes on Fibers. In: Carbon Nanotube-Reinforced Polymers. Elsevier, pp. 117–144 (2018)
129. Benko, A., Nocuń, M., Berent, K., Gajewska, M., Klita, Ł, Wyrwa, J., Błażewicz, M.: Diluent changes the physicochemical and electrochemical properties of the electrophoretically-deposited layers of carbon nanotubes. Appl. Surf. Sci. **403**, 206–217 (2017). https://doi.org/10.1016/j.apsusc.2017.01.146
130. Ata, M.S., Poon, R., Syed, A.M., Milne, J., Zhitomirsky, I.: New developments in non-covalent surface modification, dispersion and electrophoretic deposition of carbon nanotubes. Carbon **130**, 584–598 (2018). https://doi.org/10.1016/j.carbon.2018.01.066
131. Zhao, Q., Liu, X., Veldhuis, S., Zhitomirsky, I.: Sodium deoxycholate as a versatile dispersing and coating-forming agent: A new facet of electrophoretic deposition technology. Colloids Surf. A **588**, 124382 (2020). https://doi.org/10.1016/j.colsurfa.2019.124382
132. Tamrakar, S., An, Q., Thostenson, E.T., Rider, A.N., Haque, B.Z. (Gama), Gillespie, J.W.: Tailoring interfacial properties by controlling carbon nanotube coating thickness on glass fibers using electrophoretic deposition. ACS Appl. Mater. Interf. **8**, 1501–1510. https://doi.org/10.1021/acsami.5b10903
133. Ku-Herrera, J.J., May-Pat, A., Avilés, F.: An Assessment of the Role of Fiber Coating and Suspending Fluid on the Deposition of Carbon Nanotubes onto Glass Fibers for Multiscale Composites: The role of fiber coating and suspending fluid on the deposition of CNTs onto GFs. Adv. Eng. Mater. **18**, 963–971 (2016). https://doi.org/10.1002/adem.201500389
134. An, Q., Rider, A.N., Thostenson, E.T.: Hierarchical composite structures prepared by electrophoretic deposition of carbon nanotubes onto glass fibers. ACS Appl. Mater. Interf. **5**, 2022–2032 (2013). https://doi.org/10.1021/am3028734
135. Xia, Q., Mei, H., Zhang, Z., Liu, Y., Liu, Y., Leng, J.: Fabrication of the silver modified carbon nanotube film/carbon fiber reinforced polymer composite for the lightning strike protection application. Compos. B Eng. **180**, 107563 (2020). https://doi.org/10.1016/j.compositesb.2019.107563
136. Balram, A., Santhanagopalan, S., Hao, B., Yap, Y.K., Meng, D.D.: Electrophoretically-deposited metal-decorated CNT Nanoforests with High Thermal/Electric Conductivity and Wettability Tunable from Hydrophilic to Superhydrophobic. Adv. Funct. Mater. **26**, 2571–2579 (2016). https://doi.org/10.1002/adfm.201504208

137. Dilonardo, E., Penza, M., Alvisi, M., Di Franco, C., Rossi, R., Palmisano, F., Torsi, L., Cioffi, N.: Electrophoretic deposition of Au NPs on MWCNT-based gas sensor for tailored gas detection with enhanced sensing properties. Sens. Actuat. B: Chem. **223**, 417–428 (2016). https://doi.org/10.1016/j.snb.2015.09.112
138. Oakes, L., Westover, A., Mahjouri-Samani, M., Chatterjee, S., Puretzky, A.A., Rouleau, C., Geohegan, D.B., Pint, C.L.: Uniform, Homogenous Coatings of Carbon Nanohorns on Arbitrary Substrates from Common Solvents. ACS Appl. Mater. Interf. **5**, 13153–13160 (2013). https://doi.org/10.1021/am404118z
139. MacLucas, T., Suarez, S.: On the solid lubricity of electrophoretically deposited carbon nanohorn coatings. Lubricants **7**, 62 (2019). https://doi.org/10.3390/lubricants7080062
140. Zhitomirsky, I.: Cathodic electrophoretic deposition of diamond particles. Mater. Lett. **37**, 72–78 (1998). https://doi.org/10.1016/S0167-577X(98)00074-3
141. Zhai, C.X., Yun, J.N., Zhao, L.L., Zhang, Z.Y., Wang, X.W., Chen, Y.Y.: Effect of annealing on field emission properties of nanodiamond coating. Phys. B **406**, 1124–1128 (2011). https://doi.org/10.1016/j.physb.2010.12.063
142. Goto, Y., Ohishi, F., Tanaka, K., Usui, H.: Formation of diamond nanoparticle thin films by electrophoretic deposition. Jpn J. Appl. Phys. **55**, 03DD10. https://doi.org/10.7567/JJAP.55.03DD10(2016)
143. Yu, S., Chen, Z., Wang, Y., Luo, R., Xu, T., Pan, Y., Liao, J.: Fabrication of SiC/diamond composite coatings by electrophoretic deposition and chemical vapor deposition. Int. J. Appl. Ceram. Technol. **14**, 644–651 (2017). https://doi.org/10.1111/ijac.12662
144. Schmidlin, L., Pichot, V., Josset, S., Pawlak, R., Glatzel, T., Kawai, S., Meyer, E., Spitzer, D.: Two-dimensional nanodiamond monolayers deposited by combined ultracentrifugation and electrophoresis techniques. Appl. Phys. Lett. **101**, 253111 (2012). https://doi.org/10.1063/1.4772983
145. Wang, Y., Liang, Y., Cheng, X., Huang, H., Zang, J., Lu, J., Yu, Y., Xu, X.: Preparing porous diamond composites via electrophoretic deposition of diamond particles on foam nickel substrate. Mater. Lett. **138**, 52–55 (2015). https://doi.org/10.1016/j.matlet.2014.09.087
146. Huang, H., Wang, Y.H., Zang, J.B., Bian, L.Y.: Improvement of suspension stability and electrophoresis of nanodiamond powder by fluorination. Appl. Surf. Sci. **258**, 4079–4084 (2012). https://doi.org/10.1016/j.apsusc.2011.12.105
147. Affoune, A.M., Prasad, B.L.V., Sato, H., Enoki, T.: Electrophoretic deposition of nanosized diamond particles. Langmuir **17**, 547–551 (2001). https://doi.org/10.1021/la001317f
148. Yuzova, V.A., Puzyr', A.P.: Introducing utradisperse detonation diamond powder into channels of porous silicon. Tech. Phys. Lett. 34. 421–423 (2008) https://doi.org/10.1134/S1063785008050180
149. Su, S., Wang, J., Wei, J., Qiu, J., Wang, S.: Thermal conductivity studies of electrophoretically deposited nanodiamond arrays. Mater. Sci. Eng. B **225**, 54–59 (2017). https://doi.org/10.1016/j.mseb.2017.08.010
150. Li, L., Zhang, K., Qiu, J., Wang, S., Van, H.H., Zhang, M.: Solution assembly of vertically aligned diamond nanotube arrays from diamond nanocrystals. Diam. Relat. Mater. **29**, 79–83 (2012). https://doi.org/10.1016/j.diamond.2012.08.002
151. Cao, G., Liu, D.: Template-based synthesis of nanorod, nanowire, and nanotube arrays. Adv. Coll. Interface. Sci. **136**, 45–64 (2008). https://doi.org/10.1016/j.cis.2007.07.003
152. Ohtani, M., Kamat, P.V., Fukuzumi, S.: Supramolecular donor–acceptor assemblies composed of carbon nanodiamond and porphyrin for photoinduced electron transfer and photocurrent generation. J. Mater. Chem. **20**, 582–587 (2010). https://doi.org/10.1039/B916634C
153. Jeun, J.-H., Kim, W.-S., Hong, S.-H.: Electrophoretic deposition of carbon nanoparticles on dendritic Sn foams fabricated by electrodeposition. Mater. Lett. **112**, 109–112 (2013). https://doi.org/10.1016/j.matlet.2013.09.002
154. Kahradeh, Kh.H., Saievar-Iranizad, E., Bayat, A.: Electrophoretically deposited carbon micro and nanospheres thin films as superhydrophobic coatings. Surf. Coat. Technol. **319**, 318–325 (2017). https://doi.org/10.1016/j.surfcoat.2017.03.070

155. Shrestha, M., Amatya, I., Wang, K., Zheng, B., Gu, Z., Fan, Q.H.: Electrophoretic deposition of activated carbon YP-50 with ethyl cellulose binders for supercapacitor electrodes. J. Energy Storage **13**, 206–210 (2017). https://doi.org/10.1016/j.est.2017.07.015
156. Ogihara, H., Okagaki, J., Saji, T.: A Facile Fabrication of Superhydrophobic Films by Electrophoretic Deposition of Hydrophobic Particles. Chem. Lett. **38**, 132–133 (2009). https://doi.org/10.1246/cl.2009.132
157. Yang, Y.J., Zhao, G.J., Hu, S.: Direct current electrodeposition of carbon nanofibers in DMF. Electrochem. Commun. **9**, 2681–2685 (2007). https://doi.org/10.1016/j.elecom.2007.08.021
158. Tada, K., Onoda, M.: Preparation of donor–acceptor nanocomposite through electrophoretic deposition. Curr. Appl. Phys. **5**, 5–8 (2005). https://doi.org/10.1016/j.cap.2003.11.069
159. Jeon, H.-G., Ryo, S., Sugiyama, T., Oh, I., Masuhara, H., Asahi, T.: Fullerene (C60) nanostructures having interpenetrating surfaces prepared by electrophoretic deposition of C60 Nanoparticles in Water. Chem. Lett. **36**, 1160–1161 (2007). https://doi.org/10.1246/cl.2007.1160
160. Chugh, S., Biswas, C., Echegoyen, L., Kaul, A.B.: Investigation of structural morphology and electrical properties of graphene-C_{60} hybrids. J. Vacuum Sci. Tech. B, Nanotech. Microelectr. Mater. Process. Measure. Phenomena **35**, 03D111 (2017). https://doi.org/10.1116/1.4982881
161. Chugh, S., Adhikari, N., Lee, J.H., Berman, D., Echegoyen, L., Kaul, A.B.: Dramatic Enhancement of Optoelectronic Properties of Electrophoretically Deposited C_{60} –Graphene Hybrids. ACS Appl. Mater. Interf. **11**, 24349–24359 (2019). https://doi.org/10.1021/acsami.9b00603
162. Umeyama, T., Okawada, Y., Ohara, T., Imahori, H.: Spontaneous Complexation of Fullerene Aggregates on Nanodiamond Aggregates and Their Enhanced Photocurrent Generation. Chem. Asian J. **14**, 4042–4047 (2019). https://doi.org/10.1002/asia.201900835
163. Li, C.J., Wang, C.R., Wang, C., Bai, C.L.: C_{70} Nanowire Arrays Prepared by a Simple Electrophoretic Deposition Method. SSP **121–123**, 259–262 (2007). https://doi.org/10.4028/www.scientific.net/SSP.121-123.259
164. Xu, J., Chen, X., Grützmacher, P., Rosenkranz, A., Li, J., Jin, J., Zhang, C., Luo, J.: Tribochemical Behaviors of Onion-like Carbon Films as High-Performance Solid Lubricants with Variable Interfacial Nanostructures. ACS Appl. Mater. Interf. **11**, 25535–25546 (2019). https://doi.org/10.1021/acsami.9b06099
165. Bhardwaj, H., Singh, C., Kotnala, R.K., Sumana, G.: Graphene quantum dots-based nanobiointerface platform for food toxin detection. Anal. Bioanal. Chem. **410**, 7313–7323 (2018). https://doi.org/10.1007/s00216-018-1341-y
166. Tjandra, R., Liu, W., Zhang, M., Yu, A.: All-carbon flexible supercapacitors based on electrophoretic deposition of graphene quantum dots on carbon cloth. J. Power Sources **438**, 227009 (2019). https://doi.org/10.1016/j.jpowsour.2019.227009
167. Nguyen, T.D., Geuli, O., Yeo, L.P., Magdassi, S., Mandler, D., Tok, A.I.Y.: Additive-Free Electrophoretic Deposition of Graphene Quantum Dots Thin Films. Chem. Eur. J. **25**, 16573–16581 (2019). https://doi.org/10.1002/chem.201903596
168. Purcell-Milton, F., Curutchet, A., Gun'ko Y, : Electrophoretic Deposition of Quantum Dots and Characterisation of Composites. Materials **12**, 4089 (2019). https://doi.org/10.3390/ma12244089
169. Modi, S., Panwar, A., Mead, J.L., Barry, C.M.F.: Effect of Molecular Weight on the Electrophoretic Deposition of Carbon Black Nanoparticles in Moderately Viscous Systems. Langmuir **29**, 9702–9711 (2013). https://doi.org/10.1021/la401657d
170. Hong, X., Shi, W., Zheng, H., Liang, D.: Effective carbon nanotubes/graphene hybrid films for electron field emission application. Vacuum **169**, 108917 (2019). https://doi.org/10.1016/j.vacuum.2019.108917
171. Kwon, Y.J., Kim, Y., Jeon, H., Cho, S., Lee, W., Lee, J.U.: Graphene/carbon nanotube hybrid as a multi-functional interfacial reinforcement for carbon fiber-reinforced composites. Compos. B Eng. **122**, 23–30 (2017). https://doi.org/10.1016/j.compositesb.2017.04.005
172. Yan, F., Liu, L., Li, M., Zhang, M., Xiao, L., Ao, Y.: Preparation of carbon nanotube/copper/carbon fiber hierarchical composites by electrophoretic deposition for enhanced thermal conductivity and interfacial properties. J. Mater. Sci. **53**, 8108–8119 (2018). https://doi.org/10.1007/s10853-018-2115-9

173. Xiao, A.W., Evers, K., Tkaczyk, M., Jones, R.S., Saxby, C., Dragnevski, K., Grobert, N.: Electrophoretic Fabrication of Robust Carbon Nanotube "Buckyfilms" for Flexible Electronics. ACS Appl. Nano. Mater. **2**, 5190–5199 (2019). https://doi.org/10.1021/acsanm.9b01050
174. Goodwin, J.W.: Colloids and interfaces with surfactants and polymers, 2nd edn. Wiley, Chichester, U.K. (2009)
175. Robins, M., Fillery-Travis, A.: Colloidal dispersions. In: Russel, W. B., Saville, D. A., Schowalter, W. R. (eds.) Cambridge University Press, Cambridge, UK, 1989, xvii + 506 pp., price: £60.00. ISBN 0 521 34188 4. J. Chem. Technol. Biotechnol. **54**, 201–202 (2007) https://doi.org/10.1002/jctb.280540216
176. Zięba, M., Korczeniewski, E., Zięba, W., Bolibok, P., Czarnecka, J., Płońska-Brzezińska, M.E., Terzyk, A.P.: Electrophoretic deposition of spherical carbon nanoobjects—a comparison of different biocompatible surfaces. Med. Devices. Sens. mds3.10075. (2020). https://doi.org/10.1002/mds3.10075
177. Dahneke, B.: The influence of flattening on the adhesion of particles. J. Colloid Interf. Sci. **40**, 1–13 (1972). https://doi.org/10.1016/0021-9797(72)90168-3
178. Hiemenz, P.C., Rajagopalan, R.: Principles of colloid and surface chemistry, 3rd ed., rev.expanded. Marcel Dekker, New York (1997)
179. Bryk, P., Korczeniewski, E., Szymański, G.S., Kowalczyk, P., Terpiłowski, K., Terzyk, A.P.: What is the value of water contact angle on silicon? Materials **13**, 1554 (2020). https://doi.org/10.3390/ma13071554

CO_2 Adsorption in Activated Carbon Materials

S. Reljic, E. O. Jardim, C. Cuadrado-Collados, M. Bayona, M. Martinez-Escandell, J. Silvestre-Albero, and F. Rodríguez-Reinoso

Abstract Selective CO_2 capture in industrial streams is of paramount importance to decrease actual emissions to the atmosphere, and to minimize the associated environmental concerns. Among the different possibilities to capture CO_2, adsorption using nanoporous networks is one of the most promising approaches, provided that the adsorbent can be properly designed in terms of porous structure and surface chemistry. The material has to be designed to achieve a preferential interaction with CO_2 versus other competing molecules (e.g., CH_4, N_2, O_2, etc.) also present in the industrial effluents. Promising adsorbents for CO_2 capture include activated carbons, zeolites, and metal–organic frameworks (MOFs), among others. This chapter will focus on activated carbon materials and the effect of the porous structure and surface chemistry in the selective adsorption of CO_2. Although the chapter will concentrate mainly in physisorption processes, a final section will be introduced devoted to a novel approach for CO_2 capture based on natural processes, the so-called solid state storage.

1 Introduction

The increasing energy demand worldwide is unambiguously associated with an alarming increase in the amount of CO_2 emitted to the atmosphere (actual numbers from the Mauna Loa station are above 411 ppm far from the 280 ppm level in the pre-industrial era). CO_2 in the atmosphere comes preferentially from the combustion of fossil fuels (coal, natural gas, and oil), solid waste, trees and chemical reactions performed at industrial scale (e.g., manufacture of cement). Although land and ocean ecosystems act as natural buffers, these systems cannot adsorb and sequester more

In memoria of Prof. Francisco Rodríguez-Reinoso.

S. Reljic · E. O. Jardim · C. Cuadrado-Collados · M. Bayona · M. Martinez-Escandell · J. Silvestre-Albero (✉) · F. Rodríguez-Reinoso (Deceased)
Laboratorio de Materiales Avanzados, Departamento de Química Inorgánica-IUMA, Universidad de Alicante, Alicante, Spain
e-mail: joaquin.silvestre@ua.es

J. C. Moreno-Piraján et al. (eds.), *Porous Materials*, Engineering Materials,
https://doi.org/10.1007/978-3-030-65991-2_5

than 50% of all CO_2 produced in Earth. Unfortunately, these increasing CO_2 emissions to the atmosphere are unambiguously associated with a dramatic alteration in Earth´s temperature and a change in the climate cycles, thus given rise to unexpected climate episodes (e.g., heavy rains, floods, tropical cyclones, dry seasons, etc.). Taking into account that these numbers will continue increasing in the next few years, it is critical to develop technologies able to mitigate CO_2 emissions to the atmosphere [1]. Actual technology to remove CO_2 in industrial streams is based on the reactivity between the acidic CO_2 molecule and amines (e.g., monoethanolamine—MEA) to form water soluble compounds, the so-called amine scrubbing. However, this technology has significant cost and efficiency penalty. Among the alternative technologies (for instance carbonate-based systems, aqueous ammonia, membranes, etc.), one of the most promising one is the application of solid sorbents able to interact with CO_2 molecules through van der Waals forces at the gas–solid interface, the so-called physisorption. Potential sorbents for CO_2 capture are activated carbons, zeolites and metal–organic frameworks, among others [2, 3]. Huge efforts have been done to design novel sorbents with a perfectly designed porous structure and surface chemistry able to selectively adsorb CO_2, with a low cost, easy scale-up and low regeneration cost. Among these systems, this chapter will concentrate in carbon materials with special emphasis in the role played by the porous structure and surface chemistry. Carbon materials have demonstrated to be one of the most promising candidates due to their high surface area, excellent thermal and chemical stability, low cost and easy regeneration for CO_2 uptake [4–6].

2 CO_2 Physisorption Using Carbon Materials

2.1 Effect of Surface Chemistry

As described above, carbon dioxide capture and storage is considered one of the most important strategies when it comes to efficient reduction and recycling of CO_2 [4]. Among the different potential adsorbents, activated carbons exhibit some advantages such as a tunable porous structure and surface chemistry, so that it can be designed specifically to enhance CO_2 adsorption/delivery. The optimization of adsorption performance in carbon materials has been performed through the proper selection of the precursor, modification in the carbonization/activation steps, post-processing surface modification, etc. [7]. Among these modification, the incorporation of heteroatoms in the carbon skeleton is a very useful approach of tuning carbon's unique physicochemical properties. The incorporation of heteroatoms such as nitrogen, sulfur, oxygen, boron and phosphorus to a variety of carbon materials, including graphite, porous carbon, carbon nanotubes, and fullerenes, has been very useful to significantly improve the performance of carbon materials in a range of applications (e.g., batteries, fuel cells, supercapacitors, adsorption, etc.) [4, 6–10]. Incorporation of heteroatoms provides a different electronegative environment,

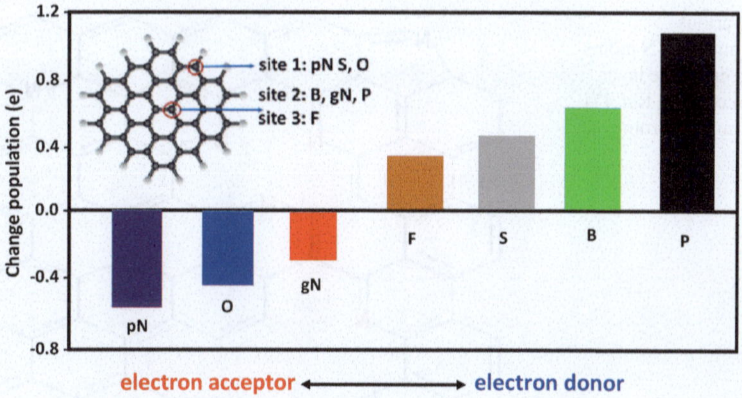

Fig. 1 Natural bond orbital (NBO) population analysis of six nonmetallic heteroatoms in graphene matrix. pN and gN represents pyridinic and graphitic type of N, respectively. Insets show the proposed doping sites for the different elements: sites 1 and 2 are the edge and center in-plane sites, respectively, and site 3 is an out-of-plane center site in graphene. Reproduced from Ref. [11] with permission from American Chemical Society

causing an alteration in the electron density map of the carbon framework and a modification in the electrical conductivity. For instance, nitrogen and oxygen heteroatoms are negatively charged, and can therefore serve as electron donors for the adjacent C atoms (Lewis base). In contrast, F, S, B and P are positively charged and act as electron acceptors (Lewis acid) (see Fig. 1). Another possibility is the co-doping with two elements with reverse electronegativity to that of carbon, e.g., B and N, so that the final composite exhibits special electron-donor properties through the so-called synergistic coupling effect between the two heteroatoms, with improved properties in the oxygen reduction reaction (ORR) [11]. Among these possibilities, N-doped carbons have been the most widely investigated for CO_2 capture.

Many approaches have been suggested when using carbon adsorbents for CO_2 adsorption to integrate heteroatoms into the carbon network, including in-situ doping and post-synthesis doping [12]. The in-situ doping method (or direct synthesis of doped carbon material) requires the carbonization of carbon precursors rich in a specific functional group (for instance, nitrogen-rich polymers such as polyacrylonitrile, melamine, etc.). This approach is very interesting since it allows to synthesize materials with a uniform distribution of heteroatoms, and a controlled porosity and surface chemistry [13]. On the contrary, the post-treatment approach involves the surface reaction between heteroatoms-rich chemicals and the carbon matrix, usually at high temperatures [14–17]. For the last approach, NH_3 has been the most popular reagent used to treat porous carbons in order to introduce nitrogen functionalities and surface basicity [18]. Ammonia can be used directly in the raw carbon material, in the carbon sample after a pre-oxidation treatment with HNO_3 (amination) or in combination with air. Other common reagents for nitrogenization of porous carbons are: HCN, HNO_3, melamine, polyaniline, urea, etc. [7].

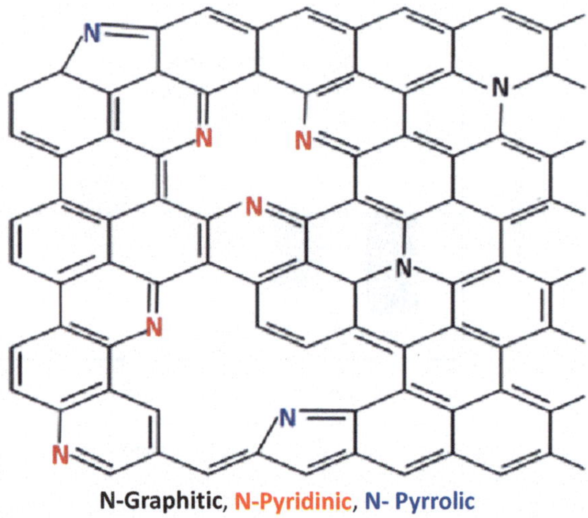

Fig. 2 Various configurations of N atoms doped in graphene layer. Reproduced from Ref. [7] with permission from Elsevier

Previous studies described in the literature have shown that the adsorption performance of carbon materials is not only defined by the incorporation of a certain functional group but also but their chemical environment. This is the case for instance for nitrogen functionalities where nitrogen can be incorporated as graphitic, pyridinic and pyrrolic, depending on the experimental conditions applied. Figure 2 shows a schematic representation of these functional groups and their environment. The incorporation of these N functionalities into the carbon network allows to enhance the basicity either due to the lone pair of electrons (for pyridinic-N and pyrrolic-N), or due to delocalized π-orbitals (for graphitic-N), with the associated changes in the adsorption properties.

Horikawa et al. demonstrated that the water adsorption capacity of porous carbons doped with nitrogen was directly related to the nitrogen content on the low pressure region due to the reduced hydrophobicity of the surface [19]. Despite the beneficial role of N for a wide range of applications, as described above, the role of these functionalities is not so clear in the case of CO_2 adsorption due to the combined effect of porosity and surface chemistry in the final uptake. To et al. prepared hierarchical carbon materials through co-assembly and polymerization of a rationally designed pyrrole monomer and a triblock copolymer soft-template [6]. Under low carbonization and activation conditions, the authors observed an optimum performance for CO_2 adsorption (adsorption capacity above 4.5 mmol/g at 0.1 MPa and 298 K). This excellent performance was attributed to the presence of ultra-small pores and, to the presence of a rich surface chemistry (nitrogen adsorption sites) able to exhibit a strong binding interaction with CO_2. The presence of these surface functional groups (mainly pyrrolic and pyridonic) could also explain the large CO_2/N_2 selectivity in these samples. A similar conclusion was achieved by Ge et al. using polyurethane-based carbon materials [4]. These studies confirm that small micropores (pore size smaller than 0.86 nm (at 273 K) and 0.70 nm (at 298 K)) play a dominant role in the

CO_2 adsorption capacity (around 80% of the adsorption arises from physical adsorption), while there is an additional 20% of the total uptake that is due to chemical adsorption originated from the interaction of basic N groups and CO_2 molecules. This observation agrees with the isosteric heat of adsorption (Q_{st}) ranging from 46 to 28 kJ/mol at CO_2 loadings of 0.01–2 mmol/g [6]. A different conclusion was drawn by Silvestre-Albero et al. for carbons materials prepared from polyaniline [20]. Despite the presence of a rich surface chemistry (around 3–8% nitrogen, mainly pyridinic and pyrrolic groups), CO_2 adsorption measurements at atmospheric (0.1 MPa) and high pressure (4.0 MPa) were in perfect agreement with the textural properties, with no clear contribution of the surface functional groups. This conclusion was also anticipated by Sevilla et al. for carbon microspheres undoped and doped with nitrogen [21]. CO_2 adsorption in these two sets of samples (with similar textural properties) confirmed that nitrogen moieties do not enhance CO_2 performance with respect to the undoped carbons. Adeniran et al. confirmed the main role of the porosity defining the CO_2 adsorption capacity using a range of microporous carbon materials with virtually identical porosity but different surface chemistry [22]. Based on these studies, small micropores in the range 0.5–1.0 nm are critical for CO_2 adsorption, while N-functionalities do not have any beneficial effect.

Other functional groups less explored are boron and sulfur. Boron is an ideal carbon dopant owing to its special and generally incomparable properties within the periodic table, such as lower electronegativity and similar atomic size to carbon [23]. Since boron has one electron less than carbon, when it replaces carbon in the graphene sheet, localized states below the Fermi level (valence band) appear due to its Lewis acid character [24]. Up to now, B-doped carbon materials have been broadly used in different research areas such as super capacitors, fuel cells, oxygen reduction reaction (ORR), among others, with very promising results [23]. However, no studies have been reported in the literature so far about the adsorption properties of these B-doped systems, except the pioneering work of Liu et al. where an improvement in the adsorption performance of porous biochar towards peroxydisulfate (PDS) was observed [24]. Compared to boron, studies dealing with sulfur doping are very limited and represent a growing field in carbon science. Although today the high potential of such materials is mainly exploited for fuel cells, supercapacitors or batteries, little is known about sulphur-doped carbons. Their understanding principally covers graphite-sulfur composites and their superconductivity performance, as well as theoretical investigations related to the effect of individual sulfur atoms on carbon nanotubes or graphene sheets [14, 23, 25–27]. In the specific case of sulfur-doped carbons, there are few studies in the literature that suggest a promoting effect of S doping in the CO_2 adsorption performance [28–30]. Seredych et al. suggest that the improved adsorption performance in the S-doped carbons is due to the presence of acid–base interactions of CO_2 in small pores with sulfur incorporated into aromatic rings of the pore walls, to polar interactions of CO_2 with sulfoxides, sulfones and sulfonic aids, and to hydrogen bonding of CO_2 with acidic groups on the surface (oxygen groups developed in a subsequent oxidation step) [30]. An important effect of the sulfur-containing functional groups was also proposed by Sun et al. for polymer-based carbon materials. Based on these studies, oxidized sulfur groups (-SO_3H) exhibit a significant role in CO_2 adsorption at low pressures, while the microporosity exhibits the predominant role at moderate pressures (0.8 MPa) [29].

Overall, actual results reported in the literature highlights the preferential role of the porosity defining the adsorption performance of carbon materials, as discussed in the next section, although some marginal role of the surface chemistry cannot be completely ruled out, preferentially for sulfur-based functional groups.

2.2 *Effect of the Porous Structure on CO_2 Adsorption*

As anticipated in the previous section, the porous architecture of inorganic materials is of paramount importance defining their adsorption properties. The presence of nanocavities with enhanced adsorption potential is a pre-requisite to promote adsorption of light molecules (e.g., CO_2) under low pressure conditions. To this end, the porous structure has to be properly designed, pore size and shape, to achieve an optimum packing density for the target molecule [31]. Based on these premises, carbon materials can be considered one of the most promising candidates due to the possibility to tune their porous network through proper synthesis routes (e.g., carbonization and activation conditions) [31]. The presence of a disordered structure with voids or cavities in between the graphene-like layers allows to adsorb a large amount of gas or liquid molecules, provided that these cavities are properly designed.

CO_2 is an acidic molecule with a kinetic diameter of 0.33 nm. CO_2 adsorption at 273 K and atmospheric pressure has been traditionally applied to complement N_2 adsorption in the characterization of carbon materials [32, 33]. Whereas N_2 adsorption at cryogenic temperatures (77 K) and atmospheric pressure provides information about the microporosity (pores below 2 nm) and mesoporosity (pores in between 2 and 50 nm), CO_2 adsorption at 273 K and atmospheric pressure provides information about the narrow microporosity (pores below 0.7 nm). This assumption anticipates that CO_2 adsorption at atmospheric pressure (partial pressure 0.03) requires the presence of narrow pores, preferentially those below 0.7–0.8 nm.

Grand canonical Monte Carlo (GCMC) simulations have shown that CO_2 forms a dense monolayer in pores wider than 0.3 nm, with a complete filling of these cavities at a pressure of 5×10^{-3} atm [34]. These simulations predict a transition from a single to a double layer of adsorbed CO_2 molecules for pores between 0.65 and 0.8 nm, these two layers remaining unchanged for wider pores, i.e. wider pores do not adsorb additional CO_2 at sub-atmospheric pressures, the central region of the pores remaining unfilled. These theoretical predictions confirm the above mentioned assumption, i.e. CO_2 adsorption at atmospheric or sub-atmospheric pressure conditions require the presence of pores below 1 nm to achieve an optimum packing density. These predictions were confirmed experimentally by different research groups [35–39]. Silvestre-Albero et al. synthesized a series of porous carbon molecular sieves using petroleum pitch as carbon precursor and KOH and activated agent [35]. The evaluation of these microporous carbon materials in CO_2 capture at atmospheric pressure suggested that the total adsorption capacity is not defined by the total surface area (BET) but rather by the volume of narrow micropores. Similarly, Martin et al. analyzed the CO_2 adsorption on different polymeric porous carbons prepared by physical activation of a phenolic resin [36]. These authors concluded that CO_2 capture at 298 K and atmospheric pressure involves exclusively smaller micropores (<1 nm). A more detailed analysis about the role of the porous structure in the adsorption performance was described by Presser et al. by using a series of carbide-derived carbons (CDCs) [39]. The adsorption performance of these CDCs for CO_2 at atmospheric pressure was correlated with the porous structure through the correlation with specific pore size regions. These correlations anticipated that the optimum pore size for CO_2 adsorption at 0.1 MPa is 0.8 nm, while smaller pores (<0.5 nm) are mandatory for processes working at 0.01 MPa. Last but not least, Sevilla et al. carried out similar analysis in carbon materials prepared by physical activation with CO_2 [21]. These studies confirmed that the adsorption potential for CO_2 is reduced with an enlargement of the micropores as the activation process proceeds, as shown in Fig. 3.

In conclusion, the above described studies confirm that small micropores (those below 1 nm) are mandatory to achieve an optimum adsorption performance for CO_2 at atmospheric pressure. Although majority of studies described in the literature are dealing with atmospheric pressure conditions, preliminary studies have shown that the scenario changes completely at high pressures. Under these conditions larger cavities are required to achieve an optimum packing density. The knowledge of the adsorption performance of a given inorganic system at high pressure is highly important for applications such as gas storage in the transportation sector. It is widely accepted in the literature that for adsorption processes at high pressure (close to CO_2 vapour pressure, i.e. 3.4 and 6.4 MPa at 273 K and 298 K, respectively), the adsorption of CO_2 takes place in cavities similar to those filled by N_2 at 77 K and 0.1 MPa [40]. Under these circumstances, CO_2 uptake can be notably enhanced because both large

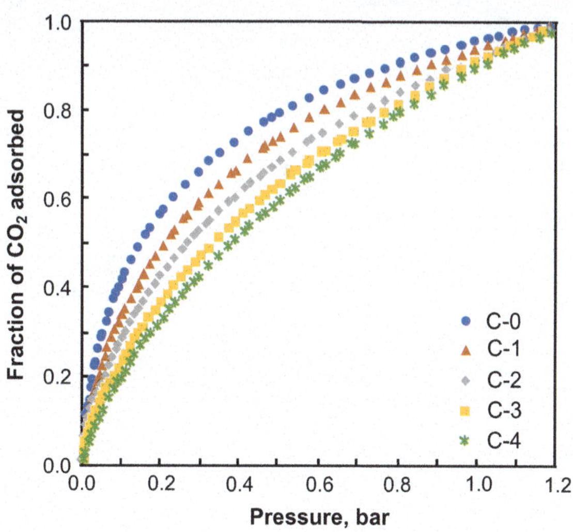

Fig. 3 Comparison of the fraction of CO_2 adsorbed by undoped carbons at different activation degrees. Adsorption temperature of 273 K. Reproduced from Ref. [21] with permission from American Chemical Society

micropores and mesopores are now available to participate in the adsorption process. For this reason, the CO_2 uptake at high pressures shows a better correlation with the total pore volume or BET surface area, as it has been already reported in the literature, rather than with the micropore volume (as stated above for atmospheric pressure conditions) [40–43]. At high pressures, the optimal threshold pore size increases with the adsorption pressure since experimental conditions are sufficient to enable capillary condensation in large cavities (mainly mesocavities). Casco et al. showed that in the specific case of petroleum-pitch based activated carbons, CO_2 adsorption at 4.5 MPa and 298 K exhibits a good correlation with pores below 2 nm, although the correlation is also quite good for larger pores (<3 nm), see Fig. 4 [44]. These results constitute an experimental evidence that large micropores and small mesopores are the optimum cavities to store CO_2 at high pressure with an optimum packing density. Under these conditions, the delivery capacity can be up to 1400 mg CO_2/g between 4.5 and 0.1 MPa. Contrary to atmospheric pressure conditions, no clear correlation could be observed at high pressure with the amount of narrow micropores (<1 nm).

Ashourirad et al. determined that for the case of benzimidazole-based KOH activated carbons, both the surface excess and the absolute CO_2 uptakes at 3.0 MPa and 298 K exhibit a very good correlation with the volume of pores smaller than 2.2 nm [45]. Similarly, He et al. studied the CO_2 adsorption performance of polyaniline-based KOH activated carbons at 3.0 MPa and 298 K [46]. Under these experimental

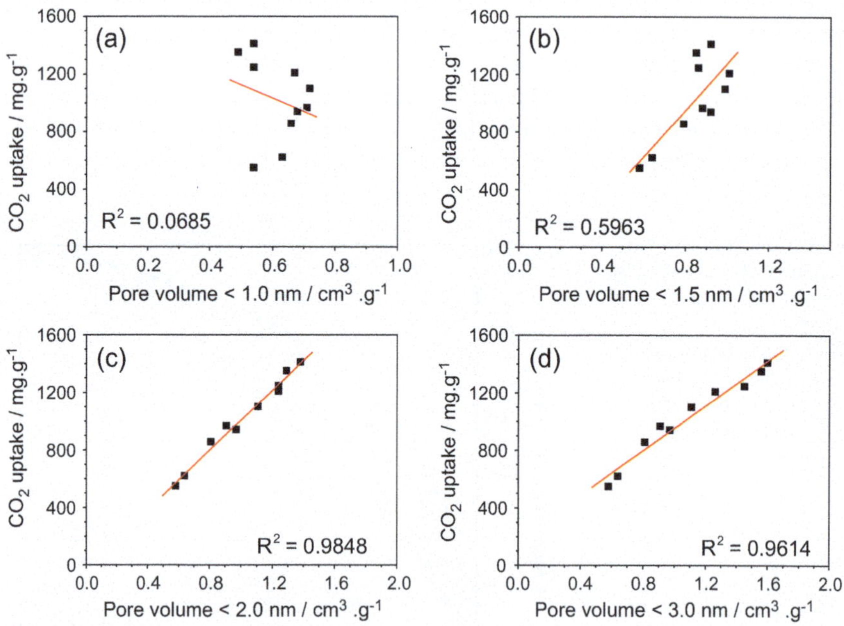

Fig. 4 The correlation between the CO_2 adsorption capacity (mg/g) and the volume of pores smaller than 3.0, 2.0, 1.5, and 1.0 nm. Reproduced from Ref. [44] with permission from Elsevier

conditions the amount adsorbed was perfectly correlated with the volume of pores smaller than 4 nm.

Last but not least it is important to highlight that in the last few years, novel materials have been proposed in the literature able to overpass the adsorption performance of carbon materials for CO_2 adsorption. These materials are the well-known metal–organic frameworks (MOFs). When activated carbons are compared with the most promising MOFs (Fig. 5), it seems that the latest offer better performance in terms of total capacity. For instance, values as high as 250 wt% at 5.0 MPa were reported for MOF-177 and MOF-200, systems with an exceptional surface area above 4500 m^2/g [47]. Based on these premises, Silvestre-Albero et al. compared the adsorption performance of one of the best performing carbon materials with the aforementioned MOFs [35]. These studies suggested that properly designed activated carbon materials (carbons prepared from petroleum pitch with KOH as activating agent) can compete, and even overpass, MOFs performance in volumetric basis. While activated carbon VR-93 was able to reach an excess uptake of 500 v/v at 5.0 MPa, MOF-177 reached an asymptotic value around 300 v/v under similar experimental conditions. At this point it is important to highlight that the volumetric data for MOFs were obtained assuming the crystallographic density, i.e. assuming a single crystal of MOF.

In summary, these results show that wider cavities are needed to store large quantities of CO_2 under high pressure conditions. Properly designed carbon materials can achieve adsorption values as high as 2000 mg/g, in gravimetric basis, or 500 v/v in volumetric basis, at 5.0 MPa and 298 K. Although MOF can compete with carbon materials in terms of total adsorption capacity for CO_2, one has to keep in mind the excellent mechanical and chemical properties of carbons as a differentiating factor to other porous sorbents, for instance several MOFs.

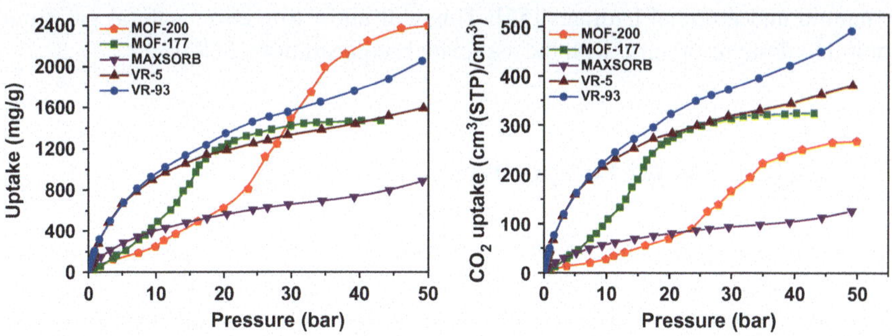

Fig. 5 CO_2 adsorption isotherms in gravimetric and volumetric basis (cm^3/cm^3) for the VR carbon molecular sieve samples at 298 K and up to 5.0 MPa. Two MOF materials (MOF-177 and MOF-200) with exceptional adsorption properties and commercial activated carbon MAXSORB are included for the sake of comparison. Adapted from Ref. [35] with permission from The Royal Society of Chemistry

3 CO_2 Adsorption in the Form of Clathrates

As described in previous sections, although carbon materials are excellent sorbents for CO_2 adsorption both at low and high pressure, at this stage it is difficult to design novel carbon materials able to outperform actual values. This is due to the presence of a highly optimized porous structure with the proper pore size and shape for a number of materials reported in the literature. However, recent studies from our research group have shown that these numbers can be highly improved by mimicking nature, though the so-called CO_2 hydrates. This new strategy consists in trapping CO_2 molecules in the cages created by water molecules in the clathrate crystal. Under these conditions large quantities of gas molecules can be captured in a very small volume [48]. Although gas hydrates are very common in nature (for instance deep under the sea or in the permafrost), their artificial synthesis is not so simple due to the presence of important kinetic limitations at the gas–liquid interface [49]. However, these kinetic restrictions can be overcome using nanoporous materials as a host structure to promote the nucleation and growth of these hydrates [50–53]. Under confined conditions, gas hydrates can be grown under milder pressure and temperature conditions compared to artificial hydrates, and more importantly, with faster kinetics. Majority of studies reported in the literature involve methane as a guest molecule, with excellent results in terms of adsorption capacity compared to the dry system [50, 51, 53]. However, recent studies performed using activated carbon as a host structure confirm that CO_2 hydrate can also be promoted under confinement conditions, at lower pressures compared to methane hydrates (1.8 MPa vs. 3.4 MPa) [54]. Under water supplying conditions, the amount of CO_2 adsorbed can be increased from 15 mmol/g up to 40 mmol/g at 10.0 MPa and 275 K (Fig. 6). Furthermore, these studies anticipate the possibility to replace CH_4 by CO_2 in these confined hydrates, thus opening the gate towards in-situ replacement or swapping in natural and artificial hydrates [55]. This will allow to recover natural gas from natural hydrate reservoirs, with the associated sequestration of CO_2.

4 Conclusions and Perspectives

The main goal of this chapter was to provide some guidelines about the role of carbon materials in CO_2 adsorption processes and the effect of the different variables (mainly pore size and surface chemistry) defining the adsorption performance. These studies described in the literature show that the design of the porous structure (size and shape of the cavities) is crucial to achieve a proper adsorption performance for CO_2 with an optimum packing density. On the contrary, the role of the functional groups present on the carbon surface is still a matter of debate. For low-pressure applications, a pore size around 0.8–1.0 nm is defined as a optimum value, while larger pores (wide micropores and small mesopores) are needed for high-pressure applications. Concerning surface chemistry, incorporation of sulfur functionalities

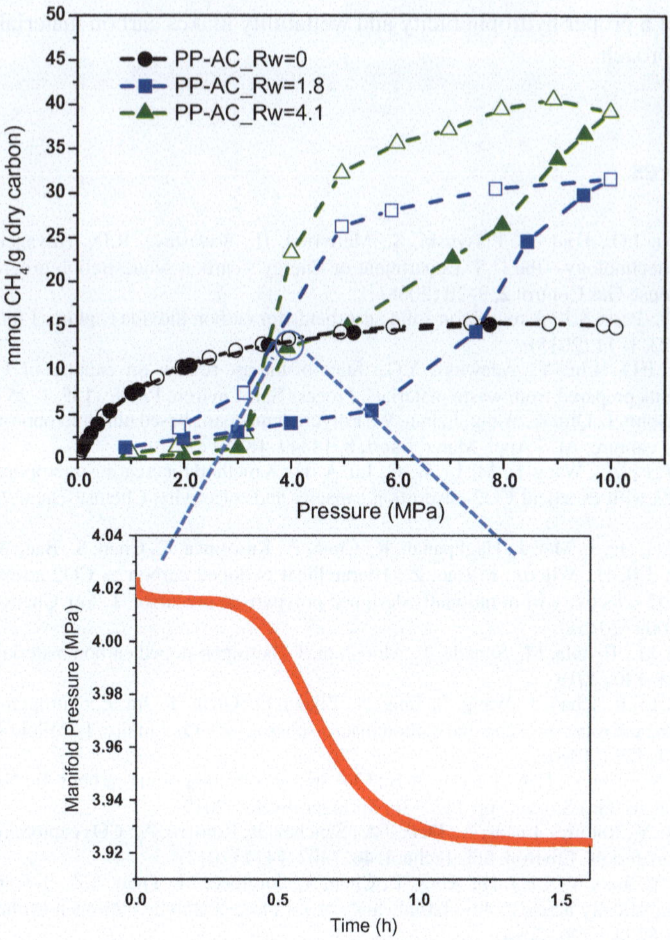

Fig. 6 High-pressure CO_2 isotherm at 275 K for a petroleum pitch derived carbon obtained by KOH activation. Reproduced from Ref. [54] with permission from John Wiley and Sons

has been proposed as a promising approach to improve the gas-framework interactions, with the associated improvement in the adsorption of CO_2. However, the situation is not so clear in the case of nitrogen functionalities. Although some studies claim a promoting effect of these surface functional groups, the predominant effect of the porous structure precludes a clear confirmation of this hypothesis. Last but not least, this chapter incorporates a new approach to store CO_2 in carbon materials based on natural processes. Under water supplying conditions and high pressure, carbon materials can promote the formation of CO_2 hydrates with a similar stoichiometry to the natural systems. Compared to the dry physisorption process, these crystal structures allow to improve the storage capacity more than 2–3 times. The presence of a proper hydrophobicity and wettability makes carbon materials unique for this approach.

References

1. Figueroa, J.D., Fout, T., Plasynski, S., McIlvried, H., Srivastava, R.D.: Advances in CO_2 capture technology—the U.S. Department of Energy's carbon sequestration program. Int. J. Greenhouse Gas Control **2**, 9–20 (2008)
2. Lee, S.Y., Park, S.J.: A review on solid adsorbents for carbon dioxide capture. J Industr. Eng. Chem. **23**, 1–11 (2015)
3. Ochedi, F.O., Liu, Y., Adewuyi, Y.G.: State-of-the-art review on capture of CO_2 using adsorbents prepared from waste materials. Process Saf. Environ. Protec. **139**, 1–25 (2020)
4. Ge, C., Song, J., Qin, Z., Wang, J., Fan, W.: Polyurethane foam-based ultramicroporous carbons for CO2 capture. ACS Appl. Mater. Interf. **8**, 18849–18859 (2016)
5. Qian, D., Lei, C., Wang, E.-M., Li, W.-C., Lu, A.-H.: A method for creating microporous carbon materials with excellent CO2-adsorption capacity and selectivity. Chemsuschem **7**, 291–298 (2014)
6. To, J.W.F., He, J., Mei, J., Haghpanah, R., Chen, Z., Kurosawa, T., Chen, S., Bae, W.-G., Pan, L., Tok, J.B.-H., Wilcox, J., Bao, Z.: Hierarchical N-doped carbon as CO2 adsorbent with high CO2 selectivity from rationally designed polypyrrole precursor. J. Am. Chem. Soc. **138**, 1001–1009 (2016)
7. Inagaki, M., Toyoda, M., Soneda, Y., Morishita, T.: Nitrogen-doped carbon materials. Carbon **132**, 104–140 (2018)
8. Sun, Y., Li, K., Zhao, J., Wang, J., Tang, N., Zhang, D., Guan, T., Jin, Z.: Nitrogen and sulfur Co-doped microporous activated carbon macro-spheres for CO_2 capture. J. Colloid Interf. Sci. **526**, 174–183 (2018)
9. Sevilla, M., Ferrero, G.A., Fuertes, A.B.: CO_2 storage on nanoporous carbons. In: Nanoporous Materials for Gas Storage, pp 287–330, Springer, Berlin (2019)
10. Jimenez, V., Ramirez-Lucas, A., Diaz, J.A., Sanchez, P., Romero, A.: CO_2 capture in different carbon materials. Environ. Sci. Technol. **46**, 7407–7414 (2012)
11. Zheng, Y., Jiao, Y., Li, L.H., Xing, T., Chen, Y., Jaroniec, M., Qiao, S.Z.: Towards design of synergistically active carbon-based catalysts for electrocatalytic hydrogen evolution. ACS Nano **8**, 5290–5296 (2014)
12. Shen, W., Fan, W.: Nitrogen-containing porous carbons: synthesis and application. J. Mater. Chem. A **1**, 999–1013 (2013)
13. Machnikowski, J., Grzyb, B., Machnikowska, H., Weber, J.V.: Surface chemistry of porous carbons from N-polymers and their blends with pitch. Microp. Mesop. Mater. **82**, 113–120 (2005)
14. Kicinski, W., Szala, M., Bystrzejewski, M.: Sulfur-doped porous carbons: Synthesis and applications. Carbon **68**, 1–32 (2014)
15. Paraknowitsch, J.P., Thomas, A.: Functional carbon materials from ionic liquid precursors. Macromol: Chem. Phys. **213**, 1132–1145 (2012)
16. Xu, Y., Yang, Z., Zhang, G., Zhao, P.: Excellent CO_2 adsorption performance of nitrogen-doped waste biocarbon prepared with different activators. J. Clean Prod. **264**, 121645 (2020)
17. Stein, A., Wang, Z., Fierke, M.A.: Functionalization of porous carbon materials with designed pore architecture. Adv. Mater. **21**, 265–293 (2009)
18. Mangun, C.L., Benak, K.R., Economy, J., Foster, K.L.: Surface chemistry, pore sizes and adsorption properties of activated carbon fibers and precursors treated with ammonia. Carbon **39**, 1809–1820 (2001)
19. Horikawa, T., Sakao, N., Sekida, T., Hayashi, J., Do, D.D., Katoh, M.: Preparation of nitrogen-doped porous carbon by ammonia gas treatment and the effects of N-doping on water adsorption. Carbon **50**, 1833–1842 (2012)

20. Silvestre-Albero, A., Silvestre-Albero, J., Martinez-Escandell, M., Molina-Sabio, M., Kovacs, A., Rodriguez-Reinoso, F.: Novel synthesis of micro-mesoporous nitrogen-doped nanostructured carbon from polyaniline. Microp. Mesop. Mater. **218**, 199–205 (2015)
21. Sevilla, M., Parra, J.B., Fuertes, A.B.: Assessment of the role of micropore size and N-doping in CO_2 capture by porous carbons. ACS Appl. Mater. Interf. **5**, 6360–6368 (2013)
22. Adeniran, B., Mokaya, R.: Is N-doping in porous carbons beneficial for CO_2 storage? Experimental demonstration of the relative effects of pore size and N-doping. Chem. Mater. **28**, 994–1001 (2016)
23. Paraknowitsch, J.P., Thomas, A.: Doping carbons beyond nitrogen: an overview of advanced heteroatom doped carbons with boron, sulphur and phosphorus for energy applications. Energy Environ. Sci. **6**, 2839–2855 (2013)
24. Liu, B., Guo, W., Wang, H., Si, Q., Zhao, Q., Luo, H., Ren, N.: B-doped graphitic porous biochar with enhanced affinity and electron transfer for efficient peroxydisulfate activation. Chem. Eng. J. **396**, 125119 (2020)
25. da Silva, R.R., Torres, J.H., Kopelevich, Y.: Indication of superconductivity at 35K in graphite-sulfur composites. Phys. Rev. Lett. **87**, 147001 (2001)
26. Terrones, M., Souza Filho, A.G., Rao, A.M.: Doped carbon nanotubes: synthesis, characterization and applications. In: Jorio, A., Dresselhaus, G., Dresselhaus, M.S. (eds.) Carbon Nanotubes. Topics in Appl. Physics, v. 111, pp 531–566, Springer, Berlin (2007)
27. Denis, P.A.: When noncovalent interactions are stronger than covalent bonds: bilayer graphene doped with second row atoms, aluminium, silicon, phosphorus and sulfur. Chem. Phys. Lett. **508**, 95–101 (2011)
28. Xia, Y., Zhu, Y., Tang, Y.: Preparation of sulfur-doped microporous carbons for the storage of hydrogen and carbon dioxide. Carbon **50**, 5543–5553 (2012)
29. Sun, Y., Zhao, J., Wang, J., Tang, N., Zhao, R., Zhang, D., Guan, T., Li, K.: Sulfur-doped millimeter-sized microporous activated carbon spheres derived from sulfonated poly(styrene-divinylbenzene) for CO_2 capture. J. Phys. Chem. C **121**, 10000–10009 (2017)
30. Seredych, M., Jagiello, J., Bandosz, T.J.: Complexity of CO2 adsorption on nanoporous sulfur-doped carbons—is surface chemistry an important factor? Carbon **74**, 207–217 (2014)
31. Marsh, H., Rodriguez-Reinoso, F.: Activated Carbon. Elsevier Science (2016)
32. Garrido, J., Linares-Solano, A., Martin-Martinez, J.M., Molina-Sabio, M., Rodriguez-Reinoso, F., Torregrosa, R.: Use of nitrogen vs. carbon dioxide in the characterization of activated carbons. Langmuir **3**, 76–81 (1987)
33. Rodriguez-Reinoso, F., Garrido, J., Martin-Martinez, J.M., Molina-Sabio, M., Torregrosa, R.: The combined use of different approaches in the characterization of microporous carbons. Carbon **27**, 23–32 (1989)
34. Vishnyakov, A., Ravikovitch, P.I., Neimark, A.V.: Molecular level models for CO_2 sorption in nanopores. Langmuir **15**, 8736–8742 (1999)
35. Silvestre-Albero, J., Wahby, A., Sepulveda-Escribano, A., Martinez-Escandell, M., Kaneko, K., Rodriguez-Reinoso, F.: Ultrahigh CO_2 adsorption capacity on carbon molecular sieves at room temperature. Chem. Commun. **47**, 6840–6842 (2011)
36. Martin, C.F., Plaza, M.G., Pis, J.J., Rubiera, F., Pevida, C., Centeno, T.A.: On the limits of CO_2 capture capacity of carbons. Separ. Purif. Techn. **74**, 225–229 (2010)
37. Zhang, Z., Zhou, J., Xing, W., Xue, Q., Yan, Z., Zhuo, S., Qiao, S.Z.: Critical role of small micropores in high CO_2 uptake. Phys. Chem. Chem. Phys. **15**, 2523–2529 (2013)
38. Lee, J.H., Kwack, K., Lee, H.J., Lim, S.Y., Jung, D.S., Jung, Y., Choi, J.W.: Optimal activation of porous carbon for high performance CO_2 capture. ChemNanoMat **2**, 528–533 (2016)
39. Presser, V., McDonough, J., Yeon, S.-H., Gogotsi, Y.: Effect of pore size on carbon dioxide sorption in carbide derived carbon. Energy Environ. Sci. **4**, 3059–3066 (2011)
40. Cazorla-Amoros, D., Alcañiz-Monge, J., Linares-Solano, A.: Characterization of activated carbon fibers by CO_2 adsorption. Langmuir **12**, 2820–2824 (1996)
41. Coromina, H.M., Walsh, D.A., Mokaya, R.: Biomass-derived activated carbon with simultaneously enhanced CO Biomass-derived activated carbon with simultaneously enhanced CO_2 uptake for both pre and post combustion capture applications. J. Mater. Chem. A **4**, 280–289 (2016)

42. Srinivas, G., Krungleviciute, V., Guo, Z.-X., Yildirim, T.: Exceptional CO_2 capture in a hierarchically porous carbon with simultaneous high surface area and pore volume. Energy Environ. Sci. **7**, 335–342 (2014)
43. Srinivas, G., Burress, J., Yildirim, T.: Graphene oxide derived carbons (GODCs): synthesis and gas adsorption properties. Energy Environ. Sci. **5**, 6453–6459 (2012)
44. Casco, M.E., Martinez-Escandell, M., Silvestre-Albero, J., Rodriguez-Reinoso, F.: Effect of the porous structure in carbon materials for CO_2 capture at atmospheric and high-pressure. Carbon **67**, 230–235 (2014)
45. Ashourirad, B., Arab, P., Islamoglu, T., Cychosz, K.A., Thommes, M., El-Kaderi, H.M.: A cost-effective synthesis of heteroatom-doped porous carbons as efficient CO_2 sorbents. J. Mater. Chem. A **4**, 14693–14702 (2016)
46. He, J., To, J.W.F., Psarras, P.C., Yan, H., Atkinson, T., Holmes, R.T., Nordlund, D., Bao, Z., Wilcox, J.: Tunable polyaniline-based porous carbon with ultrahigh surface area for CO_2 capture at elevated pressure. Adv. Energy Mater. **6**, 1502491 (2016)
47. Furukawa, H., Ko, N., Go, Y.B., Aratani, N., Choi, S.B., Choi, E., Yazaydin, A.O., Snurr, R.Q., O'Keeffe, M., Kim, J., Yaghi, O.M.: Ultrahigh porosity in metal-organic frameworks. Science **329**, 424–428 (2010)
48. Sloan, E.D.: Fundamental principles and applications of natural gas hydrates. Nature **426**, 353–359 (2003)
49. Vlasov, V.A.: Diffusion-kinetic model of gas hydrate growth along the gas-water interface. Heat Mass Transf. **55**, 3537–3545 (2019)
50. Casco, M.E., Silvestre-Albero, J., Ramirez-Cuesta, A.J., Rey, F., Jorda, J.L., Bansode, A., Urukawa, A., Peral, I., Martinez-Escandell, M., Kaneko, K., Rodriguez-Reinoso, F.: Methane hydrate formatin in confined nanospace can surpass nature. Nature Commun. **6**, 6432 (2015)
51. Borchardt, L., Casco, M.E., Silvestre-Albero, J.: Methane hydrate in confined spaces—an alternative energy storage system. ChemPhysChem **19**, 1298–1314 (2018)
52. Handa, Y.P., Stupin, D.Y.: Thermodynamic properties and dissociation characteristics of methane and propane hydrates in 70-.ANG.-radius silica gel pores. J. Phys. Chem. **96**, 8599–8603 (1992)
53. Celzard, A., Mareche, J.-F.: Optimal wetting of active carbons for methane hydrate formation. Fuel **85**, 957–966 (2005)
54. Casco, M.E., Jorda, J.L., Rey, F., Fauth, F., Martinez-Escandell, M., Rodriguez-Reinoso, F., Ramos-Fernandez, E.V., Silvestre-Albero, J.: High-performance of gas hydrates in confined nanospace for reversible CH_4/CO_2 storage. Chem: A Europ. J. **22**, 10028–10035 (2016)
55. Palodkar, A.V., Jana, A.K.: Fundamental of swapping phenomena in naturally occurring gas hydrates. Sci. Rep. **8**, 16563 (2018)

Graphene-Based Carbons of Tuned Nanoporosity and Crystallinity

Shuwen Wang and Katsumi Kaneko

Abstract This review introduces the tuning of porosity and crystallinity of graphene-based porous carbons through a series of methods like chemical reagent activation (with KOH, $ZnCl_2$ and H_3PO_4), physical activation (with CO_2), compression and thermal treatment. The KOH activation shows a superior porosity tunability compared with the chemical reagent activations with $ZnCl_2$ or H_3PO_4 and CO_2 activation. A shape-retention activation method with ice-templating can be applied to preparation of free-standing porous graphene monoliths with well-aligned textures. A long compression of highly porous graphene for 25 h unfolds the double-bent like structures of nano-sized graphenes. Heat-treatment results in the increase of crystallinity and decrease of porosity of porous graphene monolith; highly crystalline porous graphene monolith with high porosity can be obtained by the heat-treatment at 1873 K. The heat-treated porous graphene monolith shows a variable-range hopping conduction, being a promising material for electrochemical applications.

1 Introduction

We had active scientific arguments on the upper limit of surface area of porous carbons in adsorption and carbon sciences 30 years ago; many scientists believed that the surface area limit of porous carbons was 2630 m^2g^{-1} of the geometrical surface area of graphene. On the contrary, KOH activation [1] enables to produce porous carbons whose BET surface area is much larger than 2630 m^2g^{-1}, offering an active discussion for the scientific understanding. We had competitive researches for developing nanoporous carbons of larger surface area for storage of supercritical gases such as hydrogen and methane. Very recently Professor Rodríguez-Reinoso and one of the authors edit the book entitled on Nanoporous Materials for Gas Storage

S. Wang · K. Kaneko (✉)
Research Initiative for Supra-Materials, Shinshu University, Nagano 380-8553, Japan
e-mail: kkaneko@shinshu-u.ac.jp

S. Wang
e-mail: s_wang@shinshu-u.ac.jp

J. C. Moreno-Piraján et al. (eds.), *Porous Materials*, Engineering Materials,
https://doi.org/10.1007/978-3-030-65991-2_6

[2], in which representative porous carbons for methane and hydrogen storage are given. Nanoporous carbons having the BET surface area larger than 2630 m^2g^{-1} may be called "superhigh surface area carbons" [3]. Kaneko et al. clarified the contribution of nanoscale graphitic units as the component of nanoporous carbons to the surface area larger than 2630 m^2g^{-1} without overestimation; they proposed the subtracted pore effect (SPE) method for more accurate evaluation of the surface area [3–5]. Also they showed that the routine BET analysis lead to overestimation of the surface area for supermicropores and underestimation for ultramicropores.

Still we have an intensive demand for the superhigh surface area carbon, because promotion of renewable energy technologies is indispensable to the high surface area carbon electrodes for supercapacitors [6–9] in addition to efficient adsorbents for methane storage [10–12]. Outstanding confinement effects were observed with the relevance to fundamental studies on supercapcitors using porous carbons, stimulating fundamental science on nanoconfinement. Chmiola et al. [13] found an anomalous increase of capacitance in ultramicroporous carbons, agreeing with dehydration of ions in highly confined carbon pores [14]. Futamura et al. elucidated novel concentration effect of co-ions in carbon ultramicropores due to the marked image charge effect [15]. These novel scientific findings have also encouraged to develop highly designed porous carbons for better understanding of nanoconfinement effects. Thus, we must develop porous carbons having customized pore structures depending on our scientific target or industrial application. Porous carbons are mostly prepared from a variety of carbon precursors whose structures are too diverse to achieve a full understanding. Consequently, it's necessary to innovate nanoporous carbons with well-defined wall structure and tunable pore structure.

Graphene naturally becomes a versatile nanoscale building block to achieve nanoporous carbons with designed structure and properties due its unique physical and chemical properties [16–18]. Graphene-derived porous carbons show superior performance over conventional carbons due to their high thermal [19–21] and electrical conductivity [22–24], as well as good mechanical property [25–27]. However, most of the reported graphene-based carbons are of low surface area, or have N_2 isotherms of Type II or IV [28, 29], suggesting that micropores are not sufficiently developed. Consequently, a systematic route on the porosity tuning of graphene-based porous carbons, especially those with abundant micropore is highly desired. This chapter offers a review on the porosity development on the graphene-based carbon by using KOH, CO_2, $ZnCl_2$ and H_3PO_4 as an activation reagent, as well as their porosity regulation by using high-pressure compression and heat-treatment at high temperature.

2 High Surface Area Graphene Monoliths with KOH Activation

KOH activation is one of the most efficient way for developing nanopores on graphene-based carbons. It is of significant importance to provide a guideline for the bottom-up design of graphene-based porous carbons with KOH activation method [30]. The graphene-based carbons are prepared in monolith form due to their easiness for handling. The porous graphene monolith (PGM) can be obtained by KOH activation method; the ice-templated monoliths prepared from the mixture of KOH and GO are freeze-dried, heat-treated and washed repeatedly by 1 M HCl and distilled water. The PGM shows a remarkable shrinkage (see Fig. 1d) due to its microscopic structural change. For comparison, the reduced GO (RGO) monolith without KOH activation (See Fig. 1a) is prepared through heating the GO monolith in Ar atmosphere at a low heating rate (1.5 K/min). The RGO monolith without activation consists of crumpled and uniform sheets detaching from each other, which construct abundant macropores (see Fig. 1b). TEM observation indicates RGO monolith has stacking graphene sheets of large size and integrated shape (see Fig. 1c). On the other hand, PGM is constructed by irregular lamellar structure (see Fig. 1e), which should stem from the aggregation of GO on addition of the chemical reagents. The graphene sheets break into smaller sizes and become entangled after the KOH activation process, constructing the three-dimensional (3D) network with abundant nanospaces (see Fig. 1f). Different from conventional highly porous carbon that consists of graphene units of a few nanometers in scale [31–33], PGM has graphene sheets of considerably large size (but smaller than RGO) and integrated structure, being responsible for its enhanced mechanical and electrical property.

Fig. 1 Morphology of RGO monoliths (**a–c**) and PGM monoliths (**d–f**). **a**, **d** photos, **b**, **e** SEM images, **c**, **f** TEM images

The N_2 adsorption isotherms of non-activated RGO and PGMs prepared with KOH/C = 1 are all of type IV with a slight combination of type I (see Fig. 2a), indicating the presence of predominant mesoporosity with a little microporosity. The PGMs have much higher N_2 uptake amount than that of RGO in whole P/P_0 region, suggesting the significant development of nanoporosity. The PGMs prepared with KOH/C = 2 have isotherms with less characteristic of type IV but more characteristic of type I than those prepared with KOH/C = 1 (see Fig. 2b), suggesting the deterioration of large mesopores and the development of micropores. The PGMs prepared with KOH/C = 3 show a significant N_2 uptake enhancement at $P/P_0 < 0.4$ (see Fig. 2c), which comes from the substantial increase of micropores. The saturation of adsorption isotherms at $P/P_0 > 0.5$ indicates the disappearance of large mesopores. On the other hand, the continuous enlargement of micropores induced by intensive activation (1023 and 1073 K) leads to the development of small mesopores, which is evidenced by the notable hysteresises at $P/P_0 = 0.4$–0.6. Even at more intensive activation conditions with KOH/C = 5, further development of micropores and mesopores is evidenced according to the larger initial N_2 uptake amount in the low P/P_0 region and the distinct hysteresis loop at $P/P_0 > 0.4$ (see Fig. 2d).

The porosity parameters as a function of the activation temperature and KOH/C ratio are illustrated in histogram for a better comparison. The total and external

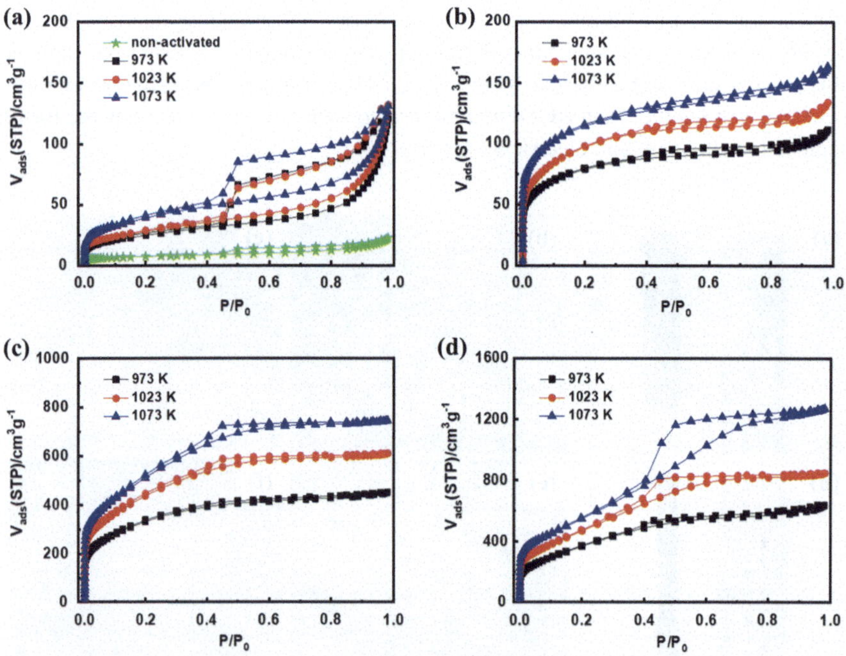

Fig. 2 N_2 adsorption isotherms (77.4 K) of PGM obtained from KOH activation with KOH/C = 1 **a**, 2 **b**, 3 **c** and 5 **d** at various temperatures. Reprinted (adapted) with permission from [30]. Copyright (2014) Elsevier

surface areas are determined by subtracting pore effect (SPE) method [5] using high resolution α_s-plot of the N_2 adsorption isotherm. The micropore volume is evaluated by the Dubinin-Radushkevich (DR) equation [34] and the mesopore volume is determined by Dollimore-Heal (DH) method [35, 36] at $P/P_0 = 0.95$, respectively. The surface area and micropore volume of PGM increase with the KOH/C ratio and activation temperature within the applied experimental conditions (see Fig. 3a, b). The most remarkable increase of surface asrea and micropore volume occurs during the KOH/C ratio increases from 2 to 3. The increase of surface area is largely caused by the development of micropores. On the other hand, the changes of external surface area and mesopore volume under KOH activation show similar tendency in general; both decrease firstly, then increase with the KOH/C ratio under each activation temperature (see Fig. 3c, d).

The porosity development of PGM with increase of the KOH/C ratio from 1 to 5 can be described by a simple activation model based on the obtained porosity parameters and TEM images. The non-activated RGO is constructed by randomly stacked graphene layers (see Fig. 4a) with abundant interspaces of micrometer scale, which cannot be determined by N_2 adsorption at 77 K. KOH activation with KOH/C = 1 breaks up large graphene layers into fine pieces and brings out smaller interspaces

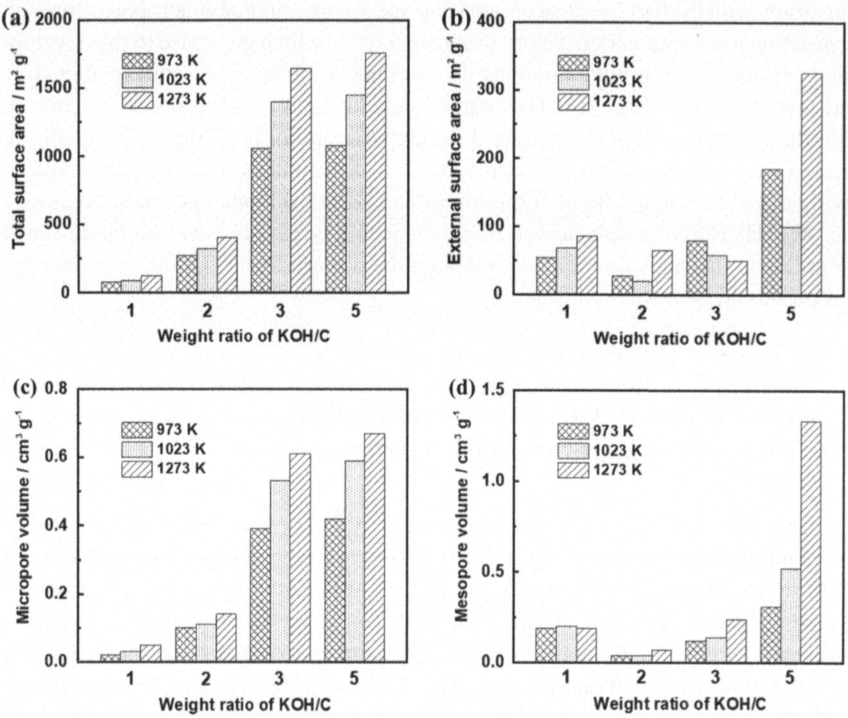

Fig. 3 Porosity variation of PGMs activated at different temperatures and KOH/C ratio. **a** Total surface area, **b** external surface area, **c** micropore volume, **d** mesopore volume

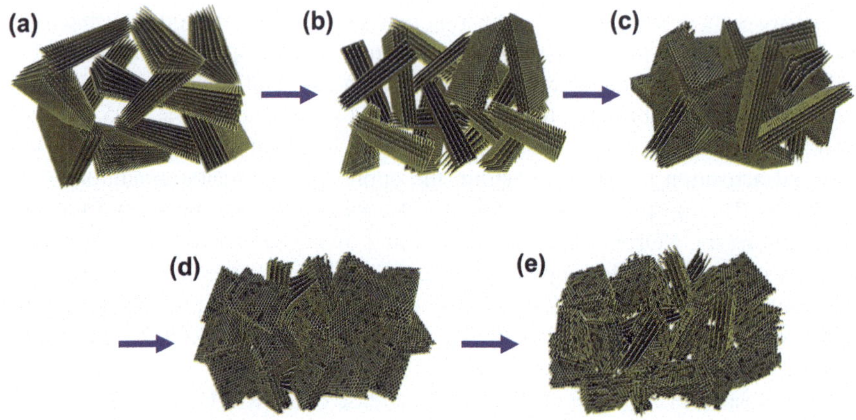

Fig. 4 Porosity development model of PGM under KOH activation. **a** Non-activated RGO; PGM obtained from KOH activation at 1073 K with KOH/C = 1 (**b**), 2 (**c**), 3 (**d**), 5 (**e**)

(see Fig. 4b), which are mainly large mesopores and/or macropores according to the remarkable N_2 uptake at high P/P_0 and the distinct hysteresises (see Fig. 2a). KOH activation with KOH/C = 2 decreases the mesopores and/or macropores through destructing the framework between graphene layers, which gives rise to the development of micropores through breaking the stacking graphene layers into smaller sizes and few layers (see Fig. 4c). The marked development of micropores is achieved with the increase of KOH/C ratio to 3 through the further breaking of the graphene layers (see Fig. 4d). Activation with KOH/C = 5 exfoliates more graphene layers, giving rise to the enlargement of micropores and the development of mesopores (see Fig. 4e). Therefore, graphene-based carbon with a target porosity can be designed according to this activation model, although the porosity contribution from the edge carbon should be further considered.

3 High Surface Area Graphene Monoliths with Shape-Retaining KOH Activation

The high surface area PGM with tunable porosity can be fabricated through a KOH activation process as introduced in Sect. 2. However, the structural deformation of PGM during the preparation process limits its application in energy and environmental issues. A shape-retaining route [29] can solve the structural deformation problem effectively through freeze-drying the sample at each drying process, giving a much improved structural designability. Firstly, a cylindrical RGO monolith with diameter and length of about 1 and 3 cm (see Fig. 5a) is prepared by using the method as introduced in Sect. 2. The RGO monolith is then filled with KOH solution of the KOH to carbon weight ratio of 10/1, and quickly frozen by liquid N_2 and freeze-dried

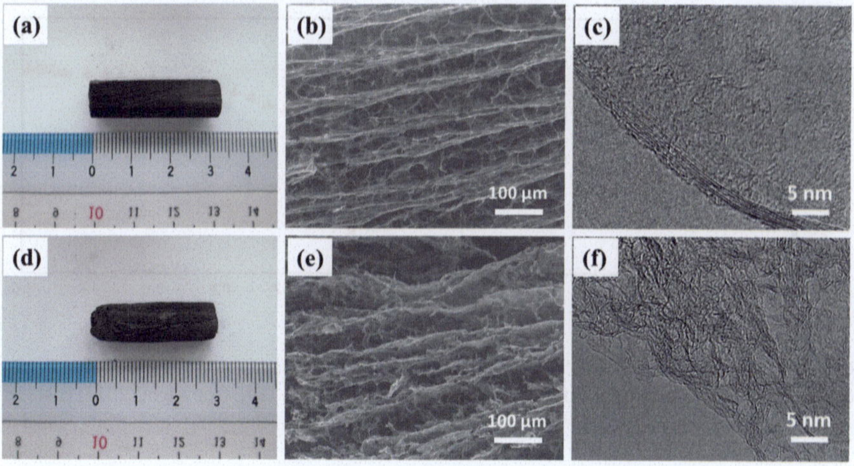

Fig. 5 Morphology of RGO monolith (**a–c**) and PGM (**d–f**) obtained after shape-retaining activation. **a**, **d** Photos, **b**, **e** SEM images, **c**, **f** TEM images. Reprinted (adapted) with permission from [29]. Copyright (2017) Elsevier

for 24 h under vacuum condition (<10 Pa). The KOH-loaded monolith is activated at 973 K followed by washing with 0.1 M HCl and deionized water. The product is frozen by liquid N_2 and freeze-dried in vacuum to avoid the structural deformation. The obtained PGM (see Fig. 5d) can largely maintain its original geometry with less than 10% volume shrinkage. The bulk density evaluated is 8 ± 0.5 mg cm^{-3}, being much lower than conventional activated carbons.

The microscopic morphology of RGO monolith can be also well-preserved after the activation process. Pristine RGO monolith has unidirectional texture structures parallel to the ice growth direction with inter-sheets bridging each other (see Fig. 5b), being similar to the reported materials produced by ice-freezing method [37, 38]. Such parallel textures are well preserved even after the activation process, although some parts are slightly distorted (see Fig. 5e). The TEM observation indicates that pristine RGO monolith consists of large and integrated graphene sheets of a few layers (see Fig. 5c), which are exfoliated into thinner and smaller defective graphene sheets after the KOH activation process (see Fig. 5f). Even so, the entangled and interconnected graphene sheets of considerably large size guarantee the free-standing nature of PGM.

The porosity development of PGM through KOH activation is evidenced by the N_2 adsorption isotherms at 77.4 K. Pristine RGO monolith has a N_2 adsorption isotherm of type IV(a) with an H4 hysteresis loop (see Fig. 6a), suggesting the presence of slit-shaped mesopores. After the shape-retaining activation, the isotherm changes into type I(b) (see Fig. 6b), being indicative of the development of large micropores. The surface area of the PGM increases from 26 to 990 m^2 g^{-1}. Thus-prepared monolith has wide application potential in the fields which require both fast mass-transfer and

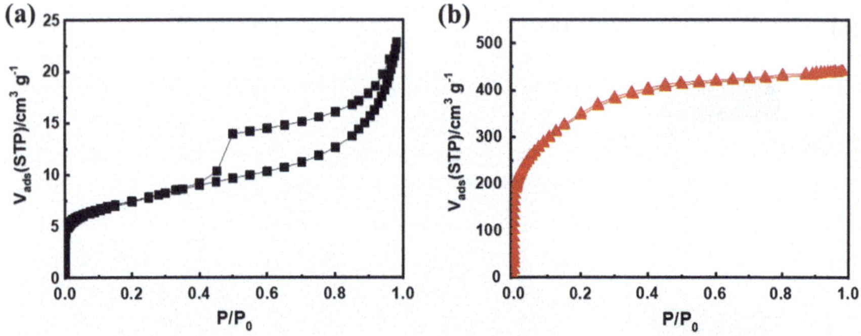

Fig. 6 N_2 adsorption isotherms (77.4 K) of RGO monolith (**a**) and PGM obtained after shape-retaining activation (**b**). Reprinted (adapted) with permission from [29]. Copyright (2017) Elsevier

efficient adsorption due to its unique hierarchical porous structure with both marked microporosity and macroporosity.

4 CO_2 Activation Effect of Graphene Colloids

Porosity development through CO_2 activation is achieved by extraction of carbon atoms from the carbon precursor during the gasification reaction. The RGO monolith introduced in Sect. 2 can be activated through heat-treatment in a 30% CO_2/Ar flow (200 mL min^{-1}) [30]. The N_2 adsorption isotherms of RGO monoliths activated at 1223 K for 1 and 4 h are both of type IV (see Fig. 7a). However, the adsorption isotherm of RGO monoliths activated for 4 h has a steeper uptake at high P/P_0 region comparing with that activated for 1 h; the hysteresis loop changes from H4 to H3, indicating that longer activation time induces the broadening of mesopores. The surface area and micropore volume of CO_2-activated RGO monoliths (see Fig. 7c) are much smaller than those of the reported works [39, 40].

Marsh and Rodríguez-Reinoso [41] reported that even trace amount of metal can play a significant catalytic effect in the gasification of carbon. The GO colloid used in this study experienced sufficient washing during the preparation process, then the metal content in RGO monoliths should be very low. In order to testify the effect of metal on the porosity development, CO_2 activation is conducted on RGO monoliths loaded with Ni or Cu at mole ratio of 1/1000 and 1/200. However, the porosity development is still small by all means tried (see Fig. 7b, d), suggesting these metal catalysts are not efficient for the porosity development. The innerness of RGO for the metal catalyst-aided gasification should be ascribed to the special structure of graphene units. It is known that at the edges of the graphene planes, CO_2 molecules undergo a reaction 10^2–10^3 order faster than those on the basal planes [41]. The RGO monoliths prepared in this work has more basal planes and less prismatic

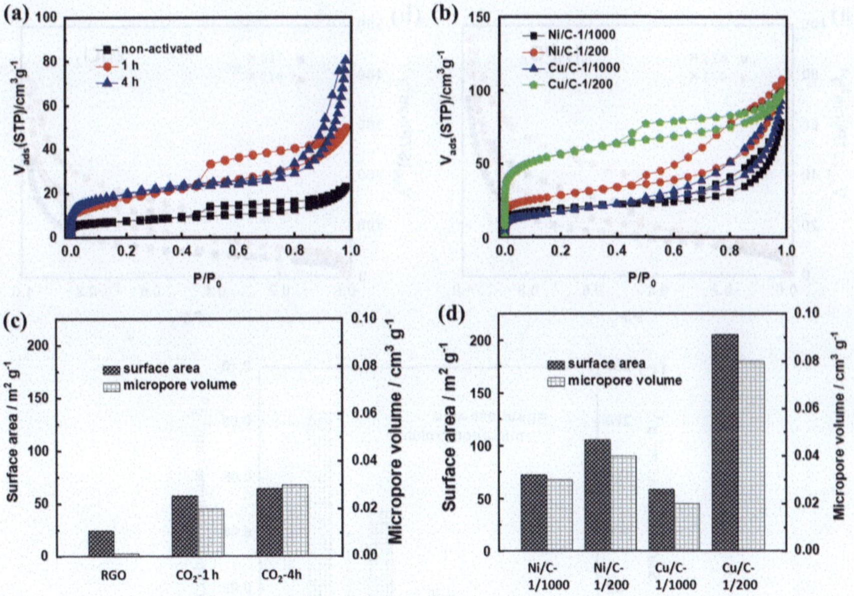

Fig. 7 N_2 (77.4 K) adsorption isotherms (**a, b**) and porosity parameters (**c, d**) of RGO monoliths treated by CO_2 activation. **a, c** RGO activated at 1223 K for 1 and 4 h, **b, d** metal-loaded RGO activated at 1223 K for 2 h. Reprinted (adapted) with permission from [30]. Copyright (2014) Elsevier

edges comparing with conventional carbons, then CO_2 activation is not an effective method for RGO.

5 $ZnCl_2$ and H_3PO_4 Activation Effects for Graphene Colloids

$ZnCl_2$ and H_3PO_4 are used for developing pores on GO monoliths [30] due to their tremendous dehydrating effect. Similar to the preparation and activation of KOH-contained GO monoliths as described in Sect. 2, the $ZnCl_2$ (or H_3PO_4) -contained GO monoliths are prepared from the mixture of GO suspension and $ZnCl_2$ (or H_3PO_4) by using the ice-templating and freeze-drying method, then heated under Ar atmosphere for 1 h. The weight ratio of carbon to chemical reagents is adjusted to be 1. The $ZnCl_2$-contained monoliths are activated at 773 and 873 K, while the H_3PO_4-contained monoliths are activated at 623 and 673 K. The amounts of chemical reagents used and the activation conditions are adopted from the optimized conditions reported [42–45].

The isotherms of both H_3PO_4- and $ZnCl_2$-activated RGO monoliths are of typical type IV according to the IUPAC classification (see Fig. 8a, b). The relatively small

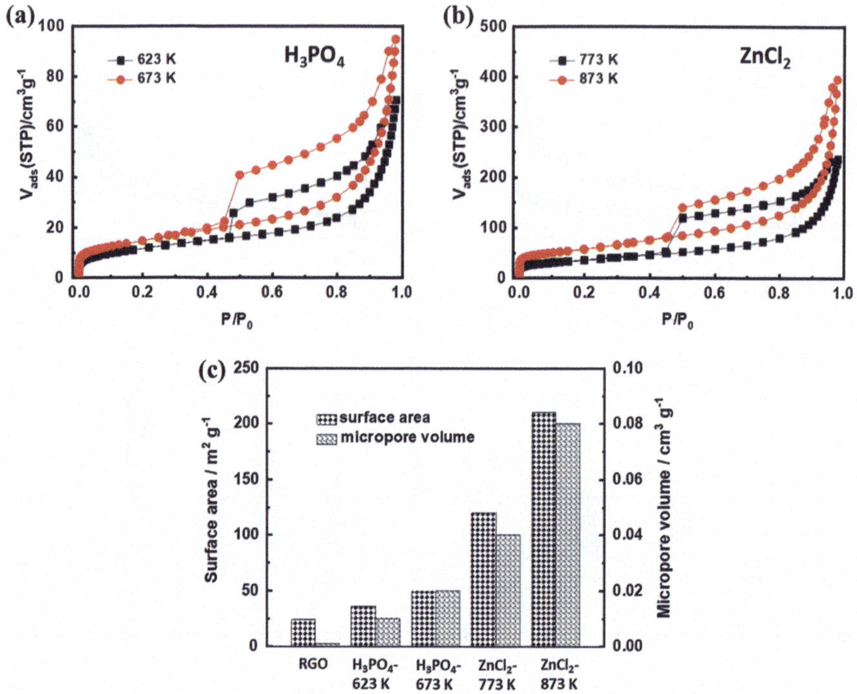

Fig. 8 N_2 (77.4 K) adsorption isotherms of RGO monolith activated by **a** $ZnCl_2$ and **b** H_3PO_4 at various temperatures for 1 h, and their corresponding porosity parameters (**c**). Reprinted (adapted) with permission from [30]. Copyright (2014) Elsevier

surface area determined from SPE method and the typical type IV isotherm suggest their micropores are not sufficiently developed. The distinct hysteresis loops with a steep uptake are observed at high P/P_0, being indicative of the presence of large mesopores. The N_2 uptake amounts of $ZnCl_2$-activated RGO monoliths are higher than those of H_3PO_4-activated RGO monoliths, suggesting that $ZnCl_2$ induces better porosity development than H_3PO_4. Nevertheless, the surface area and micropore volume obtained (see Fig. 8c) are smaller than those in the previously studies [42–45]. The porosity development of carbon through $ZnCl_2$ (or H_3PO_4) activation should largely depends on the precursors used. For vegetal precursors, the dehydration of the cellulose, hemicellulose and lignin compounds by $ZnCl_2$ or H3PO_4 leaves vacancy on the carbon matrix, giving rise to the significant porosity development. On the other hand, GO-based precursors have more ordered structure than vegetal precursors at atomic level, giving the observed stability on the activation. However, porous RGO monoliths prepared from $ZnCl_2$-activaiton have significant mesopores, being promising as catalyst supports or adsorbents for biomacromolecules.

6 Tuning Porosity Through Mechanical Compression

The porosity of PGM is effectively tuned by the mechanical compression process [46] due to the excellent elastic property of graphene [47, 48]. PGM prepared by KOH activation method (KOH/C = 10, 1073 K) as introduced in Sect. 2 is compressed at 43 MPa over different time period from 3 to 25 h to obtain graphene monoliths with various pore structures. The sample compressed for β h is nominated as PGM-βh. Non-compressed PGM has abundant macroporous structure which is constructed by the interconnected carbon layers with ruffled surfaces (see Fig. 9a). Compression for 25 h degrades the macropores and converts the disoriented carbon layers into stacked walls (see Fig. 9d). The morphology conversion of PGM at nanoscale indicates the folded graphene sheets changes to smooth and flat layers (see Fig. 9b, e). TEM examination indicate that the fundamental structure of graphene units is well preserved after the compression process (see Fig. 9c, f).

The Ar adsorption isotherms at 87 K are more appropriate for micropore characterization than N_2 isotherms at 77 K due to its smaller molecular size (the Lennard–Jones size parameter of Ar and N_2 molecules are 0.341 nm and 0.375 nm, respectively) which is suitable for narrow micropore, its non-polar feature which can avoid the underestimation of micropore size as observed in N_2 adsorption isotherms, and the higher measuring temperature which enables smooth intrapore diffusion [49, 50]. The compression process decreases both Ar and N_2 adsorbed amounts (see Fig. 10), especially in the higher P/P_0 region, and shrinks the adsorption hysteresis loop in their isotherms, suggesting the degradation of mesopores. Sample compressed more

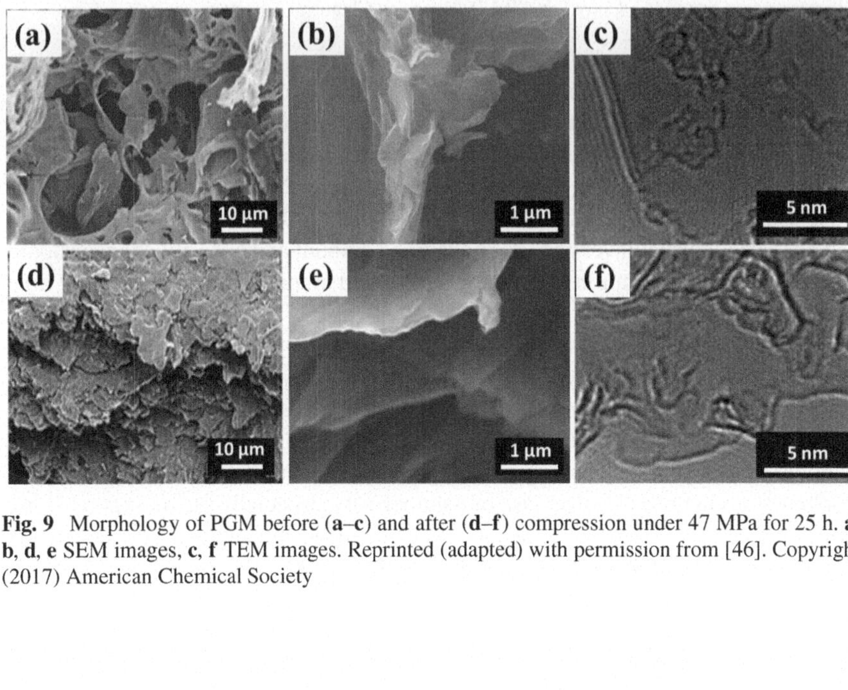

Fig. 9 Morphology of PGM before (**a–c**) and after (**d–f**) compression under 47 MPa for 25 h. **a**, **b**, **d**, **e** SEM images, **c**, **f** TEM images. Reprinted (adapted) with permission from [46]. Copyright (2017) American Chemical Society

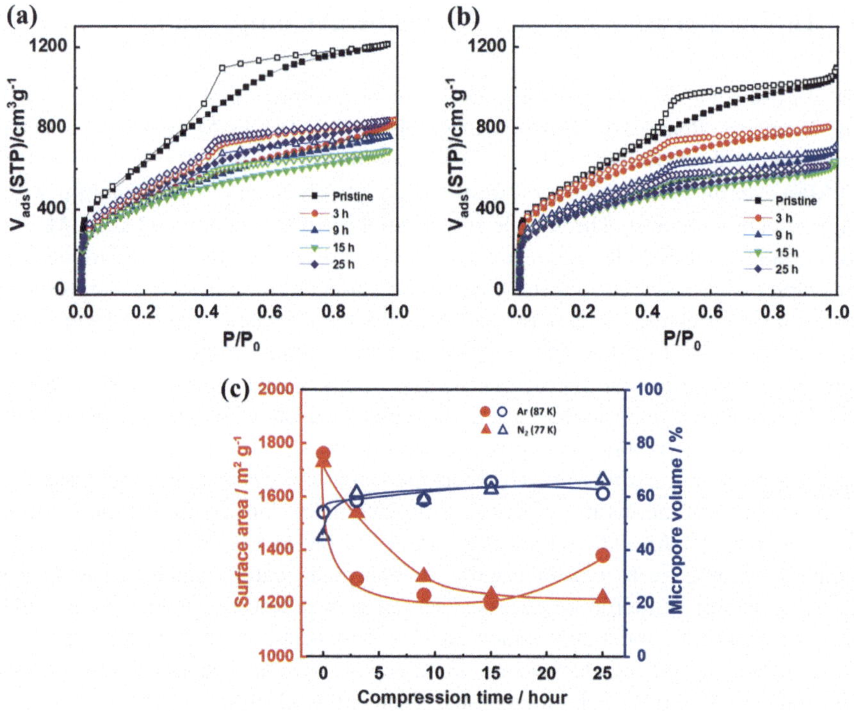

Fig. 10 Ar adsorption isotherms at 87 K (**a**), N_2 adsorption isotherms at 77 K (**b**) and variations of surface area and micropore volume (**c**) of PGM with the compression time. Reprinted (adapted) with permission from [46]. Copyright (2017) American Chemical Society

than 9 h has an adsorption isotherm close to Type I(b) according to IUPAC classification [49], indicating the predominant presence of micropores. Meanwhile, the compression induces low-pressure hysteresis which relates to the restricted diffusion in ultramicropore. Extending the compression time up to 25 h gives rise to the increase of Ar adsorption amount, which is less remarkable on the N_2 adsorption isotherm. The observed unusual inversion of Ar adsorbed amount and porosity change should associate with the relaxation of stacking nano-sized graphenes induced by long time compression.

A relaxation model of stacked nano-sized graphenes with elasticity is proposed (see Fig. 11). The interstices in the randomly aggregated nano-sized graphenes mainly offer mesopores and the intergraphene spaces provide micropores; the mesoporosity is much predominant compared with the microporosity. The initial compression leads to the bending of nanographene sheets between other nano-sized graphenes, giving rise to a decrease of nanoporosity, especially the mesopore. Long time compression restores the bent nanographene sheets into mutually stacked layers with ultramicropore where Ar molecules can, but N_2 molecules cannot access. Thus,

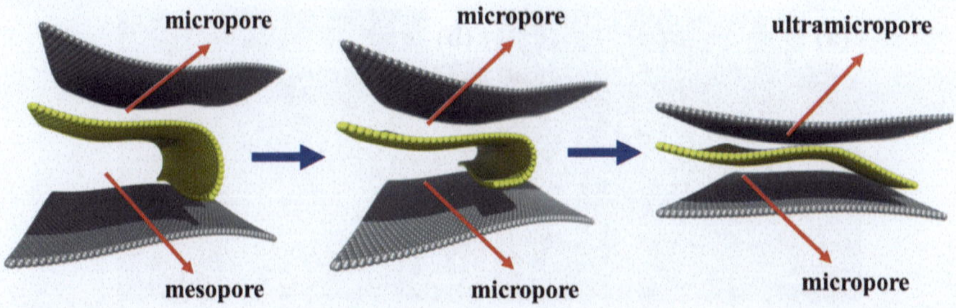

Fig. 11 Relaxation model of a bent nano-sized graphenes under long-time compression. Reprinted (adapted) with permission from [46]. Copyright (2017) American Chemical Society

the discrepancy observed between Ar and N_2 isotherms stems from the high elasticity of nano-sized graphenes.

7 Tuning of Porosity and Structure Through Enforced Crystallization at High Temperature

High temperature treatment improves the graphitization and reduces the defects of carbon material effectively, but decreases their porosity at the same time. It is quite difficult to get highly crystalline carbons having an extremely high porosity. We must search an optimum heating temperature of carbon materials having considerably high crystallinity and high surface area. The unit graphene size of PGM is much larger than that of the conventional porous carbons. The PGM has a soft carbon nature and thereby high temperature treatment of the PGM enables to give a highly porous carbon consisting of well-crystalline graphenes [51]. PGMs prepared from KOH activation (KOH/C = 8, 1073 K) are heat-treated at different temperatures under Ar atmosphere. The pristine PGM mainly consists of disoriented graphene sheets entangling with each other (see Fig. 12a). Heat-treatment at 1873 K converts the disoriented graphenes into partially stacked graphene layers with wrinkled surfaces (see Fig. 12b), which is a transitional state between distorted graphene networks and ordered graphitic structures. Heat-treatment at 2273 K produces stacking graphene layers with improved uniformity (see Fig. 12c), which further reconstruct themselves into highly crystalline graphitic structure at 2673 K with the coexistence of disordered graphitic parts (see Fig. 12d), i.e., the growth of highly ordered graphitic structure should only consume the crystallizable part, leaving the disordered structures.

The heat-treatment below 1373 K doesn't change the N_2 adsorption isotherms, but that at temperature higher than 1373 K decreases the N_2 adsorbed amount with elevation of the heating temperature (see Fig. 13a). The decrease of the N_2 uptake at the submonolayer region suggests the degradation of micropores (see Fig. 13b). The hysteresis loop changes from Type H4 to Type H2 at the heating temperature above 1473 K, indicting the heat-treatment induces narrow pore necks on PGM [49].

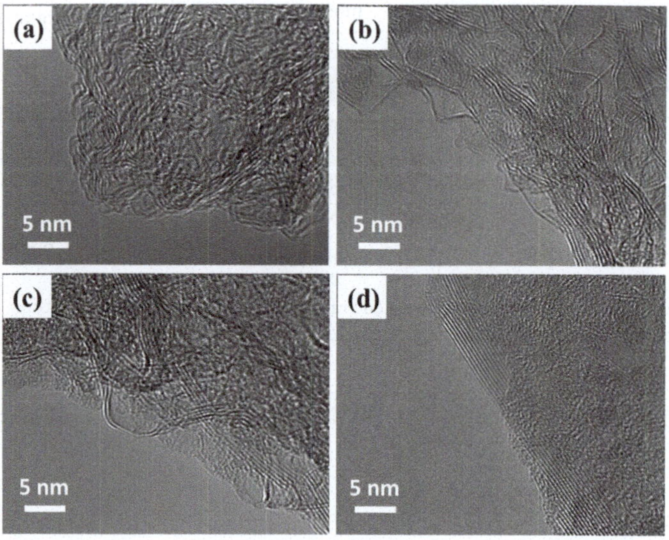

Fig. 12 HR-TEM images of **a** PGM, **b** PGM-1873, **c** PGM-2273, and **d** PGM-2673. The porous graphene monolith heated at T K is nominated as PGM-T

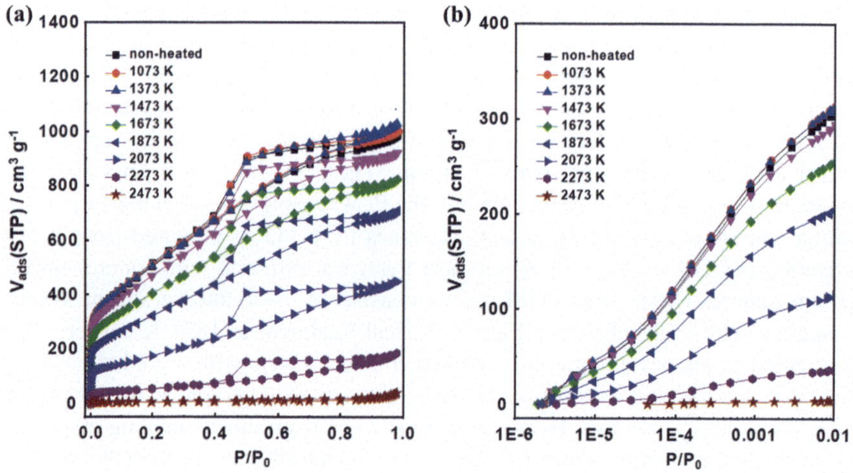

Fig. 13 N_2 adsorption isotherms (77.4 K) of heat-treated PGM represented in linear (**a**) and logarithmic (**b**) coordinates. Reprinted (adapted) with permission from [51]. Copyright (2016) Elsevier

The hysteresis loop of the sample heated at 2073 K and 2273 K is still evident. The heat-treatment at 2473 K degrades the mesopore structures, losing the adsorption hysteresis.

The evolution of the surface area and pore volume with heating temperature are illustrated in histogram for a better comparison. The surface area doesn't change significantly with heating temperature up to 1373 K, then decreases at the heating temperature higher than 1373 K (see Fig. 14a). The heat-treatment at 2473 K reduces the surface area to only 20 $m^2 g^{-1}$. The changes of total, micro- and mesopore volumes are similar to that of the surface area (see Fig. 14b). The porosity change with the heating temperature indicates a structural transformation of porous and disoriented graphene structures into nonporous stacking graphene layers structures.

The Raman spectroscopy and X-ray diffraction (XRD) evidence the structural transformation of PGM with the heating temperature (see Fig. 15). Pristine PGM has

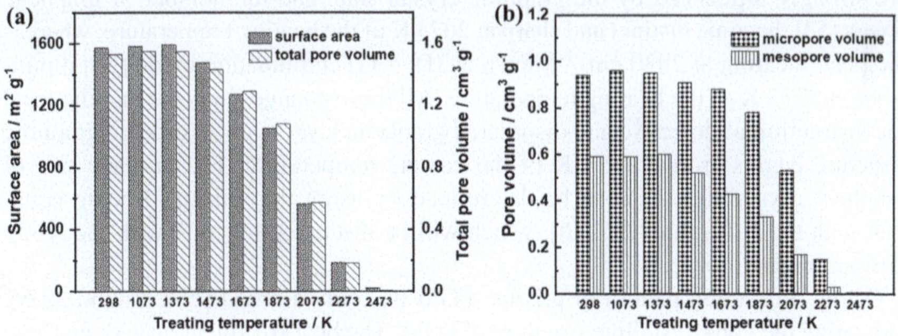

Fig. 14 Porosity variation of PGM heat-treated at different temperatures. **a** Total surface area and total pore volume, **b** micro- and mesopore volumes

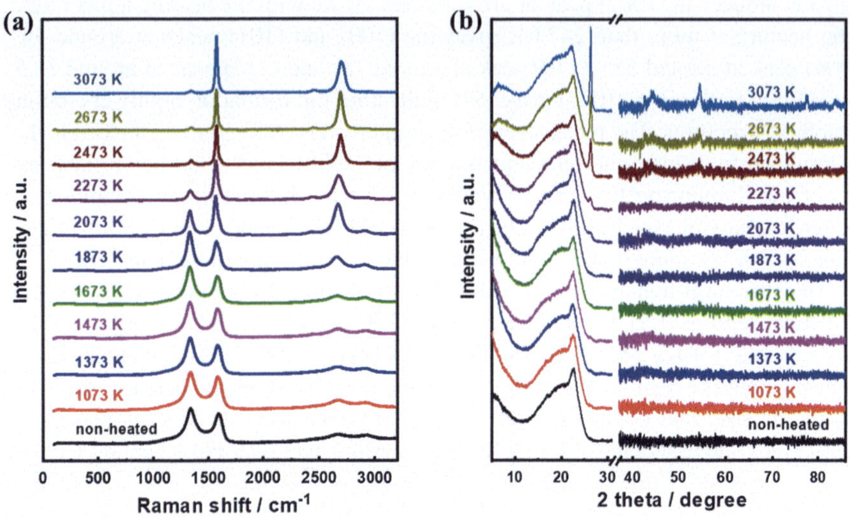

Fig. 15 Raman spectra (**a**) and XRD patterns (**b**) of PGM heat-treated at different temperatures. Reprinted (adapted) with permission from [51]. Copyright (2016) Elsevier

two overlapping broad bands locating at ~1350 cm^{-1} (D band) and ~1590 cm^{-1} (G band), which correspond to the defective part of carbon and the stretching vibration in the aromatic planes, respectively. These broad D and G bands are typical features of Raman spectrum of amorphous carbons. The heat-treatment below 1873 K doesn't change the Raman spectra significantly, while that above 1873 K leads to an evident decrease of D band and increase of G band; the I_D/I_G ratio decreases from 1.52 to 0.11 with the heating temperature increase form 2073 to 3073 K. The structural transformation from Raman spectroscopy begins around 1873 K, being higher than the temperature suggested by the porosity change. Meanwhile, the peak at 2450 cm^{-1} that corresponds to the overtone mode longitudinal optical phonons [52] and the peak at 2700 cm^{-1} (2D band, also called G′ band) whose frequency and intensity are strongly influenced by the graphitic crystal state and the number of graphene layers [53], become distinct and sharp at 2073 K of the heating temperature, whereas the peak locating at 2930 cm^{-1} known as (D + G) combination mode [54] diminishes at 2273 K of the heating temperature. All these changes indicate that the local reconstruction of defective and disordered graphene layers into an ordered graphitic structure begins around 1873 K of the heating temperature. The in-plane size of graphitic crystallites (L_a) can be determined by using the Tuinstra-Koenig equation with the I_D/I_G ratio [55, 56], which will be discussed together with the X-ray diffraction results.

X-ray diffraction pattern of pristine PGM has only a broad peak around 22.6°, indicating the non-crystalline structure of PGM. The heat-treatment doesn't give rise to a significant change in the X-ray diffraction patterns below 1873 K, suggesting the mutual ordering of graphene layers needs to overcome a high energy barrier corresponding to the thermal energy of 1873 K (~16 kJ mol^{-1}). Higher temperature above 1873 K induces the (002) peak at 26.4°, which grows with the heating temperature. The heating at more than 2473 K gives the (100) and (101) peaks at around 44°, (004) peak at around 55°, (110) peak at around 78° and (112) peak at around 83.5° in addition to the sharp (002) peak [57], indicating the formation highly crystalline graphitic structures. The mean sizes of the graphitic crystallites along the c axis (L_c) determined by using Scherrer's equation are plotted together with the in-plane size (L_a) from Raman spectroscopy as well as the micro- and mesopore volumes for a better comparison (see Fig. 16). The L_a values of PGM heated at temperature higher than 2473 K are more than 20 nm, whereas the L_c values are more than 12 nm.

The structural transformation of PGM with heating temperature is divided into 4 distinct regions: (A) $T < 1373$ K, (B) $1373 \text{ K} < T < 1873$ K, (C) $1873 \text{ K} < T < 2473$ K, and (D) $T > 2473$ K. Both porosity and crystallite size don't change significantly within the region (A). In region (B), the porosity starts to decrease, while the crystallite size does not change; the energy required for the stacking of disoriented less-crystalline graphenes is larger than the thermal energy at these heating temperatures. However, partial breaking of the bridging between the graphene units begins, collapsing part of the pores. In region (C), the decrease of porosity and increase of crystallite size occur simultaneously. The PGM having both considerably large porosity and moderate crystallinity can be obtained at this region. In region (D),

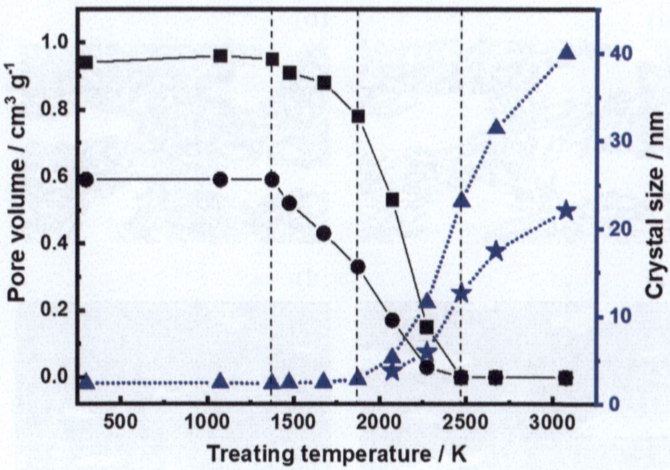

Fig. 16 Comparison between the crystallite size and pore volume of PGM heat-treated at different temperatures. Mesopore volume: ■, micropore volume: ●, in-plane size of graphite crystallites L_a: ▲, mean crystallite size along the c axis L_c: ★. Reprinted (adapted) with permission from [52]. Copyright (2016) Elsevier

almost all pores collapse due to complete losing of the three-dimensional graphene-graphene bridges, while L_a and L_c increase markedly with the heating temperature due to growth of the stacking structures of graphene through coalescence of graphene crystallites. PGM treated at 3073 K still has disordered parts according to XRD patterns and TEM images, suggesting the coexistence of graphitizable and non-graphitizable carbons in PGM.

The structural transformation model of PGM on high temperature treatment is proposed (see Fig. 17) on the basis of the experimental results shown above. The non-heated PGM consists of highly disoriented nanographene sheets which bridge each other, providing micropores and mesopores. Partial stacking of nanographene layers leads to the collapse of micropores and mesopores on heating above 1373 K. Heating at temperature higher than 1873 K induces the coalescence of nanographenes, giving a remarkable growth of the stacked graphitic layers and losing a predominant porosity. A long range ordered graphitic structure is formed at 2473 K, whereas almost all open nanoporosity disappears.

The electrical conductivity reflects sensitively the graphitic unit-graphitic unit channel structure, being useful to understand the structural transformation of PGM on the high temperature heating. The electrical conductivity of PGM is measured by using a physical property measurement system (PPMS) over the temperature range of 2–370 K (see Fig. 18a). By losing the 18% surface area through the heat-treatment, PGM-1673 gains 36, 52 and 200% increases of electrical conductivity at 273 K, 77.4 K (at liquid N_2) and 4.2 K (at liquid He), respectively. Similarly, PGM-2073 gains 56, 102 and 400% increases of the electrical conductivity at the same temperature shown above. The $\ln\sigma T^{1/2}$ versus $T^{-1/4}$ plots for pristine and heat-treated PGM

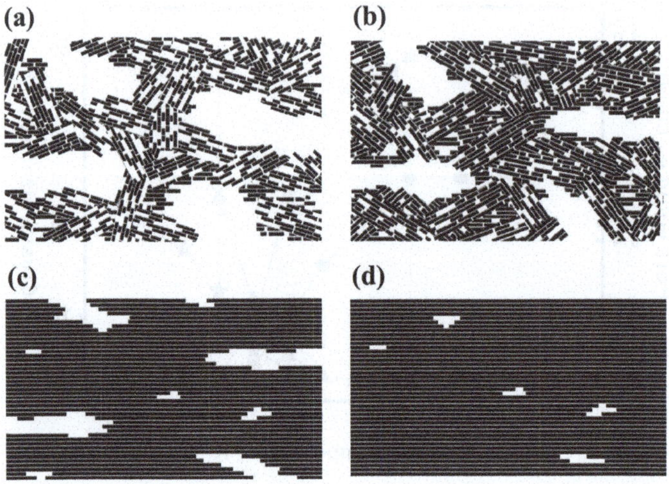

Fig. 17 Structural transformation model of PGM heat-treated at different temperature regions. **a** T < 1373 K, **b** 1373 K < T < 1873 K, **c** 1873 K < T < 2473 K, and **d** T > 2473 K. Reprinted (adapted) with permission from [51]. Copyright (2016) Elsevier

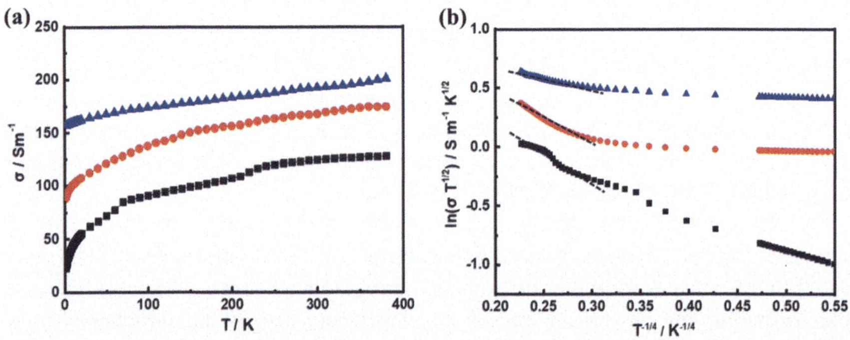

Fig. 18 The temperature dependences of electrical conductivity of pristine and heat-treated PGM represented in **a** linear and **b** $\ln\sigma T^{1/2}$ versus $T^{-1/4}$ plots. The ordinate values of PGM-1673 and PGM-2073 in **b** are upshifted by 0.3 and 0.6, respectively. PGM: ■, PGM-1673: ●, PGM-2073: ▲. Reprinted (adapted) with permission from [51]. Copyright (2016) Elsevier

samples are quite different from each other (see Fig. 18b). The $\ln\sigma T^{1/2}$ versus $T^{-1/4}$ plot of pristine PGM has no linear fit at 100–300 K, while those of PGM-1673 and PGM-2073 show more linear fit at the same temperature range, which is related to the variable-range hopping conduction [58, 59]. The electrical conductivities of PGM-1673 and PGM-2073 show similar temperature dependence to that of polycrystalline graphite, suggesting those heat-treated PGM have a semimetallic nature [60, 61]. N_2 adsorption, Raman spectroscopy and X-ray diffraction studies (see Fig. 16) show that PGM-1673 has an imperfect graphitic structure. However, the electrical conductivity data on PGM-1673 indicates that PGM-1673 has well-combined electrical channels through highly crystalline graphitic layers as well as PGM-2073. N_2 adsorption, Raman spectroscopy and X-ray diffraction give the average structure information, while the electrical conductivity over the wide temperature range is governed by the most electrically contacted nanographene channels. Therefore, we understand that even not-well crystalline PGM like PGM-1673 has highly electrical conductivity linkages though chain-like connections of nano-sized graphene units. Thus, the electrical conductivity provides a unique structural information.

8 Conclusion

A variety of methods have been proposed for the porosity development of carbon materials in last few decades, including the chemical reagent activation with KOH, $ZnCl_2$ or H_3PO_4 and physical activation with CO_2. However, when these methods are applied to graphene-derived carbon, we obtain very different nanoporosity developments depending on the activation route. For example, chemical activation with $ZnCl_2$ or H3PO_4 and physical activation with CO_2 cannot develop micropores on graphene-derived carbon effectively. On the contrary, KOH activation of graphene-based carbon is efficient for activation of graphene-derived carbon. The KOH activation procedure makes graphenes to be finely divided to provide micropores and to produce mesopores by reconstruction of the assemblies of the fine graphenes.

On the other hand, a long compression of highly porous graphene leads to an unusual restoring of the bent nano-sized graphenes due to their high elasticity. Highly porous carbon with well-crystalline graphenes are obtained by heat-treatment of the porous graphene monolith (PGM) of a soft carbon nature. The heat-treated PGM exhibits a variable-range hopping conduction as well as high surface area, being a promising material as electrodes.

Although the porosity and structure of graphene-based carbon can be considerably well tuned, the graphene-based porous carbons cannot meet sufficiently the demands from industries and fundamental studies. This review gives a useful guidepost for designing and preparation of highly customized porous carbons.

Acknowledgements KK expresses heartfelt thanks to Professor F. Rodríguez-Reinoso for long-term friendship, encouraging him. Professor F. Rodríguez-Reinoso suddenly passed way. Then, KK would like to express his thanks to Professor Juan Carlos Moreno-Piraján for editing this book in the honor of Professor F. Rodríguez-Reinoso. This book is a splendid memorial book for Professor F. Rodríguez-Reinoso and his friends. This work was supported by the OPERA project from Japan Science and Technology Agency.

References

1. Otowa, T., Nojima, Y., Miyazaki, T.: Development of KOH activated high surface area carbon and its application to drinking water purification. Carbon **35**, 1315–1319 (1997)
2. Kaneko, K., Rodríguez-Reinoso, F.: Nanoporous Materials for Gas Storage. Springer Nature, Singapore (2019)
3. Kaneko, K., Ishii, C., Ruike, M., Kuwabara, H.: Origin of superhigh surface area and microcrystalline graphitic structures of activated carbons. Carbon **30**, 1075–1088 (1992)
4. Kaneko, K.: Determination of pore size and pore size distribution 1: adsorbents and catalysts. J. Membr. Sci. **96**, 59–89 (1994)
5. Setoyama, N., Suzuki, T., Kaneko, K.: Simulation study on relationship between high resolution αs-plot and pore size distribution for activated carbon. Carbon **36**, 1459–1467 (1998)
6. Kim, T., Jung, G., Yoo, S., et al.: Activated graphene-based carbons as supercapacitor electrodes with macro-and mesopores. ACS Nano **7**, 6899–6905 (2013)
7. Zhu, Y., Murali, S., Stoller, M.D., et al.: Carbon-based supercapacitors produced by activation of graphene. Science **332**, 1537–1541 (2011)
8. Wang, Y., Xia, Y.: Recent progress in supercapacitors: from materials design to system construction. Adv. Mater. **25**, 5336–5342 (2013)
9. Salanne, M., Rotenberg, B., Naoi, K., Kaneko, K., et al.: Efficient storage mechanisms for building better supercapacitors, Nat. Energy 1, Article number: 16070 (2016)
10. Bekyarova, E., Murata, K., Yudasaka, M., Kasuya, D., et al.: Single-wall nanostructured carbon for methane storage. J. Phys. Chem. B **107**, 4681–4684 (2003)
11. Casco, M.E., Silvestre-Albero, J., Ramírez-Cuesta, A.J., et al.: Methane hydrate formation in confined nanospace can surpass nature. Nat. Commun. **6**, Article number: 6432 (2015)
12. Choi, S., Alkhabbaz, M.A., Wang, Y., et al.: Unique thermal contraction of zeolite-templated carbons enabling micropore size tailoring and its effects on methane storage. Carbon **141**, 143–153 (2019)
13. Chmiola, J., Yushin, G., Gogotsi, U., et al.: Anomalous increase in carbon capacitance at pore sizes less than 1 nanometer. Science **313**, 1760–1763 (2006)
14. Ohkubo, T., Konishi, T., Hattori, Y., et al.: Restricted hydration structures of Rb and Br ions confined in slit-shaped carbon nanospaces. J. Am. Chem. Soc. **124**, 11860–11861 (2002)
15. Futamura, R., Iiyama, T., Takasaki, Y., et al.: Partial breaking of the Coulombic ordering of ionic liquids confined in carbon nanopores. Nat. Mater. **16**, 1225–1232 (2017)
16. Novoselov, K.S., Geim, A.K., Morozov, S.V., et al.: Electric field effect in atomically thin carbon films. Science **306**, 666–669 (2004)
17. Meyer, J.C., Geim, A.K., Katsnelson, M.I., et al.: The structure of suspended graphene sheets. Nature **446**, 60–63 (2007)
18. Geim, A.K., Novoselov, K.S.: The rise of graphene. Nat. Mater. **6**, 183–191 (2007)
19. Kim, J.Y., Lee, J.H., Grossman, J.C.: Thermal transport in functionalized graphene. ACS Nano **6**, 9050–9057 (2012)
20. Malekpour, H., Chang, K.H., Chen, J.C., et al.: Thermal conductivity of graphene laminate. Nano Lett. **14**, 5155–5161 (2014)
21. Peng, L., Xu, Z., Liu, Z., et al.: Ultrahigh thermal conductive yet super flexible graphene films. Adv. Mater. **29**, 1700589 (2017)
22. Berger, C., Song, Z., Li, X., Wu, X., et al.: Electronic confinement and coherence in patterned epitaxial graphene. Science **312**, 1191–1196 (2006)
23. Heersche, H.B., Jarillo-Herrero, P., Oostinga, J.B., et al.: Bipolar supercurrent in graphene. Nature **446**, 56–59 (2006)
24. Qiao, Z., Jung, J., Niu, Q., MacDonald, A.H.: Electronic highways in bilayer graphene. Nano Lett. **11**, 3453–3459 (2011)
25. Park, S., Lee, K.S., Bozoklu, G., et al.: Graphene oxide papers modified by divalent ions-enhancing mechanical properties via chemical cross-linking. ACS Nano **2**, 572–578 (2008)
26. Rafiee, M.A., Rafiee, J., Wang, Z., et al.: Enhanced mechanical properties of nanocomposites at low graphene content. ACS Nano **3**, 3884–3890 (2009)

27. Suk, J.W., Piner, R.D., An, J., Ruoff, R.S.: Mechanical properties of monolayer graphene oxide. ACS Nano **4**, 6557–6564 (2010)
28. Nardecchia, S., Carriazo, D., Ferrer, M.L., et al.: Three dimensional macroporous architectures and aerogels built of carbon nanotubes and/or graphene: synthesis and applications. Chem. Soc. Rev. **42**, 794–830 (2013)
29. Wang, S., Wang, Z., Futamura, R., et al.: Highly microporous-graphene aerogel monolith of unidirectional honeycomb macro-textures. Chem. Phys. Lett. **673**, 38–43 (2017)
30. Wang, S., Tristan, F., Minami, D., et al.: Activation routes for high surface area graphene monoliths from graphene oxide colloids. Carbon **76**, 220–231 (2014)
31. Endo, M., Furuta, T., Minoura, F., et al.: Visualized observation of pores in activated carbon fibers by HRTEM and combined image processor. Supramolecul. Sci. **5**, 261–266 (1998)
32. Harris, P.J.F., Liu, Z., Suenaga, K.: Imaging the atomic structure of activated carbon. J. Phys.: Condens Matter **20**, 362201–362205 (2000)
33. Wang, H., Gao, Q., Hu, J.: High hydrogen storage capacity of porous carbons prepared by using activated carbon. J. Am. Chem. Soc. **131**, 7016–7022 (2009)
34. Dubinin, M.M.: Fundamentals of the theory of adsorption in micropores of carbon adsorbents: characteristics of their adsorption properties and microporous structures. Carbon **27**, 457–467 (1989)
35. Dollimore, D., Heal, G.R.: An improved method for the calculation of pore size distribution from adsorption data. J. Appl. Chem. **14**, 109–114 (1964)
36. Dollimore, D., Heal, G.R.: Pore-size distribution in typical adsorbent systems. J. Colloid Interface Sci. **33**, 508–519 (1970)
37. Mukai, S.R., Nishihara, H., Tamon, H.: Formation of monolithic silica gel microhoneycombs (SMHs) using pseudosteady state growth of microstructural ice crystals. Chem. Commun. **2004**, 874–875 (2004)
38. Zhang, N., Qiu, H., Si, Y., Wang, W., Gao, J.: Fabrication of highly porous biodegradable monoliths strengthened by graphene oxide and their adsorption of metal ions. Carbon **49**, 827–837 (2011)
39. Molina-Sabio, M., Gonzalez, M.T., Rodríguez-Reinoso, F., Sepúlveda-Escribano, A.: Effect of steam and carbon dioxide activation in the micropore size distribution of activated carbon. Carbon **34**, 505–509 (1996)
40. Turmuzi, M., Daud, W.R.W., Tasirin, S.M., et al.: Production of activated carbon from candlenut shell by CO_2 activation. Carbon **42**, 453–455 (2004)
41. Marsh, H., Rodríguez-Reinoso, F.: Activated Carbon, pp. 268, 330–331. Elsevier, London. (2006)
42. Olivares-Marín, M., Fernández-González, C., Macías-García, A., Gómez-Serrano, V.: Preparation of activated carbon from cherry stones by chemical activation with $ZnCl_2$. Appl. Surf. Sci. **252**, 5967–5971 (2006)
43. Yorgun, S., Vural, N., Demiral, H.: Preparation of high-surface area activated carbons from Paulownia wood by $ZnCl_2$ activation. Micropor. Mesopor. Mat. **122**, 189–194 (2009)
44. Teng, H., Yeh, T.S., Hsu, L.Y.: Preparation of activated carbon from bituminous coal with phosphoric acid activation. Carbon **36**, 1387–1395 (1998)
45. Suárez-García, F., Martínez-Alonso, A., Tascón, J.M.D.: Activated carbon fibers from Nomex by chemical activation with phosphoric acid. Carbon **42**, 1419–1426 (2004)
46. Chotimah, N., Putri, A.D., Ono, Y., et al.: Nanoporosity change on elastic relaxation of partially folded graphene monoliths. Langmuir **33**, 14565–14570 (2017)
47. Lee, C., Wei, X., Kysar, J.W., Hone, J.: Measurement of the elastic properties and intrinsic strength of monolayer graphene. Science **321**, 385–388 (2008)
48. Nishihara, H., Simura, T., Kobayashi, S., et al.: Oxidation-resistant and elastic mesoporous carbon with single-layer graphene walls. Adv. Funct. Mater. **26**, 6418–6427 (2016)
49. Thommes, M., Kaneko, K., Neimark, A.V., et al.: Physisorption of gases, with special reference to the evaluation of surface area and pore size distribution. Pure Appl. Chem. **87**, 1051–1069 (2015)

50. Wang, S., Minami, D., Kaneko, K.: Comparative pore structure analysis of highly porous graphene monoliths treated at different temperatures with adsorption of N_2 at 77.4 K and of Ar at 87.3 K and 77.4 K. Micropor. Mesopor. Mat. **209**, 72–78 (2015)
51. Wang, S., Morelos-Gomez, A., Lei, Z., et al.: Correlation in structure and properties of highly-porous graphene monoliths studied with a thermal treatment method. Carbon **96**, 174–183 (2016)
52. Shimada, T., Sugai, T., Fantini, C., et al.: Origin of the 2450 cm^{-1} Raman bands in HOPG, single-wall and double-wall carbon nanotubes carbon. Carbon **43**, 1049–1054 (2005)
53. Dresselhaus, M.S., Jorio, A., Hofmann, M., et al.: Perspectives on carbon nanotubes and graphene Raman spectroscopy. Nano Lett. **10**, 751–758 (2010)
54. Malard, L.M., Pimenta, M.A., Dresselhaus, G., Dresselhaus, M.S.: Raman spectroscopy in graphene. Phys. Rep. **473**, 51–87 (2009)
55. Tuinstra, F., Koenig, J.L.: Raman spectrum of graphite. J. Chem. Phys. **53**, 1126–1130 (1970)
56. Mattia, D., Rossi, M.P., Kim, B.M., et al.: Effect of graphitization on the wettability and electrical conductivity of CVD-carbon nanotubes and films. J. Phys. Chem. B **110**, 9850–9855 (2006)
57. Li, Z.Q., Lu, C.J., Xia, Z.P., et al.: X-ray diffraction patterns of graphite and turbostratic carbon. Carbon **45**, 1686–1695 (2007)
58. Zeller, H.R.: Electrical conductivity of one-dimensional conductor. Phys. Rev. Lett. **28**, 1452–1455 (1972)
59. Mott, N.F.: Localized states in a pseudo gap and near extremities of conduction and valence bands. Philos. Mag. **19**, 835–852 (1969)
60. Pedraza, D.F., Klemens, P.G.: Effective conductivity of polycrystalline graphite. Carbon **31**, 951–956 (1993)
61. Spain, I.L.: Electronic transport-properties of graphite, carbons, and related materials. Chem. Phys. Carbon **16**, 119–304 (1981)

Application of Adsorption Processes for the Treatment of Diluted Industrial Effluents

Nuno S. Graça, Ana M. Ribeiro, Alexandre Ferreira, and Alírio E. Rodrigues

Abstract Adsorption processes have been widely used in the purification of polluted streams and diluted wastewaters. Another interesting aspect of these processes is their ability to concentrate solutes. In conventional fixed bed processes, the treatment involves a saturation step followed by desorption and regeneration steps. However, these processes present low efficiency, since only a fraction of the adsorbent is used. Additionally, the use of chemicals for bed regeneration causes a new problem of waste disposal. The use of a cyclic adsorption process, such as parametric pumping, where the mobile phase percolates through the adsorption bed upwards and downwards alternately, associated with a simultaneous change in a thermodynamic variable (temperature, pressure, pH, etc.), enables the continuous purification of a polluted stream. This operation mode avoids the use of chemical regenerants and allows the recycling of the concentrated stream back to the industrial process.

1 Introduction

The first reports of the use of adsorption can be dated back to 2000 BC [1]. However, only in the 1700s, the phenomenon of adsorption, as today understood, was first described by Scheele on experiments on gases exposed to carbon [2]. The adsorption found its first industrial application a few years later by the use of wood charcoal to clarify sugar during refining [2]. In more recent times, among several applications in liquid or gas phase, the adsorption based processes have been used in the treatment of polluted streams and diluted wastewaters [3–5].

Activated carbons are one of the most used adsorbents in the removal of a variety of compounds from wastewaters. Its relatively low-cost and good removal efficiency justifies its popularity [6]. However, its reduced response to regeneration appears as a significant drawback [7]. In large scale facilities, the saturated activated carbon can be collected and regenerated at high temperatures. An energetically integrated system of heat exchangers is a typical way to reach these regeneration temperatures. However,

N. S. Graça · A. M. Ribeiro · A. Ferreira · A. E. Rodrigues (✉)
Faculty of Engineering, University of Porto, Rua Dr. Roberto Frias s/n, 4200-465 Porto, Portugal
e-mail: arodrig@fe.up.pt

J. C. Moreno-Piraján et al. (eds.), *Porous Materials*, Engineering Materials,
https://doi.org/10.1007/978-3-030-65991-2_7

on a small scale, the use of high temperatures for regeneration is not economically viable. Therefore, the activated carbon is usually disposed of and replaced by a fresh adsorbent [8]. A possible alternative to activated carbons can be polymeric resins, which can be produced with a variety of physical properties such as functionality, surface area, and porosity. This possibility of choosing some physical properties allows the customization of the polymeric resins for the selective removal of specific compounds [9].

Besides its use for water purification purposes, the adsorptive processes have the ability to concentrate solutes and have the advantage of not being destructive since they usually do not rely on extreme operating conditions. This aspect is especially important for valuable solutes. In conventional fixed-bed adsorption, the process is generally carried out in two steps: the first is the saturation or loading step, which is followed by the second step of desorption or regeneration. However, this operative scheme presents two significant drawbacks: one is the need for a chemical regenerant to get an operative bed, which causes an additional waste disposal problem, and the second is the lack of efficiency since the adsorbent is only partially used during the fixed-bed operation.

A possible alternative to the conventional fixed-bed adsorption is the cyclic process named parametric pumping, which is based on the fact that the adsorption equilibrium of the solutes changes with a thermodynamic variable (temperature, pressure, pH, etc.). The main difference between the parametric pumping and the conventional adsorption techniques is that in the conventional processes the feed solution is percolated through the column in just one direction until the column becomes saturated, and then the column is regenerated. In contrast in the parametric pumping, the evolution of the process occurs in successive changes of the flow direction coupled with the variation of a thermodynamic variable. The column is never completely saturated or regenerated.

Although different parametric pumping operation modes have been described and studied over the years, more attention has been given to the thermal mode, where the temperature is the changing variable assuming that the adsorption equilibrium of the solutes changes with the periodical change of temperature. Since thermal parametric pumping does not require the use of chemical regenerant agents, it enables the treatment of industrial wastewaters allowing the recycling of the concentrated streams back to the industrial process.

Moreover, the use of the parametric pumping process could lead to a reduction on the equipment costs and energy consumption since the process does not require phase changes (boiling or freezing), does not need heat sources or "sinks" able to reach extreme temperatures, and does not require the adding or removing of energy associated with the phase change [10].

In summary, the parametric pumping process has the potential to overcome the limitations of the conventional adsorption processes, namely in the regeneration of the adsorbents. Additionally, its ability to simultaneously treat effluents and concentrate solutes while operating as a continuous or semi-continuous process opens several possibilities for its application at the industrial level.

2 Parametric Pumping

Parametric pumping is a separation technique developed by Richard Wilhelm and coworkers in 1966 [11]. Its principle of operation is based on the stationary phase capacity to alternately retain and release specific compounds due to the periodic change of the equilibrium conditions (through the variation of a thermodynamic variable) coupled with the shift on the fluid phase flow direction, leading to the release and transport of the retained species from a region of the column to the other. Figure 1 shows the movement of a solute during the thermal parametric pumping operation. During the downwards flow, the movement of the solute is retarded by the adsorbent due to the cooling of the column (adsorption step). Then, when the flow direction is upward, the column is heated, and the solute is released by the adsorbent and carried by the fluid phase (desorption step). By repeating this process, the solute will move upwards through the column until it eventually comes out to

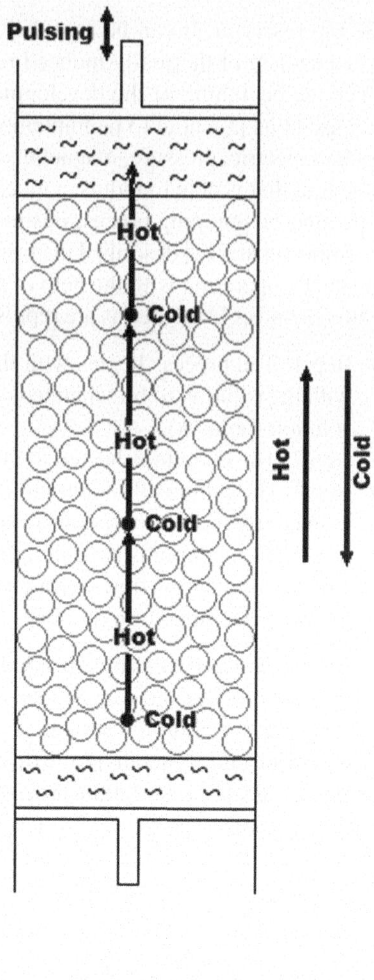

Fig. 1 Schematic representation of the parametric pumping principle

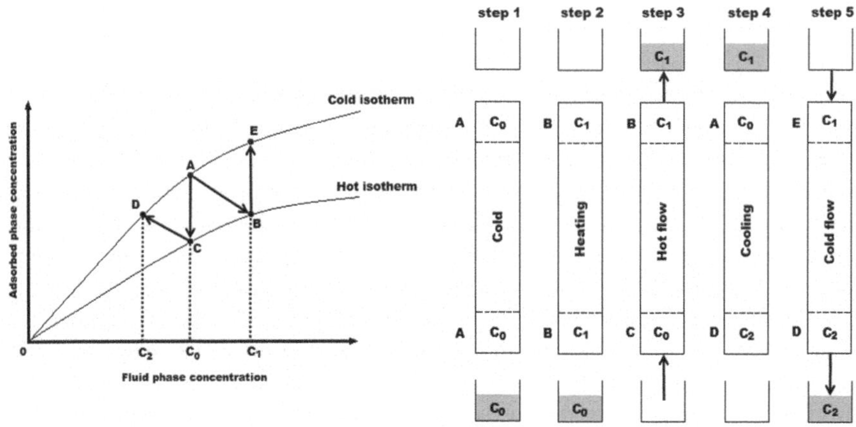

Fig. 2 Principles of parametric pumping explained based on a McCabe–Thiele type diagram

the top reservoir. It can be said that the operating principle described is based on a mechanism of thermally induced retardation-release of the solute [12]. However, since the beginning of the development of the parametric pumping process, it was suggested its potential to be implemented using different types of driving potentials such as electric, pressure, magnetic, and chemical [13]. Therefore, theoretically, any potential that is able to induce a mechanism of retardation-release can be applied in a parametric pumping process design.

Simon et al. [14] described the principles of parametric by illustrating the steps of the cyclic operation with the help of a McCabe–Thiele type diagram (Fig. 2), where both hot and cold isotherms are represented. Then the following steps are illustrated:

- step 1: The process begins with the column equilibrated at the cold temperature with the solution to be separated, and the bottom reservoir is filled with the same solution (point A).
- step 2: The column is heated to the hot temperature (point B).
- step 3: The solution flows through the column from the bottom to the top reservoir at the hot temperature (points C and B).
- step 4: The column is cooled to the cold temperature (points D and A).
- step 5: The solution flows through the column from the top to the bottom reservoir at the cold temperature (points E and D)

The process proceeds by repeating steps 2 to 5 for a determined number of cycles. During the process, the concentration of the solute increases in the top reservoir and decreases in the bottom reservoir.

Grévillot and Tondeur [15, 16] described an analogy between the parametric pumping and the distillation processes. The upwards flow during the parametric pumping hot half-cycle can be compared with the vapor phase, and the downward flow during the cold half-cycle can be compared to the liquid flow in the distillation.

The adsorption isotherms at two temperatures can be seen as analogous to the liquid–vapor equilibrium curve. Additionally, the parametric pumping can operate in total reflux (batch) or partial reflux (continuous or semi-continuous).

The concept of parametric pumping was also studied as a possible explanation for the active cellular transport mechanism [17].

2.1 Parametric Pumping Operation

In the original conception of the parametric pumping process, it was proposed an operation mode based on adsorption–desorption called recuperative mode [11]. In this operation mode, the fluid phase is heated or cooled before entering the adsorption column (Fig. 3a). Alternatively, the process can be operated using the direct mode where the temperature change is imposed through the column wall (Fig. 3b). The main advantage of using the recuperative mode is avoiding the heat-transfer difficulties present in the direct mode when applied in large scale processes [18].

Ghasem described a combined operation mode of the parametric pumping process where the temperature change is imposed in both the fluid entering the column (recuperative mode) and at the column wall through a jacket (direct mode) [19].

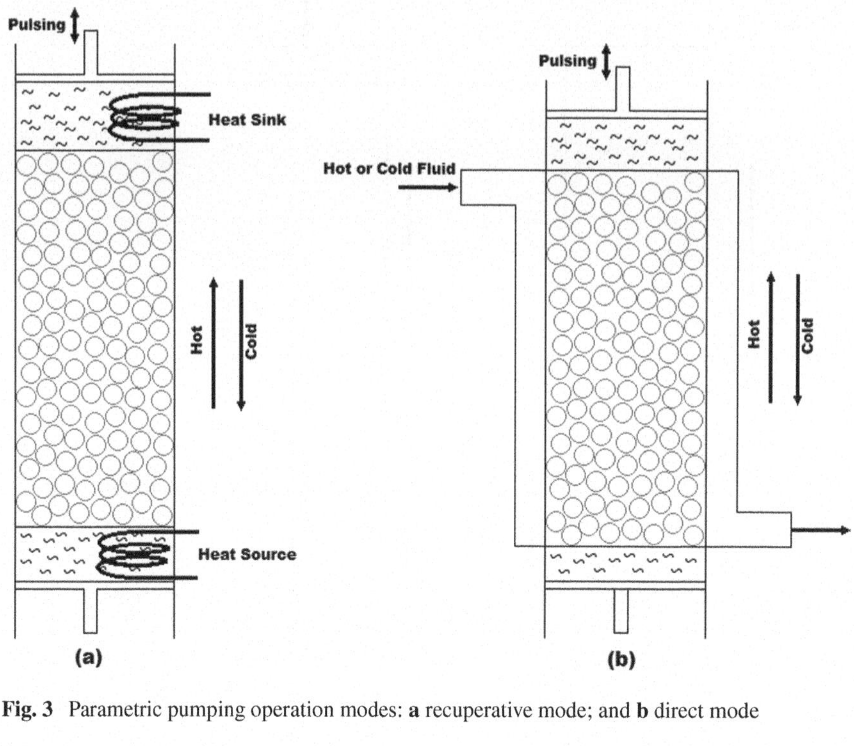

Fig. 3 Parametric pumping operation modes: **a** recuperative mode; and **b** direct mode

The parametric pumping process can also be operated as a closed or an open system. In a closed parametric pumping system (Fig. 3), there are no feed or product streams; while in an open system (Fig. 4), there is a feed reservoir and two other reservoirs to collect the top and the bottom products. In the open system, the cycle usually begins with the column in equilibrium with the feed solution at the hot temperature, and the feed fluid is placed on the bottom reservoir. During the hot half-cycle, the liquid solution flows from the bottom to the top reservoir, from where the top product is withdrawn, and the feed is added. In the cold half-cycle, the liquid solution flows from the top reservoir to the bottom reservoir, from where the bottom product is withdrawn. The duration of each half-cycle is determined by the time required to transfer all the solution contained in a reservoir, either downward or upward.

Simon et al. [14] presented an alternative operation mode called temperature swing chromatography (TSC) as a particular case of parametric pumping. In this process,

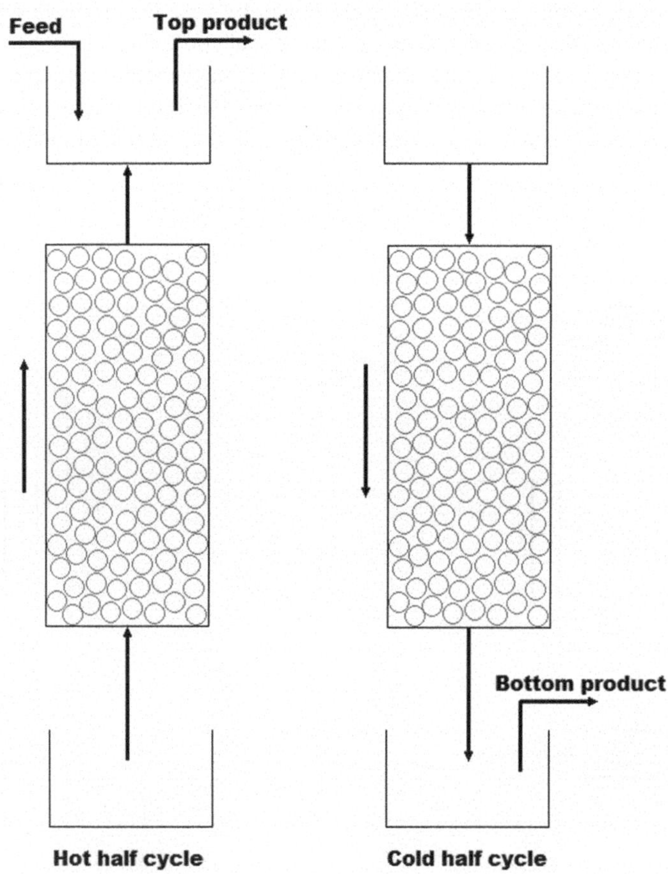

Fig. 4 Open parametric pumping cycle

there is no prior equilibration with the feed solution; and the column is filled with a fresh adsorbent. Before the beginning of the experiment, the solution to be separated is placed in the top reservoir. Then the process runs through the following steps:

- Cold flow from the top to the bottom reservoir;
- Heating;
- Hot flow from the bottom to the top reservoir;
- Cooling;

In TSC, the displaced volume during the cold flow is always higher than the displaced volume during the hot flow ($V_c > V_h$). Therefore, by repeating these steps, the top reservoir will eventually become empty. The last step of the process is the regeneration of the adsorbent, by feeding the column with pure solvent at the high temperature. Therefore, at the end of the process, it will be obtained a bottom solution containing the less adsorbed components, and a solution resulted from the regeneration containing the more adsorbed components.

An analogy between TSC and batch rectification can be made. As in batch rectification where there is no feed, and a part of the top product is withdrawn, in TSC, the same asymmetry occurs since the volume of hot flow is less than the volume of cold flow. The less adsorbed components in TSC break through the column, and the more adsorbed remain at the top end of the column. Analogously, in batch rectification, the 'light' components reach the top of the column, and the 'heavy' components remain at the bottom of the column. Simon et al. [14] showed the advantage of using TSC relatively to the simple frontal chromatography (FC) by performing experiments with arginine.

The use of a multi-column parametric pumping system with multiple adsorbents presenting different selectivities was proposed by Wankat and Tondeur to separate a feed mixture into two or more products [20]. Later Sheng and Costa studied two-column multi-adsorbent systems for the separation of binary mixtures when a selectivity reversal is present [21]. Another multi-column process based on the parametric pumping concept was proposed for the recovery of aniline. Such a process can be implemented on industrial processes where the aniline is used as a monomer or as feedstock for the preparation of a wide range of chemicals [22]. In its simplest form, the process can be operated with two columns in a three-step cycle. In step 1 (adsorption step), the aniline is removed from the feed mixture by passing the aqueous solution through column 1 at a low temperature. This step proceeds until immediately before the aniline breakthrough. During step 1 on column 1, the column 2 (previously saturated) goes through two steps (step 2 and step 3). During step 2, the adsorbent is regenerated, passing part of the purified water from step 1 (heated by a heat exchanger) through the column. After that, the column is cooled down using part of the cold water from step 1. During steps 2 and 3, the aniline is recovered. In the next cycle, the cooled regenerated column (column 2) will be used in the adsorption step (step 1), and the saturated column (column 1) will go through steps 2 and 3. The flow direction in the columns during a cycle is reversed relative to the previous cycle (Fig. 5).

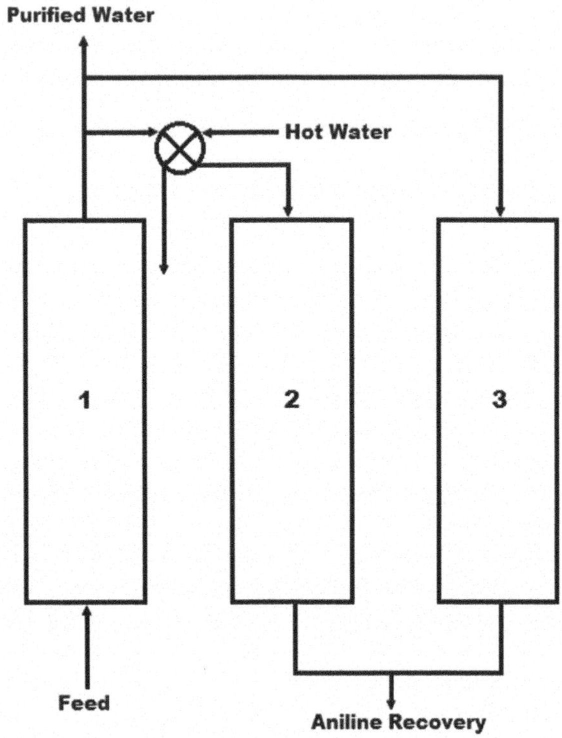

Fig. 5 Multi-column parametric pumping-based process for aniline recovery

Besides the use of temperature as the changing variable, there are alternative parametric pumping operation modes where other variables such as pH can be manipulated [23]. Nevertheless, the use of pH-induced parametric pumping processes requires that the pH has a significant impact on the adsorption equilibrium. However, in some separation problems such as heavy metals using strong acid ion-exchangers, the pH can have minimal influence since the heavy metal cations are non-preferred species. In this case, the pH is called a weak variable. This limitation can be overcome by adding a complexing agent to the liquid phase, which will convert part of the heavy metal cations in neutral species. The degree of complexation of the different heavy metal cations is dependent on the pH. Therefore, in these conditions, the manipulation of pH will impact the adsorption equilibrium, and the pH becomes a strong variable [24].

2.2 Mathematical Modeling

Different mathematical models have been used for the description of the different parametric pumping operation modes. All these models can be divided into two categories: complete models and equilibrium models. The complete models take into consideration the dispersive effects (mass transfer resistance and axial dispersion), while in the equilibrium models, these effects are neglected. The complete models are generally constituted by parabolic partial differential equations (PDEs), which can be numerically solved. The equilibrium models are formed by hyperbolic PDEs, which can be analytically solved using the method of characteristics.

An important step on the mathematical modeling of the parametric pumping process was given by Pigford et al. [25] by simplifying the model assuming linear isotherms, local equilibrium between the solid and liquid phases, constant flow, and absence of axial dispersion. This analytical solution was generalized by Arris [26] and applied to different parametric pumping processes [27–29]. Costa et al. [30] used the equilibrium model, considering non-mixed dead volumes. Besides some unrealistic assumptions, the simple equilibrium model contains the essential principles of the process and allows an insight into the separation.

The first complete model proposed by Wilhelm et al. in 1966 [11] included non-linear isotherms, mass transfer resistance on the liquid/solid interface, but considered the absence of axial dispersion. A complete model, including axial dispersion, was proposed later by Wilhelm et al. [13].

Gupta and Sweed [31] used the mixing cell theory to describe the adsorption column. Wankat [32] applied the stage equilibrium theory to describe the liquid–liquid extraction by parametric pumping. The stage theory was also used by Grevillot and Tondeur [15, 16] and Grevillot [33] making an analogy between parametric pumping and distillation, where the upward flux during the hot half-cycle is compared with the vapor phase in the distillation. While, the downward flux is compared with the liquid phase in the distillation process. Apostolopoulos [34, 35] used a quasi-equilibrium approach to describe the parametric pumping working as a chemical reactor. The effect of radial heat transfer on the separation was studied by Foo and Rice [36, 37] using frequency response solutions.

Lumped-parameter complete model

A simplified version of the complete model, which considers a lumped parameter representation for the mass transfer of the solute between phases, has been successfully applied to describe the parametric pumping process in several published works [38–41]. This model is constituted by mass balances in liquid and solid phases, a global energy balance, non-linear adsorption equilibrium isotherm, and a linear driving force (LDF) rate equation to describe the intra-particle diffusional mass transfer. Additionally, the mass balances to the reservoirs containing the product from the hot and cold half-cycles (top and bottom reservoirs) are also considered.

Additionally, the model assumes:

- perfect mixing in the reservoirs,
- radial temperature gradients in the packed bed are negligible,
- the heat transfer model is one-dimensional and pseudo-homogeneous,
- the physical properties of the fluid and adsorbent are constant,
- the heat and mass transfer parameters are not affected by the transient temperature regime,

- the film heat and mass transfer resistances are negligible,
- Equilibrium between the bulk fluid concentration and the adsorbed phase concentration at the particle surface,
- Intraparticle mass transfer rate described by the LDF model.

From the mass balance to a bed volume element results,

$$D_{ax}\frac{\partial^2 C(z,t)}{\partial z^2} \pm u_i\frac{\partial C(z,t)}{\partial z} = \frac{\partial C(z,t)}{\partial t} + \frac{1-\varepsilon}{\varepsilon}\rho f_h\frac{\partial q(z,t)}{\partial t} \tag{1}$$

where D_{ax} is the axial dispersion coefficient, u_i is the interstitial velocity, ε is the bed porosity, z is the axial position, t is the time variable, C is the concentration in the bulk fluid phase, q is the average adsorbed concentration in the solid phase, ρ is the density of the adsorbent, and f_h is the adsorbent humidity factor. Considering $z = 0$ the bottom of the column and $z = L$ its top, in the term $\pm u_i[\partial C(z,t)/\partial z]$ the "+" sign corresponds to the cold half-cycle, and the "−" sign corresponds to the hot half-cycle.

The axial dispersion coefficient can be determined by the following correlation [42],

$$\frac{u_0 d_p}{D_{ax}} = 0.2 + 0.011Re^{0.48} \tag{2}$$

with

$$Re = \frac{u_i\rho\varepsilon d_p}{\eta} \tag{3}$$

where Re is the Reynolds number, u_0 is the superficial fluid velocity, d_p is the adsorbent particle diameter, and η is the fluid viscosity.

The mass-transfer rate inside the adsorbent particles is given by the LDF model [43],

$$\frac{\partial q(z,t)}{\partial t} = k_s\left[q^*(z,t) - q(z,t)\right] \tag{4}$$

where k_s is the effective mass-transfer coefficient and q^* is the adsorbed concentration in equilibrium with the bulk concentration.

The overall effective LDF rate constants can be determined by the following correlation [44],

$$\frac{1}{k_s} = \frac{r_p}{3k_f}\rho f_h\frac{\partial q^*}{\partial C} + \frac{r_p^2}{\Omega D_{pe}}\rho f_h\frac{\partial q^*}{\partial C} = \frac{r_p}{3k_f}\rho f_h\frac{\partial q^*}{\partial C} + \frac{1}{k_{LDF}} \tag{5}$$

where D_{pe} is the effective pore diffusivity, r_p is the radius of the adsorbent particles, Ω is a factor equal to 15 for spherical particles, k_{LDF} is the LDF kinetic rate constant, k_f is the external-film mass transfer coefficient, and $\partial q^*/\partial C$ is the slope of the equilibrium isotherm.

The effective diffusivity is given by [45],

$$D_{pe} = \frac{\varepsilon_p D_m}{\tau} \tag{6}$$

where D_m is the molecular diffusivity that can be estimated using the Wilke–Chang method [46], τ is the tortuosity of the adsorbent and ε_p its porosity.

The external-film mass-transfer coefficient can be estimated using the Ranz-Marshall correlation [47],

$$Sh = \frac{k_f d_p}{D_m} = 2 + 0.6\left(\frac{\eta}{\rho D_m}\right)^{1/3}\left(\frac{u_0 \rho d_p}{\eta}\right)^{1/2} = 2 + 0.6 Sc^{1/3} Re^{1/2} \tag{7}$$

where Sh, Sc, and Re are the Sherwood, Schmidt, and Reynolds numbers, respectively.

The equilibrium adsorbed concentration, q^*, can be given by the Langmuir model,

$$q^* = \frac{q_m K_L C}{1 + K_L C} \tag{8}$$

where q_m is the adsorptive capacity. The Langmuir equilibrium constant, K_L, dependence on temperature is given by the Van't Hoff equation,

$$K_L = K_{L0} \exp\left(\frac{-\Delta H}{RT}\right) \tag{9}$$

where T is the temperature, K_{L0} is the pre-exponential factor of the Van't Hoff equation, R is the ideal gas constant, and ΔH is the enthalpy of adsorption.

The potential for the thermal separation can be evaluated by the b parameter [25],

$$b = \frac{a}{1 + \overline{m}} \tag{10}$$

where the average slope $\overline{m} = (m(T_h) + m(T_C))/2$ and the deviation $a = (m(T_h) - m(T_C))/2$, with $T_h > T_C$.

The thermal capacity factor is given by,

$$m(T) = \frac{(1 - \varepsilon)\rho f_h K(T)}{\varepsilon} \tag{11}$$

where for the Langmuir isotherm $K(T) = q_m K_L$.

The mass balances of the parametric pumping reservoirs are schematically represented in Fig. 6.

From the mass balance to the reservoirs during the hot half-cycle results,

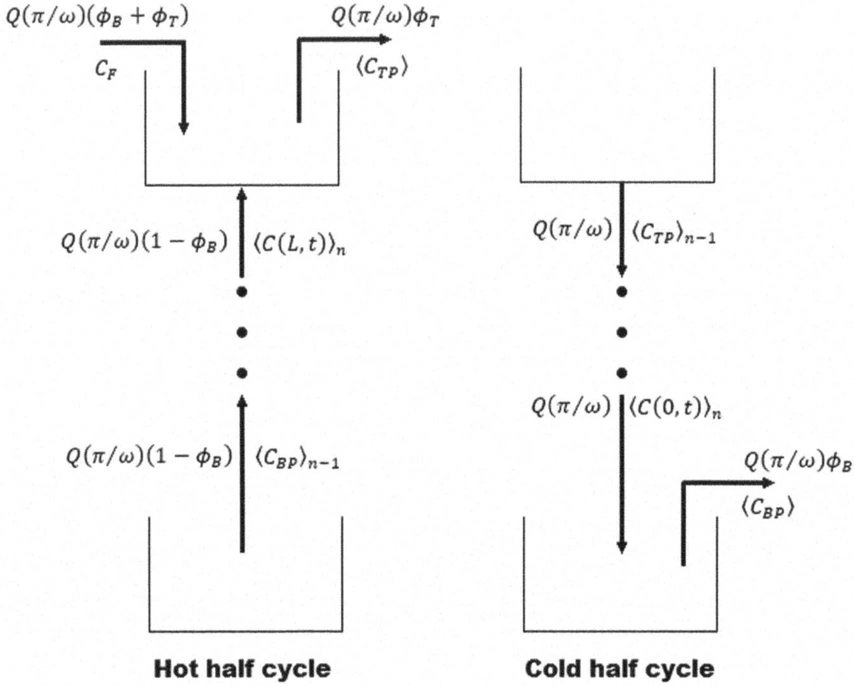

Fig. 6 Schematic representation of the parametric pumping reservoir mass balances

$$\langle C_{BP}\rangle = \langle C_{BP}\rangle_{n-1} \tag{12}$$

$$C_{TP} = \frac{(1-\phi_B)C(L,t)_n}{(1+\phi_T)} + C_F\frac{(\phi_B+\phi_T)}{(1+\phi_T)} \tag{13}$$

and during the cold half-cycle,

$$C_{BP} = C(0,t)_n \tag{14}$$

where C_{BP} and C_{TP} are the average concentration of the solute in the bottom and the top reservoirs, respectively, ϕ_B and ϕ_T the fractions of the total displaced volume $(Q(\pi/\omega))$ that are withdrawn as bottom and top product, respectively, and n is the number of cycles.

From the global energy balance to the column results,

$$\begin{aligned}\left[\rho_f C_{pf}\varepsilon + \rho C_{ps}(1-\varepsilon)\right]\frac{\partial T(z,t)}{\partial t} &= K_{ae}\frac{\partial^2 T(z,t)}{\partial z^2} \\ \pm \rho_f C_{pf} u_0 \frac{\partial T(z,t)}{\partial z} &- h_w A_w (T - T_{amb}) + (-\Delta H)\frac{\partial q(z,t)}{\partial t}\end{aligned} \tag{15}$$

where K_{ae} is the axial thermal conductivity, h_w is the heat-transfer coefficient of the wall of the column, A_w is the wall specific area, ρ_f is the density of the fluid and, and C_{pf} and C_{ps} are the heat capacities of the fluid and solid phases, respectively. The term $\pm\rho_f C_{pf} u_0[\partial T(z,t)/\partial z]$ has the " + " sign for the cold half-cycle (downward flow) and the "−" sign for the hot half-cycle (upward flow).

The boundary and initial conditions for the parametric pumping mode are the following:

For the hot half-cycle,

$$z = 0 \quad C(0,t) = \langle C_{BP} \rangle_n \tag{16}$$

$$T = T_h \tag{17}$$

$$z = L \quad \frac{\partial C(L,t)}{\partial z} = 0 \tag{18}$$

$$\frac{\partial T(L,t)}{\partial z} = 0 \tag{19}$$

for cold half-cycle,

$$z = 0 \quad \frac{\partial C(0,t)}{\partial z} = 0 \tag{20}$$

$$\frac{\partial T(0,t)}{\partial z} = 0 \tag{21}$$

$$z = L \quad C(L,t) = \langle C_{TP} \rangle_n \tag{22}$$

$$T = T_c \tag{23}$$

The initial conditions are,

$$C(z,0) = C_F \tag{24}$$

$$T(z,0) = T_0 \tag{25}$$

Detailed model

The assumption of a lumped parameter for the representation of inter-phase solute mass transfer is only valid when the intraparticle gradients can be neglected. Ferreira and Rodrigues used a complete detailed mathematical model for the recuperative mode considering the intraparticle mass transfer resistance [48]. In this model, the mass balance in a volume element described above (Eq. 1) should be replaced by,

$$D_{ax}\frac{\partial^2 C(z,t)}{\partial z^2} \pm u_i\frac{\partial C(z,t)}{\partial z} = \frac{\partial C(z,t)}{\partial t} + \frac{3}{R_p}\frac{(1-\varepsilon)}{\varepsilon}k_f\big(C(z,t) - C_p\big(R_p, z, t\big)\big) \tag{26}$$

where C_p is the concentration of the solute in the adsorbent pores, and R_p is the radius of the adsorbent particles.

A mass balance inside the adsorbent particles is included in the model,

$$\rho f_h\frac{\partial q(r,z,t)}{\partial t} + \varepsilon_p\frac{\partial C_p(r,z,t)}{\partial t} = \varepsilon_p D_p\left(\frac{2}{r}\frac{\partial C_p(r,z,t)}{\partial r} + \frac{\partial^2 C_p(r,z,t)}{\partial r^2}\right) \tag{27}$$

where r is the radial coordinate, ε_p is the particle porosity, and D_p is the pore diffusivity.

Contrary to the lumped parameter model, where is considered an average concentration inside the adsorbent particles, in this model, the adsorbed-phase concentration, q, is dependent on the radial position inside the particles.

This model also requires additional boundary conditions for the adsorbent particles,

$$r = 0 \quad \frac{\partial C_p(0,z,t)}{\partial r} = 0 \tag{28}$$

$$r = R_p \quad \varepsilon_p D_p\frac{\partial C_p\big(R_p, z, t\big)}{\partial r} = 3k_f\big[C(z,t) - C_p\big(R_p, z, t\big)\big] \tag{29}$$

and the initial condition,

$$C(z,0) = C_p(r,z,0) = C_F \quad (z > 0) \tag{30}$$

The system of parabolic partial differential equations (PDEs) that constitutes these complete parametric pumping mathematical models can be solved numerically using appropriate computer software. The numeric methods employed usually make use of finite element orthogonal collocation for the discretization of the spatial variables, which results in an ordinary differential equations (ODEs) system that can then be integrated over the time variable.

2.3 *Parametric Pumping Applications*

After its invention, the parametric pumping process was studied at lab-scale by different authors for single and multi-component separations. Some examples of single-component separations are the NaCl from water using retardation resin [49], toluene from n-heptane using silica gel [50], and phenol from water using Duolite

ES861 [30]. Some examples of multi-component separations are toluene/aniline/n-heptane and glucose/fructose/water using cation exchange systems [33, 51].

Table 1 summarizes some examples of experimental systems used in thermal parametric pumping studies.

Alternative operation modes were proposed over the years. Sabadell and Sweed [23] combined the recuperative mode with simultaneous pH control in the separation of Na^+ and K^+ ions from aqueous solutions. The concept of pH parametric pumping also showed its potential for protein separation on ion-exchange columns. The periodical change of the pH coupled with the change in the flow direction makes the separation possible due to the net charge of the protein dependence on the pH. Chen et al. used the pH parametric pumping system for separating haemoglobin and albumin on CM Sepharose [64]. Hollein et al. showed that the protein separation could be improved by applying an electrical field to a pH parametric pumping system [65].

Shaffer and Hamrin [66] combined affinity chromatography with parametric pumping to reduce the enzyme concentration of an aqueous solution. Oren and Soffer [67] extended the parametric pumping to electrochemical systems for water desalinization and isotopes separation.

In the early 1980s, parametric pumping began to be considered for wastewater purification purposes. The application of a small-scale parametric pumping unit for the purification of wastewater containing phenol using a polymeric resin was studied by Costa et al. [30], Almeida et al. [60], and Ramalho [61]. The same purification

Table 1 Summary of adsorbent and chemical systems used by thermal parametric pumping

Adsorbent	Chemical system	References
Activated carbon	Urea-water	[52]
	Oxalic acid–water	[53]
Silica gel	Toluene-n-heptane	[13, 25, 50, 54]
	Toluene-n-heptane-aniline	[51]
BioRad AG11A8	Sodium chloride-water	[49, 55]
	Sodium nitrate-water	[56]
Dowex 50X8	K^+-H^+-water	[57]
	K^+-Na^+-water	[57]
BioRad AG 50 W-X4	Glucose-fructose-water	[33]
Amberlite IR-45/IRC-50	Sodium chloride-water	[11, 13, 58]
Duolite C265	Potassium nitrate-water	[59]
Duolite ES861	Phenol-water	[30, 48, 60–62]
Sephabeads SP206	Salicylic acid–water	[39]
	Phenol-4-nitrophenol-water	[40, 41]
Sawdust carbon	Chromium-water	[63]

was later performed by Ferreira et al. using a fully automated pilot-scale adsorptive parametric pumping unit [62].

2.4 Pilot Scale Process

Phenolic wastewater purification

Phenolic compounds are considered hazardous pollutants because of their potential to harm human life. These compounds can enter water bodies through discharge from pharmaceutical, petrochemical, and other chemical manufacturing processes. Therefore, the use of suitable treatment systems to deal with these wastewaters is a priority.

The purification of wastewater containing phenol and 4-nitrophenol using thermal parametric pumping was studied by Otero et al. [40]. The phenolic wastewater purification was carried out in a fully automated parametric pumping pilot unit working in recuperative mode (Fig. 7). The adsorbent used was the polymeric resin Sephabeads SP206, packed in a borosilicate column with 0.09 m internal diameter and 0.8 m height.

The process was operated in a semi-continuous way with feed in the top reservoir and with the operating conditions presented in Table 2. The pilot unit used has a top, a bottom, and a feed reservoir. The top and the bottom reservoir receive the product during the hot and cold half-cycles, respectively. A solution of 500 mg L^{-1} of either phenol or 4-nitrophenol is pumped from the feed to the top reservoir during the hot half-cycle using the peristaltic pump 18. The solution is passed through the column either downward (cold half-cycle) or upward (hot half-cycle) using the peristaltic pumps 15 or 16, respectively. During the cold half-cycle, valve 11 is open, and valve 12 closed; the reverse occurs during the hot half-cycle. The end of each half-cycle occurs when all the solution contained in a reservoir is transferred, which is determined by pressure transducers placed on the bottom of the reservoirs. The liquid solution is heated or cooled by heat exchangers, 6 and 7. This study was carried out using a high temperature of 60 °C for the hot half-cycle and a temperature of 20 °C for the cold half-cycle. The sampling of the bottom and top products was carried out, pumping the liquid solution from the respective reservoirs to a fraction collector using pumps 19 and 17. The solution goes to the fraction collector or the drain by using three-way solenoid valves 13 and 14. The goal of this process was to obtain purified water in the bottom reservoir and concentrate the phenolic compounds in the top reservoir.

The experimental results showed that under the appropriated conditions, after ten cycles, it was possible to reduce the concentration of phenol to $C/C_F = 0.0003$ and the concentration of 4-nitrophenol to $C/C_F = 0.0013$. The experimental results were simulated and compared using the equilibrium and LDF models, showing that the LDF model provides the best prediction of the experimental results (Fig. 8).

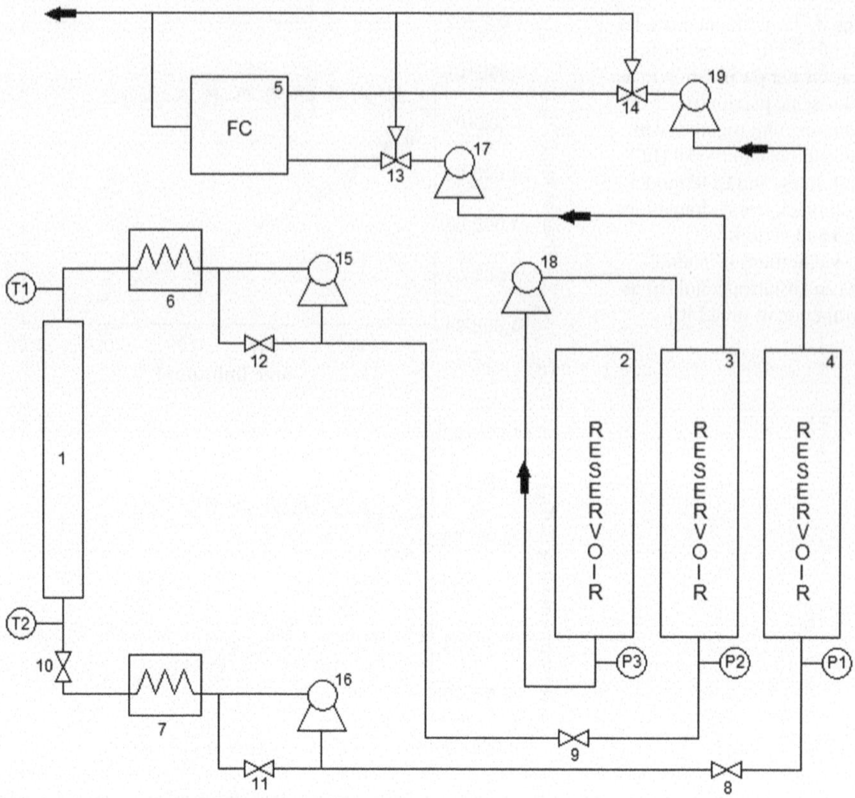

Fig. 7 Experimental set-up of the parametric pumping unit: (1) borosilicate column; (2) feed reservoir; (3) top reservoir; (4) bottom reservoir; (5) fraction collector; (6) and (7) heat exchangers; (8)–(12) two-ways solenoid valves; (13) and (14) three-ways solenoid valves; (15) and (19) peristaltic pumps; (T1) and (T2) thermocouples; (P1)–(P3) pressure transducers

Table 2 Experimental condition used for the parametric pumping in the recuperative mode

Operating variables	
$Q(\pi/\omega)(mL)$	20,000
ϕ_B	0.15
ϕ_T	0.25
$t_c(min)$	85
$t_h(min)$	100
$Q_c\left(mLmin^{-1}\right)$	200
$Q_h\left(mLmin^{-1}\right)$	200
$Q_{TP}\left(mLmin^{-1}\right)$	62.5
$Q_{BP}\left(mLmin^{-1}\right)$	32

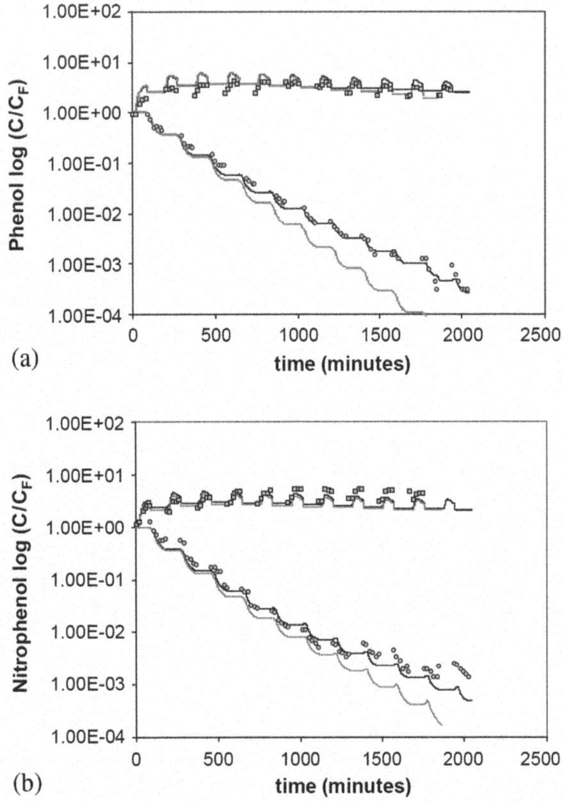

Fig. 8 Experimental results (symbols) for the phenolic wastewater purification in a pilot-scale parametric pumping unit together with the equilibrium model (full gray lines) and LDF model (full black lines) simulations. Top and bottom concentrations of phenol (**a**) and 4-nitrophenol (**b**) as a function of time [40]

Salicylic acid bearing wastewater purification

Salicylic acid is another potentially harmful compound to human health. Concentrations of salicylic acid above 800 mg/L after 6 h of post-exposure can be extremely toxic and possibly lethal [68].

The same parametric pumping pilot unit described above with the same adsorbent and operating at the same conditions (Table 2) was used in the purification of wastewater containing salicylic acid [38]. The high temperature of 60 °C was chosen for the hot half-cycle, and the temperature of 20 °C was used for the cold half-cycle. The experimental results showed that the concentration of salicylic acid in the purified stream (bottom reservoir) was reduced nearly 100 times when compared with the feed concentration ($C_F = 101$ mg L^{-1}). The comparison between the experimental and simulated results showed that the LDF model provides a good prediction of the process (Fig. 9).

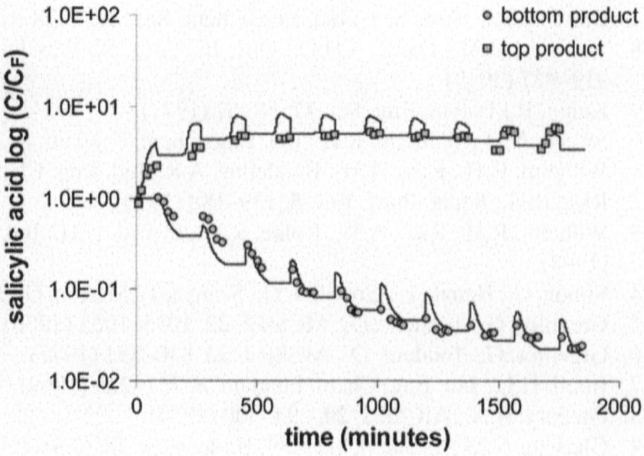

Fig. 9 Experimental results (symbols) for the operation of the parametric pumping system salicylic acid/water/Sephabeads SP206, together with the model predictions (full lines). The top and bottom product concentrations are shown as a function of time in recuperative thermal parametric pumping [38]

3 Final Remarks

The importance of the adsorptive processes in water treatment systems has been growing due to the necessity to deal with the increasing occurrence of anthropogenic compounds in the aquatic environment. Moreover, there is a need for the development of decentralized water treatment systems able to overcome the inadequacy of the conventional centralized systems to manage, treat, and provide quality water in rapidly growing urban areas [69]. However, the use of such decentralized systems can be limited by its necessity of having a constant supply of spare parts and consumables [70]. Therefore, in the case of the adsorptive processes, the ability to regenerate and reuse the adsorbent is an essential feature. Processes such as parametric pumping are good candidates to be used in decentralized water treatment systems due to its capacity for the in-situ regeneration of the adsorbent while providing continuous or semi-continuous operation.

Despite the potential showed since its invention, the parametric pumping process never found widespread industrial application. Nevertheless, the possibility of the parametric pumping-based processes to be applied in decentralized water treatment systems could undoubtedly justify the revival of the interest in the study and development of this process.

References

1. Crittenden, J.C., Trussell, D.W., Hand, K.J., Howe, G.: Tchobanoglous, MWH's Water Treatment: Principles and Design, 3rd edn. (2012)
2. Hassler, J.W., J. Am. Pharm. Assoc. (Scientific ed.) **40**, 418–419 (1951)
3. Knaebel, K.S.: Chem. Eng. **106**, 92–101 (1999)
4. Kumar, A., Kumar, S., Kumar, S.: Carbon **41**, 3015–3025 (2003)
5. Roostaei, N., Tezel, F.H.: J. Environ. Manage. **70**, 157–164 (2004)
6. Luo, Y., Guo, W., Ngo, H.H., Nghiem, L.D., Hai, F.I., Zhang, J., Liang, S., Wang, X.C.: Sci. Total Environ. **473–474**, 619–641 (2014)

7. Browne, T.E., Cohen, Y.: Ind. Eng. Chem. Res. **29**, 1338–1345 (1990)
8. Aumeier, B.M., Dang, A.H.Q., Ohs, B., Yüce, S., Wessling, M.: Environ. Sci. Technol. **53**, 919–927 (2019)
9. Kunin, R.: Polym. Eng. Sci. **17**, 58–62 (1977)
10. Sweed, N.H., Wilhelm, R.H.: Ind. Eng. Chem. Fundam. **8**, 221–231 (1969)
11. Wilhelm, R.H., Rice, A.W., Bendelius, A.R.: Ind. Eng. Chem. Fundam. **5**, 141–144 (1966)
12. Rice, R.G.: Separ. Purif. Rev. **5**, 139–188 (1976)
13. Wilhelm, R.H., Rice, A.W., Rolke, R.W., Sweed, N.H.: Ind. Eng. Chem. Fundam. **7**, 337–349 (1968)
14. Simon, G., Hanak, L., Grévillot, G., Szanya, T., Marton, G.: J. Chromatogr. A **732**, 1–15 (1996)
15. Grevillot, G., Tondeur, D.: AIChE J. **22**, 1055–1063 (1976)
16. Grevillot, G., Tondeur, D.: AIChE J. **23**, 840–851 (1977)
17. Booij, H.L.: Ind. Eng. Chem. Fundam. **8**, 231–235 (1969)
18. Gregory, R.A.: AIChE J. **20**, 294–300 (1974)
19. Ghasem, N.M.: J. Chem. Technol. Biotechnol. **78**, 666–669 (2003)
20. Wankat, P.C., Tondeur, D.: Use of multiple sorbents in pressure swing adsorption, parametric pumping and cycling zone adsorption. In: AIChE Symposium Series, pp. 74 83 (1985)
21. Sheng, P., Costa, C.A.V.: Sep. Purif. Technol. **12**, 81–95 (1997)
22. Grande, C.A.: Process for the recovery of aniline (2012)
23. Sabadell, J.E., Sweed, N.H.: Separ. Sci. **5**, 171–181 (1970)
24. Bartosch, C., Kiefer, R., Höll, W.H.: React. Funct. Polym. **45**, 197–210 (2000)
25. Pigford, R.L., Baker, B., III., Blum, D.E.: Ind. Eng. Chem. Fundam. **8**, 144–149 (1969)
26. Aris, R.: Ind. Eng. Chem. Fundam. **8**, 603–604 (1969)
27. Chen, H.T., Hill, F.B.: Separ. Sci. **6**, 411–434 (1971)
28. Butts, T.J., Gupta, R., Sweed, N.H.: Chem. Eng. Sci. **27**, 855–866 (1972)
29. Gregory, R.A., Sweed, N.H.: Chem. Eng. J. **1**, 207–216 (1970)
30. Costa, C.A.V., Rodrigues, A.E., Grevillot, G., Tondeur, D.: AIChE J. **28**, 73–85 (1982)
31. Gupta, R., Sweed, N.H.: Ind. Eng. Chem. Fundam. **12**, 335–341 (1973)
32. Wankat, P.C.: Ind. Eng. Chem. Fundam. **12**, 372–381 (1973)
33. Chen, H.T., D'Emidio, V.J.: AIChE J. **21**, 813–815 (1975)
34. Apostelopoulos, G.P.: Ind. Eng. Chem. Fundam. **14**, 11–16 (1975)
35. Apostolopoulos, G.P.: Ind. Eng. Chem. Fundam. **15**, 14 (1976)
36. Foo, S.C., Rice, R.G.: AIChE J. **21**, 1149–1158 (1975)
37. Foo, S.C., Rice, R.G.: AIChE J. **23**, 120–123 (1977)
38. Otero, M., Zabkova, M., Grande, C.A., Rodrigues, A.E.: Ind. Eng. Chem. Res. **44**, 927–936 (2005)
39. Otero, M., Zabkova, M., Rodrigues, A.: Adsorption **11**, 887–892 (2005)
40. Otero, M., Zabkova, M., Rodrigues, A.E.: Water Res. **39**, 3467–3478 (2005)
41. Otero, M., Zabkova, M., Rodrigues, A.E.: Chem. Eng. J. **110**, 101–111 (2005)
42. Guiochon, G., Shirazi, S.G., Katti, A.M.: Fundamentals of Preparative and Nonlinear Chromatography (1994)
43. Glueckauf, E.: Trans. Faraday Soc. **51**, 1540–1551 (1955)
44. Farooq, S., Ruthven, D.M.: Ind. Eng. Chem. Res. **29**, 1084–1090 (1990)
45. Reid, R.C., Prausnitz, J.M., Poling, B.E.: The Properties of Gases and Liquids (1987)
46. Wilke, C.R., Chang, P.: AIChE J. **1**, 264–270 (1955)
47. Yang, R.T.: Gas Separation by Adsorption Processes (1987)
48. Ferreira, L.M., Rodrigues, A.E.: Adsorption **1**, 213–231 (1995)
49. Sweed, N.H., Gregory, R.A.: AIChE J. **17**, 171–176 (1971)
50. Chen, H.T., Reiss, E.H., Stokes, J.D., Hill, F.B.: AIChE J. **19**, 589–595 (1973)
51. Chen, H.T., Lin, W.W., Stokes, J.D., Fabisiak, W.R.: AIChE J. **20**, 306–310 (1974)
52. Sweed, N.: Progr. Separ. Purif. **4**, 171–240 (1971)
53. Rice, R.G., Mackenzie, M.: Ind. Eng. Chem. Fundam. **12**, 486–487 (1973)
54. Chen, H.T., Rak, J.L., Stokes, J.D., Hill, F.B.: AIChE J. **18**, 356–361 (1972)
55. Gregory, R.A., Sweed, N.H.: Chem. Eng. J. **4**, 139–148 (1972)

56. Chen, H.T., Manganaro, J.A.: AIChE J. **20**, 1020–1022 (1974)
57. Butts, T.J., Sweed, N.H., Camero, A.A.: Ind. Eng. Chem. Fundam. **12**, 467–472 (1973)
58. Rolke, R.W., Wilhelm, R.H.: Ind. Eng. Chem. Fundam. **8**, 235–246 (1969)
59. Grevillot, G.: AIChE J. **26**, 120–131 (1980)
60. Almeida, F., Costa, C., Rodrigues, A., Grevillot, G.: Stud. Environ. Sci. **19**, 169–178 (1982)
61. Ramalho, E., Costa, C., Rodrigues, A., Grevillot, G.: Sep. Technol. **1**, 99–107 (1991)
62. Ferreira, L.M., Rodrigues, A.E.: Adsorption **1**, 233–252 (1995)
63. Mishra, P.K., Nanagre, D.M., Yadav, V.L.: J. Sci. Ind. Res. **66**, 79–84 (2007)
64. Chen, H.T., Yang, W.T., Pancharoen, U., Parisi, R.: AIChE J. **26**, 839–849 (1980)
65. Holleln, H.C., Me, H.C., Huang, C.R., Chen, H.T.: Ind. Eng. Chem. Fundam. **21**, 205–214 (1982)
66. Shaffer, A.G., Jr., Hamrin, C.E., Jr.: AIChE J. **21**, 782–786 (1975)
67. Oren, Y., Soffer, A.: J. Electrochem. Soc. **125**, 869–875 (1978)
68. Balali-Mood, M.: Effects of Forced Alkaline Diuresis on Salicylate Distribution and Elimination Following Over-dosage, 132–270 (1981)
69. Larsen, T.A., Hoffmann, S., Lüthi, C., Truffer, B., Maurer, M.: Science **352**, 928–933 (2016)
70. Sobsey, M.D., Stauber, C.E., Casanova, L.M., Brown, J.M., Elliott, M.A.: Environ. Sci. Technol. **42**, 4261–4267 (2008)

Activated Carbons for H_2S Capture

Sebastião Mardonio Pereira de Lucena, Daniel Vasconcelos Gonçalves, José Carlos Alexandre de Oliveira, Moises Bastos-Neto, Célio Loureiro Cavalcante Jr., and Diana Cristina Silva de Azevedo

Abstract The size and dispersion-repulsion forces of the H_2S molecule make activated carbon a first-choice adsorbent material. With a wide degree of freedom in the distribution of pores and possibilities for impregnation, the activated carbon can be carefully designed, through synthesis techniques, to meet the particular separation process. In this chapter we describe the main variables to guide the choice or synthesis of an activated carbon suitable for H_2S retention. We review the most recent studies of textural characterization and what we have learned about the structure–property relationship in H_2S capture. We present the promising molecular simulation technique, still weakly explored for this specific application, detailing how it can be useful. Additionally, we review specific details of dynamic tests to guide the dimensioning of units on an industrial scale.

1 Introduction

Activated carbons have been successfully and widely used for adsorption separations both in liquid and gas phases. One of the most common applications in the gas phase consists in pollutants concentration reduction to regulatory environmental limits. Some of these gas phase phase applications are: removal of sulfur gases (H_2S and/or SO_2), removal of volatile organic compounds (VOCs), mercury (Hg) removal from in gaseous streams, H_2 storage by adsorption, and other specific applications in the military and nuclear fields [1].

Hydrogen sulfide (H_2S) is particularly undesirable in industrial and public operational processes due to its toxicity and corrosiveness. The H_2S toxicity has been a

S. M. P. de Lucena (✉) · D. V. Gonçalves · J. C. A. de Oliveira · M. Bastos-Neto ·
C. L. Cavalcante Jr. · D. C. S. de Azevedo (✉)
Grupo de Pesquisas em Separações Por Adsorção (GPSA), Departamento de Engenharia Química (DEQ), Universidade Federal do Ceará (UFC), Campus do Pici, bl. 709, Fortaleza, CE 60760-400, Brazil
e-mail: lucena@ufc.br

D. C. S. de Azevedo
e-mail: diana@gpsa.ufc.br

J. C. Moreno-Piraján et al. (eds.), *Porous Materials*, Engineering Materials,
https://doi.org/10.1007/978-3-030-65991-2_8

health public concern since the beginning of the last century, when the rayon industrial production developed, being one of the first genuinely multinational industries [2]. The rayon production process uses carbon disulfide for cellulose treatment that is afterward contacted with sulfuric acid. H_2S is then generated as a by-product of the CS_2 reaction with sulfuric acid. H_2S concentrations in the range of 30–50 ppm were measured in this industrial sector and were then related to loss of vision by workers in that industry [2]. Because of the very low concentrations of H_2S found in the rayon industry, adsorption removal was the process of choice for this problem, with activated carbon as selected adsorbent. Not only are workers exposed to risks in industrial units where toxic gas is produced, the population in the vicinity of the plants and the environment are also penalized.

In recent years, the widespread use of biogas from landfills or waste water treatment plants, as source of clean and sustainable energy, there has been an increasing and urgent need to remove H_2S from the biogas streams. The European Biogas Association (EVA) reports a total of ca. 18,000 biogas plants producing *ca.* 65 TWh among the European Union countries in 2018. In particular, the number of biomethane plants grew from 187 plants in 2011 up to a total of 540 plants in 2018 [3].

For gas streams with high H_2S concentrations, the Claus process is largely used. However, the tail gases of the Claus process still carry *ca.* 3–5% of H_2S. For these low H_2S concentrations, adsorptive/catalytic processes are more economically feasible than reduction/absorption processes using MEA/DEA in solution [4]. The capture of H_2S using activated carbons may occur by three different mechanisms: (1)—Physical adsorption of the gas; (2)—Chemisorption and resulting formation of carbon–sulfur surface products and (3)—Selective catalytic oxidation with deposition of elemental sulfur. Depending on the mechanism involved, regeneration of the carbon may be performed using vacuum or inert gas (physical adsorption) or solvent extraction (elemental sulfur). However, removal of carbon–sulfur surface products is rather complicated and should be avoided [5].

H_2S experimental measurements are difficult and costly, which means that even today many several thermodynamic properties, under different conditions, are unknown. Furthermore, these experimental difficulties of using H_2S sums with the typically amorphous structure of activated carbons, accessed only indirectly through their textural characteristics, leaving concerns about the precise mechanisms that result in successful adsorption projects. This chapter attempts to gather the most relevant aspects related to the use of carbonaceous materials to capture H_2S.

2 Carbon Textural Properties

For engineers and synthesis chemists, it shall be important to define which textural properties of a carbon sample would optimize the capture of H_2S by adsorption. The characterization needs to be performed carefully to provide a clear correlation between the carbon structure and its H_2S adsorption properties, which is unfortunately not always reported in the open literature.

As a general rule, most authors report that a suitable carbon for this application should ideally present both micropores and mesopores. It has been accepted that these carbons should have mesopores in the range 30–80 Å [5]. Recent studies have shown that carbons for H_2S removal should have pores regularly distributed above and below 10 Å or even 4 Å [6–8].

A compilation of carbon samples that have been reported in studies for H_2S capture is listed in Table 1 with their respective pore size distribution (PSD) shown in Fig. 1. These carbons were selected for having very similar textural properties in order to provide better basis for performance comparison among them. For instance, all of them have pH approximately 10 and had been impregnated with Na and have shown high efficiency for H_2S separation. The influence of the textural parameters of the carbons in Table 1 (surface areas, pore volumes and PSD) will be discussed in this section, while other factors such as impregnation concentration ($C_{Impregnant}$), pH and flue gas composition will be evaluated in the next section. In addition to the textural parameters of area and pore volume, it is essential that PSD calculation be performed (Fig. 1). It is also helpful that the respective probe-gas isotherms be presented.

In general, we may observe that the BET areas vary from 815 to 1518 m^2/g, and total pore volumes between 0.38 and 0.69 cm^3/g. These intervals are relatively wide to be able to establish a clear correlation between the adsorption capacity and the textural properties. Nevertheless, Chen et al. [6] has shown that, for a similar series of activated carbons, both BET area and pore volume correlate well with the obtained adsorption capacity.

Because the pore size distribution is limited to narrower intervals, this would be a more adequate textural parameter to correlate with the H2S adsorption capacity. The carbon samples of Desorex K43-Na, P3 I, WSC 10 e VA 10 all have pores up to a maximum of 20 Å, except for N31, with pores up to 50 Å. Also, all pore volumes are approximately evenly distributed between pores of sizes below and above 10 Å, except for Desorex K43-Na with a higher proportion of pores below 10 Å. Furthermore, in contrary to all other carbons, Desorex K43-Na presents ultramicropores (<4 Å).

Thus we may conclude that activated carbons, in order to be preliminarly considered for H_2S capture, should meet the following conditions: (1)—balanced pore size distribution up to 20 Å; in case most pore volume is below 10 Å, ultramicropores should be present, and (2)—minimum BET area of 800 m^2/g. It should be observed that for any of the PSD's presented in Fig. 1, increasing the pore volume in the range below or above 10 Å (or 4 Å) would benefit the H_2S capture capacity. For instance, Bandosz [10], tested a series of carbons (W1, W2 and W3, from Westvaco), with pore size distributions above 10 Å that are very similar, but with different pore volumes below 10 Å. Samples W1 and W3, with similar pore volumes below 10 Å, captures 295 and 230 mg/g, respectively. On the other hand, the sample W2, with pore volumes below 10 Å much lower than for samples W1 and W3, the H2S retention was only 17 mg/g.

Since the pore size distribution is the most consistent textural property to indicate the eventual applicability of the carbon sample for H_2S adsorption, more recent kernels, in which the surface heterogeneity is taken into account, should be used for

Table 1 Textural parameters, experimental conditions and H_2S retention ($C_{Ret.}$) of selected carbons

		Raw material	Area BET	V_{tot} (cm^3/g)	$C_{Ret.}$ (mg/g)	$C_{Impregnant}$	pH	T	H_2S flue gas
1	Desorex K43-Na[a]	Bituminous + coconut	815	0.38 ($V_{<0.4\ nm} = 0.04$)	**8.16**	Na 4.7 (wt%) Fe 3.8 (wt%)	10.2	298	100 ppm 0% RH
2	Desorex K43-Na[b]	Bituminous + coconut	815	0.38 ($V_{<0.4\ nm} = 0.04$)	**1050**	Na 4.7 (wt%) Fe 3.8 (wt%)	10.2	298	1000 ppm 80% RH
3	P3 I[c]	Carbon fiber	1147	0.52 ($V_{<0.7\ nm} = 0.12$)	**810**	Na_2CO_3 (Sol. 6%)	9.7	303	1000 ppm 80% RH
4	N3 I[c]	Carbon fiber	886	0.51 ($V_{<0.7\ nm} = 0.12$)	**780**	Na_2CO_3 (Sol. 6%)	9.8	303	1000 ppm 80% RH
5	WSC 10[d]	Celulosic	899	0.43 ($V_{<1\ nm} = 0.28$)	**172**	Na 2.8 $\mu mol/m^2$	10.2	423	1000 ppm 24% RH
6	VA 10[d]	Coconut shell	1518	0.69 ($V_{<1\ nm} = 0.33$)	**208**	Na 1.7 $\mu mol/m^2$	10.3	423	1000 ppm 24% RH
7	WSC[d]	Celulosic	860	0.40 ($V_{<1\ nm} = 0.28$)	**66**	Na 0	10	423	1000 ppm 24% RH
8	VA[d]	coconut shell	1020	0.47 ($V_{<1\ nm} = 0.22$)	**26**	Na 0	9.7	423	1000 ppm 24% RH

[a]Menezes [8]
[b]Castrillon [9]
[c]Chen [6]
[d]Bashkova [7]

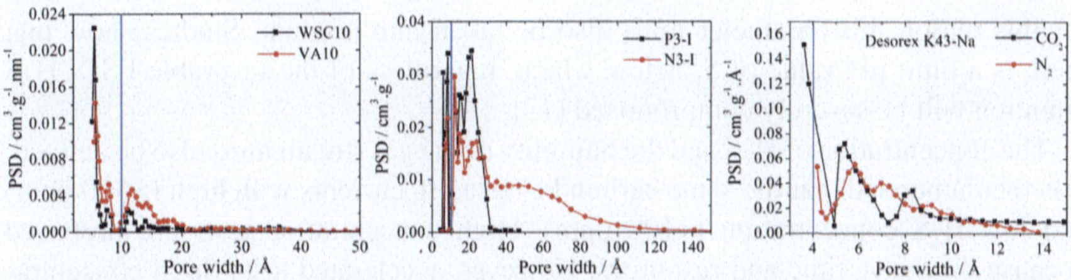

Fig. 1 PSD of selected carbons presented in Table 1. WSC10, VA10, P3-I and N3-I were obtained from N_2 isotherms at 77 K. Two PSDs from N_2 at 77 K and CO_2 at 273 K are presented to Desorex K43-Na carbon

this evaluation [11–14]. Furthermore, at least two different probe molecules should be used: N_2 at 77.4 K or Ar at 87 K for larger pores, and CO_2 at 273 K to investigate the presence of ultramicropores.

The occurrence and importance of ultramicropores have been more studied recently. A series of commercial carbons with high performance (Desorex K43) was found to show their pore volumes concentrated below 20 Å. Measurements of CO_2 isotherms at 273 K for this carbon series showed the presence of ultramicropoes in all samples. The adsorption capacity of H_2S increased with increasing ultramicropores volumes [8]. Next the micropores were blocked with CO_2 and then the same H_2S stream was evaluated. That blocking let to a drastic decrease in H_2S retention [15]. For samples in which chemisorption was the dominant adsorption mechanism, there was a reduction up to 400% in H_2S capacities.

3 Physical and Chemical Adsorption

The textural parameters must be analyzed together with the variables related to the amount of impregnation ($C_{impregnant}$), pH, H_2S concentration and moisture in the flue gas (H_2S flue gas).

Among the various possibilities of impregnating agents (Na, K, Fe), as long as their incorporation does not substantially change the pore distribution, increasing the concentration of the impregnating agent increases the retention of H_2S [7]. Most of the impregnating agents are incorporated via the wet method which has certain deficiencies regarding the distribution in the carbonaceous material. Studying KOH impregnation in coal-based carbon, Yan [16] found that in addition to differences in concentrations of approximately 15% on the surface, where bright and dark areas could be discriminated, the analysis of the interior of the pellets showed concentrations of 50–80% below those found on the surface. Very often the impregnating agent can block the pores of the activated carbon, for this reason further PSD analysis must be done after impregnation.

As the first and second ionization of H_2S occurs at pH 7.2 and 13.9 respectively, a moderately basic microenvironment, with higher pH values, favors impregnation, for this reason this parameter must also be taken into account. Studies show that there is a limit pH value of 5, below which, regardless of the favorable PSD, H_2S retention will be severely compromised [17].

The concentration of H_2S and the humidity of the gas stream must also be defined. It is recommended that the same carbon be tested in currents with high (>500 ppm) and low H_2S concentration (<100 ppm). High concentration tests are preferred because they save time and resources. However, accelerated tests (high concentration) may not adequately describe the kinetics of a low concentration H_2S stream [18]. Accelerated tests can be used in an initial screening and then, if necessary, tests at lower concentrations, close to field conditions should be done.

Moisture is a factor of high impact on the performance of an activated carbon. It is believed that a monolayer of water induces the H_2S dissociation process, increasing the capture considerably. The effect of humidity can be seen by comparing cases 1 and 2 for the carbon Desorex K43 Na (Table 1), 80% of relative humidity increased retention in more than 10 times. The lower H_2S retention for samples 5–8 is also due to the lower current humidity (24% RH). Other examples of increased retention related to moisture have been reported in the literature [10]. Not only does the humidity of the stream influence the capture of H_2S, if the sample is pre-humidified to the limit of 2 mmol of water per g of carbon, the H_2S retention can increase from 2 to 6 times [10].

3.1 Chemisorption and Oxidation

The chemisorption process involves the production of HS^- ions which is promoted by the presence of water and the impregnating agent. Once the HS^- and dissociatively adsorbed oxygen (O*) ions are available, the conversion reactions to elemental sulfur (Sx), SO_2 and H_2SO_4 takes place. Adsorbed oxygen molecules are the source of dissociative adsorbed oxygen (O*). Probably the active sites of oxygen adsorption and dissociation should be edges or defective walls that can exerted stronger interactions with O_2 molecules [6].

Below, the main proposed reaction:

HS^- ion production

Water film

$$H_2S + H_2O \rightarrow HS^- + H_3O^+$$

impregnating agent
Na_2CO_3 [6]

$$H_2S + CO_3^{2-} \rightarrow HS^- + HCO_3^-$$
$$H_2S + HCO_3^- \rightarrow HS^- + H_2CO_3$$

KOH [19]

$$H_2S(g) + KOH(q) - C \rightarrow KHS(q) + H_2O$$
$$KHS(q) \leftrightarrow K^+(q) + HS^-(q)$$
$$HS^-(q) \leftrightarrow HS^-(ads)$$
$$2xHS^-(ads) + xO^*(ads) \rightarrow 2Sx(ads) + xH_2O$$

Conversion reactions [6,10]

$$C^-O^* + HS^- \rightarrow Cf + S^* + OH^-$$
$$xS^* \rightarrow Sx$$
$$S^* + O^2 \rightarrow SO_2$$
$$SO_2 + 0.5O_2 + H_2O \rightarrow SO_4^{2-} + 2H^+$$

Proposed mechanisms for other impregnating agents and in conditions with different concentrations of HS^- can be found elsewhere [4].

In an attempt to make the microstructure of activated carbon compatible with the adsorption/chemisorption mechanism, Chen [6] proposed a schematic representation of the process (Fig. 2). In this representation, the oxidation of H_2S can be divided into 4 steps: (1) The impregnant deposits in the larger pores (>7 Å); (2) A water film or/and clusters are formed in the micropores and even mesopores, resulting from water accumulation and condensation in the presence of moisture; (3) H_2S dissolves in the water films and dissociates into HS^- ions. At the same time O_2 dissociates into dissociatively adsorbed oxygen (O*); (4) HS^- reacts with O* to form elemental sulfur or sulfur oxides. In the larger pores where the HS^- concentration is higher, the reaction favors elemental sulfur as main product. In the small pores the low HS^- concentration and high O* and surface energy shifts the equilibrium to production of SO_2 and H_2SO_4.

3.2 *Physical Adsorption*

In a H_2S capture process where the amounts retained are attributed to the processes of chemisorption, oxidation reactions and physical adsorption that occur simultaneously, it is extremely difficult to define the individual contributions of each mechanism. In an attempt to clarify the contribution of the physical adsorptive process to the H_2S accumulation, a molecular simulation model, which will be explained in detail in the next section, was used [20].

Combining the molecular simulation model with the representative pore methodology [20] it was possible to estimate the theoretical amount adsorbed exclusively by physical adsorption (Table 2). Note that the amounts adsorbed exclusively by physisorption are much lower than the total amount retained in the impregnated samples.

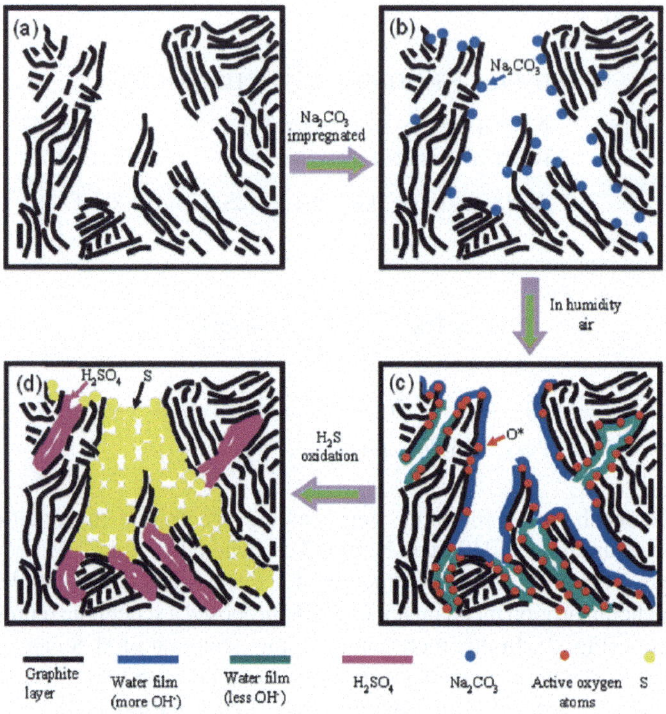

Fig. 2 Schematic representation of role of different pore sizes in the process of H_2S oxidation. *Source* Chen [6] Reproduced with permission of ACS

Two conclusions can be drawn from these results: non-impregnated samples basically adsorb by physical adsorption as shown by the proximity of the predicted value to the experimental value. Physical adsorption contributes very little to the total adsorbed when chemisorption processes are also present. However, as previously discussed, physisorption in ultramicopores is an active part of the H_2S retention process.

According to the molecular simulation model, the smaller pores of the ultramicroporous range adsorb the largest possible amounts of H_2S. The amount adsorbed falls quickly as the pores increase in size as shown in Fig. 3. These molecular simulation calculations indicate that ultramicropores are the main responsible for H_2S mobilization. Without this mobilization, which is eminently based on physical adsorption, the following processes of chemisorption and oxidation could not proceed. This mechanism was confirmed in the experiment that blocked these pores, verifying adsorption decreases of up to 400% [15]. A recent study revealed that molecules adsorbed in the pores of activated carbon has considerable mobility [21], reinforcing this hypothesis.

Table 2 Experimental H_2S retention ($C_{Ret.}$) and theoretical adsorption by physisorption of selected carbons

	Raw material	Area BET	V_{tot} (cm^3/g)	$C_{Ret.}$ (mg/g)	Theoretical adsorption by physisorption (mg/g)	$C_{Impregnant}$	pH	H_2S flue gas
Desorex K43-Na[a]	Bituminous + coconut	815	0.38 ($V_{<0.4\ nm}$ = 0.042)	**8.16**	0.3	Na 4.7 (wt%) Fe 3.8 (wt%)	10.2	100 ppm 0% RH
Desorex K43[a]	Bituminous + coconut	1003	0.48 ($V_{<0.4\ nm}$ = 0.02)	**0.34**	0.2	0	7.1	100 ppm 0% RH
Norit RB4-K[b]	Peat-base	1011	0.46 ($V_{<0.4\ nm}$ = 0.025)	**3.79**	0.36	KOH (Sol. 10%)	–	100 ppm 0% RH
Norit RB4[b]	Peat-base	957	0.50 ($V_{<0.4\ nm}$ = 0.01)	**0.22**	0.24	0	–	100 ppm 0% RH

[a]Menezes [8]
[b]Oliveira [15]

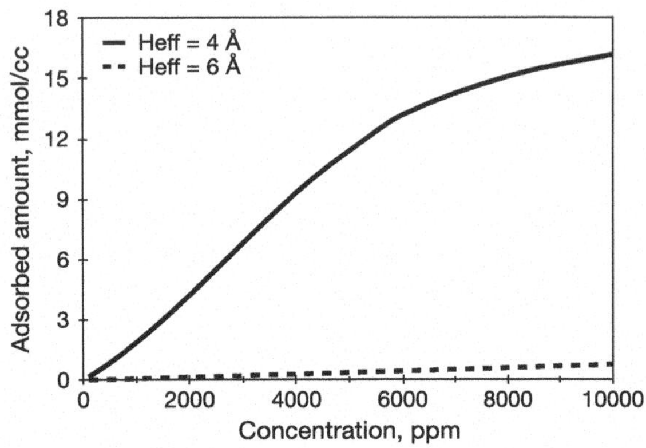

Fig. 3 Simulated adsorption isotherms of H_2S at 298 K on carbon slit-pores of 4 and 6 Å (effective pore width—H_{eff})

4 Molecular Simulation

Molecular simulation is the most efficient method when it is necessary to discover details of the molecular interaction with the surface, inaccessible by experimental techniques or to extrapolate the temperature and pressure conditions available in laboratory equipment or even to test hazardous components. Although the method has consistently contributed to several problems in the adsorption community [22], we find few investigations dedicated to the study of H_2S adsorption in carbonaceous materials.

Over the past ten years, the technique has been used to study adsorption on metalorganic structures—MOFs [23–25] and carbon nanotubes [26,27]. The first study of molecular simulation of H_2S on activated carbons [20] was carried out to predict monocomponent adsorption and mixtures with CO_2 and CH_4 using the grand canonical ensemble.

Central tools in molecular simulation are the elaboration of adsorbate and adsorbent models, the definition of the potential and the sampling method of the accessible states [28]. Activated carbons are commonly represented by slit-pores, which consist of walls with two or 3 sheets of graphene regularly spaced to represent the pores sizes (Fig. 4).

The model of the H_2S molecule that has shown best results was developed by Kristof and Liszi [29], being able to accurately reproduce the H_2S liquid–vapor equilibrium. The potential needed to predict interactions most used is that of Lennard-Jonnes (LJ):

$$U\left(r_{ij}\right) = 4\varepsilon_{ij}\left[\left(\frac{\sigma_{ij}}{r_{ij}}\right)^{12} - \left(\frac{\sigma_{ij}}{r_{ij}}\right)^{6}\right]$$

where ε_{ij} is the well depth, σ_{ij} is the diameter, and r_{ij} is the distance between the interacting atoms *i* and *j*.

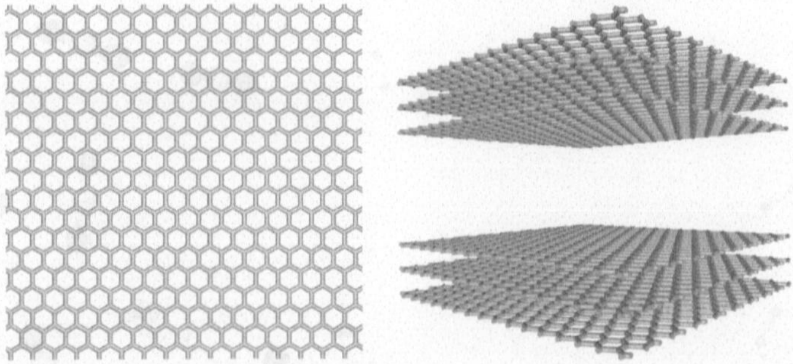

Fig. 4 View of graphene sheet and slit-pore model

In the case of the slit-pore model, the parameters ε_{ij} and σ_{ij} of the potential LJ equation are well known [30] however the interaction between the molecule and the carbon surface needs experimental data from the graphite equilibrium isotherm. The most suitable method for searching accessible states to obtain adsorption data is the Monte Carlo algorithm [31] in the canonical grand ensemble.

The value of the parameter ε_{ij} between the surface of the slit-pore model and the H_2S molecule was obtained by Gonçalves [20], from the reproduction of the isosteric heat of 40 kJ/mol at low coverage obtained experimentally for a series of activated carbons [32]. The solid–fluid interaction parameter ε_{ij} was estimated as 96.2 K. Once this parameter was defined, it was possible to predict the H_2S adsorption isotherm in activated carbon RB4 (Fig. 5) [20].

This same methodology was used to predict the amounts adsorbed exclusively by physisorption on the carbons shown in Table 2. It was also this methodology, based on molecular simulation, that made it possible to evaluate the importance of ultramicopores for H_2S adsorption. An advantage of molecular simulation is the

Fig. 5 Experimental and simulated adsorption isotherms of H_2S at 298 K in activated carbon RB4

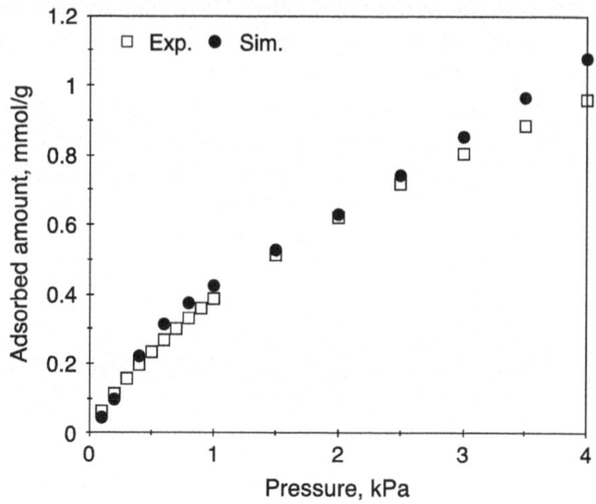

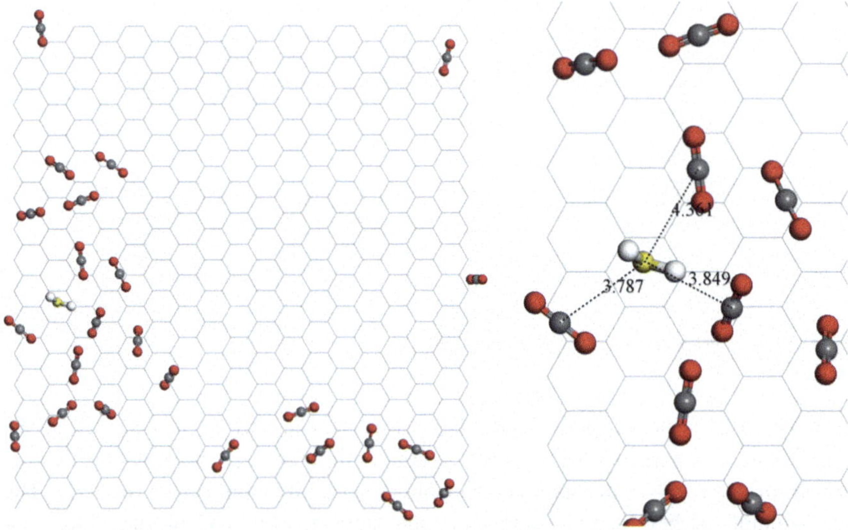

Fig. 6 Study of a mixture of 200 ppm of H_2S with 40% of CO_2 in an 8.9 Å pore showing the cooperative effect of CO_2 molecules favoring H_2S adsorption. The upper view of the graphene pore wall is shown with adsorbed CO_2 and H_2S molecules. In the detail, the distances in Å between H2S and CO_2 are highlighted. Hydrogen in white, sulfur in yellow, carbon in grey and oxygen in red. *Source* Gonçalves et al. [20]. Reproduced with permission of Elsevier

easy prediction of multicomponent mixture. For example, a cooperative effect of CO_2 molecules favoring H_2S adsorption has been recently reported by Gonçalves [20] (see Fig. 6).

In another study Huang [33] molecular dynamics was performed in a hybrid graphene-Au model with the potential for studies on activated carbons. The model was able to reproduce the adsorption energies of H_2S, H_2O and CO on the gold surface. A reference pore of 46.5 Å was prepared for the studies. One wall consisted of graphene sheets as usual in the activated carbons slit-pores models and the other wall was made up of Au (111) atoms.

Typical gases from reformer streams such as H_2O, CO_2, CO and H_2S were studied individually and in mixtures. The behavior of water in the concentration of 0.29–3.67 mol/L was examined indicating how the morphology of the water film, necessary to accelerate the formation processes of HS^- ions, could be. The study also found that H_2O and H_2S create dense H-bonding networks.

5 Breakthrough Curves and Design (Dynamic Column)

For practical applications of H_2S adsorption separations, fixed bed columns are generally used due to their easy and continuous operation. However, for an efficient process, besides a good selection of the carbon material and its properties (as seen in the previous sections), it is important to understand and model the dynamics of the fixed bed adsorption process. In this section we will present two recent studies to illustrate the important parameters for the design of an industrial column for this application. Also, the regenerability of the adsorption in conditions similar to the real process applications shall be evaluated in this continuous and dynamic process mode.

The tests are typically performed in a fixed bed unit as schematically shown in Fig. 7. The H_2S concentration in the feed stream is established via mass flow controllers prior to entering a fixed bed containing the adsorbent under temperature control. The sulfur concentration at the outlet of the adsorbent column is measured through an appropriate sulfur sensor, e.g. a Pulsed Flame Photometric Detector (PFPD) within a gas chromatograph. Details about the experimental procedures and conditions may be found in Castrillon [9], Menezes [8] and Oliveira [15].

In a typical breakthrough experiment, performed at given constant flow rate and temperature, the adsorption capacities may be estimated by integrating the area above the breakthrough curve at the breakpoint ($t = t_b$) and at saturation ($t = \infty$), using the following equation:

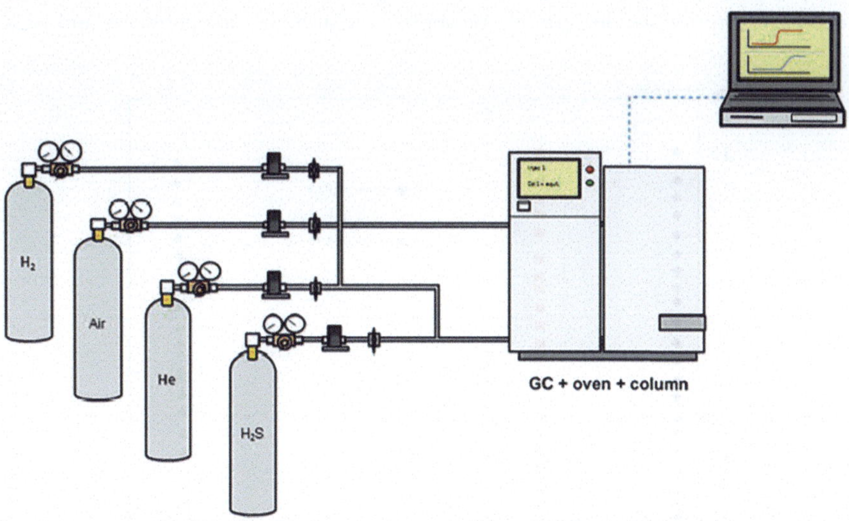

Fig. 7 Experimental scheme of a dynamic unit for measurement of breakthrough curves

$$q_0 = \frac{C_0}{M_B}\left[Q\int_{o}^{t}\left(1 - \frac{C}{C_0}\right)dt - V_C\varepsilon\right]$$

where q_0 is the adsorption capacity of the fixed bed at breakthrough or saturation, C_0 is the feed concentration, and C is the concentration at the outlet of the column as a function of time. ε is the void fraction of the bed, Q is the feed flow rate, V_C is the column volume, and M_B is the mass of adsorbent packed in the bed. The breakpoint time is usually considered as the time elapsed until the sorbate concentration at the bed outlet reaches a given fraction of the feed concentration.

Commercial samples of activated carbon (from Donau Caebon GmbH, Germany) have been evaluated through dynamic breakthrough experiments using a fixed bed unit. Experiments showed that chemical adsorption plays an important role in H_2S removal by adsorption from gaseous streams. A comparison of breakthrough curves performed under similar conditions of flow rate and temperature is shown in Fig. 8. The impregnation of the parent carbon (Desorex K43) with Fe_2O_3 (Desorex K43-Fe) or with NaOH (Desorex K43-Na) enhanced significantly the adsorption capacity for H_2S at 298 K, more specifically for the sample impregnated with NaOH (see Table 3).

The adsorption capacities obtained for breakthrough experiments using different temperatures for the same materials are reported in Table 4. An increase in sorption capacity with temperature was observed for both impregnated samples, especially for the one with NaOH. This is a clear evidence of the occurrence of chemical adsorption as a major mechanism for the H_2S retention in the carbon structure, since physisorption would certainly be suppressed at higher temperatures and such

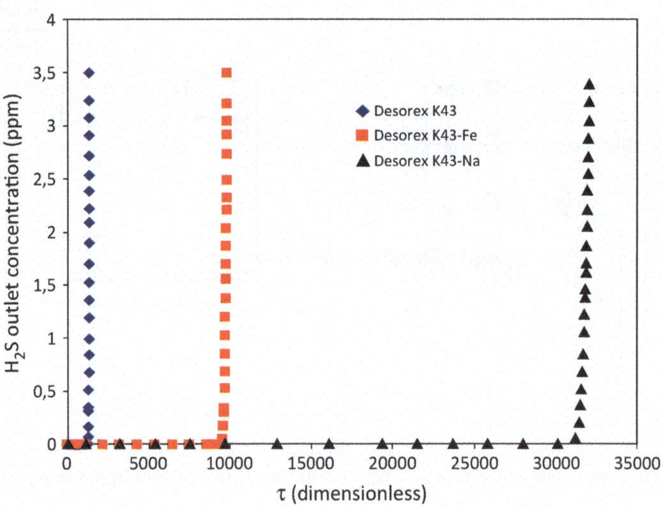

Fig. 8 Breakthrough curves of H_2S in commercial carbons, with and without chemical impregnation, at 298 K. τ (dimensionless time) is based on the feed volumetric flow rate

Table 3 Adsorption Capacities of H_2S in commercial activated carbons, with and without chemical impregnation, at 298 K

Carbon samples	q_{bt} (mg/g)	% pores < 4 Å
Desorex K43	0.34	9.3
Desorex K43-Fe	2.38	8.1
Desorex K43-Na	8.16	17.7

Table 4 H_2S adsorption capacities in commercial activated carbons, with and without chemical impregnation, at different temperatures

Carbon samples	q_{bt} (mg/g)			
	298 K	323 K	348 K	423 K
Desorex K43	0.34	0.24	0.34	1.02
Desorex K43-Fe	2.38	2.69	4.01	
Desorex K43-Na	8.16	10.20	15.64	

low partial pressures. In the case of the carbon matrix (Desorex K43), interesting results were found. The first increase in the temperature (from 298 to 323 K) led to a reduction of the H_2S adsorption capacity, as expected from a pure physical adsorption mechanism. However, as the temperature is further increased (from 323 to 348 K), the H2S capacity increases back to a value similar to 298 K. This may be related to the adsorbing molecules overcoming mass transfer resistances within the carbon framework and thus accessing ultramicropore regions in the carbon that could not be accessed at lower temperatures. This is further enhanced going to an even higher temperature (423 K) in the breakthrough experiments.

Finally, an important feature for any continuous industrial separation process is the regenerability of the adsorbent throughout consecutive cycles in operation. In Fig. 9, the breakthrough curves for four consecutive cycle, at 298 K, are displayed for each carbon material (parental and impregnated with NaOH or Fe_2O_3). Before the first cycle and between each two consecutive cycles, the adsorbent bed was regenerated at 423 K under constant flow of an inert gas (He). The experimental adsorption capacities of H_2S observed in each cycle are listed in Table 5. There is a drastic decrease in adsorption capacity between the 1st and 2nd cycles, for all samples. This trend is also observed in the following cycles, but somewhat less intensely for the nonimpregnated sample. But still, in all situations, the impregnated samples (specially the one with NaOH) show higher adsorption capacities up to the 4th cycle. This indicates a need to include in an eventual industrial process for H_2S removal a step for chemical regeneration of the carbon samples right after the physical regeneration with high temperature and inert gas flow.

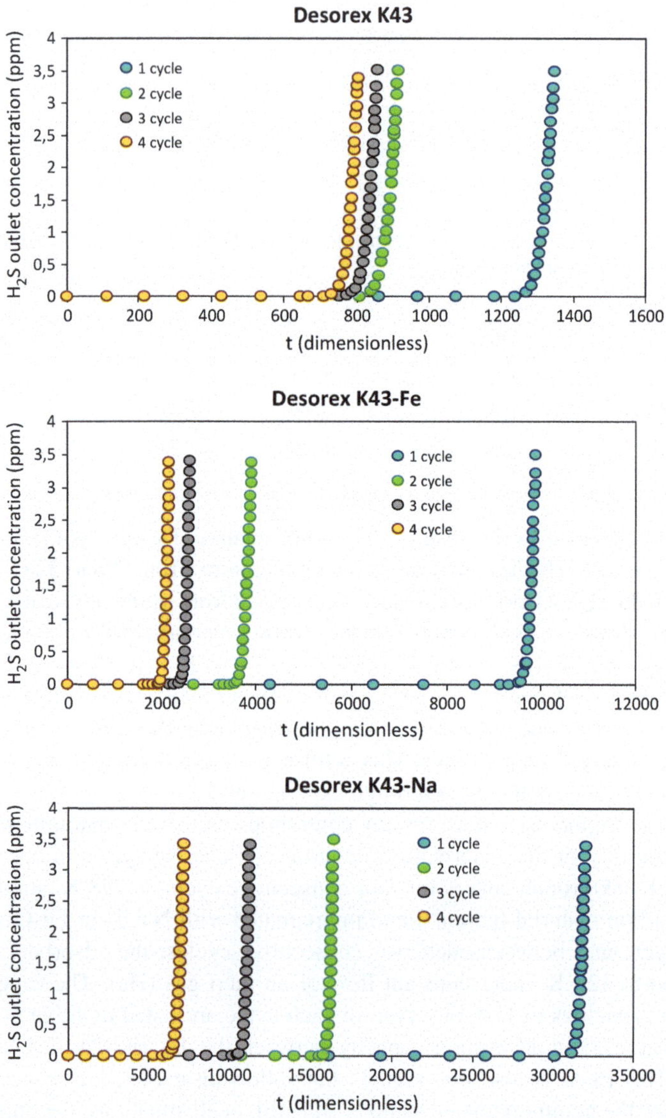

Fig. 9 Breakthrough curves of H_2S in commercial activated carbons, with and without chemical impregnation, for four consecutive adsorption/desorption cycles at 298 K. τ (dimensionless time) is based on the feed volumetric flow rate

Table 5 H_2S adsorption capacities in commercial activated carbons, with and without chemical impregnation, in four consecutive cycles of adsorption (298 K) and desorption (423 K)

Carbon samples	q_{bt} (mg/g)			
	Cycle 1	Cycle 2	Cycle 3	Cycle 4
Desorex K43	0.34	0.20	0.20	0.20
Desorex K43-Fe	2.38	0.88	0.58	0.48
Desorex K43-Na	8.16	3.74	2.72	1.73

6 Outlook and Perspectives

With the renewed interest in the capture of H_2S in biogas units, efforts have been made to improve the capture technology according to the specifics of this industry. The adsorption process is suitable for these units because they are disseminated close to urbanized areas and they have medium and small size. Despite the great competition with other adsorbent materials, activated carbon is still the material with the best cost–benefit ratio. The resumption of studies on these processes based on a better understanding of the relationship between the structure and oxidation reactions can bring gains to the entire list of processes that face the challenge of safely extracting H_2S from the process. Studies to allow better characterization of activated carbons, identify reaction mechanisms and better design strategies for PSA units are necessary to make more compact and economical units feasible. Important areas of research are still improperly explored: regeneration strategies, with evident economic and environmental benefits and molecular simulation, where important insights can be obtained from the application of quantum methods, Monte Carlo and molecular dynamics. Beginning in the first war rayon factories and passing through the oil, mining, chemical industries of barium carbonate and dyes, the capture of H_2S is not yet a fully solved problem.

Acknowledgements The authors wish to acknowledge financial support for this study from CAPES, CNPq, Petrobras and FUNCAP, and the use of the computer cluster at the Brazilian National Laboratory of Scientific Computing (LNCC/MCTI, Brazil). We are also grateful for having shared Prof. Rodriguez-Reinoso's friendship and teachings on carbon adsorption science, fundamentals and applications throughout the last two decades.

References

1. Marsh, H., Rodriguez-Reinoso, F.: Activated Carbon. Elsevier Science, London (2006)
2. Blanc, P.D.: Fake silk: The lethal history of Viscose Rayon (2016)
3. European Biogas Association (EBA): European Biogas Association Annual Report 2019 (2019)
4. Zhang, X., Tang, Y., Qu, S., Da, J., Hao, Z.: H2S-selective catalytic oxidation: Catalysts and processes. ACS Catal. **5**, 1053–1067 (2015). https://doi.org/10.1021/cs501476p
5. Bansal, R.C., Goyal, M.: Activated Carbon Adsorption. Taylor & Francis Group, New York (2005)

6. Chen, Q., Wang, Z., Long, D., Liu, X., Zhan, L., Liang, X., Qiao, W., Ling, L.: Role of pore structure of activated carbon fibers in the catalytic oxidation of H2S. Ind. Eng. Chem. Res. **49**, 3152–3159 (2010). https://doi.org/10.1021/ie901223j
7. Bashkova, S., Armstrong, T.R., Schwartz, V.: Selective catalytic oxidation of hydrogen sulfide on activated carbons impregnated with sodium hydroxide. Energy Fuels **23**, 1674–1682 (2009). https://doi.org/10.1021/ef800711c
8. Menezes, R.L.C.B., Moura, K.O., De Lucena, S.M.P., Azevedo, D.C.S., Bastos-Neto, M.: Insights on the mechanisms of H_2S retention at low concentration on impregnated carbons. Ind. Eng. Chem. Res. **57**, 2248–2257 (2018). https://doi.org/10.1021/acs.iecr.7b03402
9. Castrillon, M.C., Moura, K.O., Alves, C.A., Bastos-Neto, M., Azevedo, D.C.S., Hofmann, J., Möllmer, J., Einicke, W.D., Gläser, R.: CO_2 and H_2S removal from CH_4-rich streams by adsorption on activated carbons modified with K_2CO_3, NaOH, or Fe_2O_3. Energy Fuels **30**, 9596–9604 (2016). https://doi.org/10.1021/acs.energyfuels.6b01667
10. Bandosz, T.J.: On the adsorption/oxidation of hydrogen sulfide on activated carbons at ambient temperatures. J. Colloid Interface Sci. **246**, 1–20 (2002). https://doi.org/10.1006/jcis.2001.7952
11. Neimark, A.V., Lin, Y., Ravikovitch, P.I., Thommes, M.: Quenched solid density functional theory and pore size analysis of micro-mesoporous carbons. Carbon **47**, 1617–1628 (2009). https://doi.org/10.1016/j.carbon.2009.01.050
12. Jagiello, J., Olivier, J.P.: 2D-NLDFT adsorption models for carbon slit-shaped pores with surface energetical heterogeneity and geometrical corrugation. Carbon **55**, 70–80 (2013). https://doi.org/10.1016/j.carbon.2012.12.011
13. Lucena, S.M.P., Paiva, C.A.S., Silvino, P.F.G., Azevedo, D.C.S., Cavalcante, C.L., Jr., Cavalcante, C.L.: The effect of heterogeneity in the randomly etched graphite model for carbon pore size characterization. Carbon **48**, 2554–2565 (2010). https://doi.org/10.1016/j.carbon.2010.03.034
14. Lucena, S.M.P., Oliveira, J.C.A., Gonçalves, D.V., Silvino, P.F.G.: Second-generation kernel for characterization of carbonaceous material by adsorption. Carbon **119**, 378–385 (2017). https://doi.org/10.1016/j.carbon.2017.04.061
15. Oliveira, J.L.B., Nascimento, B.O., Gonçalves, D.V., Santiago, R.G., Lucena, S.M.P., Azevedo, D.C.S., Bastos-Neto, M.: Effect of ultramicropores on the mechanisms of H_2S retention from biogas. Chem. Eng. Res. Des. **154**, 241–249 (2020). https://doi.org/10.1016/j.cherd.2019.12.019
16. Yan, R., Liang, D.T., Tsen, L., Tay, J.H.: Kinetics and mechanisms of H_2S adsorption by alkaline activated carbon. Environ. Sci. Technol. **36**, 4460–4466 (2002). https://doi.org/10.1021/es0205840
17. Bagreev, A., Adib, F., Bandosz, T.J.: pH of activated carbon surface as an indication of its suitability for H_2S removal from moist air streams. Carbon **39**, 1897–1905 (2001). https://doi.org/10.1016/S0008-6223(00)00317-1
18. Elsayed, Y., Seredych, M., Dallas, A., Bandosz, T.J.: Desulfurization of air at high and low H_2S concentrations. Chem. Eng. J. **155**, 594–602 (2009). https://doi.org/10.1016/j.cej.2009.08.010
19. Yan, R., Chin, T., Ng, Y.L., Duan, H., Liang, D.T., Tay, J.H.: Influence of surface properties on the mechanism of H_2S removal by alkaline activated carbons. Environ. Sci. Technol. **38**, 316–323 (2004). https://doi.org/10.1021/es0303992
20. Gonçalves, D.V., Paiva, M.A.G., Oliveira, J.C.A., Bastos-Neto, M., Lucena, S.M.P.: Prediction of the monocomponent adsorption of H2S and mixtures with CO_2 and CH_4 on activated carbons. Colloids Surf. Physicochem. Eng. Asp. **559**, 342–350 (2018). https://doi.org/10.1016/j.colsurfa.2018.09.082
21. Valleroy, Z., Dos Santos, G., Lombardi, T., Wexler, C.: Adsorption of natural gas mixtures of methane, ethane, and propane in nanoporous carbon: fully atomistic numerical studies. Langmuir **36**, 3690–3702 (2020). https://doi.org/10.1021/acs.langmuir.9b03962
22. Bastos-Neto, M., Azevedo, D.C.S., Lucena, S.M.P.: Adsorption. In: Kirk-othmer encyclopedia of chemical technology (2020)

23. Hamon, L., Leclerc, H., Ghoufi, A., Oliviero, L., Travert, A., Lavalley, J.C., Devic, T., Serre, C., Férey, G., De Weireld, G., Vimont, A., Maurin, G.: Molecular insight into the adsorption of H_2S in the flexible MIL-53(Cr) and rigid MIL-47(V) MOFs: infrared spectroscopy combined to molecular simulations. J. Phys. Chem. C. **115**, 2047–2056 (2011). https://doi.org/10.1021/jp1092724
24. Ishak, M.A.I., Jumbri, K., Daud, S., Abdul Rahman, M.B., Abdul Wahab, R., Yamagishi, H., Yamamoto, Y.: Molecular simulation on the stability and adsorption properties of choline-based ionic liquids/IRMOF-1 hybrid composite for selective H_2S/CO_2 capture. J. Hazard. Mater. 399 (2020). https://doi.org/10.1016/j.jhazmat.2020.123008
25. Li, Z., Liao, F., Jiang, F., Liu, B., Ban, S., Chen, G., Sun, C., Xiao, P., Sun, Y.: Capture of H_2S and SO_2 from trace sulfur containing gas mixture by functionalized UiO-66(Zr) materials: a molecular simulation study. Fluid Phase Equilib. **427**, 259–267 (2016). https://doi.org/10.1016/j.fluid.2016.07.020
26. Darvish Ganji, M., Kiyani, H.: Molecular simulation of efficient removal of H_2S pollutant by cyclodextrine functionalized CNTs. Sci. Rep. **9** (2019). https://doi.org/10.1038/s41598-019-46816-2
27. Wang, W., Peng, X., Cao, D.: Capture of trace sulfur gases from binary mixtures by single-walled carbon nanotube arrays: A molecular simulation study. Environ. Sci. Technol. **45**, 4832–4838 (2011). https://doi.org/10.1021/es1043672
28. Gomes, V.A.M., Coelho, J.A., Peixoto, H.R., Lucena, S.M.P.: Easily tunable parameterization of a force field for gas adsorption on FAU zeolites. Adsorption **21**, 25–35 (2014). https://doi.org/10.1007/s10450-014-9647-3
29. Kristóf, T., Liszi, J.: Effective intermolecular potential for fluid hydrogen sulfide. J. Phys. Chem. B. **5647**, 5480–5483 (1997). https://doi.org/10.1021/jp9707495
30. Steele, W.: The Interactions of Nitrogen Molecules Adsorbed on Graphite. Le J. Phys. Colloq. **38**, C4-61-C4–68 (1977). https://doi.org/10.1051/jphyscol:1977410
31. Metropolis, N., Rosenbluth, A.W., Rosenbluth, M.N., Teller, A.H., Teller, E.: Equation of state calculations by fast computing machines. J. Chem. Phys. **21**:1087–1092 (1953). https://doi.org/10.1063/1.1699114
32. Bagreev, A., Adib, F., Bandosz, T.J.: Initial heats of H2S adsorption on activated carbons: Effect of surface features. J. Colloid Interface Sci. **219**, 327–332 (1999). https://doi.org/10.1006/jcis.1999.6485
33. Huang, P.H.: Molecular dynamics investigation of separation of hydrogen sulfide from acidic gas mixtures inside metal-doped graphite micropores. Phys. Chem. Chem. Phys. **17**, 22686–22698 (2015). https://doi.org/10.1039/c5cp02803e

The Immersion Calorimetry as a Tool to Study of the Adsorbate-Adsorbent Interactions on the Adsorption of Emerging Pollutants onto Activated Carbon from Water: Case Methylparaben and Paracetamol

Valentina Bernal, Juan Carlos Moreno-Piraján, Liliana Giraldo-Gutierrez, and Fernando Gómez-Granados

Abstract More and more liters of water are demanded to supply the needs of the world's population. However, many of the human activities generate water pollution. Some of the emerging pollutants, as pharmaceuticals, are hazardous for the environmental and human health. The adsorption on activated carbon is used as tertiary treatment in wastewater plants, in this step the emerging pollutants are removed when the molecules are resistant to other techniques. The immersion calorimetry is a technique used to determine the energy exchange during the formation of pollutant-activated carbon interactions, there are the most important interaction on the adsorption because the amount of pollutant adsorbed depends to the affinity with the adsorbent surface. In this work we using the immersion calorimetry to determine the affinity between the activated carbon and a pharmaceutical compound (pollutant) and its correlation with the adsorbate uptake. The tests were carried out on three activated carbons with different content of oxygenated groups and aqueous solutions of paracetamol and methylparaben. A deep thermodynamic study of the adsorbate-adsorbent interactions was carried out by calculation of two enthalpies: The immersion and interaction enthalpy. The first determines the energy associated

In memory and gratitude to Dr. Francisco Rodriguez-Reinoso.

V. Bernal · J. C. Moreno-Piraján (✉)
Grupo de Investigación en Sólidos Porosos y Calorimetría, Departamento de Química, Universidad de Los Andes, Calle 18A No. 0-03, Bogotá, Colombia
e-mail: jumoreno@uniandes.edu.co

V. Bernal
e-mail: v.bernalf2@uniandes.edu.co

L. Giraldo-Gutierrez · F. Gómez-Granados
Grupo de Calorimetría, Departamento de Química, Universidad Nacional de Colombia, Carrera 45 No. 26-85, Bogotá, Colombia
e-mail: lgiraldogu@unal.edu.co

F. Gómez-Granados
e-mail: fagomezg@unal.edu.co

J. C. Moreno-Piraján et al. (eds.), *Porous Materials*, Engineering Materials,
https://doi.org/10.1007/978-3-030-65991-2_9

with the formation of solvent-adsorbent and adsorbate-adsorbent interaction while the second, only evaluates the energy exchanges during adsorbate-adsorbent interaction. The values obtained for immersion enthalpy are between −38.9 and −69.5 J g^{-1} at solutions 6.62 mmol L^{-1}. The interaction enthalpy has values in the range of −19.2 and 48.8 J g^{-1}. The endothermic enthalpies are associated with a highest affinity solvent-activated carbon.

Keywords Activated carbon · Adsorption · Enthalpy · Interactions · Methylparaben · Paracetamol · Wetting

1 Introduction

1.1 Clean Water as an Indicator of Development: Technological Challenges and Risks of Emerging Pollutants

According to United Nations Educational, Scientific and Cultural Organization (UNESCO) in its inform "Water for a sustainable world (2015)", the world population growth is closer to 80 million persons per year and a global water deficit of nearly 40% is expected by 2030, and nowadays, the 20% of the groundwater sources on the word are being overexploited [1]. At the same time, this would imply agricultural (food security), health and energy problems.

The problem is not only related with the water supply, the pollution of water resources accents other problems related with the public health and the environmental care. According to the United Nations Committee on Economic, Social and Cultural Rights the access to safe water and sanitation is a human right but close to 745 million people do not have a good source of drinking water and other 2500 million do not have good sanitation facilities either [2].

The Word Health Organization (WHO) in the Guidelines for drinking-water quality shows that the pharmaceutical and chemical pollutants are the main sources of water contamination [3]. The increases in the pharmaceutical waste is produced by the change in the consumption habits of drugs like painkillers, antidiarrheal, antipyretic and other over the counter medicines. Also, to the increase in patients with comorbidities. This led to consider the pharmaceuticals as emerging pollutants.

According to the WHO, partially consumed drugs generate residues that contaminate water resources. Likewise, part of the drugs administered to humans and animals are not metabolized in the ADME process (Absorption, Distribution, Metabolism, and Excretion) and are eliminated through the urine and stool, reaching the sewage. Additionally, pharmaceutical molecules can accumulate in soils [3].

For instance, Paracetamol or acetaminophen is one of the most used drugs to treat the pain due to its analgesic and antipyretic properties. In countries such as Greece and Portugal the paracetamol sales are close 200 tons, 6300 tons in UK, and 10,000 tons in France. Only in UK, the paracetamol sales generated £191 million [4]. The increase in analgesic consumption is related to the appearance of molecules of this type in wastewater and other water sources [5–7].

Several of the effects of these molecules on the environment are currently known. For paracetamol, the main effects reported are endocrine disruption, oxidative stress and hepatotoxicity in male fish [8], other studies indicated intrauterine growth retardation and teratogenic effects of prenatal exposure in Zebrafish [9].

Furthermore, Methylparaben is used as an excipient in many pharmaceutical forms and is massively added to cosmetics and personal care.

products for its antimicrobial properties. Its presence in water resources around the word has been reported in literature, the concentrations found in rivers of Portugal and U.S.A are between 8.8 and 26 ng L^{-1} [10]. The studies indicate that the Methylparaben generates endocrine disruption in marine mammals, fish, birds and humans. Additionally, one of the most important problems with this compound is its accumulation in tissues which generates bioaccumulation and biomagnification [10, 11].

One of the biggest challenges for water treatment plants in respect to these contaminants is that they present physicochemical characteristics that make them persistent to many procedures. Some of these characteristics are small size, high Partition Coefficient, solubility in water higher than 30 ppm and a photolysis half-life greater than one week [12].

Particularly, in the countries that reusing water to irrigate the crops, the laws are strict and the detection limits of these pollutants are lowest. For this reason, the water treatment plants require applying more than one process to remove the pollutants. Even so, its complete removal cannot be assured.

A conventional water treatment plant includes processes such as chlorination, ozonation, adsorption, and reverse osmosis. By 2008, in the United States, a plant with these characteristics had a cost of US$ 136 million and it could process 378,000 m^3 [13]. The use of adsorption as tertiary treatment in the water plants has a lower cost than ozone, ozone/ H_2O_2 and ultrafiltration and removal results similar between them [14]. Furthermore, adsorption is less likely to form harmful compounds derived from pharmaceuticals if the adsorbent is an inert material. This is not the case of chlorination or oxidation with H_2O_2, which for example generate dioxins or organochloride compound [15].

The activated carbon (AC) is one of the most used adsorbents in water treatment plants because its physicochemical properties as largest surface area and porous structure allow the remove of inorganic and organic pollutants such pharmaceutical compounds. The international annual output of activated carbon are 500,000 tons [16], however its price depends the precursor, e.g. the activated carbon from coal has a cost of USD 1190–16,346/tons, from coconut shell between 1870 and 18,267/tons and when the precursor used is wood the prices hover around USD 2200–16,016/tons. For this reason, the investigation in new precursors and materials to prepare AC is booming [17]. To avoid the unnecessary production of activated carbon, physical and chemical processes are made in the adsorbent to modify the initial properties of a commercial AC, and in this way increase the adsorption capacity of the contaminants that are to be removed.

Until now it has been seen that access to drinking water is an indicator of economic development. However, beyond the technical aspects (costs, types of adsorbents, flow rates), the true development is visualized through the improvement of processes through the application of science and engineering.

The water purification by adsorption is a much-studied field in chemistry, chemical engineering, civil engineering and physics. The research for each of these disciplines is united in the software's development that would simulate the adsorption and thus, predict the adsorption capacities of the pollutants, the time required to complete the process, the energy released or adsorbed, and the adsorbent amount dosage [18–20].

1.2 Removing the Pharmaceutical Compounds from Water by Adsorption on Activated Carbon: A Thermodynamic Viewpoint of the Adsorbate-Adsorbent Interactions

From chemical point of view, the adsorption is the result of the compensation on the surface forces by the interaction with a molecule in gaseous or aqueous phase, the adsorbate. The interactions can be strong or weak, but this will depend to the affinity adsorbent-adsorbate. If the interactions are strong, the process is irreversible and the energy exchange is greater. The process is classified as chemisorption and implies the reaction between the chemical groups on the adsorbent and the pollutant. When the interactions are weak only involve Van der Waals and London forces and the energy exchange has values lower than the chemisorption [19, 21, 22].

In both cases, Chemisorption or physisorption, the aim of the process is the binding of drug to chemical groups on the adsorbent and in this way, to increase the remove of the pollutant. The scaling of adsorption as a chemical phenomenon on an industrial scale requires kinetic and thermodynamic favorability, otherwise the adsorbate-adsorbent interactions may not form or the process may be very slow and its application unfeasible.

The formation of the adsorbate-adsorbent interactions is a rallying-point of several steps on the process. This is largely due to the relation between thermodynamics and kinetic describes in the kinetic-molecular theory, as shown in Eq. (1) where the kinetic constant (K_c) is expressed as a function of A (the frequency or pre-exponential factor) and $e^{(-Ea/RT)}$, that represents the fraction of collisions that have enough energy to react. This mathematical model is known as Arrhenius equation.

$$K_c = Ae^{\frac{-Ea}{RT}} \tag{1}$$

The Arrhenius equation applied to the adsorption is explained in Fig. 1.

In the first step of Fig. 1, the reactants form a heterogeneous system (solid–liquid) and is comprised of activated carbon and the pollutant dissolved in the water. The reactants have a higher level of energy than the products, because the activated carbon has a high surface energy. In the second step, the transition state is forming.

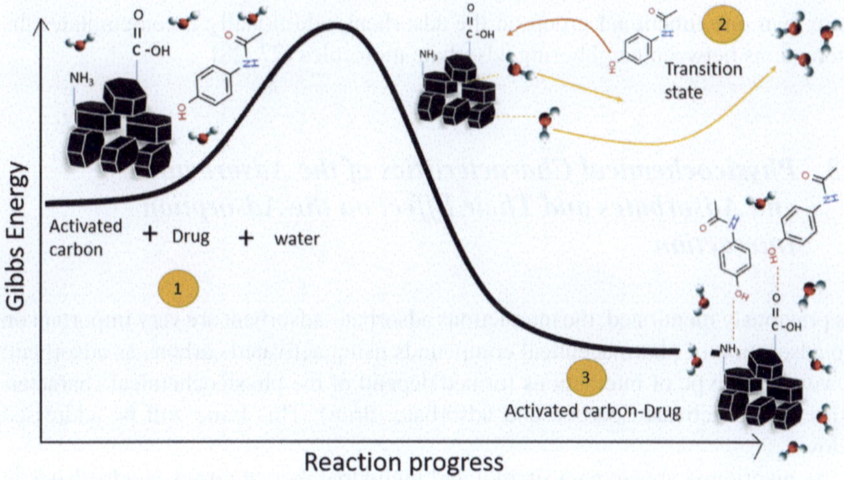

Fig. 1 Application of the kinetic-molecular theory to adsorption of drugs in aqueous solution on activated carbon

In the system, the solvent molecules are present in a greater proportion than the pollutant, therefore the interaction solvent-activated carbon is formed to compensate the surface energy on the adsorbent, then the pollutant (drug) must be displaced the solvent from the surface to form the adsorbate-adsorbent interactions. The energy required to displace the solvent will depend to the force of this interaction and the affinity drug-activated carbon [23–25]. Finally, in the third step the energy in the system decreases by the formation of adsorbent-adsorbate interactions [26].

The energy exchanged during the formation of the adsorbate-adsorbent interactions does not correspond to the total energy of the process; Additionally, the kinetic energy derived from the diffusion of the adsorbate through the porous matrix and the potential energy associated with the distance between the adsorbate and the pore walls must be taken into account.

From the molecular point of view, it is assumed that the formation of adsorbate-adsorbent interactions on the adsorption follows the kinetic-molecular theory; therefore, as the temperature increases, the collision factor with the functional groups on the adsorbent surface decreases.

At the same time, the interactions adsorbate-adsorbent are related with the adsorption models, that is, mathematical models used to describe how much drug (adsorbate) is adsorbed based on the initial amount of the compound in the solution once the chemical equilibrium is reached at a certain temperature. Some of the most used models to describe the adsorption of pharmaceutical compounds in activated carbon is the Langmuir and Freundlich model, these models classify the adsorption as homogeneous or heterogeneous, respectively. In the first case, Langmuir's model assumes that only interactions between adsorbate and specific groups are formed on activated carbon, therefore the energy exchanged in the process remains constant. The Freundlich's model allows the formation of interactions between the adsorbate and more than one functional group on the adsorbent, additionally it contemplates the interactions between neighboring adsorbate molecules [27, 28].

1.3 Physicochemical Characteristics of the Adsorbent and Adsorbates and Their Effect on the Adsorption Interaction

As previously mentioned, the interactions adsorbate-adsorbent are very important on the adsorption of pharmaceutical compounds using activated carbons as adsorbent; however, the type of interactions formed depend of the physicochemical characteristics on the activated carbon and adsorbate (drug). This issue will be addressed below.

As mentioned above, paracetamol and methylparaben are used as adsorbates in this study because these compounds are widely employed in pharmaceutical and personal care products, and they are considered pollutants by their appearance in water resources. The chemical structures and physicochemical characteristics are presented in Table 1.

The adsorption on activated carbon requires that the adsorbates have a small size because the adsorption is carried out on the micropores (pores with internal width

Table 1 Chemical structures and physicochemical characteristics of methylparaben and paracetamol

	Methylparaben	Paracetamol
	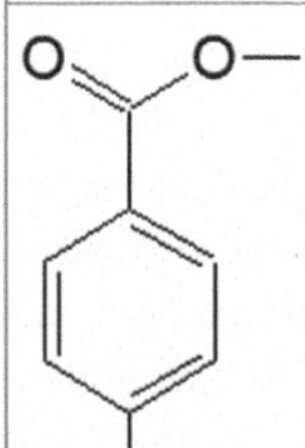	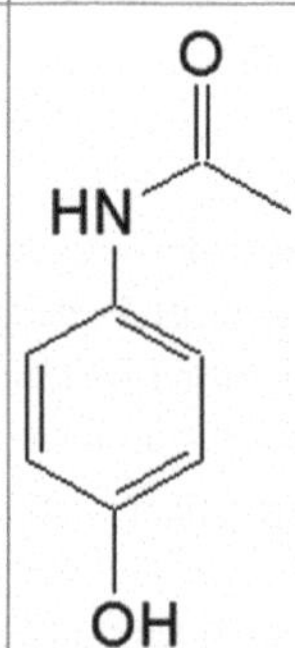
Molecular size (Å^2)	8.69 × 5.02	8.50 × 5.70
Molecular weight	152	151
pKa	8.50	9.46
Log P	2.17	0.91
Solubility in water at 298 K (M)	0.024	0.027
Hydrogen acceptor count	2	2
Hydrogen donor count	1	2

lower than 2 nm); thus, if the both adsorbates have molecular sizes lower than this value is expected that the diffusion do not a problem [29].

Moreover, the chemical equilibriums present in the adsorption system must be taken into account, in this point the solution chemistry plays a very important role in the adsorption performance. The equilibriums constants are the representation of acid–base and solubility equilibria; therefore, at solution pH values higher than 10 both adsorbates will have negative electric charges associated with the deprotonation of phenol group. Depending on the type of functional groups present on the activated carbon surface, the electrical charges can induce the formation of ionic and ion–dipole interactions or, conversely, favor the formation of repulsion forces. Additionally, electric charges increase the polarity of the adsorbate which disadvantages the affinity with activated carbon.

On the other hand, the solubility is an indicator of the solvent-adsorbate affinity, if its value is very large, the adsorption process is expected to make a higher energy consumption to separate the adsorbate molecules from the solvent and thus be able to adsorb on the activated carbon. The Lundelius rules is an empirical rule that correlate the solubility and adsorption performance for a homologous series of adsorbates (similar chemical structures), the rule indicated that adsorption from solution increases when the molecules become more hydrophobic; then, their solubility in water decrease [30].

From the environmental point of view other chemical properties are relevant, e.g. the partition coefficient allows determining if a compound can accumulate in fatty tissues or other matrices with non-polar or amphipathic characteristics. A high Log P value indicates that the compound can bioaccumulate in animal body fat and pass to humans through the trophic chain, as well as can be strongly retained from organic matter present in soils, sediments and biota.

The chemical groups present in the structure of the adsorbate make it possible to determine the behavior of the molecule in the acid–base balance and to establish, for example, the number of possible hydrogen bonds that can be formed in the adsorbate-adsorbent interactions.

Activated carbons are defined as amorphous solids composed primarily of carbon, although they may contain heteroatoms such as oxygen, nitrogen, sulfur, among others, from the raw materials used to prepare the activated carbon, almost always lignocellulosic and industrial residues, polymers, anthracitic or bituminous carbon. Additionally, the content of heteroatoms can be modified once the activated carbon is obtained through physical and chemical processes.

From the structural point of view, activated carbon is made up of graphene layers that are located randomly giving rise to the characteristic porosity of this solid. Chemical groups are usually located on the adsorbent surface, but some groups such as oxygenated functions can form at the pore edges. This will also have practical implications in the adsorbate diffusion if the molecule does not have small size.

The effect of the chemical groups on the activated carbon interactions will be addressed during this work but the effect of precursors on the characteristics of activated carbon will not be explained.

Currently, adsorption on activated carbon remains a relevant topic in research due to its industrial uses. One of the areas most studied on pharmaceutical adsorption in aqueous solution is the development of adsorbents whose surface chemistry has been modified to increase the adsorbate-adsorbent affinity.

In the literature it is found that the main methods to modify the surface chemistry of activated carbons that will be used to adsorb paracetamol and methylparaben are: immersion in strong acids or bases, salts and heat treatment at high temperatures.

Table 2 summarizes some of the main results in the literature about the use of activated carbons with modified chemical surface on the adsorption of methylparaben and paracetamol. Some of the treatments done on the samples (commercial activated carbons or lignocellulosic materials) were performed before activation to observe their effect in the physical properties of the adsorbent.

The electrostatic interactions between the adsorbate and the adsorbent depend not only on the pKa of the adsorbate, but also on the electric charge on the adsorbent, which is quantified by the point of zero charge (pH_{pzc}), that corresponds at a pH value where the number of electric positive charges is equal to the negative charges on adsorbent surface. At pH values above the pH_{pzc} the electric charge of the surface is negative, thus if the adsorbate is an weak acid will have the same charge and the repulsive forces increase in the system disfavoring the adsorption [29].

To determine the effect of adsorbate-adsorbent interactions on the adsorption of Paracetamol and methylparaben on activated carbon, three activated carbons with different surface chemistry were used. Table 3 summarizes the methods used to modify the surface chemistry of a commercial activated carbon (CAC).

The methods used to characterize the activated carbons CAC, $AC\text{-}H^+$ and AC-T are summarized in Table 4.

The results of the physicochemical characterization of the activated carbon are reported in Table 5.

The effect of the treatments carried out on the activated carbons in both physical and chemical properties has been reported in previous works. The oxidation with HNO_3 increases the acid groups present in the starting activated carbon (mainly carboxylic acids); However, the formation of these groups occurs at the edge of the micropores, which decreases the adsorption of Nitrogen due to steric effects, thus, the surface area and micropore volume on $AC\text{-}H^+$ decrease [29].

The heat treatment decreased the content of phenol groups on the starting sample, however the content of carboxylic acids increased due to reactions between free radicals (formed by the breaking of carbonaceous chains) and environmental CO_2. The temperature generated the collapse of carbonaceous structures that generated new porosity, this favored the nitrogen adsorption, therefore, an increase in the surface area [42].

The work is focused in thermodynamic issues of the interaction adsorbate-adsorbent on the adsorption of emerging pollutants, as pharmaceuticals, from aqueous solution. it was studied the energy exchange when the activated carbons are putted in contact with water and drug solutions at different concentrations using calorimetric techniques. The experimental data was correlated with the physicochemical properties of the activated carbons and the solution chemistry.

Table 2 Methods to modify the surface chemistry of activated carbons and their effect on adsorbate-adsorbent interactions

Modification method	Reagent	Effect on chemical surface groups	References
Immersion of AC in strong acids after carbonization	**HNO_3, H_2O_2** Paracetamol Adsorption capacity: 11.09 mg g^{-1} Methylparaben Adsorption capacity: 2.20 mg g^{-1}	It increases the formation of phenolic and carboxylic groups. pH_{pzc}: 3–5 The interactions adsorbent-adsorbate are favored by the increase of carboxylic acid groups	[29, 31, 32]
Impregnation of precursor in strong bases before the carbonization	**NaOH, KOH** Paracetamol Adsorption capacity: 7.57 mg g^{-1} Methylparaben Adsorption capacity: not reported	It Induces the formation alcohols, phenols and esters and an increasing in the total basicity. pH pH_{pzc}: 8.74	[31]
Impregnation of precursor in salts before the carbonization	**NH_4Cl-** Paracetamol % removal: 88%	It Increases the formation of Hydroxyl, aliphatic, carboxylic, and carbonyl groups. pH_{pzc}: 7.8. The paracetamol adsorption decreases at highest values of pH	[33]
	$Cu(NO_3)_2$- Methylparaben % removal: 52%	It generates an increase in the total acidity and phenolic groups. The uptake of methylparaben increases by the formation of specific interactions between the adsorbate and acidic groups	[34]
Thermal treatments at high temperatures on AC	**1073 K, 1173 K, 1273 K** Paracetamol Adsorption capacity at 1073 K: 1.38 mg g^{-1} Methylparaben Adsorption capacity at 1073 K: 1.53 mg g^{-1}	The temperature decreases the content of acidic groups by thermal instability of them	[32]

Table 3 Methods for obtaining activated carbons CAC, AC-H^+ and AC-T

Activated carbon	Modification method
CAC	Commercial activated carbon (Carbochem brand GS50). The sample was washed with HCl to eliminate the impurities and then was washed with distilled water until to reach pH constant. The activated carbon was dried at 383 K and stored in amber glass containers leaving minimal headspace
AC-H^+	A part of the activated carbon CAC was taken and immersed in a concentrated solution of HNO_3 at boiling point for 24 h. The sample was filtered and washed with distilled water until to reach pH constant. The activated carbon was dried at 383 K and stored in amber glass containers leaving minimal headspace. The aim was modifier the surface chemistry of CAC increasing the acidic chemical groups by oxidation
AC-T	Other part of the activated carbon CAC was subjected to heat treatment at 1073 K for 2 h in a furnace Thermolyne™ with cell in quartz an inert atmosphere (N_2). The activated carbon was stored in amber glass containers leaving minimal headspace. The aim was modifier the surface chemistry of CAC decreasing the chemical groups which thermal stability are lower than 1073 K (thermal reduction)

2 Calorimetry: Fundamentals and Methods

2.1 Calorimeters

The heat is the energy transfer between two bodies with different temperatures. The exchanged energy is contained in the bodies and is called internal energy, during physicochemical processes it can take different forms: kinetic, potential, thermal, electrical energy, among others. On adsorption, the uncompensated internal energy at the activated carbon surface is compensated through adsorbate-adsorbent interactions, therefore the new complex is more stable because it has lower potential energy.

In aqueous solution, energy transfer can not only be attributed to the formation of adsorbate-adsorbent interactions, since the solvent is found in a greater proportion and also interacts with the functional groups present on the adsorbent surface, one of the challenges of adsorption on activated carbon is to decrease its affinity for the solvent without affecting the affinity for the adsorbate. However, the magnitude of the energy associated with the solvent-adsorbent interactions is not the same associated with the solvent-adsorbate interactions, in this work these energy differences are reported and based on the results, the affinity of the adsorbent for one of the components of the system is determined.

The instruments used to measure the heat flow are the calorimeters. Current studies in calorimetry are based on the results obtained by Tian in the period between 1923 and 933. During this period Tian developed a more sensitive calorimeter, that is still in use today, and allows the measurement of energy outputs of the order of a

Table 4 Methods used to characterize the activated carbons CAC, AC-H^+ and AC-T

Method	Used to determine	Brief description	References
N_2 adsorption/desorption	Surface area (BET), pore volume (microporous-DR), pore size distribution (DFT)	The process was carried out in an apparatus Autosorb 3B (Quantachrome instruments, Anton Paar) at 77 K in a cell 9 mm. The weight of the sample was approximately 0.3 g. The activated carbon was previously desorbed at 393 K for two hours	[35–37]
Boehm Titration	Acidic functional groups: carboxylic acids, phenols and lactones Total basicity	The content of functional groups is known through an acid–base back titration. A quantity of 500 mg of each activated carbon were weighed in amber glass containers and added to 50 mL of the base solution to determine the content of acidic groups (0.1 M NaOH, Na_2CO_3 or $NaHCO_3$). To determine the total basicity, we used as immersion liquid 50 mL of 0.1 M HCl solution and the same quantity of activated carbon. The containers were stored at 298 K under constant agitation for 5 days. After this, aliquots of the supernatant were taken from each container and the sample was titrated with NaOH or HCl solution, respectively	[38, 39]
Mass titration	pH_{pzc}	50–500 mg of each activated carbon was weighed and placed in glass containers with 10 mL of 0.1 M NaCl. The samples were stored under agitation (150 rpm) and constant temperature (293 K) for 48 h. Finally, the pH of each solution was measured	[40, 41]

Table 5 Physicochemical characteristics of activated carbons CAC, $AC\text{-}H^+$ and AC-T

Activated carbon	Surface area (m^2 g^{-1})	Micropore volume (cm^3 g^{-1})	Content of phenol (μmol g^{-1})	Content of lactones (μmol g^{-1})	Content of carboxylic acids (μmol g^{-1})	Total basicity (μmol g^{-1})	pH_{pzc}
CAC	864	0.34	46.6	21.8	22.2	742	5.40
$AC\text{-}H^+$	469	0.18	53.9	490	106	735	3.40
AC-T	1127	0.42	6.36	21.2	66.1	1210	11.1

microwatt (μW). In the International System of Units (SI), heat is expressed in the unit of energy Joule (J) equivalent to W s.

The calorimeters do not measure the quantity of heat produced directly but records the rate of heat production (calorific power) expressed as $\frac{\partial Q}{\partial t}$. The experimental data obtained with the apparatus correspond to a continuous curve that represent the change of calorific power with the time. The heat change is calculated from the area under the curve for time interval.

The calorimeters can be classified through the parameter called thermal conductivity (p), which is directly associated with heat flux (θ) and temperature as shown in Eq. (2).

$$\theta = p(Ti - Te) \tag{2}$$

Ti and Te represent the temperatures of the internal and external boundary in the calorimeter. The internal boundary corresponds to the calorimeter container "cell" and the external boundaries are the walls of the surrounding cavity. Therefore, the energy exchanged during a chemical process, as the adsorption, increases the temperature in the cell and it generates a temperature difference between the two boundaries and the production of the heat flow.

According with the values of calorific power, the calorimeters are classified in three types adiabatic, isothermal and Tian-Calvet. The differences between them are summarizes in Table 6.

The apparatus used to determine the heat exchange generated during the immersion of activated carbon in water or in aqueous drug solutions was the Tian-Calvet calorimeter.

As indicated above, in Tian-Calvet calorimeters, the heat generated in the cell is measured through thermopiles, therefore it is governed by the principles of thermoelectricity, that is differences the heat is equivalent to the electric charge, the heat flow rate to the electric current, the temperature to the voltage, the thermal resistance to the electric resistance, and the heat capacity to the electric capacitance. The thermoelectric measures depend the several effects such as: seebeck, Peltier and Joule.

These physical phenomena occur in thermocouples, which are the structural units of thermopiles, they are made from two different metals (A and B) and their union allows the flow of the electrons.

Table 6 Types of calorimeters

Calorimeters	Characteristics
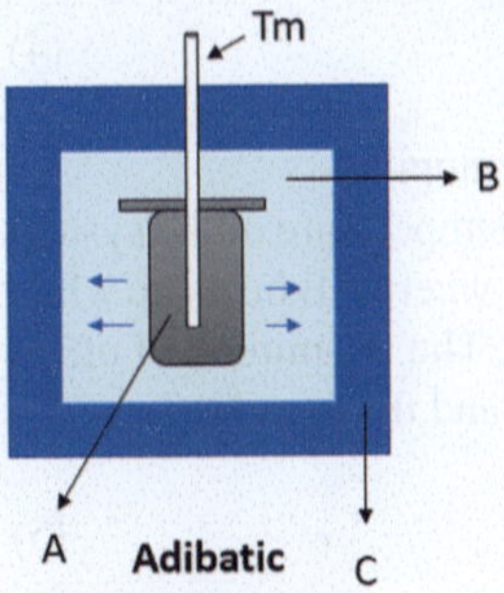	The thermal conductivity is zero or the values are smallest. Thus, it does not exchange heat with the environment because the internal (A) and external (C, good conductor of heat) limits are separated by an insulator (B) The temperature is measured with a thermometer adapted to the calorimeter cover
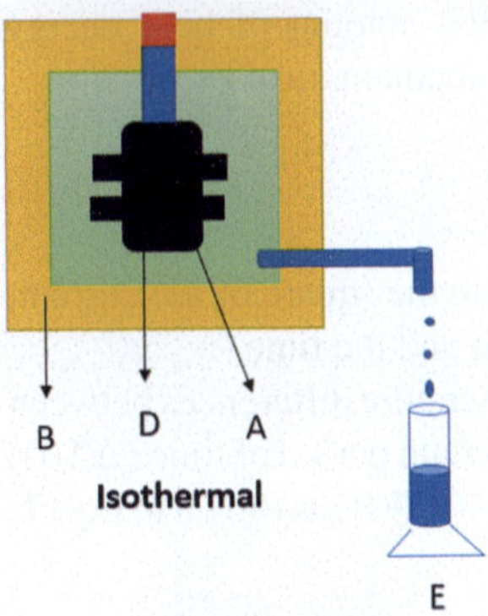	In this calorimeter the heat generated in cell (A) is used to induce the fusion of a solid (it is putted in D) that has previously been weighed. The generated liquid is collected in the test tube (E), its weight is proportional to the heat produced The letter B is still an insulating wall The thermal conductivity is very large
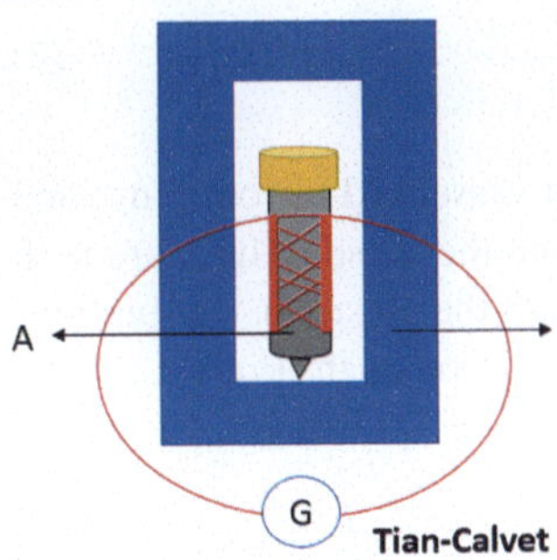	The heat generated in the cell (A) is conducted to the external jacket (C) through the thermopiles that surround the cell (red squares). The electromotive force produced by the thermopiles is proportional to the calorific power transferred G represents the galvanometer

The seebeck effect is the formation of an electric current (E) associated with the difference in temperatures at the junctions of metals AB and BA. An electromotive force can also occur, its magnitude will depend on the nature of the metals and the difference in temperatures (T2-T1). In Eqs. (3) and (4) are represented the thermoelectric power (e.m.f) for a thermocouple (tc) and thermopile (tp), respectively.

$$e.m.f = \frac{\partial E}{\partial T = (T_2 - T_1)} \tag{3}$$

$$E = n(e.f.m)(\partial T) \tag{4}$$

where n represent the number of thermocouples in the thermopile.

If a current is passing through the entire system, and the temperature of the system is uniform, there is a temperature change in the metal joints, where AB heats up while BA is cooled. This phenomenon is known as Peltier effect. The magnitude of effect will depend to the peltier coefficient (Pc), the temperature and the current (i)

$$calorific\ power = P_c.i \tag{5}$$

Very similar is the Joule effect, which quantifies the amount of heat energy produced by an electric current. Equation (6) shows the mathematical expression.

$$E = i^2.R.T \tag{6}$$

It can be seen that the calorific power is proportional to the square of the current, to the resistance of the material to the flow of the current and the time.

The Tian's equation permits the calculation of the temperature differences between the container of the sample (cell) and surroundings in a certain period of time ($\Delta T(t)$) from the heat production rate during the same period of time ($\Phi(t)$) as shown in Eq. (7) or Tian equation.

$$\Phi_{(t)} = -\frac{\Delta T(t)}{R_{th2}} - C_{SC}.\frac{d\Delta T(t)}{\delta t} \tag{7}$$

where R_{th2} is the thermal resistance between calorimeter vessel and surroundings and Csc is the heat generated in the sample and transferred to the vessel. These are both instrumental parameters and they are determined through the apparatus calibration.

To integrate up Eq. (7), the mathematical expression becomes in (Eq. 8):

$$\int_{t_{initial}}^{t_{final}} \Delta T(t)\delta t = -R_{th2} \cdot \int_{t_{initial}}^{t_{final}} \Phi_{(t)}(t)\delta t + C_{SC}.R_{th2} \cdot \int_{t_{initial}}^{t_{final}} \delta\Delta T \tag{8}$$

From Eq. (8), the calorimetric curve (Fig. 2) represents the energy exchange in the system as a function of time, thermal resistance, the temperature difference and the heat production rate.

There are two peaks in Fig. 2, one corresponds to the heat flow generated in the immersion and the other to the electrical calibration; in both cases, the peaks formation are related with the heat production rate, which increases to a maximum and subsequently decays as a function of time. The heat generated in each process is associated with the area under the curve of each peak. The last step is necessary in

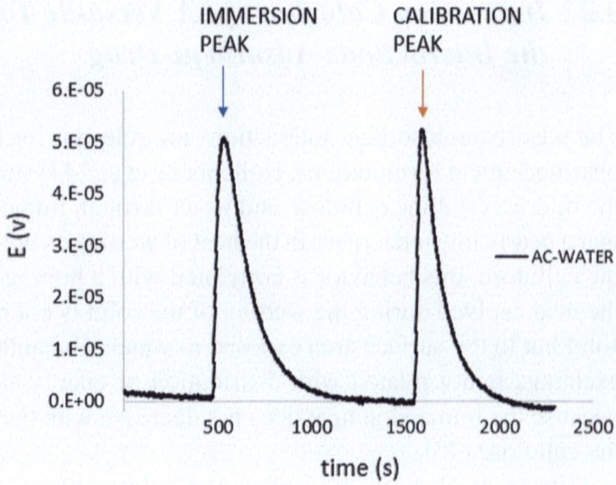

Fig. 2 Calorimetric curve obtained after the immersion of the activated carbon in water

calorimeters that involve an electrical compensation of the thermal effect to determine the sensitivity of the apparatus. Additionally, the calibration is done by measuring variables such as voltage and current, which can be measured with great accuracy.

In this work, Eqs. (9)–(11) were used to calculate the heat transferred during the immersion of the activated carbon in the aqueous solutions of methylparaben and paracetamol.

$$K_{cal} = \frac{volt\,(V) * current\,(A) * time\,(s)}{Area\,under\,curve\,(calibration\,peak)} = Watts/volt \tag{9}$$

$$Q_{imm} = K_{cal} * Area\,under\,curve\,(immersion\,peak) \tag{10}$$

$$\Delta H_{imm} = \frac{Q_{imm}}{activated\,carbon\,mass(g)} \tag{11}$$

Kc represents the calibration constant of the calorimeter, Q_{imm} is the heat of immersion and ΔH_{imm} is the immersion enthalpy.

The enthalpy and heat are related through Eq. (12):

$$\Delta H = Q_p \tag{12}$$

Equation (12) indicates that the enthalpy change is equal to the heat at constant pressure.

The internal energy (U) of the system is equal to the sum of the heat and work (Eq. 13), assuming that solids and liquids are incompressible, thus the term ΔV is zero and the internal energy is equal to heat.

$$U = Q - P\Delta V; \quad \Delta U = Q = \Delta H \quad when \Delta V = 0 \tag{13}$$

2.2 Immersion Calorimetry: A Versatile Tool to Understand the Interactions Adsorbent-Drug

The adsorbate-adsorbent interactions are relevant for the studies of the stability of pharmaceutical formulations. Hollenbeck et al. [43] studied the interactions between the microcrystalline cellulose and water through immersion calorimetry, they found that a nearly linear decrease in the heat of wetting as the moisture content increases in the cellulose, this behavior is correlated with a homogeneous surface. Additionally, the heat evolved during the wetting of the solid is not related to the total area on the solid but to the surface area exposed to water. The authors indicated that the energy exchange is not related with distribution of energy sites on the adsorbent surface because the immersion heat does not decrease with the occupation of active sites on the cellulose [43].

Gibson et al. used the immersion calorimetry to study the adsorption of the polymer (pigments and opacifiers) on the additive surface. They also analyzed the interaction between some solid additives used in tablet film coating and polymers as hydroxypropyl methylcellulose. According to the authors, the immersion enthalpies in water of titanium dioxide, red iron oxide and Talc A is −751.2, −997.4 and + 28.3 mJ g^{-1}, respectively. The exothermic values indicate highest affinity to the solids for the water, however, when the immersion liquid is a solution of hydroxypropyl methylcellulose (2%) the values of enthalpies increased in all systems because the polymer reduced the surface tension [44].

In 1991, Buckton et al. worked in the relation of wetting in other pharmaceutical issues such as: dispersion of powders to form suspensions, the effect of binders in the process of wet granulation and dissolution rate of solid dosage forms in the gastrointestinal tract. Thus, immersion calorimetry is used as a technical to improve the formulations because the role of surface energies and polarities has been investigated as a method of predicting interactions between the components of a formulation when the contact angle measurement cannot be performed [45].

Landin et al. used the immersion calorimetry to compare the characteristics of microcrystalline cellulose from different countries and different moisture contents, the aim of the authors was studied the water-cellulose interaction. The authors indicate that the immersion enthalpy increases with the content of –OH groups, however the major contribution to the immersion enthalpy is attributed to the –OH groups from the amorphous component of the cellulose. In other words, celluloses with highest crystalline content have fewer interactions with the water [46].

Hogan and Buckton investigated the effect of moisture in the crystallization of α-Lactose monohydrate. The studied the water adsorption on lactose and their interactions by immersion calorimetry; the authors indicated that the energy exchanged during calorimetry is related with the binding of water on the lactose surface, and this is the greatest part of the overall wetting response. Therefore, as amorphous lactose can adsorb more amounts of water than crystalline structures its enthalpy values are greatest [47].

Ridder et al. using the immersion calorimetry to determine the hydrophobicity of different modified zeolites used to remove pharmaceuticals compounds from demineralized and surface water. The results reported in the study show that the heat of immersion have values between 7.30 and 32.6 mJ/m^2, these values increased with the H2O vapor uptake which are between 0.45 and 0.66 mg m^2 [48].

Terzyk and Rychlicky studied the adsorption of paracetamol on activated carbon, they determined that the adsorbed amount decrease with the surface affinity by the water. The interaction AC-water was determined by immersion calorimetry, they reported that immersion enthalpies have values between −53.5 and −95.6 J g^{-1} and −85.8 and −120.5 J g^{-1} in water and paracetamol solution, respectively [24].

Moreno-Marenco et al. reported the thermodynamic and adsorption data for methylparaben on activated carbons impregned with metallic salts before the activation. The authors shown the immersion enthalpy at 100 mg L^{-1}, the enthalpy values are between −22.45 and −45.23 J g^{-1}, when the concentration of methylparaben is doubled the enthalpy remains constant [49].

2.3 Methodology

Immersion enthalpies were determined for activated carbons AC-H$^+$, CAC and AC-T in methylparaben and paracetamol solutions at 293 K. A heat conduction microcalorimeter Tyan-type was used to determine the enthalpies changes (Fig. 3). First, 10 mL of drug solution with concentration between 0.07 and 6.62 mmol L^{-1} were placed in a cell inside the calorimeter (Fig. 3 item 2). On the other hand, 0.1 g of CAC was weighed in a glass ampoule which adapts to the top of calorimeter (Fig. 3

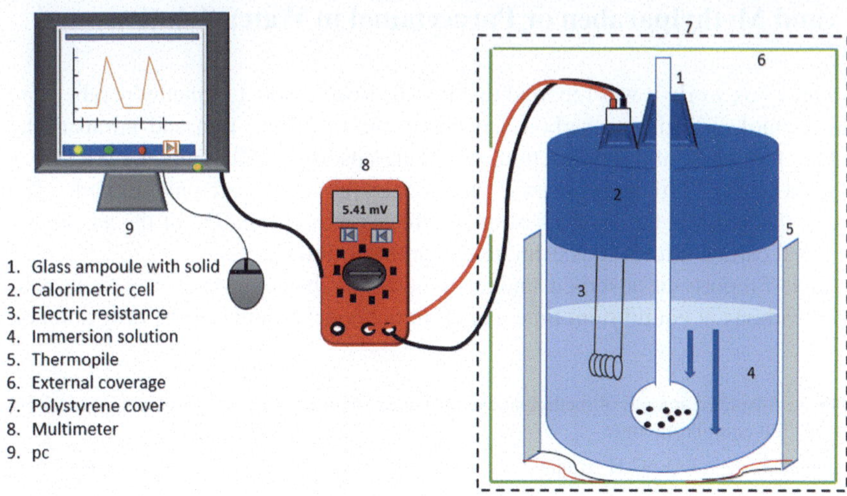

Fig. 3 Calorimeter type Tian-Calvet

item 1). The change in temperature between the system and the surroundings when the wetting of the solid occurs is determined through thermopile sensors (Fig. 3 item 5). The data capture is made with a multimeter (Fig. 3 item 8) and begins with the turn on of apparatus and ending with the calibration. This includes the time to reach the calorimeter's thermal equilibrium (baseline) and the electrical potential changes due to the wetting of the activated carbon in the drug solution or solvent. The calorimeter requires electrical calibration after the electrical potential returns to baseline.

The procedure was repeated with activated carbons AC-H^+ and AC-T and using also water as immersion liquid. To determine the enthalpic effect of substituents containing in the phenolic ring of the adsorbates, it determines the immersion enthalpy in phenol solutions at the same concentration that the drug solutions used in the study.

The Hess Law was used to determine the enthalpic contribution (ΔH_{int}; Eq. 14) of the methylparaben or paracetamol to immersion enthalpy and in this way determine the effect of the adsorbent-adsorbate interactions eliminating the interactions solvent-adsorbent. As the adsorbates share the phenolic ring in their structures the contribution of the substituents to the adsorbent-adsorbate interactions (ΔH_{sub}; Eq. 15) can be evaluated in the same form.

$$\Delta H_{int} = \Delta H_{imm} drug\, solution - \Delta H_{imm} water \tag{14}$$

$$\Delta H_{sub} = \Delta H_{int} - \Delta H_{imm} phenol\, solution \tag{15}$$

3 Insight the Interactions Between Activated Carbon and Methylparaben or Paracetamol in Water Solution

In a previous work, it was reported that the adsorbed amount of methylparaben and paracetamol on activated carbon depends to the diffusion: film and intraparticle; and the formation of the interactions adsorbate-adsorbent [50]. Only the last step is affected by the solution chemistry (solute concentration, pH and ionic strength). The aim of this work is the analysis of these effects on the formation of the adsorbate-activated carbon interactions using immersion calorimetry.

Table 7 reports the adsorbed amount of methylparaben or paracetamol on the activated carbons at equilibrium time and an initial concentration of the drug solution

Table 7 Adsorbed amount of methylparaben or paracetamol on activated carbons CAC, AC-H^+ and AC-T at equilibrium time

Compound	Q AC-H^+ (mmol g^{-1})	Q CAC (mmol g^{-1})	Q AC-T (mmol g^{-1})
Methylparaben	1.10	1.43	1.52
Paracetamol	0.76	1.21	1.38

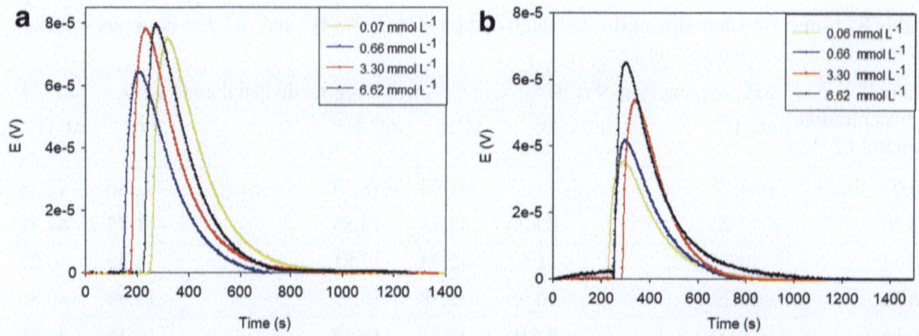

Fig. 4 Calorimetric curves obtained by the immersion of the activated carbon AC-H^+ in paracetamol (**a**) or methylparaben (**b**) solution at 293 K

of 6.62 mmol L^{-1}. According to the data, the adsorption is favored on the activated carbon AC-T for both adsorbates, they also show that the adsorbed amount of methylparaben is greater than the achieved by the paracetamol, despite the fact this adsorbate has smallest molecular size.

One of the most notable result is the adsorbed amount of methylparaben on activated carbon AC-H^+, which is a 45% higher than the adsorbed amount of Paracetamol achieved on the same activated carbon, this result is correlated with the effect of surface chemistry in the formation of activated carbon-adsorbate interactions.

Then, it is shown the calorimetric curves obtained from the immersion of the activated carbon AC-H^+ in the paracetamol (Fig. 4a) and methylparaben (Fig. 4b) solution. In both figures (a, b) the electrical potential increases with the time due to immersion of activated carbon in the solution, the area under the immersion peak is a balance between the formation of the interactions water-activated carbon, their breakdown, and finally, the formation of the adsorbate-activated carbon complex. Thus, the height of the peak does not only correlate with the concentration of pharmaceutical. However, Fig. 4a, b show that the highest immersion peaks are achieved at high concentration of the adsorbates.

In Fig. 4 only is shown the immersion peak due to the calibration peak depends to the electric work provided in the calibration but not with the interactions in the system.

The immersion enthalpy was calculated from the immersion and calibration data, the data are reported in Table 8.

The behavior of the immersion enthalpies for some system is similar; indeed, it does not depend to the adsorbate. For AC-H^+ in methylparaben solution, as well as AC-H^+ and CAC in paracetamol solution the values of the immersion enthalpy only increases from a concentration of pharmaceutical solution close to 0.66 mmol L^{-1}. For the immersion of CAC in Paracetamol solution the drug concentration necessary to increase the enthalpy is higher than the above systems, being 1.32 mmol L^{-1}.

In AC-T, the immersion enthalpies values do not vary significantly with the increasing in the adsorbate concentration.

Table 8 Immersion enthalpy of the activated carbons CAC, AC-H^+ and AC-T in aqueous solutions of paracetamol or methylparaben

Initial concentration (mmol L^{-1})	ΔH_{imm} paracetamol (J g^{-1})			ΔH_{imm} methylparaben (J g^{-1})		
	AC-H^+	CAC	AC-T	AC-H^+	CAC	AC-T
0.07	−34.72	–	−41.94	−18.34	−8.330	−42.28
0.20	−32.90	−8.820	−43.15	−18.85	−11.73	−44.32
0.33	−30.75	−11.62	−42.43	−17.81	−15.43	−40.27
0.46	−33.25	−10.95	−32.38	−18.37	−13.99	−40.86
0.66	−29.09	−8.610	−35.31	−18.17	−10.16	−38.45
1.32	−46.14	−9.710	−45.43	−30.65	−39.68	−42.53
3.30	−55.50	−31.71	−29.17	−30.35	−53.97	−46.61
6.62	−65.79	−54.32	−44.26	−69.47	−59.31	−38.86

These behaviors are related with the affinity of the activated carbon by the solvent and the number of adsorbate molecules required to displace the adsorbed solvent.

Overall, it is observed that the values of immersion enthalpies are higher for paracetamol than methylparaben in AC-H^+, in CAC at high concentrations of adsorbate the highest values are achieved in the methylparaben solution and in AC-T the values are close for the both adsorbates.

The interactions between the water and activated carbon, specifically the effect of the different surface chemistry on the affinity with the solvent, were also evaluated with immersion calorimetry. The results are reported in Table 9.

The calorimetric curves and immersion enthalpies show that the affinity of the activated carbons for the water increases following the trend AC-H^+ > CAC > AC-T; indeed, it is the same trend for the content of phenols on the activated carbon. Thus,

Table 9 Calorimetric curves and immersion enthalpies of CAC, AC-H^+ and AC-T in water at 293 K

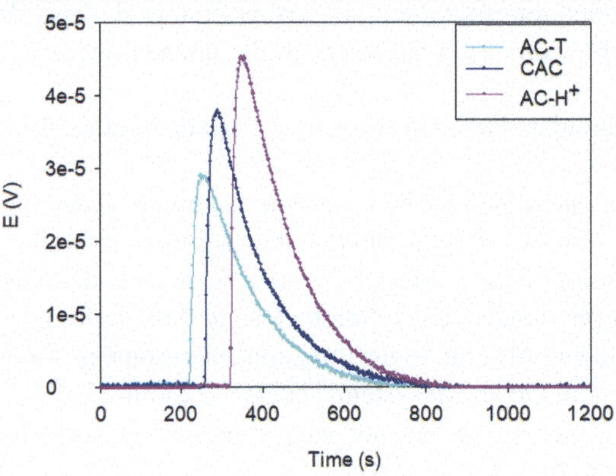

CAC (J g^{-1})	AC-H^+ (J g^{-1})	AC-T (J g^{-1})
−49.7	−66.6	−27.4

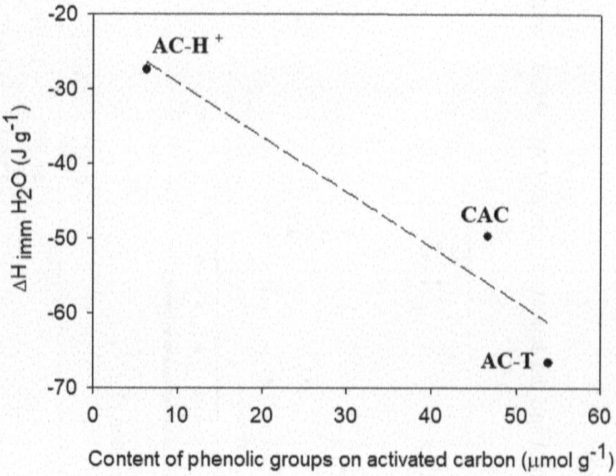

Fig. 5 Relationship between immersion enthalpy of the activated carbons CAC, AC-H⁺ and AC-T in water and their content of phenolic groups

the formation of the activated carbon–water interactions is favored with the increasing on the phenol groups on the adsorbent surface (Fig. 5). This is understandable, given that more water molecules will be attracted to the adsorbent surface, if it is polar.

The values of the immersion enthalpy of the activated carbons in water have values higher than the determine in drug solutions except for the activated carbon AC-T, this is an indicative of the affinity between the last-mentioned activated carbon and pharmaceutical compounds, which leads to a greater adsorption capacity for them.

To evaluate the interactions adsorbent-adsorbate we used the Hess law subtracting the immersion enthalpy in water to the immersion enthalpy in drug solution as shown in Eq. (14), the results are reported in Table 10.

Most of the interaction enthalpies values between the both adsorbates and the activated carbons CAC and AC-H^+ are positive, this indicate that the formation of the adsorbate-adsorbent interactions is a endothermic process, which could be related with the energy required to displace the water-adsorbent interaction because the interaction is very strong (Fig. 6), or an increase in the potential energy in the adsorbate-adsorbent complex, hence, it will be most unstable that the solvent-adsorbent interaction. This leads to, a decrease on the adsorption capacity of the two adsorbates.

In addition, only in the drug solution with the largest concentration the values are negative, therefore, to displace the adsorbent-solvent interaction it requires a greater quantity of adsorbate molecules.

On the contrary is the adsorption in AC-T, the both adsorbates present negative values for the interaction enthalpy in the all concentration range, thus, the formation of the interaction are favored mainly because the affinity of the activated carbon by the water is lowest than that for the other activated carbons as it is shown in Fig. 6.

At low concentration of the pharmaceuticals, the interaction enthalpies in CAC and AC-H^+ show a difference between the interactions formed by each adsorbate. For the Paracetamol case, the interaction enthalpies have values close to each other, therefore, the content of acidic groups (different to phenol) on the activated

Table 10 Interaction enthalpy calculated by Hess law for the interactions between methylparaben or paracetamol and the activated carbons CAC, AC-H^+ and AC-T at 293 K

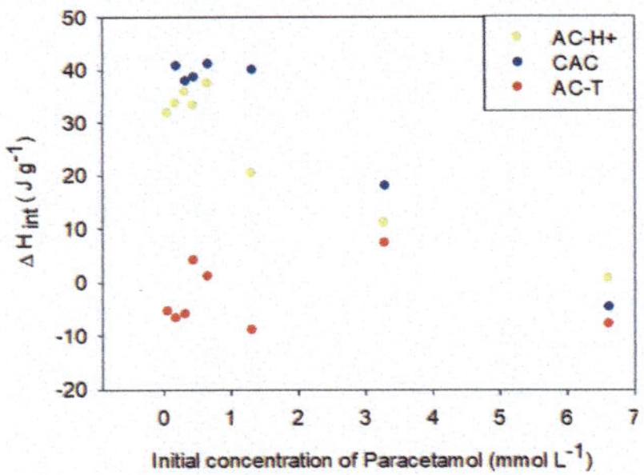

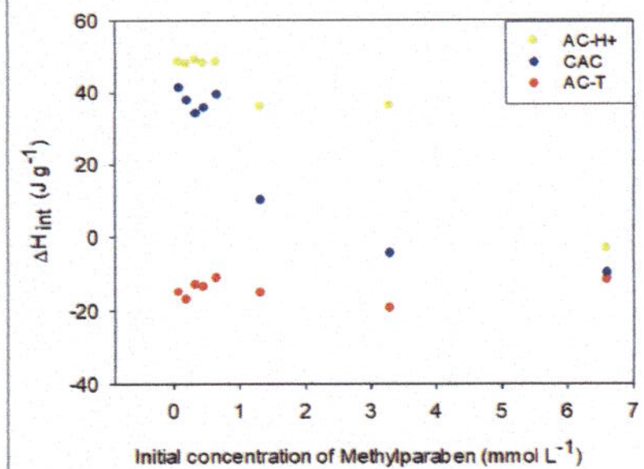

Initial concentration ($mmol\ L^{-1}$)	ΔH_{int} Paracetamol ($J\ g^{-1}$)			ΔH_{int} Methylparaben ($J\ g^{-1}$)		
	AC-H^+	CAC	AC-T	AC-H^+	CAC	AC-T
0.07	31.88	–	−5.443	48.26	41.37	−14.88
0.20	33.70	40.88	−6.648	47.75	37.97	−16.92
0.33	35.85	38.08	−5.933	48.79	34.27	−12.87
0.46	33.35	38.75	4.120	48.23	35.71	−13.46
0.66	37.51	41.09	1.190	48.43	39.54	−11.05
1.32	20.46	39.99	8.934	35.95	10.02	−15.13
3.30	11.10	17.99	7.328	36.25	−4.270	−19.21
6.62	0.81	−4.62	−7.759	−2.870	−9.610	−11.46

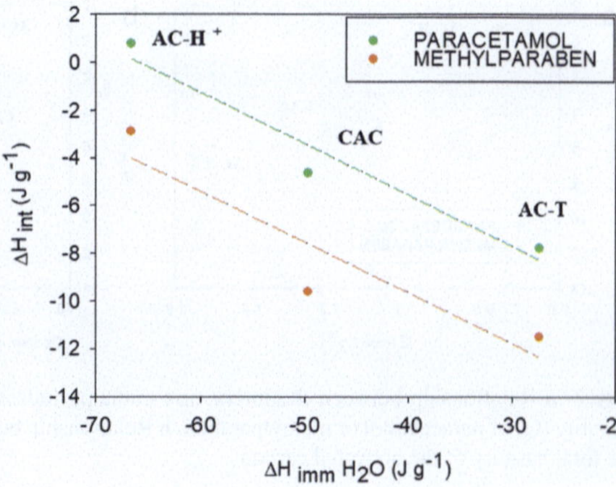

Fig. 6 Relationship between the immersion enthalpy of the activated carbons in water and the interaction enthalpy (adsorbent-adsorbate interaction)

carbon does not affect the formation of the paracetamol-activated carbon interaction. However, the enthalpies values for methylparaben-activated carbon interactions decrease when the content of acidic groups on the adsorbent also decreases. This behavior can be explained making some assumptions about the chemistry of the adsorbent surface and the chemical structures of adsorbates. In the case of activated carbon, it assumes that the only chemical groups on the surface were determined by the Boehm titration, therefore, if the phenol groups formed the interaction with the water, the active sites available will be the carboxylic acids and lactones, which are carbonyl groups or Hydrogen acceptor groups. On the other hand, paracetamol has two Hydrogen acceptor groups and methylparaben one, therefore, the interactions acid–base according to Brønsted-Lowry theory can be formed, as well as the formation of additional Hydrogen bonds. Thus, by probability, paracetamol-activated carbon interactions are more likely to form.

In all concentration range, the pH of the solutions has values between 3–6, the lowest values were found in AC-H$^+$ and CAC ($pH_{AC\text{-}H+} = 3$; $pH_{CAC} = 5.5$), therefore, the electric charge on the adsorbent surface is negative for the both activated carbons, this leads an increase in the polarity on the surface and greater affinity for the water. At the same time, the negative electric charge can decrease the interactions with the adsorbent by repulsion forces.

In Fig. 7a is shown that the adsorbed quantity of paracetamol or methylparaben on the activated carbons is directly related to the interaction enthalpies.

The adsorbed quantity of the adsorbates has a magnitude in mmol g^{-1} between 0.8 and 1.6, therefore, it is not possible that this quantity is only related to the formation of acid–base interactions and hydrogen bonds because the content of carbonyl groups is very low since its values are less than 90 μmol g^{-1}, except in AC-H$^+$, but its polarity is highest and the adsorbate affinity decreases. Figure 7b shows that the interaction enthalpy decreases with the content of basic groups on the activated carbon, which include amines, pyrones, and π-electrons, the latter are considered as Lewis bases. Currently, many investigations indicated that the π−π or π-stacked

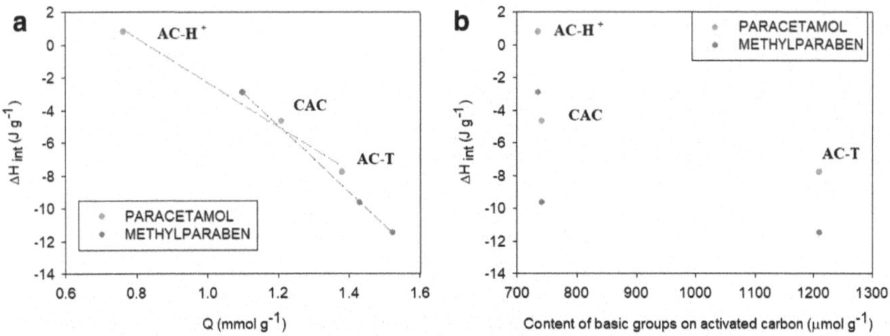

Fig. 7 **a** Relationship between the interaction enthalpy (adsorbent-adsorbate) and the adsorbed quantity (Q) of paracetamol or methylparaben. **b** Relationship between the interaction enthalpy and the total basicity of the activated carbons

are the most important interactions on the adsorption of aromatic compounds, like the pharmaceuticals used in this work, on the activated carbon. However, these interactions also depend to the solution chemistry because at extreme pH values, other interactions can be formed first such as π-H^+ in acidic medium or π-OH in basic solutions.

Up to this point, this work has shown the importance of the surface groups on the activate carbon on the adsorption of paracetamol and methylparaben, however the chemical groups in the adsorbate's structures also contributed to interactions with the adsorbent, thus, to the enthalpy change.

Both adsorbates share the phenolic ring as a core of the structure, in Table 11 it reports the immersion enthalpy od the activated carbon in phenol solutions at different concentration to know the contribution of this group in the adsorbate-adsorbent interactions.

The immersion enthalpies of the activated carbons in aqueous solution of phenol increase proportionally with the concentration of solute. Additionally, the values reported in Table 11 are lower than the values determined in water, except for the immersion of AC-T in solutions with higher concentration of solute. In conclusion, the adsorbent has a low affinity by the phenol because this adsorbate has a lower polarity than water and it requires more energy to displace the solvent on the surface. For this reason, the immersion enthalpies are positive in most systems, even to increasing the concentration of phenol, therefore, the hydrogen bonds formed between phenol groups on the adsorbent surface and water are stronger than the formed between phenol groups in both system components.

The contribution to the adsorbent-adsorbate interactions by the substituents around the phenolic ring also was determined using the interaction enthalpy, the results were reported in Table 12.

It is observed in Table 12 that the interactions enthalpies associated with the contribution of the substituents around the phenolic ring on the adsorbate-adsorbent interaction. To calculate the contribution of the substituents in the enthalpy changes only were used the interaction enthalpies at high concentration of adsorbate because

Table 11 Calorimetric curves for the immersion of the activated carbons CAC, AC-H$^+$ and AC-T in phenol solutions at 10.9 mmol L^{-1} and the immersion or interaction enthalpies calculated at different concentration of solute

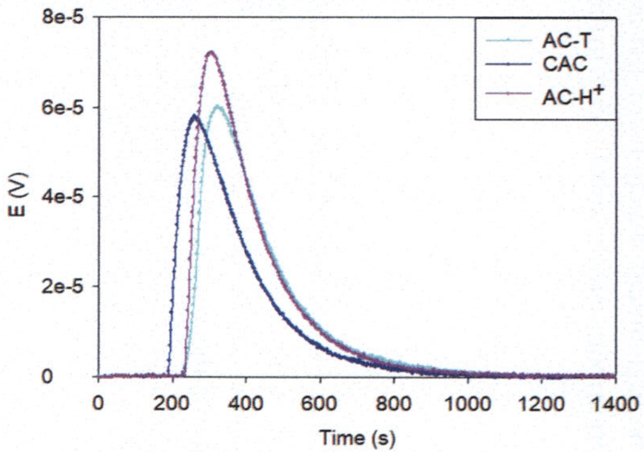

Phenol concentration (mmol L^{-1})	$-\Delta H_{imm}$ AC-H$^+$ (J g^{-1})	$-\Delta H_{imm}$ CAC (J g^{-1})	$-\Delta H_{imm}$ AC-T (J g^{-1})
1.09	19.3	7.71	12.9
5.43	33.6	39.4	35.4
10.9	57.0	39.3	39.3
Phenol concentration (mmol L^{-1})	**$-\Delta H_{int}$ AC-H$^+$ (J g^{-1})**	**$-\Delta H_{int}$ CAC (J g^{-1})**	**$-\Delta H_{int}$ AC-T (J g^{-1})**
1.09	47.3	41.9	14.5
5.43	33.0	10.2	−8.00
10.9	9.60	10.3	−11.9

Table 12 Contribution to the interaction enthalpy by the substituents on the chemical structures of methylparaben and paracetamol

Initial concentration (mmol L^{-1})	ΔH_{int} AMIDE (J g^{-1})			ΔH_{int} ESTHER (J g^{-1})		
	AC-H^+	CAC	AC-T	AC-H^+	CAC	AC-T
1.32	−26.8	−1.9	−23.4	−11.1	−31.9	−29.6
3.30	−21.9	7.80	15.3	−35.9	−14.5	−11.2
6.62	−8.80	−14.9	4.10	−9.6	−19.9	0.4

under conditions increase the interaction between the adsorbates and the activated carbon. The enthalpies are negative in most systems, it indicates the formation of interactions with the adsorbent; in the case of the interaction between AC-T and the amide group, to increase the concentration of paracetamol the enthalpy also increases, this trend can be explained by two ways, in the first case, the number of paracetamol molecules is greater than the content of available chemical groups on the surface and is necessary the displacement of the solvent to generate new active sites. In the second case, to increase the number of molecules in the carbonaceous matrix the space is reduced and steric hindrance can affect the organization of the molecules in the system, thus, their interactions. The positive value for the interaction enthalpy between AC-T and methylparaben can be explained by the second case.

Assuming that the interactions solvent-adsorbent only involve Hydrogen bonds and each chemical group on the surface only interact with one molecule of water, the immersion enthalpies values in water correspond to the energy exchange by formation of Hydrogen bonds. These values are greater than the values reported in Table 12, it suggests that the main interaction does not correspond to hydrogen bond formation but π-stacked and acid–base interactions. Additionally, the interaction enthalpies in phenol solution are positive indicating low affinity adsorbent-adsorbate.

Because of similarities between both chemical structures of the adsorbates, it could not be found a difference in the contribution to the substituents to the interaction enthalpy.

4 Conclusion

The immersion calorimetry is a tool useful to determine the energy exchanged during the wetting of the adsorbent with the solvent or aqueous solutions of pollutants present in water resources, the values determinate can be used to define the affinity of the adsorbent surface by the components of the system. For this reason, the immersion calorimetry provides experimental data that can be used to developing adsorbents with specifics characteristics for the adsorption of pollutants that resist other treatments carried out in the water treatment plants.

References

1. Connor, R.: The United Nations world water development report 2015: water for a sustainable world. United Nations Educational, Scientific and Cultural Organization, Paris, France (2015). https://sustainabledevelopment.un.org/content/documents/1711WaterforaSustainableWorld.pdf
2. Fantini, E.: An introduction to the human right to water: law, politics, and beyond. WIREs Water. **7** (2020). https://doi.org/10.1002/wat2.1405.
3. World Health Organization: Guidelines for Drinking-water Quality, 3rd edn. WHO Library Cataloguing-in-Publication Data, Geneve (2008)
4. Moore, R.A., Moore, N.: Paracetamol and pain: the kiloton problem. Eur. J. Hosp. Pharm. **23**, 187–188 (2016). https://doi.org/10.1136/ejhpharm-2016-000952
5. Seiler, R.L., Zaugg, S.D., Thomas, J.M., Howcroft, D.L.: Caffeine and pharmaceuticals as indicators of waste water contamination in wells. Ground Water **37**, 405–410 (1999). https://doi.org/10.1111/j.1745-6584.1999.tb01118.x
6. Kasprzyk-Hordern, B., Dinsdale, R.M., Guwy, A.J.: Illicit drugs and pharmaceuticals in the environment—forensic applications of environmental data, part 2: pharmaceuticals as chemical markers of faecal water contamination. Environ. Pollut. **157**, 1778–1786 (2009). https://doi.org/10.1016/j.envpol.2009.02.019
7. Park, S., Oh, S.: Activated sludge-degrading analgesic drug acetaminophen: acclimation, microbial community dynamics, degradation characteristics, and bioaugmentation potential. Water Res. **182**, 115957 (2020). https://doi.org/10.1016/j.watres.2020.115957
8. Guiloski, I.C., Ribas, J.L.C., Piancini, L.D.S., Dagostim, A.C., Cirio, S.M., Fávaro, L.F., Boschen, S.L., Cestari, M.M., da Cunha, C., Silva de Assis, H.C.: Paracetamol causes endocrine disruption and hepatotoxicity in male fish Rhamdia quelen after subchronic exposure. Environ. Toxicol. Pharmacol. **53**, 111–120 (2017). https://doi.org/10.1016/j.etap.2017.05.005
9. David, A., Pancharatna, K.: Effects of acetaminophen (paracetamol) in the embryonic development of zebrafish, Danio rerio. J. Appl. Toxicol. **29**, 597–602 (2009). https://doi.org/10.1002/jat.1446
10. Comeche, A., Martín-Villamil, M., Picó, Y., Varó, I.: Effect of methylparaben in Artemia franciscana. Comp. Biochem. Physiol. Part C Toxicol. Pharmacol. **199**, 98–105 (2017). https://doi.org/10.1016/j.cbpc.2017.04.004
11. Kang, H.-M., Kim, M.-S., Hwang, U.-K., Jeong, C.-B., Lee, J.-S.: Effects of methylparaben, ethylparaben, and propylparaben on life parameters and sex ratio in the marine copepod Tigriopus japonicus. Chemosphere **226**, 388–394 (2019). https://doi.org/10.1016/j.chemosphere.2019.03.151
12. Challis, J.K., Parajas, A., Anderson, J.C., Asiedu, E., Martin, J.W., Wong, C.S., Ross, M.S.: Photodegradation of bitumen-derived organics in oil sands process-affected water. Environ. Sci. Process. Impacts. (2020). https://doi.org/10.1039/D0EM00005A
13. Plappally, A.K., Lienhard, J.H.: Costs for water supply, treatment, end-use and reclamation. Desalin. Water Treat. **51**, 200–232 (2013). https://doi.org/10.1080/19443994.2012.708996
14. Plumlee, M.H., Stanford, B.D., Debroux, J.-F., Hopkins, D.C., Snyder, S.A.: Costs of advanced treatment in water reclamation. Ozone Sci. Eng. **36**, 485–495 (2014). https://doi.org/10.1080/01919512.2014.921565
15. Mansor, N.A., Tay, K.S.: Potential toxic effects of chlorination and UV/chlorination in the treatment of hydrochlorothiazide in the water. Sci. Total Environ. **714**, 136745 (2020). https://doi.org/10.1016/j.scitotenv.2020.136745
16. Li, K., Xu, Z., Liu, Q., Li, Z., Wang, X., Zhang, J., Zhu, J.: Space-confined construction of nitrogen-rich cobalt porphyrin-derived nanoparticulates anchored on activated carbon for high-current lithium thionyl chloride battery. Electrochim. Acta. **353**, 136543 (2020). https://doi.org/10.1016/j.electacta.2020.136543
17. Yunus, Z.M., Al-Gheethi, A., Othman, N., Hamdan, R., Ruslan, N.N.: Removal of heavy metals from mining effluents in tile and electroplating industries using honeydew peel activated carbon:

a microstructure and techno-economic analysis. J. Clean. Prod. **251**, 119738 (2020). https://doi.org/10.1016/j.jclepro.2019.119738
18. Guediri, A., Bouguettoucha, A., Chebli, D., Chafai, N., Amrane, A.: Molecular dynamic simulation and DFT computational studies on the adsorption performances of methylene blue in aqueous solutions by orange peel-modified phosphoric acid. J. Mol. Struct. **1202**, 127290 (2020). https://doi.org/10.1016/j.molstruc.2019.127290
19. Jiang, N., Erdős, M., Moultos, O.A., Shang, R., Vlugt, T.J.H., Heijman, S.G.J., Rietveld, L.C.: The adsorption mechanisms of organic micropollutants on high-silica zeolites causing S-shaped adsorption isotherms: an experimental and Monte Carlo simulation study. Chem. Eng. J. **389**, 123968 (2020). https://doi.org/10.1016/j.cej.2019.123968
20. Ling, Y., Klemes, M.J., Steinschneider, S., Dichtel, W.R., Helbling, D.E.: QSARs to predict adsorption affinity of organic micropollutants for activated carbon and β-cyclodextrin polymer adsorbents. Water Res. **154**, 217–226 (2019). https://doi.org/10.1016/j.watres.2019.02.012
21. Jia, Y., Zhang, H., Khanal, S.K., Yin, L., Lu, H.: Insights into pharmaceuticals removal in an anaerobic sulfate-reducing bacteria sludge system. Water Res. **161**, 191–201 (2019). https://doi.org/10.1016/j.watres.2019.06.010
22. Yamamoto, N., Nakakuki, I., Matubayasi, N.: Free-energy analysis of physisorption on solid-liquid interface with the solution theory in the energy representation. J. Chem. Phys. **149**, 014504 (2018). https://doi.org/10.1063/1.5027861
23. Considine, R., Denoyel, R., Pendleton, P., Schumann, R., Wong, S.-H.: The influence of surface chemistry on activated carbon adsorption of 2-methylisoborneol from aqueous solution. Colloids Surf. A Physicochem. Eng. Asp. **179**, 271–280 (2001). https://doi.org/10.1016/S0927-7757(00)00647-6
24. Terzyk, A.P., Rychlicki, G.: The influence of activated carbon surface chemical composition on the adsorption of acetaminophen (paracetamol) in vitro. Colloids Surf. A Physicochem. Eng. Asp. **163**, 135–150 (2000). https://doi.org/10.1016/S0927-7757(99)00298-8
25. Yu, P., Wurster, D.E.: Thermodynamic estimate of the number of solvent molecules displaced by a solute molecule for enthalpy-driven adsorption: phenobarbital and activated carbons as the model system. J. Pharm. Sci. **107**, 1055–1062 (2018). https://doi.org/10.1016/j.xphs.2017.11.017
26. Bernal, V., Giraldo, L., Moreno-Piraján, J.C.: Insight into adsorbate–adsorbent interactions between aromatic pharmaceutical compounds and activated carbon: equilibrium isotherms and thermodynamic analysis. Adsorption. **26**, 153–163 (2020). https://doi.org/10.1007/s10450-019-00057-x
27. Chung, H.-K., Kim, W.-H., Park, J., Cho, J., Jeong, T.-Y., Park, P.-K.: Application of Langmuir and Freundlich isotherms to predict adsorbate removal efficiency or required amount of adsorbent. J. Ind. Eng. Chem. **28**, 241–246 (2015). https://doi.org/10.1016/j.jiec.2015.02.021
28. Lima, D.R., Hosseini-Bandegharaei, A., Thue, P.S., Lima, E.C., de Albuquerque, Y.R.T., dos Reis, G.S., Umpierres, C.S., Dias, S.L.P., Tran, H.N.: Efficient acetaminophen removal from water and hospital effluents treatment by activated carbons derived from Brazil nutshells. Colloids Surf. A Physicochem. Eng. Asp. **583**, 123966 (2019). https://doi.org/10.1016/j.colsurfa.2019.123966
29. Noh, J.S., Schwarz, J.A.: Effect of HNO_3 treatment on the surface acidity of activated carbons. Carbon **28**(5), 675–682 (1990)
30. Karanfil, T.: Activated carbon adsorption in drinking water treatment, pp. 345–373 (2006) (Chapter 7). https://doi.org/10.1016/S1573-4285(06)80016-5.
31. Yanyan, L., Kurniawan, T.A., Zhu, M., Ouyang, T., Avtar, R., Dzarfan Othman, M.H., Mohammad, B.T., Albadarin, A.B.: Removal of acetaminophen from synthetic wastewater in a fixed-bed column adsorption using low-cost coconut shell waste pretreated with NaOH, HNO_3, ozone, and/or chitosan. J. Environ. Manage. **226**, 365–376 (2018). https://doi.org/10.1016/j.jenvman.2018.08.032
32. Bernal, V., Giraldo, L., Moreno-Piraján, J.: Physicochemical properties of activated carbon: their effect on the adsorption of pharmaceutical compounds and adsorbate-adsorbent interactions. C. **4**, 62 (2018). https://doi.org/10.3390/c4040062

33. Mashayekh-Salehi, A., Moussavi, G.: Removal of acetaminophen from the contaminated water using adsorption onto carbon activated with NH_4 Cl. Desalin. Water Treat. **57**, 12861–12873 (2016). https://doi.org/10.1080/19443994.2015.1051588
34. Moreno-Marenco, A.R., Giraldo, L., Moreno-Piraján, J.C.: Parabens adsorption onto activated carbon: relation with chemical and structural properties. Molecules **24**, 4313 (2019). https://doi.org/10.3390/molecules24234313
35. Rodriguez-Reinoso, F., Martin-Martinez, J.M., Prado-Burguete, C., McEnaney, B.: A standard adsorption isotherm for the characterization of activated carbons. J. Phys. Chem. **91**, 515–516 (1987). https://doi.org/10.1021/j100287a006
36. Thommes, M., Kaneko, K., Neimark, A.V., Olivier, J.P., Rodriguez-Reinoso, F., Rouquerol, J., Sing, K.S.W.: Physisorption of gases, with special reference to the evaluation of surface area and pore size distribution (IUPAC technical report). Pure Appl. Chem. **87**, 1051–1069 (2015). https://doi.org/10.1515/pac-2014-1117
37. Jagiello, J., Thommes, M.: Comparison of DFT characterization methods based on N_2, Ar, CO_2, and H_2 adsorption applied to carbons with various pore size distributions. Carbon N. Y. **42**, 1227–1232 (2004). https://doi.org/10.1016/j.carbon.2004.01.022
38. Boehm, H.P.: Some aspects of the surface chemistry of carbon blacks and other carbons. Carbon N. Y. **32**, 759–769 (1994). https://doi.org/10.1016/0008-6223(94)90031-0
39. Goertzen, S.L., Thériault, K.D., Oickle, A.M., Tarasuk, A.C., Andreas, H.A.: Standardization of the Boehm titration. Part I. CO_2 expulsion and endpoint determination. Carbon N. Y. **48**, 1252–1261 (2010). https://doi.org/10.1016/j.carbon.2009.11.050
40. Bourikas, K., Vakros, J., Kordulis, C., Lycourghiotis, A.: Potentiometric mass titrations: experimental and theoretical establishment of a new technique for determining the point of zero charge (PZC) of metal (hydr)oxides. J. Phys. Chem. B **107**, 9441–9451 (2003). https://doi.org/10.1021/jp035123v
41. Noh, J.S., Schwarz, J.A.: Estimation of the point of zero charge of simple oxides by mass titration. J. Colloid Interface Sci. **130**, 157–164 (1989). https://doi.org/10.1016/0021-9797(89)90086-6
42. Bernal, V., Erto, A., Giraldo, L., Moreno-Piraján, J.: Effect of solution ph on the adsorption of paracetamol on chemically modified activated carbons. Molecules **22**, 1032 (2017). https://doi.org/10.3390/molecules22071032
43. Hollenbeck, R.G., Peck, G.E., Kildsig, D.O.: Application of immersional calorimetry to investigation of solid-liquid interactions: microcrystalline cellulose-water system. J. Pharm. Sci. **67**, 1599–1606 (1978). https://doi.org/10.1002/jps.2600671125
44. Gibson, S., Rowe, R., White, E.: Quantitative assessment of additive-polymer interaction in pigmented hydroxypropyl methylcellulose formulations using immersion calorimetry. Int. J. Pharm. **48**, 113–117 (1988). https://doi.org/10.1016/0378-5173(88)90253-0
45. Buckton, G., Russell, S.J., Beezer, A.E.: Pharmaceutical calorimetry: a selective review. Thermochim. Acta. **193**, 195–214 (1991). https://doi.org/10.1016/0040-6031(91)80184-K
46. Landín, M., Martínez-Pacheco, R., Gómez-Amoza, J.L., Souto, C., Concheiro, A., Rowe, R.C.: Effect of country of origin on the properties of microcrystalline cellulose. Int. J. Pharm. **91**, 123–131 (1993). https://doi.org/10.1016/0378-5173(93)90331-9
47. Hogan, S.E., Buckton, G.: The quantification of small degrees of disorder in lactose using solution calorimetry. Int. J. Pharm. **207**, 57–64 (2000). https://doi.org/10.1016/S0378-5173(00)00527-5
48. de Ridder, D.J., Verberk, J.Q.J.C., Heijman, S.G.J., Amy, G.L., van Dijk, J.C.: Zeolites for nitrosamine and pharmaceutical removal from demineralised and surface water: mechanisms and efficacy. Sep. Purif. Technol. **89**, 71–77 (2012). https://doi.org/10.1016/j.seppur.2012.01.025

49. Moreno-Marenco, A.R., Giraldo, L., Moreno-Piraján, J.C.: Dataset of the immersion enthalpy of activated carbons chemically modified in methylparaben aqueous solution: relation with adsorption. Data Br. **25**, 104100 (2019). https://doi.org/10.1016/j.dib.2019.104100
50. Bernal, V., Giraldo, L., Moreno-Piraján, J.C.: A new methodology to determine the effect of the adsorbate-adsorbent interactions on the analgesic adsorption onto activated carbon using kinetic and calorimetry data. Environ. Sci. Pollut. Res. (2020). https://doi.org/10.1007/s11356-020-09725-w

Conversion of Nonconventional Aluminosiliceous Sources into Microporous Adsorbents for Water Remediation

M. R. Gonzalez, A. M. Pereyra, and E. I. Basaldella

Abstract Human activity has led to a high degree of industrialization, which has drastically increased in the last decades. Industrial growth has meant more waste generation and, at the same time, a negative environmental impact. Consequently, the need of searching for novel processes related to waste minimization and reuse arises, as is reflected in the increasing number of research works related to the reuse of industrial wastes in recent years. The present survey deals with the general evolution of the preparative methodologies employed for the recycling of aluminosiliceous wastes, particularly by hydrothermal zeolitilization. The study focuses on the reutilization of two distinctive industrial residues: the discarded catalysts coming from the cracking unities used in the oil industry and the fly ashes generated in coal energy production. Both materials are high volume wastes produced by the industrial energy sector. Their use as a negative-cost raw material in the synthesis processes of zeolite-containing microporous materials and their potential applications as adsorbents in environmental remediation technologies will be emphasized.

1 Zeolite Based Adsorbent Obtained from Aluminosiliceous Wastes

1.1 Aluminosiliceous Industrial Wastes: Origin, Features, and Environmental Implications

Huge amounts of inorganic solid aluminosilicate residues from the industrial sector are generated each year. The management of those enormous volumes of waste increasingly requires improved solutions for their treatment or disposal to reduce or eliminate their negative impact on the environment. This situation becomes relevant considering the perspective that in the near future, the disposal of this kind of waste

M. R. Gonzalez · A. M. Pereyra · E. I. Basaldella (✉)
Centro de Investigación y Desarrollo en Ciencias Aplicadas Dr. J.J. Ronco (CINDECA) (CONICET-CIC-UNLP), 47 N° 257, B1900 AJK La Plata, Argentina
e-mail: eib@quimica.unlp.edu.ar

J. C. Moreno-Piraján et al. (eds.), *Porous Materials*, Engineering Materials,
https://doi.org/10.1007/978-3-030-65991-2_10

will be limited because of more severe environmental regulations. So, among the industrial wastes that are generated every year, we will place special emphasis on those from power generation such as spent catalysts discarded from oil cracking units and fly ash produced in thermal power plants, both materials predominantly composed of aluminosilicates or silicates.

Fluidized catalytic cracking (FCC) is one of the main processes in the refinery industry. Although it has been used for more than 75 years, it is still the main technology used nowadays to obtain gasoline. The catalytic cracking units annually generate approximately 840,000 tons of exhausted FCC catalyst worldwide [1]. The enormous amount of spent catalyst generated each year constitutes a solid waste that is difficult to handle and dispose of. Several papers dealing with the reuse of this particular residue can be found in the literature, its incorporation as a non-contaminant additive in cement formulations being particularly highlighted [2, 3].

On the other hand, electric power generation from coal combustion in thermoelectric plants is still present in a large number of countries, especially in the developing world. A high amount of fly ash generated by the combustion of coal is produced each year. Worldwide, this type of waste represents approximately 750 million tons [4] per year. Due to the amount generated and its physical and chemical properties, this waste also constitutes an environmental problem in terms of its final disposal. Different papers have reported its use as an additive in concrete [5, 6], as filler for the construction of road bases [7] or as pozzolanic material in cement mortar formulations [2, 3].

The above considerations encouraged the study of possible recycling methods based on the chemical and mineralogical characteristic of both wastes, Fig. 1. For

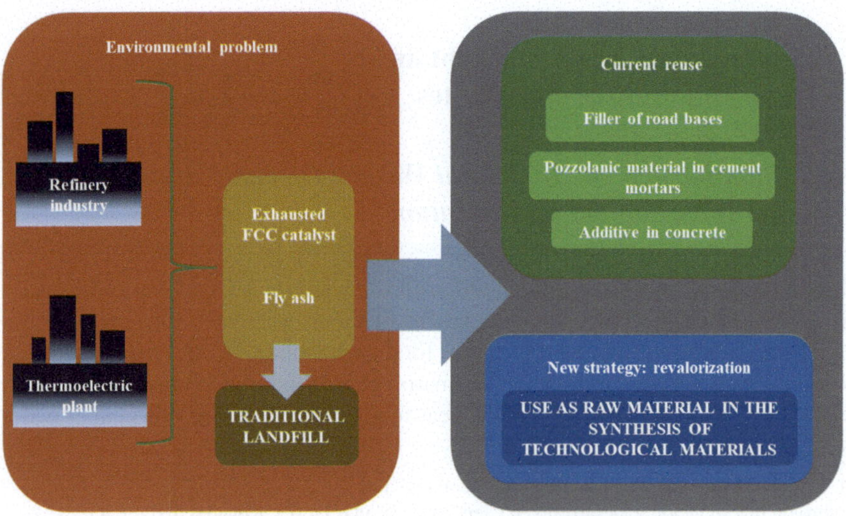

Fig. 1 Current and new strategies for the management of aluminosiliceous wastes

Fig. 2 SEM micrographs of spent FCC catalyst from the YPF refinery located in Ensenada, Argentina. **a** X100 magnification and **b** X500 magnification

this purpose, the study and characterization of the discarded material is an important step on the way to establish accurate reusing methodologies.

In the case of FCC catalysts, the main components are synthetic zeolite (15–50 wt%), kaolin clay (60–25 wt%), an aluminosiliceous matrix (ca. 5 wt%), and binder (ca. 20 wt%, usually Ludox). The zeolite present in FCC catalysts corresponds to the Y type (FAU group), usually fully hydrogen exchanged (USY) [8, 9]. The chemical composition of spent FCC catalysts turns out to be 50% of SiO_2 and 45% of Al_2O_3, the remaining 5% is distributed among other minor components. The FCC catalyst is usually shaped as smooth surface, rounded pellets 60–100 μm in diameter. Figure 2 shows the scanning electron microscopy (SEM) micrograph of a spent FCC catalyst collected at the largest of four refineries run by YPF, located in Ensenada, Buenos Aires, Argentina.

In the case of fly ashes, it is known that most of them contain mullite ($Al_{4.52}Si_{1.48}O_{9.74}$) and quartz ($SiO_2$) as the main crystalline compounds, along with an important fraction of amorphous phase. Hematite (Fe_2O_3) and calcite ($CaCO_3$) are also present as traces. The chemical analysis of fly ashes frequently reveals the presence of silica and alumina in a SiO_2/Al_2O_3 molar ratio > 1, and depending on the coal origin and type of combustion, a small fraction of other components such as Na_2O, Fe_2O_3, TiO_2, MgO, K_2O and CaO [10]. Fly ashes are composed of spherical-shaped particles, differing in size, which can reach up to 300 μm in diameter. The particle surface is not uniform, smooth spherical entities covered with small deposits or presenting areas with different roughness are commonly observed. Figure 3 shows the SEM image of the fly ash coming from AES Central Thermal San Nicolás Power Plant, located in San Nicolás, Buenos Aires, Argentina.

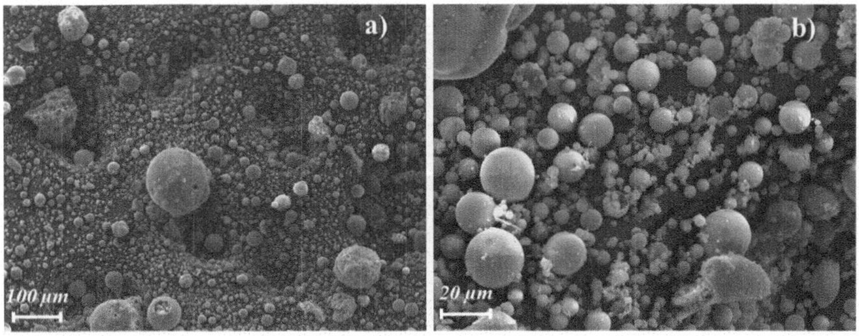

Fig. 3 SEM micrographs of fly ash provided by AES Central Thermal San Nicolás Power Plant, Argentina. **a** X100 magnification and **b** X500 magnification

1.2 Recycling Alternatives: Zeolite-Based Adsorbents

The zeolitic structures exhibit valuable features that convert them in indispensable materials for many technological applications. Based on their ionic exchange capacity, selective adsorption, ordered structure, high surface area, etc., they are widely used in cation exchange, adsorption and catalytic industrial processes.

The chemical formula of a zeolite is $M_{2/n}O$: Al_2O_3: y SiO_2: w H2O, where "n" is the valence of the M cation, "y" (the silica to alumina ratio) a number ≥ 2, and "w" the water content in the zeolite [11].

Zeolites are microporous crystalline aluminosilicates with well-defined structures that consist of $[SiO_4]^{4-}$ and $[AlO_4]^{5-}$ tetrahedra linked by oxygen atoms. The framework contains channels and cavities with molecular dimensions from 3 to 10 Å occupied by alkali and alkaline earth metal cations (Na, K, Ca, Mg) and molecules of water or other adsorbates and salts. The cations compensate the negative charge of the zeolitic framework and can be easily removed and exchanged for other cations [12]. The microporosity of zeolitic materials provides high surface areas reaching 500–900 $m^2\ g^{-1}$.

Particularly, the NaA framework (stoichiometry: $Na_{12}[(AlO_2)_{12}(SiO_2)_{12}]$:$27H_2O$) has a specific arrangement in which the unit cell contains 24 tetrahedra, 12 $[SiO_4]^{-4}$, and 12 $[AlO_4]^{-5}$. When fully hydrated, there are 27 water molecules per unit cell and there is also one sodium cation for one aluminum in the network. These sodium ions are loosely held together, so one of the main uses of this material is based on its cation exchange properties.

This zeolite has a cubic structure with a three-dimensional network of pores 4.2 Å in diameter. The structural arrangement shown in Fig. 4 consists of the union of three polyhedra (Fig. 4): a cube, a truncated octahedron cube, which is the sodalite cage or β-cage, and a 26-sided polyhedron representing the supercage (11.4 Å in diameter). The unit cell of zeolite NaA is cubic, consisting of eight α-cages parallel to the [1 0 0] surface. This cage possesses the inner symmetry of an octahedron and shares 6- and 8-membered rings with the β-cages. The β-cage (or sodalite cage) is the pseudo

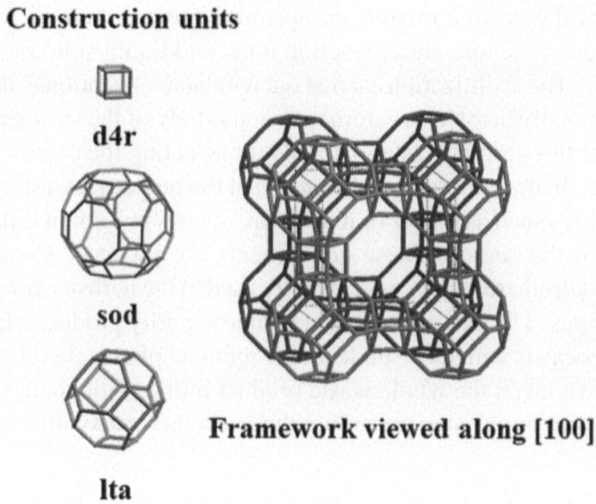

Fig. 4 Polyhedral composition of the structural arrangement of zeolite A, according to the International Zeolite Association (IZA) [13]

unit cell consisting of 24 T atoms (six 4-rings, four 6-rings, three 6–2 units or four 1–4–1 units) linked through double 4-rings for one cube face [13].

The thermal stability, high selectivity, non-toxicity and good mechanical resistance of zeolite A make it a versatile material for numerous applications such as catalytic processes [14], molecular sieve in gas separation [15], in wastewater treatment [16, 17], and for antimicrobial purposes [18–20]. Along with zeolite Y, it is the most commercialized microporous solid worldwide. The main feature arises from the Si/Al ratio = 1, which provides the highest cation exchange capacity.

As can be seen, the synthesis of zeolites requires two main precursor species based on silica and alumina. These reactants are involved in a hydrothermal process, occurring in closed vessels at controlled conditions of dilution, pH, and temperature.

Taking into account the chemical composition of both the discarded FCC and the fly ash, the production of synthetic zeolites may be a potential alternative for their reuse. The high aluminosilicate content of the mentioned wastes allows their use as a negative-cost raw material in the synthesis of zeolites. These materials offer the possibility of producing different zeolites by varying the previous treatments and altering the reaction conditions used for their reordering in zeolitic networks. In parallel, this fact takes enormous importance since it is a way of obtaining high technology value products from a residue that presents difficulties for its final disposal.

1.3 Key Steps in the Synthesis of Adsorbent Materials: Activation and Hydrothermal Crystallization

Many factors play an important role in the hydrothermal transformation of traditional raw materials into zeolite materials. The zeolite structure as well as its textural properties, which in turn define the ion exchange and adsorption properties, is determined by the starting mixture composition (silica, alumina, water, and alkali source molar ratios), temperature, reaction time, and liquid/solid ratio.

The zeolitization carried out with nonconventional aluminosiliceous sources additionally involves a careful previous study of the structural and chemical composition of the starting waste materials for selecting the correct synthesis procedure.

In the mineralogical analysis of the industrial wastes under study, inorganic structures such as quartz or mullite are always present in different proportions, depending on the source. These components do not react along the usual low temperature hydrothermal process currently used in the industry for getting the LTA or FAU structures. The consequence is a final low-purity product where the desired zeolitic phase coexists with those undesirable remnant phases. In this way, different methodologies to convert the whole waste product into zeolitic materials are based on including an "activation" step preceding the hydrothermal synthesis.

1.3.1 Activation Process

During the activation process, the operating methodology is aimed to improve the reactivity of the residues towards the generation of zeolite by increasing the availability of silica and alumina components. This objective is extremely important, because not only the selected activation procedure should allow high conversions, but also avoid extended reaction times and the use of high synthesis temperatures or pressures in the next step, the hydrothermal transformation.

In the search of methodologies to modify or dissolve the unalterable phases contained in impure raw materials, the first approach was the use of a calcination treatment as activation step, similar to the effective process used to synthesize zeolites from mining materials, such as kaolinite products [21].

The traditional activation for these clays involves a careful calcination at above the dehydration temperature and at values not as high as those required for the occurrence of new structural compromises. The desired transformation leads to the damage of the kaolinitic structure, producing metakaolinite as the activated, reactive amorphous phase. The zeolitilization of metakaolinite by hydrothermal synthesis proceeds easily using the suitable reaction parameters, leading to the crystallization of the most aluminum-rich zeolite A. If the activation temperature is increased about 900 °C, the generation of an amorphous phase, known as pre-mullite, is observed. The pre-mullite formation reduces the alumina reactivity, the zeolitization process is delayed, and the clay, under suitable reaction conditions, is converted into the more siliceous FAU structure, a material characterized by a Si/Al molar ratio higher than 1 and lower in aluminum content compared to zeolite A.

This methodology of zeolite synthesis from kaolins is well known [8] and was patented in due course [22] for the FCC catalyst manufacture. Some drawbacks related to kaolin quality are found in this process, because very pure, micronized kaolin particles are required. Iron oxides and quartz, usually found as kaolin impurities and undesirable in the catalyst formulation, are not transformed under the patented hydrothermal synthesis conditions, remaining in the final product.

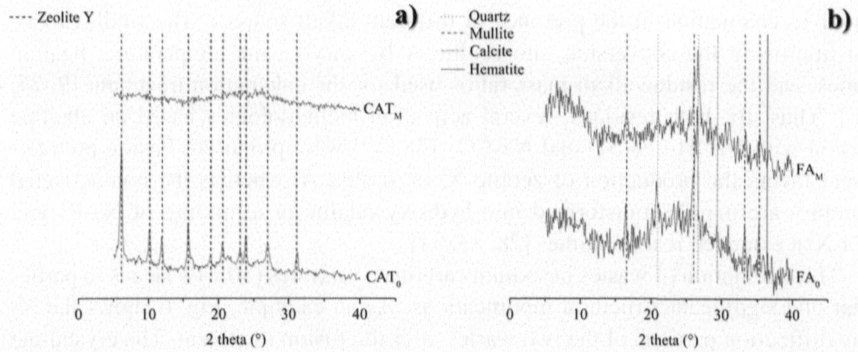

Fig. 5 X-ray diffraction patterns of mechanically activated wastes. **a** Exhausted catalyst: CAT_0 (original sample), CAT_M (after 1200 s of milling). **b** Fly ash: FA_0 (original sample), FA_M (after 3600 s of milling) (adapted from [26])

Extensive grinding was also studied as an alternative activation method, in which the kaolin clay was ground using an oscillating mill. This particular activation produces efficient conversions of kaolin clays to zeolites, similar to those obtained by heat activation. The kaolinite structure was tribologically transformed into an amorphous phase whose hydrothermal conversion in zeolites yielded acceptable results, but the effective grinding times for activation were too long [23].

Based on the experience gained from working with clays, similar methodologies were considered for obtaining zeolitic materials from the discarded FCC catalyst. In the first experiences on the transformation of the exhausted catalyst into zeolitic structures, the milling procedure was applied as an activation step. The decrease in catalyst crystallinity and particle size after 1200 s of mechanical treatment allowed better conversion into zeolite compared to those obtained with the untreated catalyst [24]. As can be seen in Fig. 5a, the zeolite Y initially present in the exhausted catalyst completely disappeared after milling.

On the contrary, when mechanical milling was applied to activate the fly ash, the results were quite different. Short milling times (from 60 up to 1200 s) produced no changes in the ash crystalline structure, and zeolite P was generated after hydrothermal synthesis. Nevertheless, a reduction in the diffraction peaks was observed from milling times of 3600 s (Fig. 5b). This change in crystallinity is related to the size reduction and particle shape modification of the mineral phases originally present in the fly ash, such as hematite, quartz, and mullite. A wide enlargement of the diffraction peaks corresponding to quartz and mullite was clearly detected, indicative of the formation of nanocrystalline particles by milling [25]. Even so, the reactivity of the ash towards zeolite A was very low and could be improved by using longer milling times. At this point, special attention should be paid to avoid contamination of the sample due to Fe from the attrition of the mill cylinders [23].

Based on more exhaustive studies, an alternative method of aluminosiliceous waste activation by means of alkaline fusion was developed. This methodology involves calcination in the presence of different alkali sources. The studies focus on improving the conversion into zeolite A by varying the temperature, heating times, and the residue/alkali mass ratios used for the calcination treatment [9, 27, 28]. Thus, for both residues, several activation methodologies based on alkaline fusion with NaOH [29–31] and Na_2CO_3 [28, 32] were proposed. Fusion pretreatment favors the production of zeolite X, or zeolite A, whereas the non-activated samples are mainly transformed into hydroxysodalite or a mixture of Na-P1 and Na-X at extended reaction times [28, 33, 34].

The calcination of wastes in sodium carbonate mixtures (50 wt%) leads to particular and significant structural modifications. As an example, Fig. 6 shows the X-ray diffraction patterns of the two wastes after the fusion treatment. The crystalline phases originally present in the residues (zeolite Y, mullite, quartz, etc., see Sect. 1.1) completely disappeared after the alkaline fusion step, meanwhile, the growth of new crystalline compounds was observed. It is interesting to note that the diffraction patterns obtained after fusion are identical for both samples, regardless of the waste type. The two emerging $NaAlSiO_4$ crystalline polymorphs were identified as nepheline and low-carnegieite.

Depending on the temperature and time of pretreatment, other minor crystalline phases were formed, due to the possible interaction with Na_2CO_3 (Fig. 6). At the pretreatment conditions studied, phases such as sodium calcium aluminum oxide silicate hydrate, sodium aluminum silicate, and sodium aluminum silicate hydroxide hydrate were found in the solid obtained after fly ash calcination. For the discarded catalyst, sodium silicate, sodium aluminum silicate hydroxide hydrate and sodium silicate appeared.

Fig. 6 X-ray diffraction patterns of exhausted catalyst (CAT) and fly ash (FA) after heating in the presence of Na_2CO_3 (800 °C, 1 h). SASH: sodium aluminum silicate hydroxide, SC: sodium carbonate, L: low-carnegieite, N: nepheline (adapted from [9])

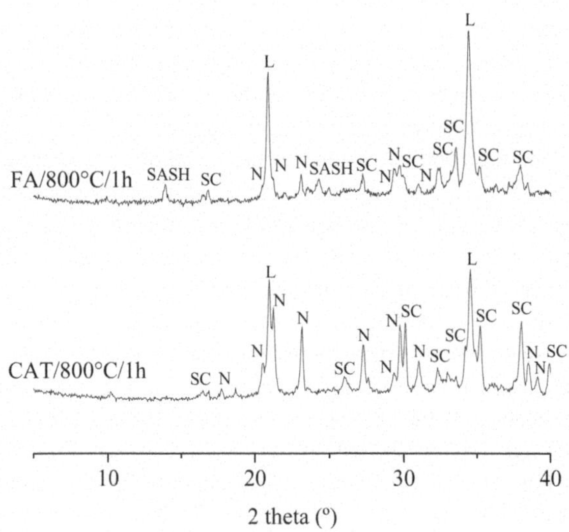

The total amount of $NaAlSiO_4$ polymorphs generated was strongly dependent on the operating conditions defined for the pretreatment step and on the structural nature of the aluminosiliceous waste.

It is also important to determine the alkali amount required for enhancing the waste transformation. After heating, residue/sodium carbonate weight ratios higher than 50% induce nepheline and low-carnegieite crystallization, but in this case, a fraction of non-reacted original waste remained. Lower ratios produce a solid having a crystalline composition identical to that obtained using 50% by weight. Therefore, from a technological and economic point of view, a weight ratio equal to 1 is recommended.

Depending on the nature of the waste, there is a temperature at which the total amount of polymorphs is maximum. The amounts of low-carnegieite and nepheline can be estimated by considering the sum of the intensities of their characteristic reflections on X-ray diffraction patterns. Thus, as can be seen in Fig. 7, for a preset calcination time of 1 h, the diffraction patterns obtained for the pretreated wastes at 800, 830 and 850 °C, show the highest intensities at the working temperature of 800 °C for the catalyst (Fig. 7a) and at 830 °C for the fly ash (Fig. 7b).

For each waste, the relative composition of the two polymorphs after heating at a preset temperature depends on the activation extent. At a constant selected temperature, the solid is preferentially rearranged into one of the polymorph structures as the heating time increases. In general, after Na_2CO_3 calcination, the spent FCC catalyst produces a large amount of nepheline compared to low-carnegieite. On the contrary, for the fly ash, a greater amount of low-carnegieite is observed compared to nepheline.

As seen in Fig. 8, fly ash activation produces the maximum percentage of low-carnegieite at 830 °C and 1 h of calcination time. In the case of the discarded catalyst, low-carnegieite formation was favored during the first hours at 800 °C, not exceeding 38%. The spent catalyst was preferentially transformed into nepheline, which in most

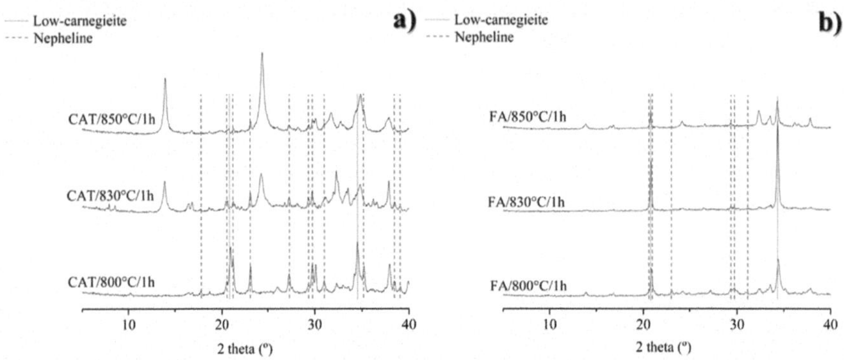

Fig. 7 X-ray diffraction patterns of products formed after 1 h of calcination in the presence of Na_2CO_3 (50%) at different temperatures. **a** Exhausted FCC catalyst. **b** Fly ash (adapted from [9])

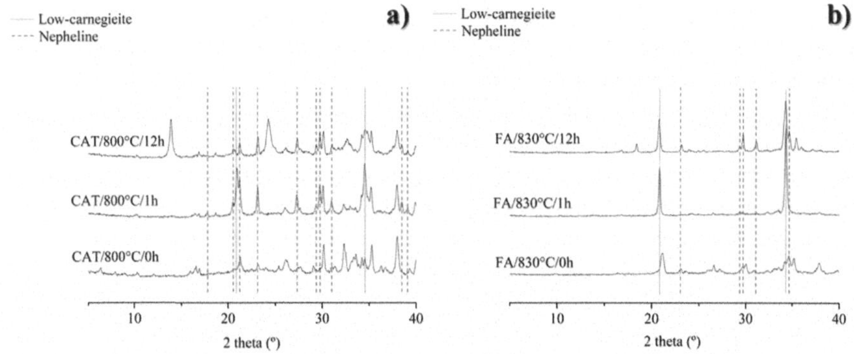

Fig. 8 X-ray diffraction patterns of the fusion activated solids obtained at different heating times. **a** Exhausted FCC catalyst treated at 800 °C. **b** Fly ash treated at 830 °C (adapted from [9])

of the cases achieved 100%. This transformation could be induced by the particular chemical composition of the waste [35–38].

For these reasons, a careful study of the residue transformation during activation allows establishing the lowest temperatures and the shortest times compatible with appreciable zeolite conversions.

The fusion treatment involves the generation of new phases by a Na_2CO_3-fluid-mediated reaction. The heating at 800–830 °C in the alkaline media allows the breakdown of all the ordered structures present in the starting materials such as zeolite, mullite, and quartz, increasing the Si and Al availability for being rearranged as $NaAlSiO_4$ polymorphs. In the wastes, the aluminum and silica atoms were linked to oxygen atoms forming octahedral or tetrahedral units, which in turn were linked among them through the oxygen atoms. The aluminosilicate chains were decomposed in [AlO_4] and [SiO_4] free tetrahedral units and transformed into the new crystalline phases low-carnegieite and nepheline. The presence of $NaCO_3$ promotes the free tetrahedra generation by the chemical bond-breaking effect of Na^+, which occurs along with CO_2 formation [39].

Furthermore, the noncrystalline fraction could involve a similar route: the amorphous aluminosilicate dissolution takes place during alkaline fusion, followed by the polymerization of active groups towards three-dimensional networks made of various units of connected SiO_4 and AlO_4 tetrahedra, and the incorporation of Na^+ as compensation ions [40, 41].

1.3.2 Hydrothermal Synthesis

The formation of zeolite microporous structure consists of nucleation and crystal growth processes occurring under hydrothermal conditions. During crystallization, the alkalinity and the cation species present in the reacting system control the formation of the polyhedral building units giving a specific framework type. Besides, the time needed for obtaining the optimum conversion into a pure A-type zeolite is very sensitive not only to the chemical composition of the synthesis media but also to working temperature. The presence of Na ions strongly influences the formation of D4R units, sodalite and α-cages. Thus, when the silica to alumina ratio is maintained constant, the increase of pH or reaction temperature accelerates the nucleation and zeolite growth, promoting the conversion into smaller sized crystals.

Considering the aforementioned statements, when using unconventional raw material for the synthesis, the real composition of the starting components, i.e., the available or true reacting sources, must be experimentally determined, and the composition of the solution synthesis must be adjusted accordingly.

At this point it is very important to highlight that in general, the SiO_2/Al_2O_3 ratio in the wastes is higher than that commonly used for synthesizing NaA, and an extra alumina source is needed. Additionally, the effects of changes in the chemical compositions of the starting mixtures on the conversion must be analyzed because the effective molar ratios vary during the crystallization. The solubilization of nutrients originally present in the solid phase, i.e., the silica and alumina fractions emerging from the solid to the liquid phase, is a function of the activation treatment and of the Na_2O concentration. In our studies, $NaAlO_2$ commercial solution was selected as the extra aluminum source optionally added at the beginning of the synthesis.

Once the synthesis batch is formed, the mixture is aged at room temperature for 48 h and placed afterwards in a conventional air oven at 90–100 °C. The course of the reaction was followed by taking aliquots at different times. The obtained solid products were washed and dried in air at 110 °C.

The particular interactions among the ionic species in the alkaline solution and the aluminosilicate frameworks of $NaAlSiO_4$ polymorphs generate a new equilibrium in the phase composition at the end of the aging [9]. In the case of the activated spent catalyst, a low dissolution of polymorphs was observed. By comparing the reflections in X-ray diffraction patterns of polymorphs before and after aging (Fig. 9a), a very small decrease in the heights of nepheline and low-carnegieite diffraction peaks occurred. On the contrary, all the activated fly ash showed preferential dissolution of nepheline, while low-carnegieite remained almost constant. The diffraction patterns presented in Fig. 9b show a fly ash sample where the initial characteristic peaks of nepheline were significantly diminished as a consequence of 48 h of aging, meanwhile the low carnegieite content remained almost unaltered.

The higher resistance to nepheline dissolution in alkaline media can be supported considering the linkage among SiO_4 tetrahedra. The nepheline phase is based on a tridymite-like structure, while the low-carnegieite is based on a cristobalite-like structure, so this crystalline arrangement determines their dissolution behavior. For all the hydrothermal synthesis experiments carried out with alkaline activated and aged wastes, as the reaction time increased, the NaA zeolite seemed to be gradually formed from the soluble silica and alumina species present in the liquid phase and also from the solid polymorphs of $NaAlSiO_2$ previously formed by alkaline fusion.

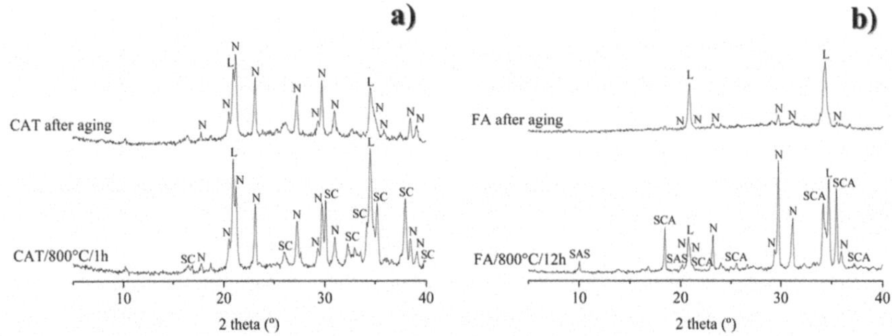

Fig. 9 Changes in the X-ray diffraction patterns of the samples after 48 h aging. **a** CAT/800 °C/1 h and aged sample. **b** FA/800 °C/12 h and aged sample. L: low-carnegieite, N: nepheline, SAS: sodium aluminum silicate hydroxide hydrate, SCA: sodium calcium aluminum oxide silicate hydrate, and SC: sodium carbonate (adapted from [9])

The evolution of the X-ray diffraction patterns of the aged phases during the course of the hydrothermal crystallization is shown in Fig. 10. As the reaction time increases, the reflections corresponding to the polymorphs decrease, whereas the characteristic peaks of NaA zeolite increase. It is interesting to note that the low-carnegieite is always completely transformed, regardless of the nature of the reacting solid. Table 1 summarizes the synthesis results obtained for both residues.

During the synthesis, the alkalinity of the batch solution contributes to the dissolution of Si^{4+} and Al^{3+} present in the activated wastes. Considering for both wastes a reaction batch as reference, changes in the initial chemical composition (changes in

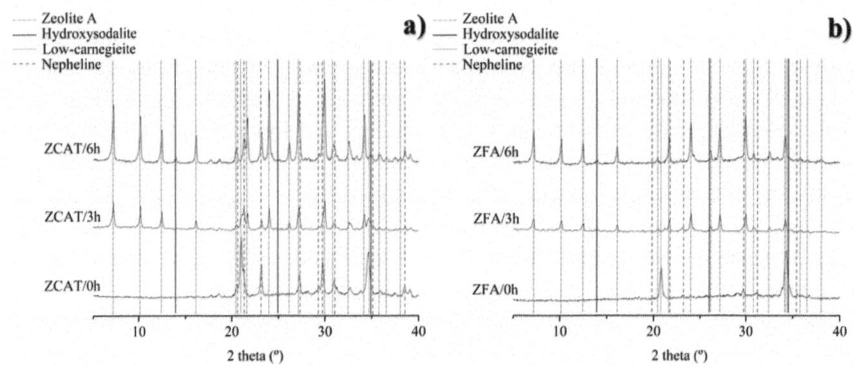

Fig. 10 X-ray diffraction patterns of the solids obtained at different synthesis times starting from the activated wastes. **a** Catalyst at 0 h (ZCAT/0 h), 3 h (ZCAT/3 h), and 6 h (ZCAT/6 h). **b** Fly ash at 0 h (ZFA/0 h), 3 h (ZFA/3 h), and 6 h (ZFA/6 h) (adapted from [9])

Table 1 Pretreatments and maximum conversion into zeolite A

Waste type	Pretreatments			Hydrothermal synthesis			
	Milling time (min)	Calcination Na_2CO_3					
		Temperature (°C)	Time (h)	Reaction time (h)	Conversion into NaA (%)	Zeolite product	Ref.
CAT	–	–	–	3	18	X + A	[28]
FA	–	–	–	23	–	HS + P	[26]
CAT	20	–	–	16	–	X	[24]
FA	10	–	–	6	–	P	[26]
FA	60	–	–	48	20	A + Tr HS	[26]
CAT	–	800	4	6	79	A + TrP/HS	[9]
FA	–	800	12	6	74	A + Tr HS	[9]

A: zeolite NaA, X: zeolite X, HS: hydroxysodalite, P: zeolite P and Tr: traces

sodium aluminate, sodium hydroxide, water concentrations) influence the percentage of NaA obtained.

Samples without any activation prior to the hydrothermal synthesis lead to the formation of different microporous solids according to the structural nature of the residue. In similar crystallization conditions, the inactivated catalyst induces NaA and NaX co-crystallization in low amount (18% of NaA). On the contrary, the results in the case of the fly ash indicated that even at higher reaction times, only HS + NaP1 are observed, and inert phases such as quartz, mullite remain in the final product.

As can be seen in Table 1, the mechanical treatment applied for shorter times leads to zeolite X for the catalyst and to zeolite P for fly ash. Long milling times are required to produce structural changes in the fly ash that improve the silica and alumina reactivity. Even so, the levels of conversion into NaA were very low.

In summary, the synthesis conditions favored the transformation of polymorphs into zeolite A by a complete topotactic process, mainly by the contribution of the low-carnegieite phase. The process could involve the fold of the planar arrangements of rings present in aluminosilicates (as a single six-membered ring, S6R) into an α-cage. The dissolved fraction of polymorphs during the aging time could also favor zeolite A formation due to the enrichment of active silicates in the synthesis batch. The results show that the highest reactivity towards zeolite crystallization could be obtained by selecting suitable aging conditions for getting the nepheline fraction entirely available before inducing zeolite growth.

In addition to structural changes, morphological differences in the surface of the particles were evidenced during the activation treatment and hydrothermal synthesis. Figures 11 and 12 show, respectively, the SEM images of the catalyst and fly ash particles after alkaline fusion pretreatment and after being aged for 48 h at room

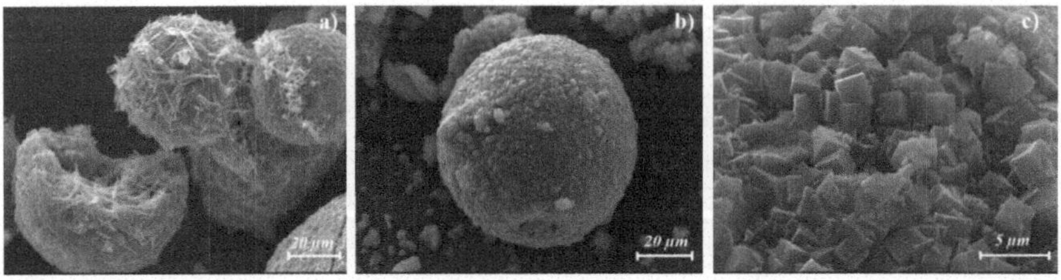

Fig. 11 Morphological transformation of exhausted catalyst. **a** After alkaline fusion pretreatment and being aged for 48 h, X1000. **b** After 6 h of synthesis, X1000. **c** After 6 h of synthesis, X5000

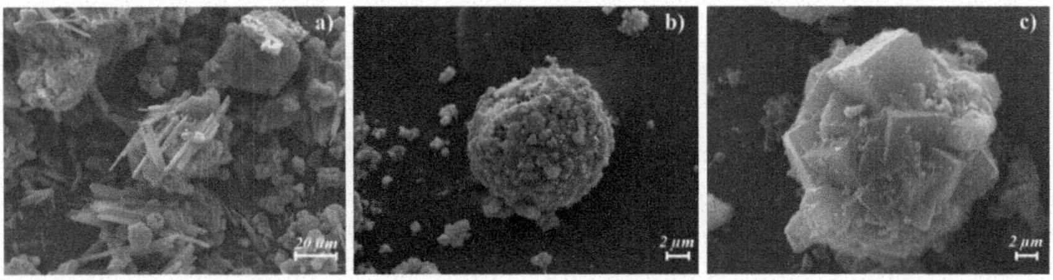

Fig. 12 Morphological transformation of fly ash. **a** After alkaline fusion pretreatment and being aged for 48 h, X500. **b** After 6 h of synthesis, X2500. **c** After 6 h of synthesis, X2500

temperature in alkaline solution (Figs. 11a and 12a) and after 6 h of hydrothermal synthesis (Figs. 11b, c, and 12b, c).

The SEM observations allow verifying the similarity in the surface transformation of both residues. The fusion activated solids consisted of rounded particles with a surface covered by needle-shaped crystals associated with low-carnegieite polymorphs.

During the synthesis, the growth of cubic crystals on the particle surface was increasingly detected. SEM analysis revealed that the starting spherical morphology was maintained and the surface was completely covered by cubic crystals. As cubic crystals are the typical morphology of NaA, these results confirm those obtained by X-ray analysis. It is interesting to note that the conservation of the spherical morphology of ash particles along the whole process indicates a reordering of the solid phase. As mentioned, the transformation of low-carnegieite present in the solid phase into NaA involves the reordering of the orthorhombic low-carnegieite network to the cubic lattice of NaA, i.e., the conversion of low-carnegieite to NaA takes place by a direct solid–solid transformation.

2 Heavy Metal Contamination

For many years, industrial activities have caused an increase in the contamination of waterways. The presence of toxic metals and anionic pollutants in aqueous courses is a problem that affects the environment and public health [42].

It is very common to find heavy metals such as Cr, Cu, Pb, Zn, Cd, Ni, etc., in the wastewater generated by metallurgical and mineral processing industries. These metallic contaminants are carcinogenic and also accumulate dangerously in living organisms. Toxic metals can be found in high concentrations (even up to 500 mg L^{-1}), but the maximum levels are limited by governmental regulations. Particularly, in the case of chromium, lead, cadmium, and nickel in water for human consumption, the World Health Organization established limits of 50, 10, 3, and 70 μg/L respectively.

Hexavalent chromium compounds are more toxic and carcinogenic than trivalent ones [43, 44] due to their facility to break through the cell membrane [45]. The enzymatic and nonenzymatic reactions occurring inside the cells produce the Cr^{6+} reduction to Cr^{5+}, Cr^{4+} and Cr^{3+} [46]. These reactive chromium intermediates are capable of generating a wide spectrum of reactive oxygen species, which cause damage to proteins, lipids, and DNA, an effect known as oxidative stress [47].

Therefore, it is important to develop technologies for controlling the pollutant concentrations in discharged wastewater. In recent years, various methodologies have been proposed to solve this problem, including ion exchange processes [48], membrane separation [49] and adsorption [50]. Currently, the processes based on exchange and adsorption are the most studied since they are the easiest to implement and generally have high efficiency [51]. The materials such as active carbons [52], carbon nanotubes [53], polymers [54], and especially zeolites [55] represent one of the best options in terms of efficiency and cost-effective issues.

The use of reconverted residues having a large percentage of zeolite NaA as microporous adsorbent is a valid alternative for the removal of heavy metal cations in water purification processes. In this way, it is necessary to carry out a wide range of physicochemical and structural characterizations, kinetic studies, equilibria at different pH, etc., to establish the ability of these synthetic materials to remove the toxic cations present in aqueous solutions. In the following section, our studies related to the effectiveness of the previously obtained zeolitized residues as water purifiers are presented.

2.1 *Chromium Adsorption by Using Zeolitized FCC Catalyst*

The sorption of Cr^{3+} in aqueous media was studied in a discontinuous system, by contacting the zeolitized material with the chromium-containing aqueous solution, under stirring. The effect of the initial chromium concentration (C_i), the working temperature (T), and the solid/liquid weight ratio (S/L) were varied in order to study their influence on the kinetic aspects of the sorption process and on the sorption capacity. Some selected results showing the reduction of chromium concentration in the exchange solution as a function of time are shown in Figs. 13 and 14, for C_i = 42 and 135 mg L^{-1}, at two temperatures (25 and 50 °C) and two solid/liquid ratios (1 and 3).

It was noted that the temperature increase accelerated the cation exchange process. Also, the removal of Cr^{3+} proceeded faster at higher solid/liquid ratios. It was evident that a system for removing 100% of chromium cation could be designed. At pH values at which the precipitation of chromium compounds is avoided, the sorption procedure

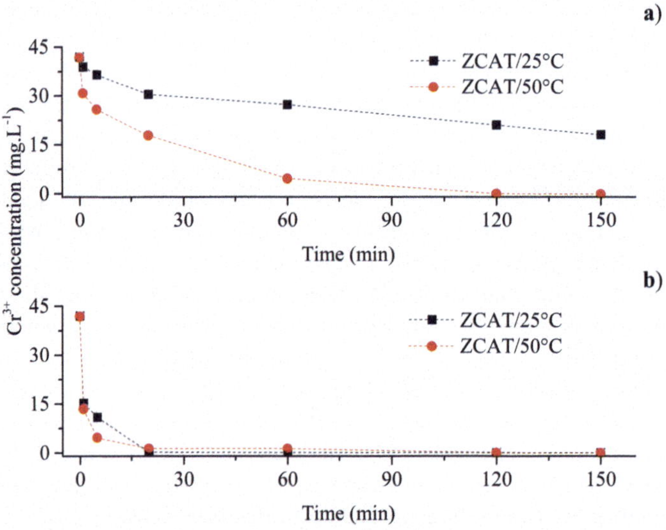

Fig. 13 Cr^{3+} concentration in liquid phase by using **a** 1 g or **b** 3 g of zeolite A-rich catalyst (ZCAT) at T = 25 °C and T = 50 °C (C_i = 42 mg L^{-1})

Fig. 14 Cr^{3+} concentration in liquid phase by using 3 g of zeolite A-rich catalyst (ZCAT) at T = 25 °C and T = 50 °C (C_i = 135 mg L^{-1})

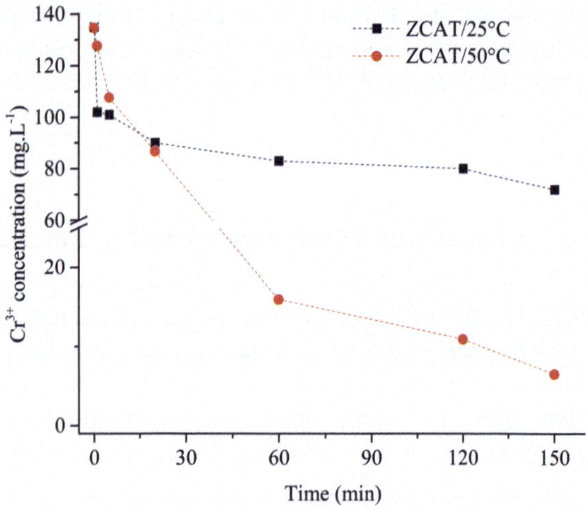

involves hydrated chromium cations replacing sodium cations, occupying exchange positions and counteracting the negative charge of the zeolite A network.

A notable pH increase was produced along the Cr^{3+} sorption process, starting from 3.5 to nearly 6.5 at the end of the exchange. It could be attributed to the occupation of the cation exchange sites by the hydrated Cr^{3+} species and the corresponding desorption of sodium cations. Competition for exchange sites with hydrogen also decreased rapidly as the pH was raised in the range 1.5–5.0 and contributed to the increased sorption.

The SEM and EDS studies of the sorbent phase after the highest Cr^{3+} sorption are shown in Fig. 15. It was observed that the initial microspheroidal catalyst morphology was maintained. Additionally, EDS mapping showed that the presence of chromium was significantly higher in the outer microsphere surface, where the zeolite crystals were mostly situated, than in the interior region (Fig. 15).

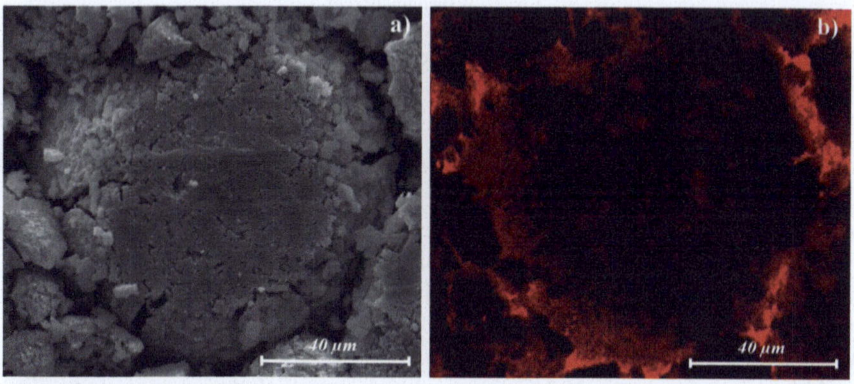

Fig. 15 SEM–EDX analyses for zeolite A-rich catalyst after the highest Cr^{3+} sorption. **a** SEM micrograph of the polished catalyst particle. **b** Chromium EDX mapping corresponding to the area shown in **a**

Fig. 16 X-ray diffraction patterns for zeolite A-rich catalyst before and after chromium cation exchange (adapted from [56])

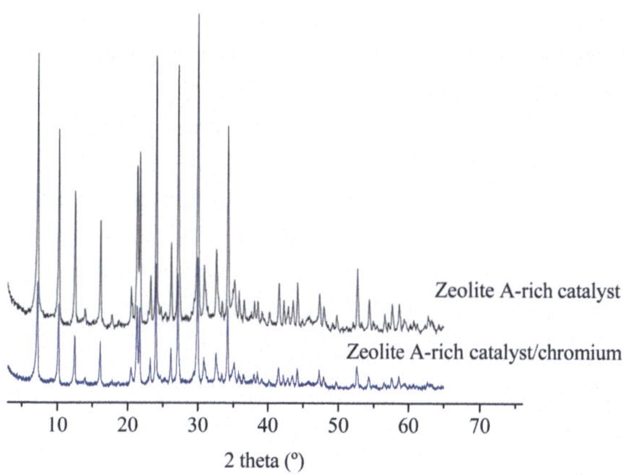

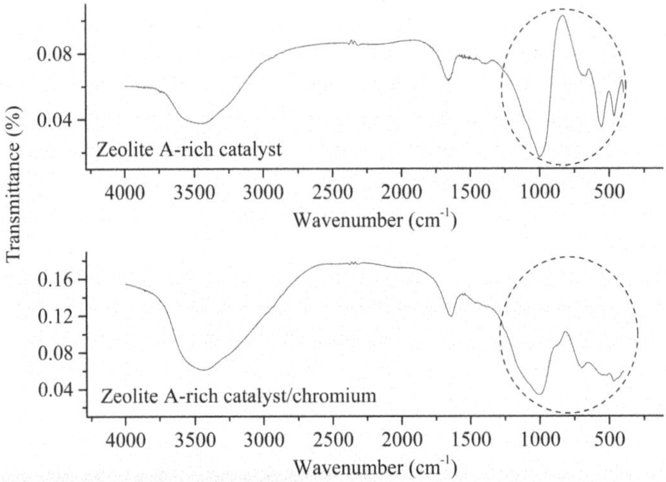

Fig. 17 FT-IR spectrum of zeolitic material before and after contact with a solution of 135 mg L^{-1} Cr^{3+} (adapted from [56])

Additionally, the structural analyses carried out by XRD (Fig. 16) and FTIR (Fig. 17) showed that the incorporation of Cr^{3+} into zeolite crystals reduced the height or intensities characteristic of the zeolite structure.

The X-ray diffraction patterns before and after cation exchange show no difference in the positions of the peaks associated with the zeolite A structure, indicating that the network parameters are preserved and reflections of oxides or hydroxides are absent.

Although the exchange Na^+/Cr^{3+} in zeolite A has a high degree of irreversibility [57], the obtained results show that chromium incorporation has not modified the crystal structure. Additionally, an appreciable decrease in peak intensities after chromium exchange was observed.

After the exchange with 135 mg L^{-1} of Cr^{3+} solution, the FT-IR absorption bands associated with the D4R zeolite units (between 450 and 600 cm^{-1}) decreased markedly. A new band between 800 and 900 cm^{-1} appeared which may be ascribed to the onset of zeolite amorphization [58].

The zeta potential versus pH curves corroborated the absence of a net charge change in the range 2–7 (Fig. 18). In fact, Cr^{3+} cations predominantly occupied exchange sites in the zeolite framework, as inferred from XRD analysis. These results are in complete agreement with the Cr^{3+} adsorption on bentonite [59].

The incorporation of Cr^{3+} cations within the crystal structure might be considered beneficial for cation inertization, because electrostatic attraction contributes to avoiding leaching.

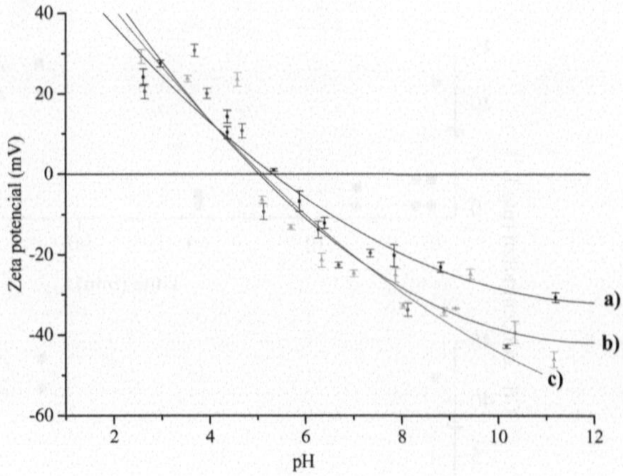

Fig. 18 Zeta potential curves of corresponding to **a** zeolite A-rich catalyst, **b** zeolite A-rich catalyst/Cr^{3+}/42, **c** zeolite A-rich catalyst/Cr^{3+}/135 (adapted from [56])

2.2 *Cadmium Sorption by Using the Zeolitized Wastes*

The removal of Cd^{2+} from aqueous solutions was carried out separately by using the zeolite A-rich materials obtained from FCC catalyst and fly ash. The cationic exchange procedure was the same used for capturing the Cr^{3+}, as described in 2.1. The influence of parameters such as temperature, initial concentration, and solid/liquid weight ratio on the adsorption capacity were also studied. Besides, for the kinetic studies, samples of the solid phases collected at predetermined exchange times were analyzed. The decrease of Cd^{2+} in the exchange solutions as a function of time, using the zeolite A-rich solids at 25 °C, is shown in Figs. 19, 20 and 21, respectively.

In the present conditions, the exchange reaction using both materials showed a very fast depletion of the cation concentration in the exchange solution, Figs. 19 and 20. Particularly, the zeolitized fly ash exhibited a better Cd^{2+} retention.

Additionally, the increase in the pH value recorded along the exchange reaction encouraged $Cd(OH)_2$ precipitation. The removal of Cd^{2+} cation seemed to be always higher than 93%, being the best results obtained from the zeolitized fly ash.

It was also noted that when the initial concentration of Cd^{2+} increased, the effectiveness in eliminating the cation largely depended on the S/L ratio. Figure 21 shows the evolution of cation concentration in the liquid phase using 1 g or 3 g of both zeolitized solids when the initial concentration of Cd^{2+} increased to 500 mg L^{-1}. Thus, for more concentrated Cd^{2+} solutions (for example $C_i = 500$ mg L^{-1}, Fig. 21), sorption improved at an increased S/L ratio. It is important to note that the zeolite A-rich fly ash still has a better performance compared to the zeolite obtained from the FCC catalyst.

Furthermore, it seems to be very important to consider the pH changes produced by changes in the S/L ratio or the initial Cd concentration. As an example, the Cd^{2+} sorption by using zeolitized fly ash notably decreased when C_i increased (Table 2).

As previously mentioned, depending on the working pH values, the removal of cadmium could occur by two mechanisms:

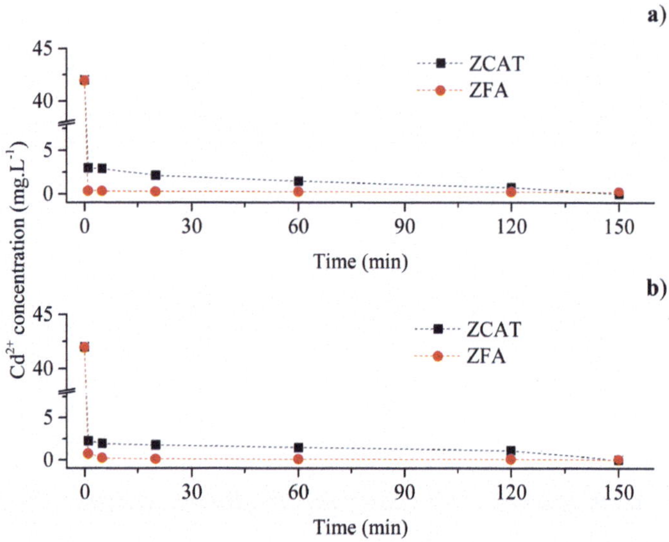

Fig. 19 Cd^{2+} concentration in liquid phase by using **a** 1 g or **b** 3 g of zeolitized catalyst (ZCAT) and zeolitized fly ash (ZFA) (T = 25 °C, C_i = 42 mg L^{-1})

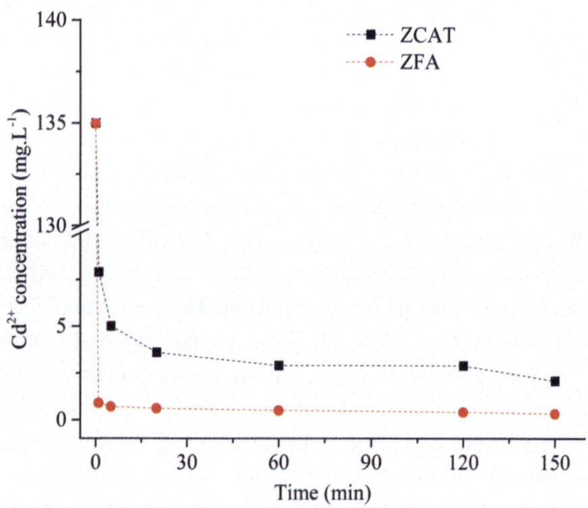

Fig. 20 Cd^{2+} concentration in liquid phase by using 3 g of zeolite A-rich catalyst (ZCAT) and zeolite A-rich fly ash (ZFA) at T = 25 °C (C_i = 135 mg L^{-1})

a) the capture process where the Cd^{2+} replaces two Na^+ cations, occupying exchange positions in the zeolite structure, and b) the $Cd(OH)_2$ precipitation at high alkalinities, mainly observed using zeolitized fly ash.

In general, the removal effectiveness could be associated with hydrolysis phenomena and coordination changes of the ionic species, together with the adsorption, exchange, and precipitation processes that are likely to occur when the ionic concentration, temperature or pH parameters are changed.

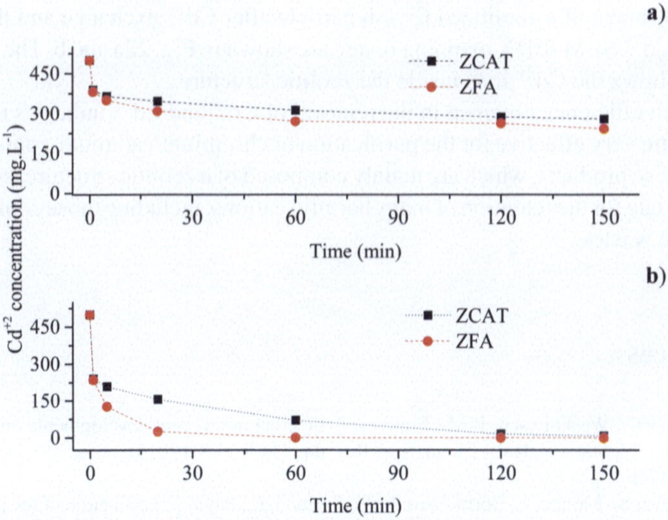

Fig. 21 Cd^{2+} concentration in liquid phase by using **a** 1 g or **b** 3 g of zeolite A-rich catalyst (ZCAT) and zeolite A-rich fly ash (ZFA) at T = 25 °C (C_i = 500 mg L^{-1})

Table 2 Cd^{2+} sorption by using zeolitized fly ash as a function of time

Time (min)	C_i = 400 mg L^{-1}		C_i = 600 mg L^{-1}	
	pH	S/L = 1 C (mg L^{-1})	pH	S/L = 1 C (mg L^{-1})
0	6.92	400	6.78	600
5 h	7.16	93	7.19	260
24 h	7.34	81	7.21	230

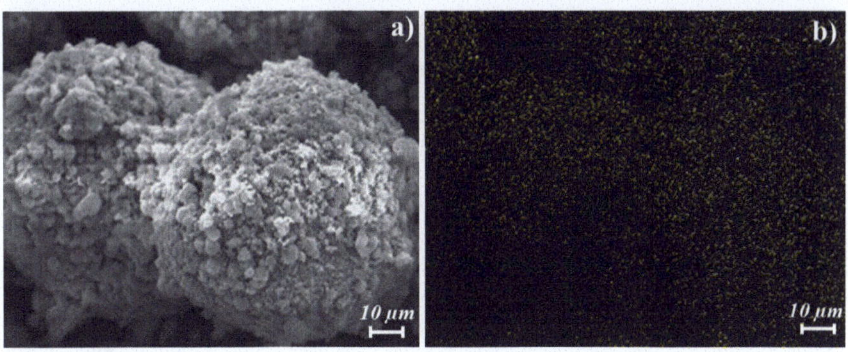

Fig. 22 SEM–EDX analyses for zeolite A-rich fly ash after the highest Cd^{2+} sorption. **a** SEM micrograph of the solid obtained after cadmium exchange. **b** Cadmium EDX mapping corresponding to the micrograph shown in **a**

A SEM image of a zeolitized fly ash particle after Cd^{2+} exchange and the corresponding Cd^{2+} SEM–EDX mapping image are shown in Fig. 22a and b. The Cd EDX mapping shows the Cd^{2+} ions inside the zeolitic structure.

The high efficiency obtained in the removal of Cr^{3+} and Cd^{2+} indicates that these materials are very effective for the purification of chromium/cadmium-contaminated waters. These products, which are mainly composed of a zeolitic structure, could also be appropriate for the retention of many harmful cations, including those coming from radioactive wastes.

References

1. Vogt, E.T.C., Weckhuysen, B.M.: Fluid catalytic cracking: recent developments on the grand old lady of zeolite catalysis. Chem. Soc. Rev. **44**, 7342–7370 (2015). https://doi.org/10.1039/c5cs00376h
2. Velázquez, S., Monzó, J., Borrachero, M.V., Soriano, L., Payá, J.: Evaluation of the pozzolanic activity of spent FCC catalyst/fly ash mixtures in Portland cement pastes. Thermochim. Acta **632**, 29–36 (2016). https://doi.org/10.1016/j.tca.2016.03.011
3. Soriano, L., Payá, J., Monzó, J., Borrachero, M.V., Tashima, M.M.: High strength mortars using ordinary Portland cement-fly ash-fluid catalytic cracking catalyst residue ternary system (OPC/FA/FCC). Constr. Build. Mater. **106**, 228–235 (2016). https://doi.org/10.1016/j.conbuildmat.2015.12.111
4. Yao, Z.T., Ji, X.S., Sarker, P.K., Tang, J.H., Ge, L.Q., Xia, M.S., Xi, Y.Q.: A comprehensive review on the applications of coal fly ash. Earth-Sci. Rev. **141**, 105–121 (2015). https://doi.org/10.1016/j.earscirev.2014.11.016
5. Bicer, A.: Effect of fly ash particle size on thermal and mechanical properties of fly ash-cement composites. Therm Sci Eng Prog **8**, 78–82 (2018). https://doi.org/10.1016/j.tsep.2018.07.014
6. Hefni, Y., El, Z.Y.A., Wahab, M.A.: Influence of activation of fly ash on the mechanical properties of concrete. Constr. Build. Mater. **172**, 728–734 (2018). https://doi.org/10.1016/j.conbuildmat.2018.04.021
7. Sobolev, K., Flores Vivian, I., Saha, R., Wasiuddin, N.M., Saltibus, N.E.: The effect of fly ash on the rheological properties of bituminous materials. Fuel **116**, 471–477 (2014). https://doi.org/10.1016/j.fuel.2013.07.123
8. Breck, D.W.: Zeolite molecular sieves—structure, chemistty and use. Wiley, New York (1974)
9. Monzón, J.D., Gonzalez, M.R., Mardones, L.E., Conconi, M.S., Pereyra, A.M., Basaldella, E.I.: The role of alkaline activation in the structural transformations of aluminosiliceous industrial wastes towards zeolite production. Mater. Today Commun. **21**, 100624 (2019). https://doi.org/10.1016/j.mtcomm.2019.100624
10. Vassilev, S.V., Vassileva, C.G.: Methods for characterization of composition of fly ashes from coal-fired power stations: A critical overview. Energy Fuels **19**, 1084–1098 (2005). https://doi.org/10.1021/ef049694d
11. Kuhl, G.H., Kresge, C.T.: Molecular sieves. In: Van Nostrand's Encyclopedia of Chemistry, First. Wiley, New York (2005)
12. Tahraoui, Z., Nouali, H., Marichal, C., Forler, P., Klein, J., Daou, T.J.: Influence of the compensating cation nature on the water adsorption properties of zeolites. Molecules **25**, 944–969 (2020). https://doi.org/10.3390/molecules25040944
13. Association, Structure Commission of the International Zeolite—Database of zeolite structures—Framework Type LTA. https://asia.iza-structure.org/IZA-SC/framework.php?STC=LTA. Accessed 13 Aug 2020
14. Ryu, T., Ahn, N.H., Seo, S., Cho, J., Kim, H., Jo, D., Park, G.T., Kim, P.S., Kim, C.H., Bruce, E.L., Wright, P.A., Nam, I.S., Hong, S.B.: Fully Copper-exchanged high-silica LTA zeolites as

unrivaled hydrothermally stable NH3-SCR catalysts. Angew Chemie Int. Ed. **56**, 3256–3260 (2017). https://doi.org/10.1002/anie.201610547

15. Li, X., Li, K., Tao, S., Ma, H., Xu, R., Wang, B., Wang, P., Tian, Z.: Ionothermal synthesis of LTA-type aluminophosphate molecular sieve membranes with gas separation performance. Microporous Mesoporous Mater. **228**, 45–53 (2016). https://doi.org/10.1016/j.micromeso.2016.03.026
16. Noroozi, R., Al-Musawi, T.J., Kazemian, H., Kalhori, E.M., Zarrabi, M.: Removal of cyanide using surface-modified Linde Type-A zeolite nanoparticles as an efficient and eco-friendly material. J. Water Process. Eng. **21**, 44–51 (2018). https://doi.org/10.1016/j.jwpe.2017.11.011
17. Li, J., Li, M., Song, Q., Wang, S., Cui, X., Liu, F., Liu, X.: Efficient recovery of Cu(II) by LTA-zeolites with hierarchical pores and their resource utilization in electrochemical denitrification: Environmentally friendly design and reutilization of waste in water. J. Hazard Mater. **394**, 122554 (2020). https://doi.org/10.1016/j.jhazmat.2020.122554
18. Pereyra, A.M., Gonzalez, M.R., Rosato, V.G., Basaldella, E.I.: A-type zeolite containing Ag^{+}/Zn^{2+} as inorganic antifungal for waterborne coating formulations. Prog. Org. Coat. **77**, 213–218 (2014). https://doi.org/10.1016/j.porgcoat.2013.09.008
19. Pereyra, A.M., Gonzalez, M.R., Rodrigues, T.A., Soares Luterbach, M.T., Basaldella, E.I.: Enhancement of biocorrosion resistance of epoxy coating by addition of Ag/Zn exchanged a zeolite. Surf. Coat. Technol. **270**, 284–289 (2015). https://doi.org/10.1016/j.surfcoat.2015.02.044
20. Machado, G.E., Pereyra, A.M., Rosato, V.G., Moreno, M.S., Basaldella, E.I.: Improving the biocidal activity of outdoor coating formulations by using zeolite-supported silver nanoparticles. Mater. Sci. Eng. C **98**, 789–799 (2019). https://doi.org/10.1016/j.msec.2019.01.040
21. Basaldella, E.I., Kikot, A., Pereira, E.: Synthesis of zeolites from mechanically activated kaolin clays. React Solids **8**, 169–177 (1990). https://doi.org/10.1016/0168-7336(90)80017-E
22. Dight, L.B., Bogert, D.C., Leskowicz, M.A.:Ultra high zeolite content FCC catalysts and method for making same from microspheres composed of a mixture of calcined kaolin clays. US Patent 5,023,220 (1991)
23. Torres Sánchez, R.M., Basaldella, E.I., Marco, J.F.: The effect of thermal and mechanical treatments on kaolinite: characterization by XPS and IEP measurements. J. Colloid Interface Sci. **215**, 339–344 (1999). https://doi.org/10.1006/jcis.1999.6241
24. Basaldella, E.I., Paladino, J.C., Solari, M., Valle, G.M.: Exhausted fluid catalytic cracking catalysts as raw materials for zeolite synthesis. Appl. Catal. B Environ. **66**, 186–191 (2006). https://doi.org/10.1016/j.apcatb.2006.03.013
25. Hamzaoui, R., Bouchenafa, O., Guessasma, S., Leklou, N., Bouaziz, A.: The sequel of modified fly ashes using high energy ball milling on mechanical performance of substituted past cement. Mater. Des. **90**, 29–37 (2016). https://doi.org/10.1016/j.matdes.2015.10.109
26. Monzón, J.D., Pereyra, A.M., Conconi, M.S., Basaldella, E.I.: Phase transformations during the zeolitization of fly ashes. J. Environ. Chem. Eng. **5**, 1548–1553 (2017). https://doi.org/10.1016/j.jece.2017.02.022
27. Basaldella, E.I., Torres Sánchez, R.M., Conconi, M.S.: Conversion of exhausted fluid cracking catalysts into zeolites by alkaline fusion. Appl. Clay Sci. **42**, 611–614 (2009). https://doi.org/10.1016/j.clay.2008.03.016
28. González, M.R., Pereyra, A.M., Basaldella, E.I.: Trivalent chromium ion removal from aqueous solutions using low-cost zeolitic materials obtained from exhausted FCC catalysts. Adsorpt. Sci. Technol. **29**, 629–636 (2011). https://doi.org/10.1260/0263-6174.29.7.629
29. Belviso, C., Cavalcante, F., Lettino, A., Fiore, S.: Effects of ultrasonic treatment on zeolite synthesized from coal fly ash. Ultrason. Sonochem. **18**, 661–668 (2011). https://doi.org/10.1016/j.ultsonch.2010.08.011
30. Sangita, K., Prasad, B., Udayabhanu, G.: Synthesis of zeolite from waste fly ash by using different methods. Asian J. Chem. **28**, 1435–1439 (2016). https://doi.org/10.14233/ajchem.2016.19682

31. Zhang, Y., Dong, J., Guo, F., Shao, Z., Wu, J.: Zeolite synthesized from coal fly ash produced by a gasification process for Ni^{2+} removal fromwater. Minerals **8**, 1–14 (2018). https://doi.org/10.3390/min8030116
32. Deng, H., Ge, Y.: Formation of NaP zeolite from fused fly ash for the removal of Cu(II) by an improved hydrothermal method. RSC Adv. **5**, 9180–9188 (2015). https://doi.org/10.1039/c4ra15196h
33. Musyoka, N.M., Petrik, L., Hums, E.: Synthesis of zeolite A, X and P from a South African coal fly ash. Adv. Mater. Res. **512–515**, 1757–1762 (2012). https://doi.org/10.4028/www.scientific.net/AMR.512-515.1757
34. Aldahri, T., Behin, J., Kazemian, H., Rohani, S.: Synthesis of zeolite Na-P from coal fly ash by thermo-sonochemical treatment. Fuel **182**, 494–501 (2016). https://doi.org/10.1016/j.fuel.2016.06.019
35. Guzmán-Carrillo, H.R., Pérez, J.M., Romero, M.: Crystallisation of nepheline-based glass frits through fast-firing process. J. Non Cryst. Solids **470**, 53–60 (2017). https://doi.org/10.1016/j.jnoncrysol.2017.04.043
36. Buchwald, A., Vicent, M., Kriegel, R., Kaps, C., Monzó, M., Barba, A.: Geopolymeric binders with different fine fillers - Phase transformations at high temperatures. Appl. Clay. Sci. **46**, 190–195 (2009). https://doi.org/10.1016/j.clay.2009.08.002
37. Kuenzel, C., Grover, L.M., Vandeperre, L., Boccaccini, A.R., Cheeseman, C.R.: Production of nepheline/quartz ceramics from geopolymer mortars. J. Eur. Ceram. Soc. **33**, 251–258 (2013). https://doi.org/10.1016/j.jeurceramsoc.2012.08.022
38. Dimitrijevic, R., Dondur, V., Vulic, P., Markovic, S., Macura, S.: Structural characterization of pure Na-nephelines synthesized by zeolite conversion route. J. Phys. Chem. Solids **65**, 1623–1633 (2004). https://doi.org/10.1016/j.jpcs.2004.03.005
39. Yan, K., Guo, Y., Fang, L., Cui, L., Cheng, F., Li, T.: Decomposition and phase transformation mechanism of kaolinite calcined with sodium carbonate. Appl. Clay Sci. **147**, 90–96 (2017). https://doi.org/10.1016/j.clay.2017.07.010
40. Tait, K.T., Sokolova, E., Hawthorne, F.C., Khomyakov, A.P.: The crystal chemistry of nepheline. Can. Mineral. **41**, 61–70 (2003). https://doi.org/10.2113/gscanmin.42.3.797
41. Kumar, A., Dhoble, S.J., Peshwe, D.R., Bhatt, J.: Structural and Photoluminescence properties of nepheline-structure NaAlSiO4:Dy3+ nanophosphors. J. Alloys Compd. **609**, 100–106 (2014). https://doi.org/10.1016/j.jallcom.2014.04.153
42. Ke, F., Qiu, L.G., Yuan, Y.P., Peng, F.M., Jiang, X., Xie, A.J., Shen, Y.H., Zhu, J.F.: Thiol-functionalization of metal-organic framework by a facile coordination-based postsynthetic strategy and enhanced removal of Hg 2+ from water. J. Hazard Mater. **196**, 36–43 (2011). https://doi.org/10.1016/j.jhazmat.2011.08.069
43. Connett, P.H., Wetterhahn, K.E.: Metabolism of the carcinogen chromate by cellular constituents (2007). Inorg. Elem. Biochem. 93–124. https://doi.org/10.1007/bfb0111319
44. De, F.S., Bagnasco, M., Serra, D., Zanacchi, P.: Genotoxicity of chromium compounds. A Review. Mutat. Res. Genet. Toxicol. **238**, 99–172 (1990)
45. De Flora, S., Wetterhahn, K.: Mechanism of chromium (VI) metabolism and genotoxicity. Life Chem. Rep. **7**, 169–244 (1989)
46. Shi, X., Chiu, A., Chen, C.T., Halliwell, B., Castranova, V., Vallyathan, V.: Reduction of chromium (VI) and its relationship to carcinogenesis. J. Toxicol. Environ. Heal. Part B Crit. Rev. **2**, 87–104 (1999). https://doi.org/10.1080/109374099281241
47. Nordberg, J., Ismael, T.Y.: Reactive oxygen species, antioxidants, and the mammalian thioredoxin system. Free Radic. Biol. Med. **31**, 1287–1312 (2001). https://doi.org/10.1016/S0891-5849(01)00724-9
48. Plazinski, W., Rudzinski, W.: Modeling the effect of surface heterogeneity in equilibrium of heavy metal ion biosorption by using the ion exchange model. Environ. Sci. Technol. **43**, 7465–7471 (2009). https://doi.org/10.1021/es900949e
49. Hasan, M.A., Selim, Y.T., Mohamed, K.M.: Removal of chromium from aqueous waste solution using liquid emulsion membrane. J. Hazard Mater. **168**, 1537–1541 (2009). https://doi.org/10.1016/j.jhazmat.2009.03.030

50. Yu, B., Zhang, Y., Shukla, A., Shukla, S.S., Dorris, K.L.: The removal of heavy metals from aqueous solutions by sawdust adsorption—removal of lead and comparison of its adsorption with copper. J. Hazard Mater. **84**, 83–94 (2001). https://doi.org/10.1016/S0304-3894(01)00198-4
51. Li, G., Zhao, Z., Liu, J., Jiang, G.: Effective heavy metal removal from aqueous systems by thiol functionalized magnetic mesoporous silica. J. Hazard Mater. **192**, 277–283 (2011). https://doi.org/10.1016/j.jhazmat.2011.05.015
52. Reed, B.E., Arunachalam, S., Thomas, B.: Removal of lead and cadmium from aqueous waste streams using granular activated carbon (GAC) columns. Environ. Prog. **13**, 60–64 (1994). https://doi.org/10.1002/ep.670130123
53. Perez-Aguilar, N.V., Muñoz-Sandoval, E., Diaz-Flores, P.E., Rangel-Mendez, J.R.: Adsorption of cadmium and lead onto oxidized nitrogen-doped multiwall carbon nanotubes in aqueous solution: equilibrium and kinetics. J. Nanoparticle Res. **12**, 467–480 (2010). https://doi.org/10.1007/s11051-009-9670-6
54. Khajeh, M., Heidari, Z.S., Sanchooli, E.: Synthesis, characterization and removal of lead from water samples using lead-ion imprinted polymer. Chem. Eng. J. **166**, 1158–1163 (2011). https://doi.org/10.1016/j.cej.2010.12.018
55. Anielak, A.M., Schmidt, R.: Sorption of lead and cadmium cations on natural and manganese-modified zeolite. Polish J. Environ. Stud. **20**, 15–19 (2011)
56. Gonzalez, M.R., Pereyra, A.M., Sánchez, R.M.T., Basaldella, E.I.: Chromium removal by zeolite-rich materials obtained from an exhausted FCC catalyst: Influence of chromium incorporation on the sorbent structure. J. Colloid Interface Sci. **408**, 21–24 (2013). https://doi.org/10.1016/j.jcis.2013.06.055
57. Wiers, B.H., Grosse, R.J., Cilley, W.A.: Divalent and trivalent ion exchange with zeolite A. Environ. Sci. Technol. **16**, 617–624 (1982). https://doi.org/10.1021/es00103a016
58. Basaldella, E.I., Vázquez, P.G., Iucolano, F., Caputo, D.: Chromium removal from water using LTA zeolites: effect of pH. J. Colloid Interface Sci. **313**, 574–578 (2007). https://doi.org/10.1016/j.jcis.2007.04.066
59. Chakir, A., Bessiere, J., Kacemi, K.E.L., Marouf, B.: A comparative study of the removal of trivalent chromium from aqueous solutions by bentonite and expanded perlite. J. Hazard Mater. **95**, 29–46 (2002). https://doi.org/10.1016/S0304-3894(01)00382-X

Removal of Pollutants from Water by Adsorbents Prepared from Animal Bone Wastes

Nahum Andres Medellin-Castillo, Miguel Mauricio Aguilera-Flores, and Bridinette Thiodjio Sendja

Abstract The disposal of water for human use and consumption, as well as its pollution by anthropogenic activities, is an issue of worldwide concern, mainly because of the amount of diseases and deaths that they generate in the population. In this chapter, an analysis is presented about the importance of water, the types of pollutants present in water and their sources of pollution. Likewise, it highlights the importance of adsorption processes as a technology of easy implementation and low cost in the treatment of water which uses various adsorbent materials such as bone char. Animal bone chars represent a versatile adsorbent alternative because of the diversity of pollutants that can be removed from aqueous solutions, among which are heavy metals, anions and organic contaminants. These materials can be synthesized from animal bones such as pork, chicken, cow, fish, sheep, among others. Similarly, this study includes a bibliometric analysis on the study of bone chars, characteristics, uses, synthesis methods, characterization, removal mechanisms and some case studies on their environmental application in the removal of pollutants from water such as fluoride, cadmium and triclosan.

N. A. Medellin-Castillo (✉)
Faculty of Engineering, Graduate Studies and Research Center, Autonomous University of San Luis Potosi, 78290 San Luis Potosi, Mexico
e-mail: nahum.medellin@uaslp.mx

N. A. Medellin-Castillo · M. M. Aguilera-Flores
Multidisciplinary Graduate Program in Environmental Sciences, Autonomous University of San Luis Potosi, 78210 San Luis Potosi, Mexico
e-mail: maguileraf@ipn.mx

M. M. Aguilera-Flores
Interdisciplinary Professional Unit of Engineering Zacatecas Campus, Instituto Politécnico Nacional, Bioengineering Academy, 98160 Zacatecas, Mexico

B. T. Sendja
Department of Mathematic and Physical Science, National Advanced School of Engineering, University of Yaounde I, 8390 Yaounde, Cameroon
e-mail: sbridine@yahoo.fr

J. C. Moreno-Piraján et al. (eds.), *Porous Materials*, Engineering Materials,
https://doi.org/10.1007/978-3-030-65991-2_11

1 Introduction

Water is one of the most abundant natural resources on the planet and is indispensable for the development of life and human activities. Due to its importance, it is necessary to develop efficient and low-cost technologies for its use and reuse. Similarly, it is necessary to produce water of adequate quality for consumption or use. There are various physicochemical and microbiological parameters that determine the quality of this resource and that are important to evaluate. Among these parameters are heavy metals, organic compounds and anionic pollutants such as fluoride. These compounds, even at low concentrations, represent a risk to human health and living organisms, so the development of affordable technologies for their application in the removal of the pollutants from water is an area of interest.

Bone char is considered a promising green material, since it is synthesized from biomass (animal bones) generated as waste, which is usually not valorized. Its use contributes with a dual purpose, the valorization of the waste and its application in the development of environmental technologies, so it is also seen as one of the environmentally friendly and cost-effective ways for removing heavy metals, anionic pollutants and emerging organic compounds from water.

In this chapter, a review is conducted of bone char and its environmental applications, including its characteristics, uses, synthesis methods, characterization, removal mechanisms, as well as a bibliometric analysis obtained from various sources that show interest in this material due to its capacity to remove various pollutants from aqueous solutions and other environmental applications. In addition, some studies focused on the application in the removal of fluorides, heavy metals and some organic compounds in aqueous solutions using bone chars from fish, cattle, pork and chicken, are presented.

2 The Importance of Water

Water is one of the fundamental resources for the conservation of biodiversity on the planet since it is essential for living organisms. Approximately, 97.5% of all water on Earth is salt water and is found in the oceans. This cannot be used without the application of expensive desalination techniques, which require a high energy demand. Water is the natural substance with the largest territorial surface, covering 70% of the planet, and only 1.75% is frozen in form of glaciers and ice caps, which are freshwater depositories not easily available for an immediate application. Only a small proportion of water (around 0.75%) is found in rivers, lakes and aquifers, and it is available for the support and development of human life in its various activities, mainly domestic, agricultural and industrial (Fig. 1) [1, 2].

Figure 2 shows the average water consumption in the world as extraction percentage of water (%) for the agricultural, domestic and industrial sector, during the 2003–2007 and 2013–2017 periods. It can be observed that the highest consumption

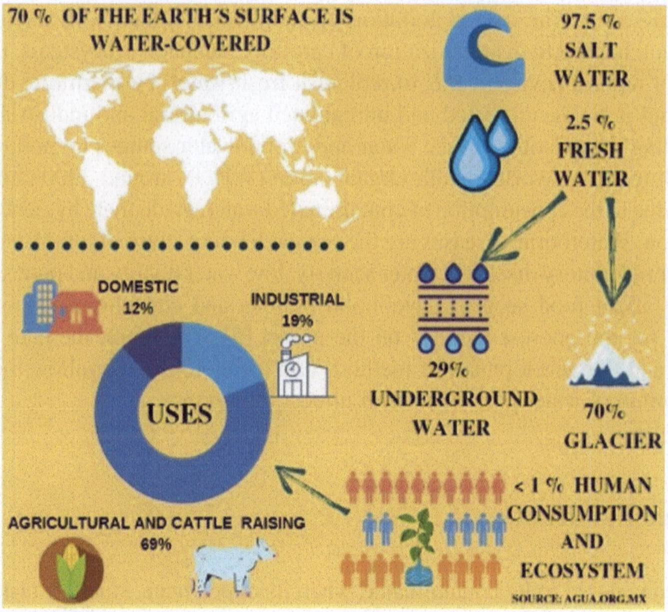

Fig. 1 Water distribution on Earth

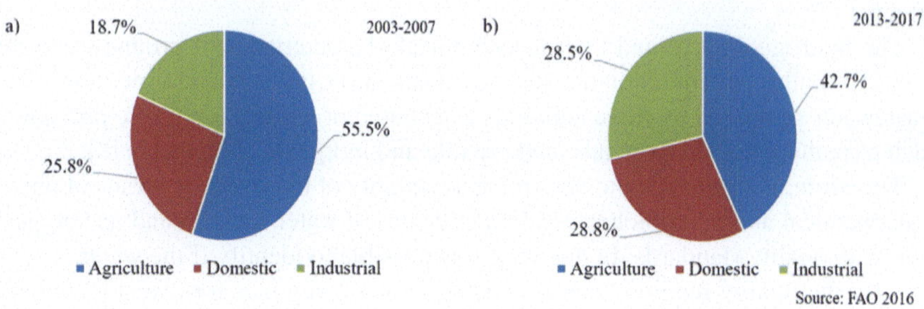

Fig. 2 Average extraction percentage of water consumption in the world for the agricultural, domestic and industrial sector, **a** 2003–2007 period, including 100 countries, **b** 2013–2017 period, including 75 countries

corresponds to the agricultural sector in both periods. However, it is noted that the extraction percentage of water in the industrial sector increased considerably during the 2013–2017 period compared to 2003–2007 period [3]. This reflects a worldwide trend of the growth of industrialization in the countries and therefore of a greater demand in the extraction of water for industrial uses, so its contribution of water extraction will probably continue increasing in the next years [4].

Global drinking water demand has increased at a considerable rate due to the exponential population growth. This leads to high water demand in industrial and agricultural sector. The overexploitation of aquifers combined with undiminished deforestation has led to disorganization of carefully balanced ecosystems, making it difficult for natural environments to replenish fresh water. Furthermore, due to the lack of strict policies, untreated and unregulated agricultural and industrial runoffs are degrading the existing surface water and groundwater sources by pollution [2].

According to the World Health Organization (WHO), around 3900 children die every day due to the consumption of contaminated water, inadequate hygiene and lack of sanitation. Waterborne diseases are the second leading cause of infant death in the world after respiratory diseases. Water scarcity, low water quality and poor sanitation negatively affect food security, livelihood options and educational opportunities, especially for the poorest families on the planet [5]. Therefore, the care of water resources and the development of methods and technologies that allow the control and prevention of water pollution are of utmost importance.

3 Water Pollution

Water is considered to be contaminated when its appearance, smell or taste change and its physical, chemical or biological elements produce adverse effects on the environment and living organisms [6]. The identification of contaminated water become complicated when pollution is intermittent or its physical characteristics are not affected [7].

The hydrogeological and hydrogeochemical characteristics of groundwater are altered by anthropogenic activities such as urbanization, industry and agriculture. The wastewater generated by these activities is a source of various types of c pollutants such as pathogens, heavy metals, and organic and inorganic compounds [8].

Therefore, it is necessary to determine the quality of the water by means of physical, chemical and microbiological characteristics of water samples and its comparison with quality standards. In this way, it is possible to identify if the water is suitable for the quality requirements associated with a given use, for example, human consumption, and if applicable, the probable purification processes required for the removal of undesirable substances [5].

The most important diseases in the healthcare sector are largely attached to the use and consumption of contaminated water, which represent a high risk on human health, causing gastrointestinal damage such as diarrhea, hepatitis, vomiting, and skeletal and endocrine systems affectations. Diarrhea is the second leading cause of death for children, which means that approximately half a million children die from this disease every year. In addition, heavy metals such as lead and nickel affect the nervous system, liver, and skin [9–11].

The WHO estimates that more than 663 million people are exposed to diseases due to the consumption of contaminated water, causing more than 502,000 deaths from diarrhea per year. This population is mainly exposed as a result of non-existent, insufficient or inadequately managed water services and sanitation. The lack of a treatment system for agricultural, domestic and industrial wastewater means that the water sources used for consumption are contaminated, mainly by chemical compounds that cause severe damage on human health, even at low concentrations, such as heavy metals, anions and emerging organic compounds [12].

3.1 *Heavy Metals*

The water pollution by heavy metals is an urgent problem that is increasing worldwide [13]. These pollutants are naturally present in soil and water, but they are also normally emitted into the environment by various anthropogenic activities that make use of a wide variety of chemical products with these compounds [14, 15]. A heavy metal is a metal or metalloid that can have a density greater than water (3.5–7 times greater); and that, even at trace levels, represents a threat to human, plant and animal life and health due to its persistence in the environment and bioaccumulation through the food chains [15, 16].

Heavy metals include arsenic (As), cadmium (Cd), copper (Cu), chromium (Cr), tin (Sn), mercury (Hg), nickel (Ni), lead (Pb), selenium (Se), thallium (Tl), vanadium (V) and zinc (Zn). These can be found in the environment in the form of organic compounds, phosphates, hydroxides, oxides, silicates, sulfates and sulphides. The natural sources of these pollutants are sea salt sprays, volcanic eruptions, forest fires, rock weathering, biogenic sources, and windborne soil particles [15].

However, anthropogenic sources contribute more to environmental pollution by the emission of heavy metals. The discharge of wastewater of agricultural activities, metallurgical-mining, electroplating, electronic and metal finishing process, and leachates generated at waste disposal sites are the most representative [15, 16]. In addition, metals emitted in the form of dust, such as in the emission of Pb by automobile exhaust, emission of Cu and Zn in smelting processes, emission of Hg, Ni, Se, Sn and V by burning fossil fuels, are carried by the wind and runoff, reaching surface water and later underground water [15]. Worldwide, drinking water has been identified as the main source of As poisoning, and the population can also be exposed to this metalloid when they consume food grown on contaminated soil with As or when is irrigated with contaminated water with As, being the use of insecticides and herbicides, or the manufacturing industry the main sources of emission of this pollutant into the environment [17]. Se is a non-metallic mineral that is used in some supplements or multivitanimics for human health, however at high concentrations in the long term it can cause nausea, vomit and diarrhea. This element is found naturally in the Earth's crust, interacting with the different elements of the environment (water, air, soil and sediments). Moreover, its presence in industrial waste water, mainly from oil and metal refineries and agriculture, is a double issue of prevention, firstly because it is a difficult element to eliminate and secondly because of the effects on human health [18]. Table 1 shows the main sources of anthropogenic emissions of metals and metalloids as well as their effects on human health.

Derived from the toxic effects on human health, heavy metals have to be removed from wastewater to prevent pollution of surface water and groundwater, and to avoid

Table 1 Main sources of anthropogenic emissions of metals and metalloids and their effects on human health (Adapted from [16, 19, 20])

Heavy metals and metalloids	Anthropogenic emission sources	Reference value [21] (mg/L)	Effects on human health
As	Agricultural applications products such as insecticides, herbicides, fungicides, among others Smelting and refining of minerals Glass manufacturing	0.01	Carcinogenesis Effects on the cardiovascular, dermatologic, nervous, hepatobiliary, renal, gastrointestinal, and respiratory system
Cd	Mining and smelting Production of alloys, pigments, stabilizers and batteries	0.003	Decreases in bone mineral density and osteoporosis Gastrointestinal tract erosion Pulmonary, hepatic, or renal injury Coma
Cr	Fossil fuels burning Manufacture of plastics and electroplating of metals Leather tanning Wood preservation	0.05	Liver and kidney damage Skin ulceration Central nervous system disorders
Cu	Mining and metallurgical processes	2	Damage to the kidney and stomach Nausea and diarrhea
Hg	Extensive use in industrial applications (batteries and steam lamps) Extraction processes	0.006	Physiological stress, abortion and tremors Brain toxicity, blindness, mental retardation, and kidney damage
Ni	Cosmetics, batteries and spark plugs production	0.07	Reduced cell growth Cancer Effects on the nervous system
Pb	Fossil fuels burning, mining, and manufacturing Batteries and metal products production Oxides for paint, glass, pigments and chemicals	0.01	Effects on the central nervous system and gastrointestinal tract Damage to the kidneys, fetus and reproductive systems
Se	Oil discharges from refineries Mining	0.01	Effects on the circulatory system Hair loss Nausea and diarrhea

(continued)

Table 1 (continued)

Heavy metals and metalloids	Anthropogenic emission sources	Reference value [21] (mg/L)	Effects on human health
Zn	Mining and metallurgical processes	–	Anemia and cholesterol

problems of intoxication and public health. The removal of heavy metals from aqueous solutions is generally accomplished by precipitation, sedimentation, and filtration. However, there are drawbacks with these treatments since they tend to generate an excessive amount of sludge which is hazardous and must be treated afterwards. An inexpensive alternative to removing heavy metals from aqueous solution is adsorption, which uses adsorbent materials such as activated carbon, biosorbents, activated alumina, and minerals (clays, zeolites, others). Recently, the removal of heavy metals from wastewater through biosorption processes using biomass has been of great interest due to the abundance of these materials, their economic availability, and the adsorption capacity of metals [22–24]. Some materials have been tested as biosorbents for metal removal, including soybean hulls [25], cottonseed [26], *Capsicum annuum* seeds [27], *Tamarindus indica* seeds [28], moss [29, 30], banana peels [31], apple peels [32], sawdust [33], sunflower stalk [34], pine bark [35], carrot and tomato waste [36], peanut shells [37], raw wheat bran [38], orange and lemon peels [39], grape stalks [40], tea waste [41], seaweeds [42], corncob [43], animal bone chars [44, 45], among others [46].

3.2 Emerging Organic Compounds

Emerging compounds are organic compounds of anthropogenic origin that are present in surface and groundwater. They are not regulated as international agencies have not yet established water quality standards for these pollutants [47]. More than 1000 emerging compounds have been identified and are classified into groups, including pharmaceuticals, personal care products, UV filters, endocrine disruptors, hydrocarbons, illicit drugs, food additives, metabolites, fire retardants, pesticides, among others [48].

The presence of these pollutants in the environment is mainly attributed to the discharge of wastewater and treated wastewater. Sources of pollution with emerging organic compounds in surface water and groundwater correspond to agricultural activities due to the application of various pesticides and the introduction of varios drugs of veterinary use (mainly antibiotics and hormones) in cattle; industrial activities where are processed or used these compounds, domestic wastewater; and landfills [49, 50].

Conventional secondary processes, as activated sludge and trickling filters are widely used in wastewater treatment; however, both are not designed to remove emerging organic compounds, so treated wastewater is discharged to surface waters with trace concentrations of these pollutants [51].

Treated and untreated wastewater discharged directly into surface waters (rivers, lagoons, seas…) is an environmental pollution problem which represents a risk to human health and aquatic organisms [50]. This problem could lead to lethal effects on aquatic organisms and human life even at very low concentrations [52].

Some of the treatment methods developed to the removal of emerging organics compounds from water are activated carbon adsorption, filtration, micro and ultrafiltration, advanced oxidation process, photolysis, ozonation, gamma radiolysis, sonolysis, reactors and bioreactors, membrane process, among others. However, these methods are not efficient to remove these pollutants, have high energy demands, operational and maintenance, and may cause secondary pollution. Therefore, there is a need to develop effective and alternative methods for removing emerging organic pollutants from the environment [53–57].

Adsorption is a promising method worldwide for removing emerging organic pollutants since it is low initial cost for implementation, has simple operating design and could be highly-efficient. The application of different adsorbents such as, activated carbons, composite adsorbents, modified biochars, bone chars and nanoadsorbents have been reported, being developed mainly for removing pharmaceutical products [52, 58, 59].

In recent years, application of biochar has attracted increasing attention for being an effective, inexpensive and environment friendly. Animal biochar or bone char has been of interest for application in the removal of emerging organic compounds however, its use has been a slightly explored area for solving this issue [60].

3.3 Fluoride

Fluoride is the lightest and most electronegative element in the halogen group. It is the 13th natural mineral in nature, is widely distributed on the earth's crust (625 mg/kg) and has and great mobility at higher temperatures [61, 62]. Its concentration is in the range of 1–35 mg/L in groundwater, around 1 mg/L in sea water and nearby 0.5 mg/L in lakes and rivers [63]. Fluoride is a pollutant in water which has been a problem of global concern for the adverse effects on human health since few decades [61]. Erosion of fluoride-rich minerals from granites and volcanic rocks is the natural source of fluoride in waters. However, some anthropogenic activities contribute significantly to environmental pollution with fluoride as glass manufacturing, ceramic industries, mining and metallurgical proccess, coal burning, electric and electronic industries, production of pesticides and fertilizers, among others [62].

Fluoride intake has always been considered a key factor for prevention of dental caries and improvement of human dental health. A concentration of fluoride in water of at least 0.7 mg/L is recommended for preventing dental caries [64]. The WHO recommends a minimum of 0.5 mg/L fluoride in the drinking water and 1.5 mg/L fluoride by intake, considering the volume of water consumed and from other sources

[21]. However, prolonged period of water intake containing high levels of fluoride may lead to a number of injurious effects, from slight dental fluorosis to skeletal fluorosis [62, 65]. Other negative impacts have been reported on human health as decreased red blood cells and thyroid function, hypertension, decreased fertility, nervous system impairment, osteoporosis, oxidative stress, renal disease, decreased intelligence in children, among others, through bioaccumulation of this pollutant [66, 67]. Different treatment methods have been developed to solve this issue of environmental pollution and human health risks.

The widely applied technologies for the removal of fluoride from water include coagulation, ion exchange, precipitation, membrane processes and adsorption, or a combination of these technologies. However, these have some limiting factors for their use in water defluoridation as use of toxic compounds, high operating cost, waste production and strict experimental conditions. Adsorption techniques have been quite popular in recent years for the removal of fluoride from water due to their simplicity and a wide range of adsorbents [63, 68]. The cost of adsorption process is associated to the unit price of the sorbent, so in last years, researches are focused on the production of low-cost adsorbents from cheaper materials [69]. Some materials tested as adsorbents for fluoride removal are: activated alumina [70], and modified activated alumina [71]; calcium-based sorbents [72], and iron-based sorbents [73]; synthetic sorbents using two or multi-metal oxides/hydroxides [74, 75]; natural materials as natural zeolites [76], synthetic zeolites [77, 78], modified attapulgite [79], bentonite clay [80], kaolinite clay [81], laterite [82], among others; biosorbents using microalgal biomass [83], and chitosan beads [84]; agricultural waste as corn cobs [85], and nut shell [86]; biochars [87, 88], and bone char [45, 89–91]. Bone char has gained significant attention as a green sorbent due to its low cost, facility of preparation and biocompatibility [63].

4 Bone Char

Millions of tons of bone waste are generated annually from the meat industry. According to the Organization for Economic Cooperation and Development, an increase of 40 million tons in meat production is expected in the next 10 years [92]. Consequently, there will be a considerable increase in generated bone waste worldwide. These materials require a proper solid waste management, so its use for the synthesis of bone char, could be used with dual-purpose, the first one as adsorbent material to remove pollutants from water and the second one to provide a better solution in waste management [93].

Animal bones contain bone apatite which is a carbonate apatite with a composition of 6–9% carbonate in the apatite structure. The carbonization process leads to change the form of mineral in the bone from apatite to hydroxyapatite $Ca_{10}(PO_4)_6(OH)_2$ [94]. The compounds found in bone char are hydroxyapatite (70–76%), amorphous carbon content (9–11%), calcium carbonate (7–9%), calcium sulfate (0.1–0.2%), iron as Fe_2O_3 (<0.3%), acid soluble ash (<3%), and have a density 0.65 g/cm^3 [93, 95, 96].

Bone char is considered as one of the adsorbents with low negative impact on the environment and human health, compared to other adsorbents such as carbon obtained from wood residues [97]. Furthermore, its regenerative capacity makes it a promising green adsorbent. However, several factors such as the initial concentration of the adsorbate and the adsorption capacity could determinate the life time of bone char [98].

4.1 Bibliometric Analysis

Globally, there is an interest in bone char for its easy preparation and implementation, low cost and its wide variety of removal of pollutants from aqueous solutions. In order to investigate world scientific publications related to research on bone char, a bibliometric analysis was performed. Web of Science database was used as the data source and the data presented in this study were mined on July 8, 2020 using the search terms "bone chars", "bone char", "pyrolyzed bones", pyrolyzed bone", "carbonizado de hueso" and "carbonizados de hueso" as Title. The timespan was 2000–2020.

Table 2 shows the first 15 most cited papers according to bibliometric analysis. It can be observed that no relevant papers were published in 2002 and 2003, being excluded, considering the 2000–2010 period. In addition, three of the most cited papers were published in 2007, two were published for each year in 2000, 2001, 2006 and 2009. For the rest of the years, only one paper was published. Papers from 2011 onwards are not presented in this list due to their recent publication, having a low number of citations.

Authors from China are predominantly the most cited, followed by those from India. The most cited papers are distributed in 10 journals, the Journal of Colloid and Interface Science being the most frequently cited (4), and number of citations (820), followed by Water Research with 2 papers and 717 citations.

Figure 3 shows the categories of study field where the papers related to the use of bone char are located. It is evident that most of the papers is focused on topics of environmental or chemical applications, however, this field can be slightly imprecise by the type of journal in which it has been published rather than by the publication's topic.

4.2 Raw Materials

Cow, chicken, fish, sheep and pork bones have been used as precursors for the production of bone char. One of the reported advantages in the use of these raw materials for production of bone char is their high availability as they are waste from the meat industry [63]. In the 2008–2017 period, the global per capita consumption of poultry grew strongly, being the type of meat most consumed, perhaps because it is cheaper;

Table 2 Top-cited bone char related papers published between 2000 and 2020

Rank	Year	Title	Lead author	Journal	Country	No. citations
1	2007	Intraparticle diffusion processes during acid dye adsorption onto chitosan	Cheung WH	Bioresour Technol	China	538
2	2001	Sorption kinetic analysis for the removal of cadmium ions from effluents using bone char	Cheung CW	Water Res	China	448
3	2007	Identification of selective ion-exchange resin for fluoride sorption	Meenakshi S	J Colloid Interface Sci	India	300
4	2006	Sorption of Zn(II), Pb(II), and Co(II) using natural sorbents: Equilibrium and kinetic studies	Al-Degs YS	Water Res	Jordan England	269
5	2000	Sorption kinetics for the removal of copper and zinc from effluents using bone char	Cheung CW	Sep Purif Technol	China	252
6	2005	Thermodynamic behaviour and the effect of temperature on the removal of dyes from aqueous solution using modified diatomite: A kinetic study	Al-Ghouti M	J Colloid Interface Sci	England	232
7	2000	Optimised correlations for the fixed-bed adsorption of metal ions on bone char	Ko DCK	Chem Eng Sci	China	199
8	2008	A conceptual overview on sustainable technologies for the defluoridation of drinking water	Ayoob S	Crit Rev Env Sci Technol	India England	187
9	2001	Adsorption of dyes from aqueous solution - the effect of adsorbent pore size distribution and dye aggregation	Walker GM	Chem Eng J	England New Zealand	180

(continued)

Table 2 (continued)

Rank	Year	Title	Lead author	Journal	Country	No. citations
10	2009	Adsorption behaviour of methylene blue onto Jordanian diatomite: A kinetic study	Al-Ghouti MA	J Hazard Mater	Jordan England	176
11	2006	Mechanisms of competitive adsorption of Pb, Cu, and Cd on peat	Qin F	Environ Pollut	China USA	168
12	2004	Adsorption of fluoride ions onto carbonaceous materials	Abe I	J Colloid Interface Sci	Japan	152
13	2007	Adsorption of fluoride from water solution on bone char	Medellin-Castillo NA	Ind Eng Chem Res	Mexico	147
14	2009	Reactive Black dye adsorption/desorption onto different adsorbents: Effect of salt, surface chemistry, pore size and surface area	Ip AWM	J Colloid Interface Sci	China	136
15	2010	A comparative study on the kinetics and mechanisms of removal of Reactive Black 5 by adsorption onto activated carbons and bone char	Ip AWM	Chem Eng J	China	120

while the per capita consumption of beef and veal decreased in the same period [92]. However, as they are types of meat more consumed, the availability of their bones as waste is higher, and therefore it is more common to find researches based on the use of bone char from cow or chicken. In addition, the textural properties and hydroxyapatite content of bone char are attractive for different technological applications, although it has been reported that the bone structure may differ with age and parity of animal species, and even from the body part of the species, so it affects its effectiveness [63].

Bone chars from animal bones have been studied in the removal of pollutants from water, mainly fluoride and heavy metals (As, Cd, Co, Cr, Cu, Hg and Pb). They have also been tested in the water treatment of emerging organic compounds such as pesticides, dyes, pharmaceutical products and surfactants, and endotoxins; obtaining appropriate adsorption capacities (Table 3). Most experiments maintain a pH close to neutral (7 and 8) for evaluating adsorption capacity of these adsorbents,

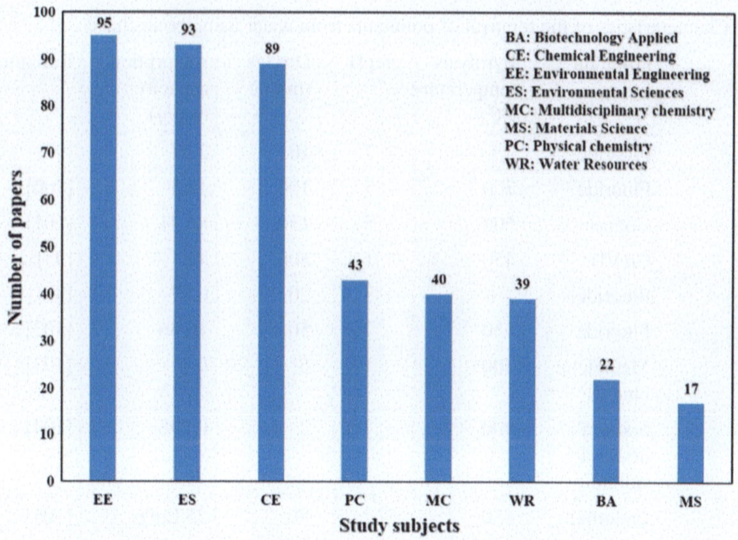

Fig. 3 Number of bone char papers published between 2000 and 2020, classified by study field

however this may also depend on the physicochemical and textural properties of the material or the nature of the water. Pyrolysis temperature for the synthesis of bone char ranges between 350 and 1000 °C, however, greater effectiveness has been found of 500–800 °C. This last argument will be further discussed in the Sect. 4.3.

4.3 Synthesis Methods

The properties of bone char depend on raw materials and the synthesis method employed. Bone chars were pyrolyzed at 500 °C from bovine bones (ribs, scapulae, vertebrae, and legs), performing a comparison of each part and their copper sorption behavior in aqueous solutions. The rib bone char showed a larger specific surface area (172 m^2/g), smaller average pore diameter (7.7 nm), and more basic functional groups than the other types of bone char. Furthermore, the rib bone char showed the best adsorption capacity 83.71 g/kg, finding that the adsorption amount increases with improving specific surface area and decreases with increasing average pore diameter [101]. Researchers [120] prepared bone char at 800 and 1000 °C in a nitrogen gas atmosphere from cow, pork, chicken and fish bones for use in the removal of fluoride from water. It was observed that the amount of fluoride adsorbed from cow and pork bones was greater than from chicken and fish bones. On the other hand, the amount adsorbed on pyrolysed bones at 800 °C was higher than at 1000 °C.

Animal bones are treated by pyrolysis at temperatures above 500 °C and below 700 °C in an oxygen-limited atmosphere [109]. The temperature of 700 °C caused the

Table 3 Characteristics of the removal of pollutants from water using bone char

Bone	Pollutant	Pyrolysis temperature (ºC)	pH	Dosage (mg/L)	Adsorption capacity (mg/g)	References
Bovine	Fluoride	–	7	10	2.37	[99]
Bovine	Fluoride	700	7	100	18.5	[100]
Bovine	Copper	500	5	150	83.71	[101]
Cattle	Cr(VI)	450	1	800	4.8	[93]
Cattle	Fluoride	–	3	20	6.77	[89]
Cattle	Fluoride	350	7	50	10.56	[102]
Cattle	Methylene blue	400	10	5	0.7	[103]
Cattle	Sodium dodecyl sulfate	900	7	2	0.298	[104]
Cattle	Endotoxin	850	7	80 Eu/mL	25 Eu/g	[105]
Chicken	Fluoride	600	5	10	11.2	[106]
Chicken	Fuchsine	800	7	25	260.8	[107]
Com-mercial	Copper nickel cadmium	–	3	2000	69.45 44.60 50.92	[108]
Com-mercial	Cadmium	–	4.8	337	64	[109]
Com-mercial	Arsenic(V)	–	10	0.5	0.08	[110]
Com-mercial	Fluoride	–	7	20	1.65	[91]
Com-mercial	Naproxen	–	7	100	3.2	[111]
Cow	Arsenic (V)	500	4	10	0.335	[112]
Cow	Pesticide	800	8	5000	–	[113]
Cow femur	Fluroride	700	7	60	7.32	[114]
Fish	Diclofenac Fluoxetine Lead	1000 1000 600		150 150 4200	43.29 55.87 714.24	[115]
Fish	Fluoride	500	7	>10	4.42	[116]
Fish	Fluoride Cadmium	500	5	60 600	24.65 175.2	[45]
Pork	Methylene blue	450	6	25	29.1	[117]
Pork	Cobalt	800	7	500	18.37	[118]
Sheep	Mercury	800	3	80	8	[119]

degradation of functional groups and decreased the efficiency in removing pollutants from the water. On the other hand, the temperature of 500 °C resulted in the addition of more organic matter to the treated water due to the incomplete removal of this in the bone structure [120]. Researchers [121] evaluated the effect of the pyrolysis temperature of bovine bones on the type of apatite obtained, finding that bone char obtained at 600 °C was composed of carbonate apatite, while in bone chars synthesized at 900 and 1200 °C, the main constituent was hydroxyapatite. Therefore, they inferred that the amount and type of apatite in bone char is directly related to the pyrolysis temperature. Authors [114] determinated that the best pyrolysis temperature for the synthesis on bone char is at 700 °C, since higher temperatures cause dehydroxylation of hydroxyapatite, reducing its adsorption capacity. Other authors [122] also studied the adsorption capacity of bone char from cow bone synthesized at temperatures between 400 and 800 °C, reporting that its capacity decreases with increasing pyrolysis temperature due to decreased specific area and pore volume.

The gas used during pyrolysis is an important factor in the quality of bone char. Researchers [123] examined the effect of the type of gas (N_2 or CO_2) in the pyrolysis of cattle bone. Bone char obtained via CO_2 showed competitive fluoride adsorption capacities compared with the results of bone char via N_2, reporting values of 5.92 and 7.32 mg/g, respectively. However, bone char obtained at 700 °C via N_2 presented the greatest specific area with a value of 85 m^2/g under the same conditions of pyrolysis, showing a minor value of 69 m^2/g for bone char via CO_2, so the adsorption process is favored on bone char obtained via N_2. Other authors [124] studied the difference between bone char with and without N_2 gas purging, finding an insignificant effect on pH, surface charge and morphology, while a slight decrease in ash content, crystallinity, cation exchange capacity and dissolved organic carbon was obtained.

The residence time during the pyrolysis process to obtain bone char is also important. Some studies report that the removal of pollutants from water is favored in materials synthesized between 400 and 700 °C and 1–2 h [114, 125]. Although this will also depend on the other factors mentioned and the type of pollutant to be eliminated.

Activation methods are employed for improving the adsorption capacity of carbonaceous materials via increasing the specific area and pore sizes and volume, or altering surface functional groups for greater selectivity of the adsorbent toward specific pollutants [63]. Authors [122] evaluated the fluoride adsorption capacity of bone char modified using nitric acid solutions at distinct concentrations, showing that its capacity was enhanced when monetite and hydroxyapatite with low crystallinity were present in the material. The highest adsorption capacity was obtained at 400 °C and when was treated with a 1.0 M nitric acid solution.

Pork bone was treated under chemical activation using sodium hydroxide, potassium hydroxide, potassium carbonate, sulfuric acid, and phosphoric acid as agents, being evaluated in methylene blue adsorption tests. Results showed significant improvement as a result of chemical activation with sulfuric acid, increasing methylene blue adsorption kinetics, caused by an intense development of microporosity, probably owing to its strong dehydration capacity. The treatment with potassium carbonate allowed an increase of the equilibrium adsorption capacity (>50%), achieving also a maximum increase in a specific area of about 28% [117].

4.4 Applications

Traditionally, bone char has been applied in the sugar industry to discolor sugar solutions in the refining process [126]. However, in recent years, it has become a novel and advantageous material due to its applications in electrochemistry, catalysis and environmental remediation.

4.4.1 Electrochemistry

Animal bone chars have been studied as electrode materials for other energy storage and conversion system such as Li–ion batteries and fuel cells. Pork bone char was tested as a novel cathode for lithium-sulfur batteries, improving the cycle stability in the batteries, showing a higher initial capacity than that of the normal cathodes with compact structures [127]. High-surface-area carbon monoliths from bovine bone char were elaborated to study their performance as supercapacitor electrodes, being a novel method for producing conductive carbon monoliths through the pyrolysis of bovine bone [128]. Cow bone char was evaluated as electrode material in two electrode symmetric supercapacitor system in non-aqueous electrolyte, exhibiting high specific capacitance with excellent retention at high current density and for long term operation [129]. Cow bone char was employed as electrode in bioelectrochemical systems for biofilm growth to form bioanodes, showing not only electroactive properties but also capacitive behavior, indicating a new direction for research on microbial biocapacitors [130].

4.4.2 Catalysis

Animal bone chars have been tested as catalysts for biodiesel production. Waste animal bones were employed as a cost effective catalyst for the transesterification of palm oil, finding the best biodiesel yield when the catalyst was calcined at 800 °C due to an increase in surface area, leading to better catalytic activity [131]. Bovine bone char was also used to catalyze the transesterification of soybean oil, obtaining an optimum yield of biodiesel using bone calcined at 750 °C and at certain reaction conditions [132]. Bone char from fish bone was tested as a heterogeneous catalyst for synthesis of biodiesel from soybean oil, finding that this material could effectively catalyze the methanolysis of refined soybean oil to yield biodiesel and could reemployed up to six times in synthesis process [133]. Catalysts from chicken bone char were tested in the transesterification reaction of waste cooking oil, obtaining a good biodiesel yield (89.33% at 5.0 g of catalyst loading), when the bone char was calcined at 900 °C, attributing its high catalytic activity to the presence of optimal number of active basic sites on its surface. This catalyst was successfully reused four times for biodiesel production [134].

Also, animal bone chars have been studied as catalyst in Advanced Oxidation Processes. Pork bone char was used as a catalyst in the persulfate activation system to effectively degrade acetaminophen, reporting that degradation of this compound could be completed within 60 min [135]. The catalytic oxidation of 2,6-diisopropylnaphthalene (2,6-DIPN) to 2,6-naphthalene dicarboxylic acid (2,6-NDCA) was studied using a catalyst of immobilizated Co/Mn nano-hybrid particles over cow bone char, finding to be a effective catalyst in the oxidation process [136]. A composite photocatalyst ZnO/bone char (ZnO/BC) was tested in photocatalytic degradation of alkaline methylene blue dye under simulated sunlight irradiation, determining that the increase of bone char in ZnO/BC composites enhanced the absorbance of visible light and can fully utilize visible light to decompose methylene blue [137].

4.4.3 Environmental Remediation

Bone chars have been studied in the environmental remediation, mainly in the removal of pollutants from water. These materials have been widely studied in the removal of heavy metals and fluoride. For example, sheep bone char with zinc chloride activation was studied for the removal of Hg(II) from aqueous solution. Auhors [112] used bone char for the removal of As(V) from wastewater, reporting an adsorption capacity of 0.335 mg/g at pH 4, which is greater than the aluminum-coated limestone, the iron-coated sand and granular activated carbon (0.150, 0.029 and 0.038 mg/g, respectively). Bovine bone char was used for heavy metal removal from aqueous solution at batch reactors, finding a removal trend mainly with cadmium, then zinc, and finally niquel. However, the multicomponent removal of these three ions in both binary and ternary mixtures was a strong antagonistic adsorption process [44]. The results of a study reveal the possibility of designing a filtration unit with bone char as an adsorbent in a large-scale water and wastewater treatment plant for the removal of Cr(VI) from contaminated water [93].

The removal of fluoride from water has been studied for a long time. Generally, it has been reported that the fluoride adsorption capacity on bone char varies between 1.33 and 11.9 mg/g, depending on the initial concentration of fluoride in solution and the type of bone char used [89, 99, 122, 138, 139]. A comparision of the fluoride adsorption capacity on activated carbon and bone char was studied. Results revealed that the highest percentage of removal was 82 and 13% for bone char and activated carbon, respectively. The high removal efficiency of bone char was attributed to hydroxyapatite of bone char [138]. Reserachers [140] studied the adsorption of fluorides on bone char, tourmaline and activated alumina, finding that the adsorption capacity of the three materials increased regard fluoride concentration. The fluoride adsorption capacity on bone char (18.5 mg/g) was higher than that of tourmaline (15 mg/g) and activated alumina (13 mg/g). These authors inferred that on the surface of bone char, calcium active sites can bind with fluoride ions; in addition, some anions contained in bone char such as PO_4^{3-} and OH^- can be exchanged for fluoride in the solution. Other Authors [122] studied the fluoride adsorption capacity on cow bone char, defining that this capacity is dependent on the pH which is increased when the pH of the solution decreased from 9 to 5, due to the electrostatic attraction between the fluoride in solution and the surface charge of the bone char.

Animal bone chars have also been tested in the removal of emerging organic compounds. One paper reported the use of bovine bone chars for removing naproxen in water, the results showed a feasible and effective process in batch conditions [111]. Defatted pork bone char was applied for the removal of 2,4-dichlorophenol (2,4-DCP). The results showed that more than 85% of 2,4-DCP can be eliminated in 30 min and almost 100% in 2 h [60]. Cow bone was analysed for the elimination of five pesticides (hexazinone, diuron, ametrine, sulfometuron-methyl and azoxystrobin) from drinking water. A near 100% elimination was achieved for the five pesticides with the highest dose of bone char (1 g), which represents a low-cost method to be used in water treatment plants or in domestic filters [113]. In addition, the elimination of endotoxins from the aqueous solution has also been studied through the use of cattle bone char by pyrolysis in a furnace at 850 °C. The results revealed that bone char can be used as an effective adsorbent for the removal of endotoxins, since its adsorption is fast. In addition, this material could regenerate and become endotoxin-free by being heated to 350 °C for 30 min [105].

5 Removal of Pollutants from Water onto Bone Chars

The most relevant results of some studies on the application of bone chars from different animals in the removal of pollutants from water are analyzed in this section. For these studies, several characterization techniques are included to relate the textural and physicochemical properties of the materials to the adsorption process. Also, data on the adsorption equilibrium and/or adsorption capacities of the bone chars are shown, as well as the mechanisms of removal of the analyzed pollutants.

5.1 Fluoride Removal on Bone Char from Different Animals

The bone chars used in this study were prepared from chicken, pork and cow bones obtained from a butchery. Each type of bone was placed in boiling water for 2 h, then washed with an alkaline detergent solution, cleaned manually to remove any organic matter and left to dry at room temperature. The bones were placed in containers in an oven used for brick production, which is constructed of the same type of material and uses dried cactus as fuel. The oven is located within the municipality of Villa de Arriaga in the state of San Luis Potosi, Mexico (Fig. 4). After 24 h, the bone chars were removed from the oven, ground to a particle size of 0.74 mm, then washed with deionized water and left to dry in the oven at 110 °C for 24 h. Finally, the prepared bone chars were stored in polypropylene containers for later use.

The total active sites of the bone chars were determined by the acid–base titration method proposed by Boehm. The determination of the point of zero charge pH_{PZC} and surface charge were determined using potentiometric titration methods.

The specific area, pore volume and average pore diameter were determined by means of a physysorption equipment, Micromeritics, model ASAP 2020. The operation of this equipment is based on the method of nitrogen adsorption at a temperature close to the boiling point of N_2 (77 K). The analysis of the specific area is based on the theory of Brunauer, Emmett and Teller (BET).

The surface of the adsorbent materials was analyzed by means of a scanning electron microscope, FEI, model QUANTA 200, equipped with an energy-dispersive EDAX microanalysis system, to perform a qualitative elemental analysis of the surface. The identifying of the crystalline species present in the bone charring samples was conducted with an X-ray diffractometer, Rigaku, model DMAX 2000.

The experimental data of the fluoride adsorption equilibrium were obtained in 500 mL batch adsorbers in which a volume of well water with an initial known fluoride concentration of 4.8 mg/L was added, the solution pH was adjusted and a 20 mL sample was taken and later analyzed to corroborate its initial concentration.

Fig. 4 Brick oven used in bone char synthesis

The solution was transferred to the batch adsorber which was partially immersed in a bath at constant temperature. A nylon bag containing a certain mass of the adsorbent material ranging from 0.25 to 2.5 g was then added. The solution and the adsorbent were left in contact until they reached equilibrium. The solution was sampled from time to time and the concentration of fluoride in the samples was determined. The mass of adsorbed fluoride q in mg/g was calculated by means of a mass balance which is mathematically represented by Eq. 1:

$$q = \frac{V\left(C_0 - C_f\right)}{m} \tag{1}$$

where V is the volume of the solution, C_0 and C_f is the initial and final fluoride concentration in mg/L, respectively, and m is the mass of the bone char in g.

The carbonization yield of chicken, pork and cow bones was 20, 24 and 38% in weight in relation to uncalcinated bones, respectively. In addition, the colouration of the bone chars obtained was grey, black and white due to oven temperatures of 500–700 °C, which were verified with a digital infrared thermometer. Figure 5 shows images of the bone chars obtained. Authors [141] reported that the grey and white bone chars are generated in calcination processes with a continuous supply of atmospheric oxygen. Other authors [142] reported that the gray and white chars are prepared at 450–600 °C and have fluoride adsorption capacities of 2.4 and 0.3 mg/g, respectively.

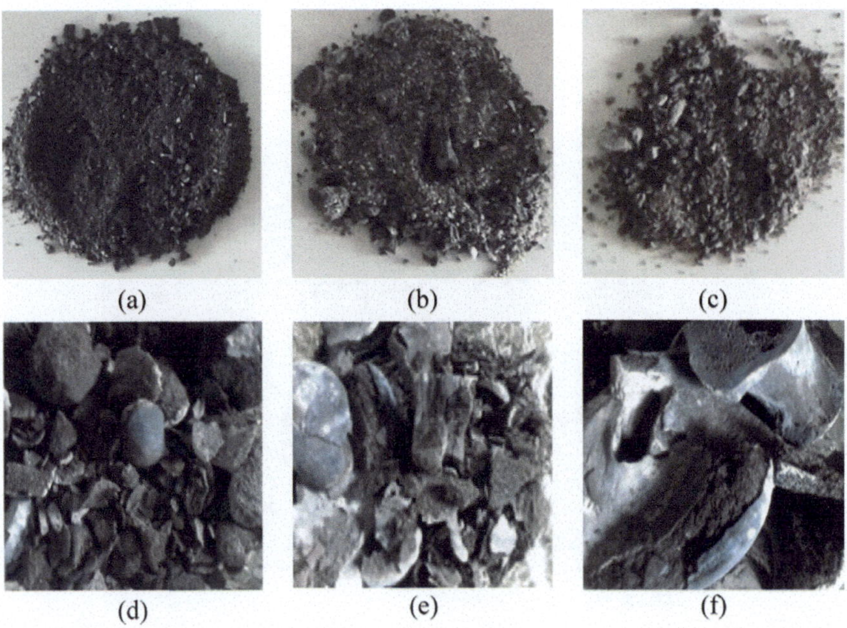

Fig. 5 Images of bone chars of **a**, **d** chicken, **b**, **e** pork and **c**, **f** cow

The physicochemical and textural properties of the bone chars synthesized in this study and a commercial sample are reported in Table 4. The results show that the specific areas obtained are slightly smaller than those of the commercial bone char, however, the pore volume of the materials is very similar and they can be considered mesoporous. Similarly, the carbonized materials are basic materials since the concentration of basic sites is higher than that of acidic sites and was confirmed with the pH_{PZC} value. Figure 6 shows the distribution of the surface charge of the bone chars where the pH_{PZC} values can be seen. Therefore, it can be seen that when the pH $< pH_{PZC}$ the surface charge of the materials is positive and the removal of anions can be favoured when the predominant mechanism is by electrostatic attraction. On the other hand, when the pH $> pH_{PZC}$ the surface is positive and there can be a repulsion by the anions but a greater attraction by the cations.

Table 4 Physicochemical and textural properties of chicken, pork, cow and commercial bone chars

Bone char	Acidic sites (meq/g)	Basic sites (meq/g)	Specific área (m^2/g)	Pore volumen (cm^3/g)	Average pore diameter (nm)	pH_{PZC}
Pork	0.10	0.33	58.5	0.32	21.9	9.75
Chicken	0.03	0.48	82.7	0.37	23.8	8.20
Cow	0.08	0.39	78.3	0.38	19.6	9.50
Commercial apelsa	0.30	0.62	104	0.30	11.1	8.00

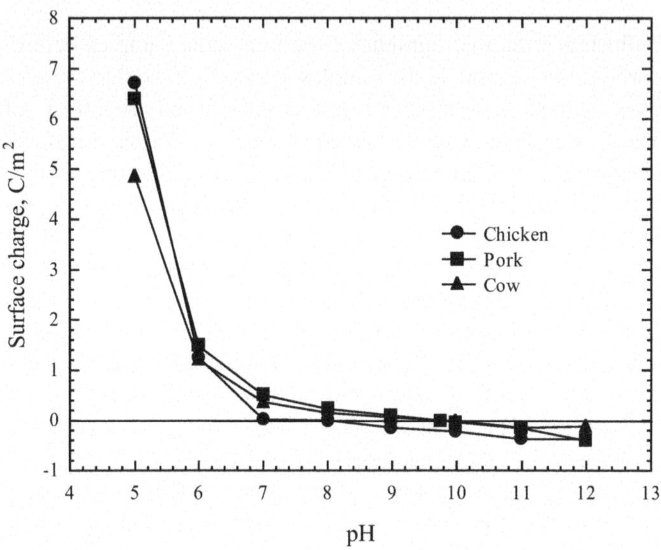

Fig. 6 Surface charge distribution of bone chars

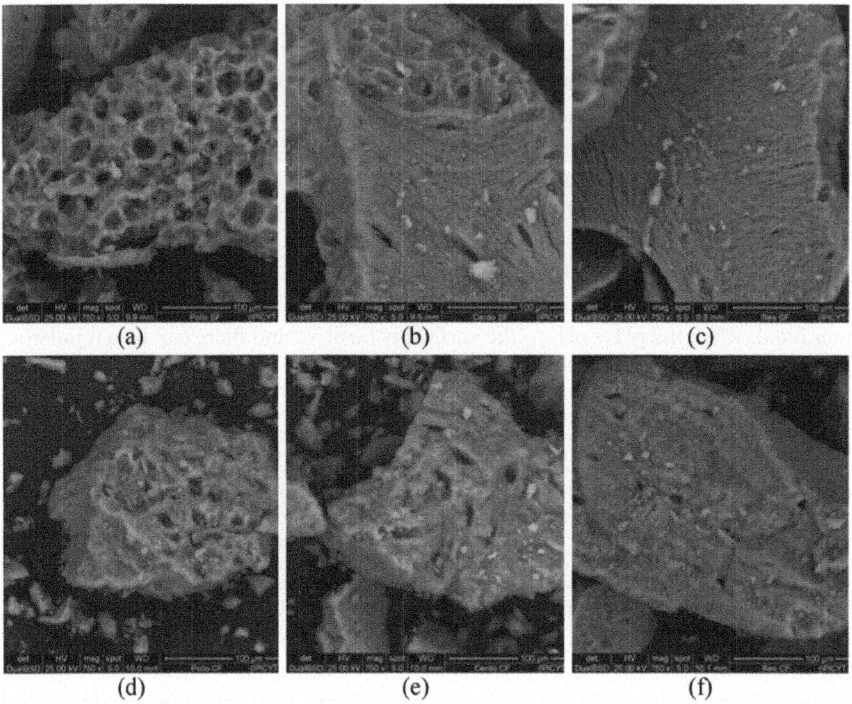

Fig. 7 Micrographs of bone chars unsaturated with fluorides **a** chicken, **b** pork, **c** cow and saturated with fluoride **d** chicken, **e** pork and **f** cow at 750×

Figure 7 illustrates the micrographs of the bone char samples before and after fluoride adsorption. In general, in the samples it was observed that the surface of the materials has no defined shape, it is heterogeneous in size and structure. Characteristic morphologies of bone chars without defined geometry, and irregular shape and pore sizes are also reported by other authors [44, 45, 143, 144]. Unfortunately, it was not possible to visualize any change in the surface of the bone chars that were used in fluoride adsorption.

X-ray fluorescence analysis revealed that the adsorbents are composed of Na, Mg, Ca, P, O and C, and this elemental chemical composition did not show significant changes between unsaturated and saturated bone chars. In the case of fluoride-saturated bone chars, no surface fluoride was identified and this may be due to the fact that the amount of fluorides adhering to the surface of the bone chars is below the detection limit. Figure 8 shows the fluorescence spectrum of cow bone char. Similar results were obtained for chicken and pork bone chars.

Figure 9 shows the diffractograms of fluoride-saturated and unsaturated chicken, pork and cow bone chars. For all materials, characteristic peaks of hydroxyapatite were identified and for fluoride-saturated materials the presence of fluoroapatite was noted. Several authors have proposed as an adsorption mechanism the ion exchange

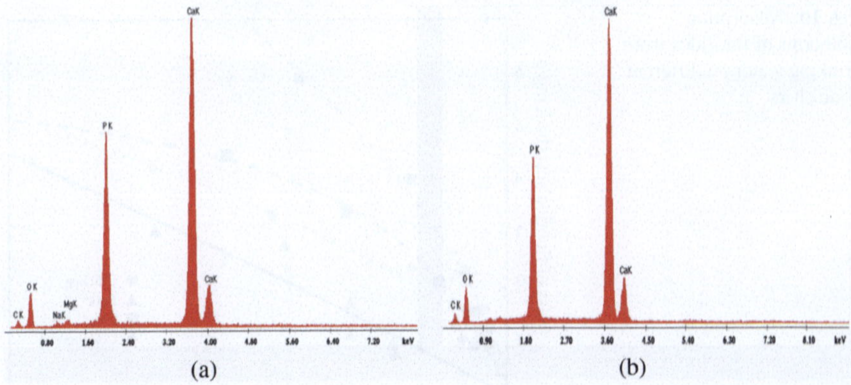

Fig. 8 X-ray fluorescence microanalysis on a cow bone char sample **a** unsaturated with fluoride and **b** saturated with fluoride

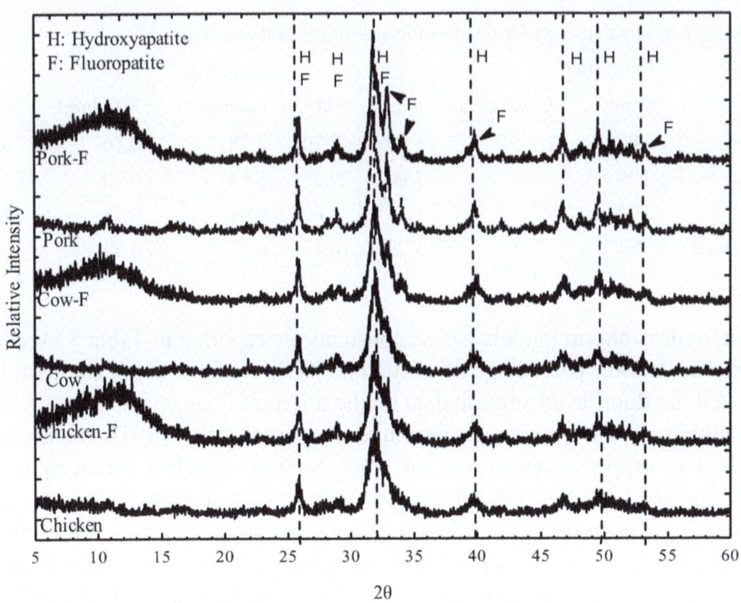

Fig. 9 Diffractograms of fluoride saturated and unsaturated bone chars

reaction of the fluoride ion in solution by the OH^- groups of the hydroxyapatite, which can be represented as [45]:

$$Ca_{10}(Po_4)_6(OH)_2 + 2F^- \rightarrow Ca_{10}(PO_4)_6F_2 + 2OH^-$$

The experimental data of the fluoride adsorption isotherms in the Laborcilla well water on the bone chars are shown in Fig. 10 and were adjusted by the Freundlich

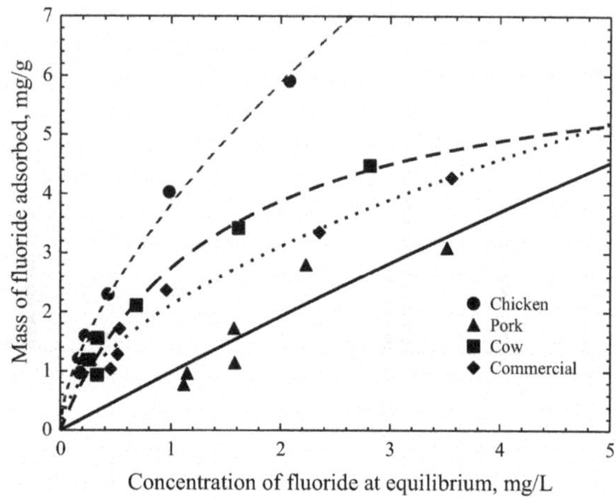

Fig. 10 Adsorption isotherms of fluorides from drinking water on different bone chars

Table 5 Isotherms parameters for the fluoride adsorption on bone chars

Bone chars	Freundlich			Langmuir		
	k ($mg^{1-1/n}$ $L^{1/n}mg^{-1}$)	n	%D	q_m (mg/g)	K (L/mg)	%D
Chicken	3.82	1.60	7.70	9.58	0.76	8.71
Pork	0.96	1.00	19.5	6.83	0.20	22.1
Cow	2.52	1.75	6.60	6.62	0.71	7.23
Commercial	3.84	1.79	10.1	6.21	0.57	10.7

and Langmuir isotherm models whose constants are reported in Table 5. According to a criteria of lower percentage of deviation (%D), the Freundlich isotherm better interpreted the fluoride adsorption data on the different bone chars.

In addition, the adsorption equilibrium results revealed that chicken bone char has the highest adsorption capacity over the other materials which is, for an equilibrium concentration of 1.5 mg/L, 1.4, 3.4 and 1.9 times higher than cow, pork and commercial bone char, respectively. In relation to the materials prepared in this study, the adsorption capacity decreased in the following order chicken > cow > pork and this was attributed to the number of basic sites and the specific area whose values also decreased in the same order. Concerning the commercial bone char, the adsorption capacity values did not show a trend in relation to the properties of the bone chars. This was attributed to the preparation method followed in the synthesis as it is done under better controlled conditions and in a controlled N_2 atmosphere. Finally, it was determined that the necessary mass of the bone chars to reduce the current concentration of fluoride in one m^3 of water from the Laborcilla well from 4.8 to 1.5 mg/L and thus comply with Mexican regulations is 0.67, 0.97, 2.26 and 1.24 kg for chicken, cow, pork and commercial bone char, respectively.

5.2 Fluoride and Cd(II) Removal on Pleco Fish Bone Char

The synthesis of bone chars from the bones of an invasive aquatic species was first performed by Medellin-Castillo et al. [45]. Pleco or devilfish of the family Loricariidae are native from the Amazon basin in South America but are considered invasive species in different regions of the world. Due to the lack of uses and their negative environmental impacts, in this study their use for the preparation of adsorbent materials was proposed as an alternative control to eradicate their population. The potential application of pleco fish in the preparation of bone char for use in the removal of fluoride and Cd(II) from aqueous solutions was evaluated. The fish used in this study were collected from the Usumacinta River in Tabasco, Mexico and the flesh was removed by separating the bones of the fins and the backbone which were designated as HA and HC, respectively. These bones were treated with aqueous hydrogen peroxide solutions and calcined at 500 °C for 3 h in a muffle furnace. The carbonized samples HA and HC were designated as CHA and CHC, respectively.

Surface area, pore volume and average pore diameter of the samples were determined by N_2 physisorption, the concentrations of acidic and basic sites were determined by the acid-base titration method proposed by Boehm and the the point zero charge pH_{PZC} were determined by a titration method. These properties are shown in Table 6 and it is observed that the HA and HC bone samples have low porosity due to the specific area values of 13.72 and 0.605 m^2/g, respectively. These values are increased by carbonization at 500 °C as well as pore volume values and are attributed to the decomposition of organic matter and exposure of the hydroxyapatite pores. Both the raw material and the samples CHA and CHC are considered as mesoporous materials because of the average pore diameter values. The pH_{PZC} values of HA and HC are slightly acidic and when being synthesized, the pH_{PZC} increases which makes the bone char to be of basic character.

The CHA sample obtained the highest adsorption capacity of fluoride and Cd(II) and was chosen for individual and multicomponent adsorption tests.

The effect of the solution pH on the adsorption capacity of the CHA sample to adsorb fluorides and Cd(II) is shown in Fig. 11a, b in which it can be seen that the capacity is pH dependent. In the case of fluoride adsorption, it increases with decreasing pH and for Cd(II) an opposite effect was observed. The effect of pH on the adsorption capacity is explained by the electrostatic interactions between

Table 6 Physical and chemical properties of adsorbents

Sample	Specific surface (m^2 g^{-1})	Pore volume (cm^3 g^{-1})	Average pore diameter (nm)	Basic sites (meq g^{-1})	Acidic sites (meq g^{-1})	pH_{PZC}
HA	13.72	0.0358	8.11	0.53	0.83	6.67
HC	0.605	0.0053	19.8	0.62	0.76	6.68
CHA	107.2	0.39	12.1	0.81	0.05	8.16
CHC	109.4	0.37	11.3	0.97	0.06	8.06

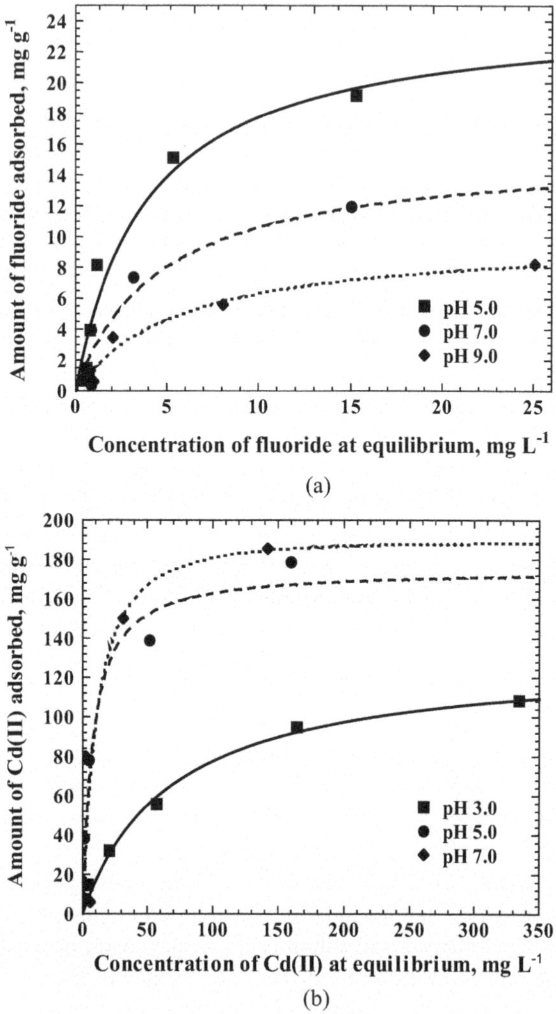

Fig. 11 Effect of pH on the adsorption isotherm on CHA at T = 25 °C. **a** Adsorption isotherms of fluoride and **b** adsorption isotherms of Cd(II)

the surface charge of the bone char and the ions in solution. At pH values lower than the pH_{PZC}, the surface of this material is positively charged and under these conditions, fluoride adsorption was favored but the Cd(II) adsorption is unfavorable by the repulsion of these ions and the surface charge of CHA. At pH higher than the pH_{PZC}, the surface of the CHA is negatively charged, so that the electrostatic repulsion does not favor the adsorption of fluoride but this one favors the adsorption of Cd(II).

Additionally, the binary adsorption of fluorides and Cd(II) on CHA sample was studied and the data was successfully interpreted by the Extended Freundlich Multi-component Isotherm (EMFI) as shown in Fig. 12 where it can be seen that the adsorption of fluorides decreases as the concentration of Cd(II) varies (Fig. 12a). On

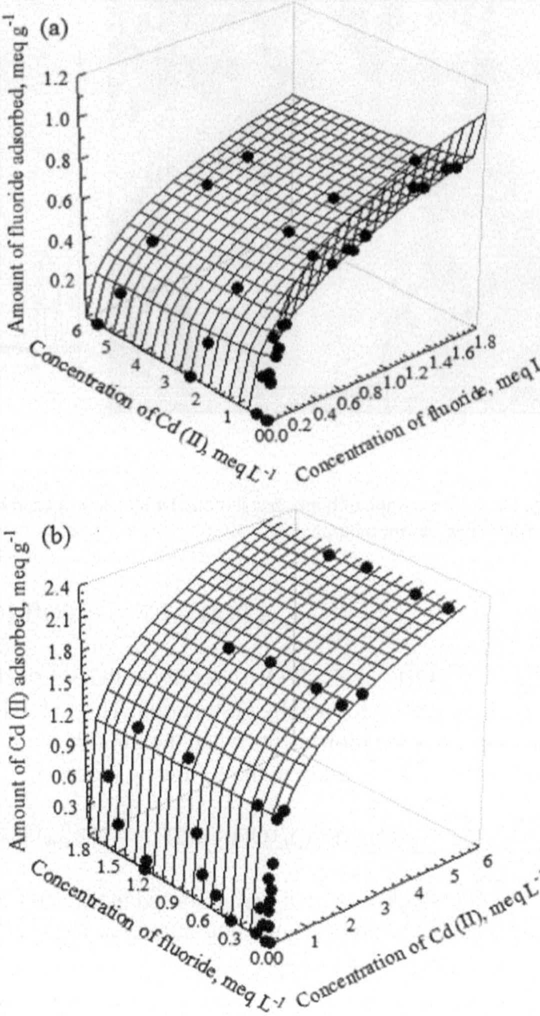

Fig. 12 The surfaces are predicted with the EFMI model. **a** Fluoride uptake and **b** Cd(II) uptake at pH = 5 and T = 25 °C

the other hand, Fig. 12b shows that the adsorption of Cd(II) does not change when the concentration of fluorides changes. The decrease of the fluoride adsorption capacity in the presence of Cd(II) was attributed to the possible formation of cadmium precipitates on the surface. This study is of importance since it is demonstrated that bone carbonization has the potential to simultaneously retain both ions in solution which is of interest in water treatment.

In relation to the adsorption mechanism it was established that the adsorption of fluorides occurs mainly by electrostatic forces according to the following reactions:

$$\text{Ca-OH}_2^+ + \text{F}^- \rightarrow \equiv \text{Ca-OH}_2\text{F}$$

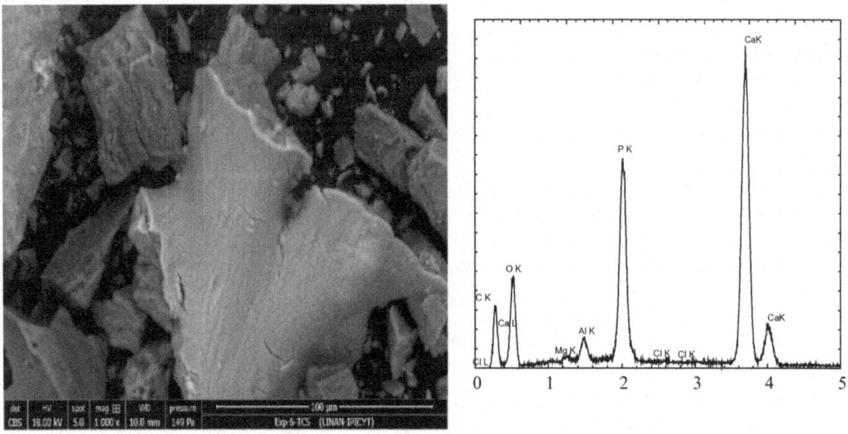

(a) (b)

Fig. 13 **a** Micrograph of bone char saturated with triclosan and **b** X-ray fluorescence microanalysis on bone char sample saturated

$$\equiv POH_2^{+} + F^{-} \rightarrow \equiv POH_2F$$

where $\equiv$Ca-OH_2^+ and $\equiv$P-OH_2^+ represent the positively charged sites on the surface of BCs, which are formed by protonation. Through XRD analysis it was shown that fluoroapatite formation occurs when the fluoride ion replaces the OH^- groups of the hydroxyapatite according to the following reaction:

$$Ca_{10}(PO_4)_6(OH)_2 + 2F^- \rightarrow Ca_{10}(PO_4)_6F_2 + 2OH^-$$

For Cd(II) adsorption, it was proposed that Cd(II) adsorption on the char bone be carried out by means of electrostatic attraction at the active sites of the hydroxyapatite. This mechanism can be expressed as:

$$\equiv POH + Cd^{2+} \rightarrow \equiv POCd^+ + H^+$$

$$\equiv PO^- + Cd^{2+} \rightarrow \equiv POCd^+$$

$$\equiv CaOH + Cd^{2+} \rightarrow \equiv CaOCd^+ + H^+$$

In addition, adsorption can be carried out by ion exchange, in which the metal ion is exchanged for the calcium ion present in CHA hydroxyapatite, which is expressed as:

$$\equiv Ca^{2+}_{(s)} + Cd^{2+}_{(aq)} \rightarrow \equiv Cd^{2+}_{(s)} + Ca^{2+}_{(aq)}$$

5.3 Triclosan Removal on Pleco Fish Bone Char

Bone chars from animal bones have also been studied in the removal of organic compounds as dyes, mainly methylene blue [103, 117]; pesticides [113]; surfactants as sodium dodecyl sulfate [104]; and pharmaceutical products as naproxen [111], diclofenac and fluoxetine [115], due to its structural properties and chemical composition.

The mechanism by which organic compounds are removed from water on bone chars has been described by some authors. For the case of dyes adsorption, it was attributed to hydrophobic interactions on the surface of the material [103]. Authors [104] reported that sodium dodecyl sulfate is adsorbed on bone char by common mass transfer mechanisms in porous materials. Other authors [111] indicated that the naproxen adsorption could include a complexation process via phosphate and naproxen, hydrogen bonding and the possibility of hydrophobic and interactions π-π.

Triclosan, or 5-chloro-2-(2,4-dichlorophenoxy)phenol or 2,4,4′-trichloro-2′-hydroxydiphenyl ether, with chemical formula $C_{12}H_7Cl_3O_2$ and molecular weight 289.54 g/mol, is a common broad-spectrum anti-microbial agent used in a wide range of personal care products (deodorants, hand soaps, toothpastes, and cleaning products), household items, medical devices, and clinical settings [145, 146]. There is a possibility that humans in all age groups are exposed to triclosan for life because of its widespread use, in fact, it has been detected in human tissues and the environment [145]. Triclosan can be introduced into the aquatic environment through numerous pathways, including the discharge of industrial effluents and wastewater treatment plants. This pollutant has been detected frequently in wastewater, sediment, and aquatic environments, and is a serious concern [147].

The removal of triclosan from water using bone chars has not been reported in the literature. This section includes some relevant results on the adsorption of this pollutant on a bone char from pleco fish or devilfish which were described in Sect. 5.2.

A micrograph of bone char saturated with triclosan and its X-ray fluorescence spectrum is shown in Fig. 5.13. It can be seen that the particles have irregular shapes and sizes, with certain porosity on the surface. Elemental analysis revealed the presence of Ca, Cl, P, Al, Mg, O and C with a content of 37.77, 0.12, 14.21, 1.1, 0.15, 37.32 and 9.33% w/w, respectively. This confirmed the presence of triclosan on the surface of the bone char due to the chlorine content.

The effect of the solution pH on the adsorption capacity of the bone char to remove triclosan is shown in Fig. 14. The adsorption experiments were carried out by varying the concentration of triclosan in the range of 5–30 mg/L and using a mass of 0.1 g of bone char in a volume of solution of 50 mL, and regulating the pH of the solution with HCl solutions until equilibrium was reached. The results of the effect of pH show that the adsorption capacity of the triclosan increases as the pH of the solution decreases. According to these results, it can be established that the triclosan is removed from the bone char by interactions π-π with the surface of the carbon contained in the bone char, since at solution pH values of 5 and 7, the triclosan is in

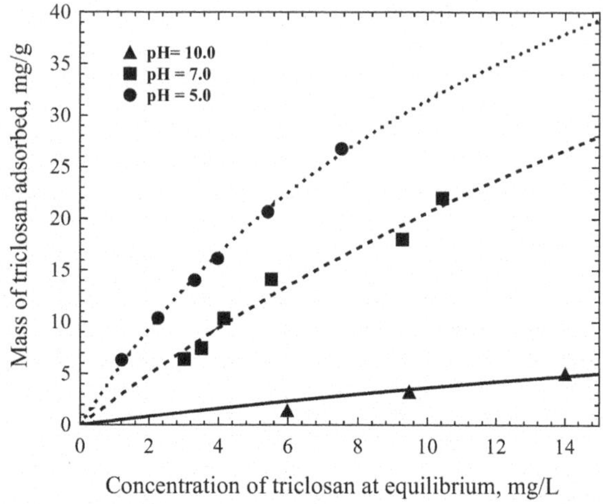

Fig. 14 Effect of the solution pH on the adsorption capacity of bone char for removal of triclosan at 25 °C

its neutral or undissociated form. Furthermore, its pKa is between 7.9 and 8.1, and under these conditions, the surface charge of the bone char is positive because the pH_{PZC} is 8.16 and however, there is an increase in the adsorption capacity as the pH of the solution decreases.

5.4 *Thermal and Acid Modification of Bone Chars for Fluoride Removal*

Authors [122] synthesized bone chars (BC) from cattle bones at different calcination temperatures and modified them by using HNO_3 solutions with different concentrations in order to increase the capacity to remove fluoride from the water. Authors observed that the calcination temperature and the acid modification of BCs affected the crystalline phases, the size of the crystals and the content of hydroxyapatite, monetite and other calcium phosphates, which are the main constituents of BC. Similarly, they determined that BC modified with HNO_3 solutions increased its adsorption capacity when monetite and low crystallinity hydroxyapatite are present in these materials. XRD, FTIR and XPS studies confirmed that the mechanism of adsorption on the modified BC is due to electrostatic interactions and chemisorption between fluoride in solution and calcium phosphates from hydroxyapatite and monetite.

The cattle bones used in this study were calcined in a muffle at temperatures of 400–800 °C and the materials were named BC400, BC500, BC600, BC700, and BC800, respectively. These materials were modified with 0.5, 1.0 and 1.5 M HNO_3 acid solutions and were identified by adding L, M and H, respectively.

Of the relevant results of this study, those of the XRD analysis can be mentioned. For the materials synthesized at different temperatures, BC400-BC800 samples, peaks were identified corresponding to the crystalline phase of hydroxyapatite, $Ca_{10}(PO_4)_6(OH)_2$. These results were expected since BC is mainly composed of this mineral. In addition, it was found that the samples synthesized at lower temperatures presented lower crystallinity (less than 15%) and this increased as the temperature increased to values close to 90%. The diffractograms of the materials modified with acid solutions are shown in Fig. 15 which indicated that the treatment with 0.5 M acid solutions (BC400L) allows the removal of impurities and amorphous material. It is also observed that by increasing the concentration of the acid solution, two crystalline phases corresponding to monetite ($CaHPO_4$) and hydroxyapatite with greater crystallinity than BC400L are presented. The XRD pattern for BC400H sample shows the complete conversion of hydroxyapatite to monetite. Figure 15 shows the XRD

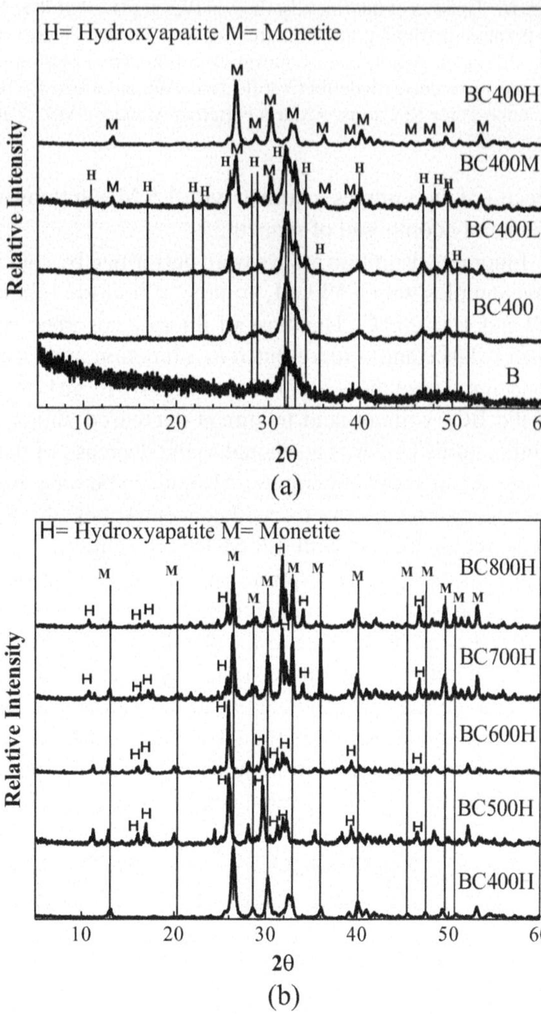

Fig. 15 XRD patterns of BCs calcined at different temperature and modified with nitric acid solutions. **a** B, BC400-BC400H. **b** BC400H-BC800H. Reprinted by permission from Springer Nature Customer Service Centre GmbH: on behalf of Cancer Research UK: Springer, Adsorption, Removal of fluoride from aqueous solution using acid and thermally treated bone char, Medellin-Castillo NA, Padilla-Ortega E, Tovar-García LD, Leyva-Ramos R, Ocampo-Pérez R, Carrasco-Marín F, Berber-Mendoza MS, © 2016

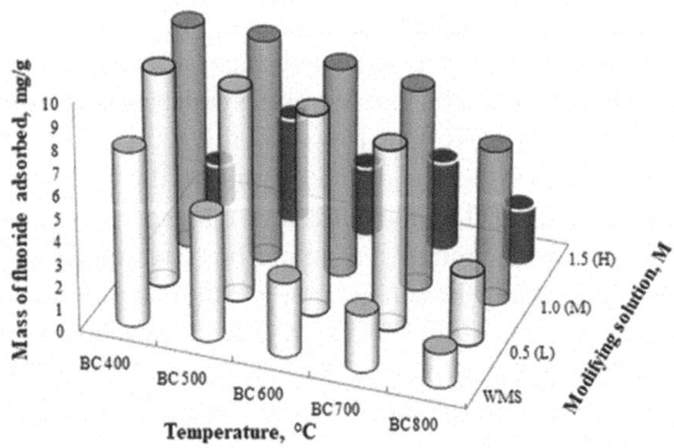

Fig. 16 Uptakes of fluoride adsorbed on BCs at pH = 7.0, T = 25 °C and C_0 = 20 mg/L. Reprinted by permission from Springer Nature Customer Service Centre GmbH: on behalf of Cancer Research UK: Springer, Adsorption, Removal of fluoride from aqueous solution using acid and thermally treated bone char, Medellin-Castillo NA, Padilla-Ortega E, Tovar-García LD, Leyva-Ramos R, Ocampo-Pérez R, Carrasco-Marín F, Berber-Mendoza MS, © 2016.

spectra of the samples modified with 1.5 M solutions which reveals that the samples are mainly composed of monetite.

Fluoride adsorption assays were performed by contacting a 1.0 g mass of the bone char samples with a 500 mL volume of fluoride solution of C_0 = 20 mg/L at pH = 7.0 and T = 25 °C. The mass of fluoride adsorbed was determined by means of a mass balance and is represented as a function of the calcination temperature and the concentration of the acid solution in Fig. 16. It can be seen that the adsorption capacity of the BCs without acid treatment decreased almost linearly with the calcination temperature. This was attributed to the decrease of the specific area by the increase of the calcination temperature. Also, it can be seen that the acid treatment increased the adsorption capacity for acid concentrations of 0.5 and 1.0 M, but decreased for BC samples treated with 1.5 M HNO_3 solution. For these materials treated with acid solutions, the adsorption capacity depended on the content of monetite and hydroxyapatite of low crystallinity.

For this study, it was reported that the adsorption of fluorides on thermally treated BCs and with acid solutions occurs by electrostatic attraction between the fluoride ion and the main constituents in these materials such as hydroxyapatite, monetite and other calcium phosphates. Also, it was established by XPS analysis that fluoride ions react with calcium from monetite and hydroxyapatite so the mechanism can be established as a chemisorption according to the following reactions:

$$\equiv Ca_{10}(PO_4)_6(OH)_2 + 20F^- \rightarrow 10 \equiv CaF_2 + 3PO_4^{3-} + 2OH^-$$

$$\equiv Ca_{10}(PO_4)_6(OH)_2 + 2F^- \rightarrow \equiv Ca_{10}(PO_4)_6F_2 + 2OH^-$$

$$\equiv CaHPO_4 + 2F^- \rightarrow \equiv CaF_2 + HPO_4^{2-}$$

6 Conclusion

The bones of different animals such as cow, pork, chicken, fish, among others, are highly available materials because they are waste generated in large quantities in the food industry. Therefore, their recovery through the synthesis of bone chars for the development of environmental technologies contributes to the mitigation of environmental impacts with a double purpose.

Bone chars are promising materials for use in electrochemistry, catalysis, environmental remediation and other fields. However, they have been mostly studied as adsorbent materials in the removal of heavy metals, anionic contaminants such as fluoride and emerging organic compounds from water as demonstrated in the bibliometric analysis presented in this study.

The physicochemical and textural properties of these materials such as specific area, amount of acidic and basic active sites, pH_{PZC} and surface charge distribution as well as the chemical composition depend on the type of bone, the conditions of synthesis such as the type of furnace, temperature, time and thermal and/or chemical modification.

According to the cases presented in this study, bone chars are basic materials with a $pH_{PZC} > 8.0$ whose base site concentration is higher than that of acidic sites and therefore they are usually of interest especially for use in the removal of anions such as fluoride from aqueous solutions but with potential use for the removal of heavy metals and organic compounds. Specific area values usually vary in the range of 50 to 180 m^2/g and are mainly due to the bone used as a precursor and the temperature of synthesis and are classified as mesoporous. According to SEM analysis, the morphology is irregular and heterogeneous which is characteristic of bone chars.The main constituents are hydroxyapatite, calcium carbonate, amorphous carbon content, calcium sulfate, iron as Fe_2O_3 and acid soluble ash which contribute to the removal of various pollutants and determine the mechanism of removal which may be by electrostatic attractions, ion exchange or chemisorption. In the case of fluoride and heavy metals such as Cd(II), hydroxyapatite can be considered responsible for the removal process and for organic compounds it is attributed to the amorphous carbon content.

Finally, it was demonstrated that bone chars are materials that are easy to synthesize and highly feasible to use in the treatment of water in areas with problems of pollution by fluorides or other contaminants, since they can be synthesized from any type of animal bone, such as those of invasive species, and in simple ovens. Once produced, they can be modified with heat treatments and acid solutions to increase their removal capacity.

Acknowledgements This work was supported by Consejo Nacional de Ciencia y Tecnología, CONACyT, Mexico, under grants Nos. CB-2016-286990 and PN-2016-3947.

References

1. Tyler Miller, G.: Introducción a la Ciencia Ambiental. In: Thomson (eds.) Madrid, España (2002)
2. Li, D., Liu, S.: Drinking water detection. Water Qual. Monitor. Manage. (2019). https://doi.org/10.1016/b978-0-12-811330-1.00010-7
3. Food and Agriculture Organization of the United Nations (FAO): AQUASTAT Main Database. https://www.fao.org/nr/water/aquastat/data/query/index.html?lang=es (2016). Accessed 09 Jul 2020
4. United Nations Educational, Scientific and Cultural Organization (UNESCO): The United Nations world water development report 2019: leaving no one behind. https://unesdoc.unesco.org/ark:/48223/pf0000367306 (2019). Accessed 08 July 2020
5. Naciones Unidas – Departamento de Asuntos Económicos y Sociales: Decenio Internacional para la acción 'El agua fuente de vida' 2005–2015. https://www.un.org/spanish/waterforlifedecade/quality.shtml (2014). Accessed 10 July 2020
6. Schweitzer, L., Noblet, J.: Water contamination and pollution. Green Chem.: Inclus. Approach (2017). https://doi.org/10.1016/B978-0-12-809270-5.00011-X
7. Singh, G., Kaushik, S.K., Mukherji, S.: Revelations of an overt water contamination. Med. J. Armed Forces India. **73**(3), 250–255 (2017). https://doi.org/10.1016/j.mjafi.2016.08.013
8. Abdalla, F., Khalil, R.: Potential effects of groundwater and surface water contamination in an urban area, Qus City. Upper Egypt. J. Afr. Earth Sci. **141**, 164–178 (2018). https://doi.org/10.1016/j.jafrearsci.2018.02.016
9. Khan, K., Lu, Y., Saeed, M.A., Bilal, H., Sher, H., Khan, H., Ali, J., Wang, P., Uwizeyimana, H., Baninla, Y., Li Q, Liang, R.: Prevalent fecal contamination in drinking water resources and potential health risks in Swat, Pakistan. J. Environ. Sci. (2017). https://doi.org/10.1016/j.jes.2017.12.008
10. Mohammadzadeh Pakdel, P., Peighambardoust, S.J.: A review on acrylic based hydrogels and their applications in wastewater treatment. J. Environ. Manage. **217**, 123–143 (2018). https://doi.org/10.1016/j.jenvman.2018.03.076
11. Vilela, C.L.S., Bassin, J.P., Peixoto, R.S.: Water contamination by endocrine disruptors: Impacts, microbiological aspects and trends for environmental protection. Environ. Pollut. **235**, 546–559 (2018). https://doi.org/10.1016/j.envpol.2017.12.098
12. World Health Organization (WHO): Drinking-water. https://www.who.int/news-room/fact-sheets/detail/drinking-water (2020). Accessed 10 July 2020
13. Hussein, F.H.: Environmental chemistry is the importance of chemistry to the environment. Arabian J. Chem. **7**(1), 1–4 (2014). https://doi.org/10.1016/j.arabjc.2013.11.007
14. Bradl, H.B.: Heavy Metals in the Environmental: Origin Interaction and Remediation. London UK (2005)
15. Masindi, V., Muedi, K.L.: Environmental contamination by heavy metals. Heavy Metals. (2018). https://doi.org/10.5772/intechopen.76082
16. Gautam, R.K., Sharma, S.K., Mahiya, S., Chattopadhyaya, M.C.: Contamination of heavy metals in aquatic media: transport, toxicity and technologies for remediation. In: The Royal Society of Chemistry (ed.) Heavy Metals in Water: Presence, Removal and Safety, pp. 1–24. Cambridge, UK (2014)
17. Chung, J.Y., Yu, S.D., Hong, Y.S.: Environmental source of arsenic exposure. J. Prev. Med. Public Health **47**(5), 253–257 (2014). https://doi.org/10.3961/jpmph.14.036
18. Stefaniak, J., Dutta, A., Verbinnen, B., Shakya, M., Rene, E.R.: Selenium removal from mining and process wastewater: A systematic review of available technologies. J. Water Supply Res. Technol. AQUA. (2018). https://doi.org/10.2166/aqua.2018.109
19. Tchounwou, P.B., Yedjou, C.G., Patlolla, A.K., Sutton, D.J.: Heavy metal toxicity and the environment. Mol. Clin. Environ. Toxicol. **101**, 133–164 (2012). https://doi.org/10.1007/978-3-7643-8340-4_6

20. Vu, C.T., Lin, C., Shern, C.C., Yeh, G., Le, V.G., Tran, H.T.: Contamination, ecological risk and source apportionment of heavy metals in sediments and water of a contaminated river in Taiwan. Ecol. Indic. **82**, 32–42 (2017). https://doi.org/10.1016/j.ecolind.2017.06.008
21. World Health Organization (WHO): Guidelines for Drinking-water Quality. https://apps.who.int/iris/bitstream/handle/10665/204411/9789241547611_eng.pdf?sequence=1 (2016). Accessed 10 July 2020
22. Volesky, B.: Detoxification of metal-bearing effluents: biosorption for the next century. Hydrometallurgy **59**(2–3), 203–216 (2001). https://doi.org/10.1016/s0304-386x(00)00160-2
23. Chandra Sekhar, K., Kamala, C., Chary, N., Anjaneyulu, Y.: Removal of heavy metals using a plant biomass with reference to environmental control. Int. J. Miner. Process **68**(1–4), 37–45 (2003). https://doi.org/10.1016/s0301-7516(02)00047-9
24. Ahluwalia, S.S., Goyal, D.: Microbial and plant derived biomass for removal of heavy metals from wastewater. Bioresour. Technol. **98**(12), 2243–2257 (2007). https://doi.org/10.1016/j.biortech.2005.12.006
25. Marshall, W.E., Wartelle, L.H., Boler, D.E., Toles, C.A.: Metal ion adsorption by soybean hulls modified with citric acid: a comparative study. Environ. Technol. **21**(6), 601–607 (2000). https://doi.org/10.1080/09593332108618075
26. Srinivasan, K., Sathiya, E.: Bimetal adsorption by cottonseed carbon: equlibrium and kinetic studies. E-J Chem. **6**(4), 1167–1175 (2009). https://doi.org/10.1155/2009/106127
27. Özcan, A., Özcan, A.S., Tunali, S., Akar, T., Kiran, I.: Determination of the equilibrium, kinetic and thermodynamic parameters of adsorption of copper(II) ions onto seeds of *Capsicum annuum*. J. Hazard. Mater. **124**(1–3), 200–208 (2005). https://doi.org/10.1016/j.jhazmat.2005.05.007
28. Agarwal, G.S., Bhuptawat, H.K., Chaudhari, S.: Biosorption of aqueous chromium(VI) by *Tamarindus indica* seeds. Bioresour. Technol. **97**(7), 949–956 (2006). https://doi.org/10.1016/j.biortech.2005.04.030
29. Gonzalez, A.G., Pokrovsky, O.S.: Metal adsorption on mosses: toward a universal adsorption model. J. Colloid Interface Sci. **415**, 169–178 (2014). https://doi.org/10.1016/j.jcis.2013.10.028
30. Koz, B., Cevik, U.: Lead adsorption capacity of some moss species used for heavy metal analysis. Ecol. Indic. **36**, 491–494 (2014). https://doi.org/10.1016/j.ecolind.2013.08.018
31. Vilardi, G., Di Palma, L., Verdone, N.: Heavy metals adsorption by banana peels micro-powder: Equilibrium modeling by non-linear models. Chin. J. Chem. Eng. **26**(3), 455–464 (2018). https://doi.org/10.1016/j.cjche.2017.06.026
32. Enniya, I., Rghioui, L., Jourani, A.: Adsorption of hexavalent chromium in aqueous solution on activated carbon prepared from apple peels. Sustain. Chem. Pharm. **7**, 9–16 (2018). https://doi.org/10.1016/j.scp.2017.11.003
33. Lim, J., Kang, H.M., Kim, L.H., Ko, S.O.: Removal of heavy metals by sawdust adsorption: equilibrium and kinetic studies. Environ. Eng. Res. **13**(2), 79–84 (2008). https://doi.org/10.4491/eer.2008.13.2.079
34. Jalali, M., Aboulghazi, F.: Sunflower stalk, an agricultural waste, as an adsorbent for the removal of lead and cadmium from aqueous solutions. J. Mater. Cycles Waste Manage. **15**(4), 548–555 (2013). https://doi.org/10.1007/s10163-012-0096-3
35. Khokhotva, A.P.: Adsorption of heavy metals by a sorbent based on pine bark. J. Water Chem. Technol. **32**(6), 336–340 (2010). https://doi.org/10.3103/s1063455x10060044
36. Changmai, M., Banerjee, P., Nahar, K., Purkait, M.K.: A novel adsorbent from carrot, tomato and polyethylene terephthalate waste as a potential adsorbent for Co(II) from aqueous solution: kinetic and equilibrium studies. J. Environ. Chem. Eng. **6**(1), 246–257 (2018). https://doi.org/10.1016/j.jece.2017.12.009
37. Wilson, K., Yang, H., Seo, C.W., Marshall, W.E.: Select metal adsorption by activated carbon made from peanut shells. Bioresour. Technol. **97**(18), 2266–2270 (2006). https://doi.org/10.1016/j.biortech.2005.10.043
38. Ogata, F., Kangawa, M., Iwata, Y., Ueda, A., Tanaka, Y., Kawasaki, N.: A study on the adsorption of heavy metals by using raw wheat bran bioadsorbent in aqueous solution phase. Chem. Pharm. Bull. **62**(3), 247–253 (2014). https://doi.org/10.1248/cpb.c13-00701

39. Husoon, Z.A., Al-Azzawi, M.N.A., Al-Hiyaly, S.A.K.: Investigation biosorption potential of copper and lead from industrial waste-water using orange and lemon peels. J. Al-Nahrain Univ. Sci. **16**(2), 173–179 (2013). https://doi.org/10.22401/JNUS.16.2.27
40. Miralles, N., Martínez, M., Florido, A., Casas, I., Fiol, N., Villaescusa, I.: Grape stalks waste as low cost biosorbents: an alternative for metal removal from aqueous solutions. Solvent Extr. Ion Exch. **26**(3), 261–270 (2008). https://doi.org/10.1080/07366290802053660
41. Thakur, L.S., Parmar, M.: Adsorption of heavy metal (CU^{2+}, Ni^{2+} and Zn^{2+}) from synthetic waste water by tea waste adsorbent. Int. J. Phys. Sci. **2**(6), 6–19 (2012)
42. Nigro, S.A., Stirk, W.A., van Staden, J., Pienaar, R.N.: Optimising heavy metal adsorbance by dried seaweeds. S. Afr. J. Bot. **68**(3), 333–341 (2002). https://doi.org/10.1016/s0254-6299(15)30395-1
43. Leyva-Ramos, R., Landin-Rodriguez, L.E., Leyva-Ramos, S., Medellin-Castillo, N.A.: Modification of corncob with citric acid to enhance its capacity for adsorbing cadmium(II) from water solution. Chem. Eng. J. **180**, 113–120 (2012). https://doi.org/10.1016/j.cej.2011.11.021
44. Rojas-Mayorga, C., Mendoza-Castillo, D., Bonilla-Petriciolet, A., Silvestre-Albero, J.: Tailoring the adsorption behavior of bone char for heavy metal removal from aqueous solution. Adsorp. Sci. Technol. **34**(6), 368–387 (2016). https://doi.org/10.1177/0263617416658891
45. Medellin-Castillo, N.A., Cruz-Briano, S.A., Leyva-Ramos, R., Moreno-Piraján, J.C., Torres-Dosal, A., Giraldo-Gutiérrez, L., Labrada-Delgado, G.J., Pérez, R.O., Rodriguez-Estupiñan, J.P., Lopez, S.Y.R., Berber Mendoza, M.S.: Use of bone char prepared from an invasive species, pleco fish (*Pterygoplichthys* spp.), to remove fluoride and Cadmium(II) in water. J. Environ. Manage. **256**, 109956 (2020). https://doi.org/10.1016/j.jenvman.2019.109956
46. De Gisi, S., Lofrano, G., Grassi, M., Notarnicola, M.: Characteristics and adsorption capacities of low-cost sorbents for wastewater treatment: a review. Sustain. Mater. Technol. **9**, 10–40 (2016). https://doi.org/10.1016/j.susmat.2016.06.002
47. Stuart, M.E., Lapworth, D.J.: Transformation products of emerging organic compounds as future groundwater and drinking water contaminants. Transform. Prod. Emerging Contam. Environ. (2014). https://doi.org/10.1002/9781118339558.ch02
48. Peña-Guzman, C., Ulloa-Sanchez, S., Mora, K., Helena-Bustos, R., Lopez-Barrera, E., Alvarez, J., Rodriguez-Pinzon, M.: Emerging pollutants in the urban water cycle in Latin America: a review of the current literature. J. Environ. Manage. **237**, 408–423 (2019). https://doi.org/10.1016/j.jenvman.2019.02.100
49. Kümmerer, K.: Pharmaceuticals in the environment. Annu. Rev. Environ. Resour. **35**(1), 57–75 (2010). https://doi.org/10.1146/annurev-environ-052809-161223
50. Greskowiak, J., Hamann, E., Burke, V., Massmann, G.: The uncertainty of biodegradation rate constants of emerging organic compounds in soil and groundwater—A compilation of literature values for 82 substances. Water Res. **126**, 122–133 (2017). https://doi.org/10.1016/j.watres.2017.09.017
51. Petrie, B., Barden, R., Kasprzyk-Hordern, B.: A review on emerging contaminants in wastewaters and the environment: Current knowledge, understudied areas and recommendations for future monitoring. Water Res. **72**, 3–27 (2015). https://doi.org/10.1016/j.watres.2014.08.053
52. Sophia, A.C., Lima, E.C.: Removal of emerging contaminants from the environment by adsorption. Ecotox. Environ. Safe **150**, 1–17 (2018). https://doi.org/10.1016/j.ecoenv.2017.12.026
53. Rubio Clemente, A., Chica Arrieta, E.L., Peñuela Mesa, G.A.: Wastewater treatment processes for the removal of emerging organic pollutants. Rev. Ambient. Agua **8**(3), 93–103 (2013). https://doi.org/10.4136/ambi-agua.1176
54. Gonzalez, O., Bayarri, B., Aceña, J., Pérez, S., Barcelo, D.: Treatment technologies for wastewater reuse: fate of contaminants of emerging concern. In: The Handbook of Environmental Chemistry (2015). https://doi.org/10.1007/698_2015_363
55. Chopra, S., Kumar, D.: Ibuprofen as an emerging organic contaminant in environment, distribution and remediation. Heliyon **6**(6), e04087 (2020). https://doi.org/10.1016/j.heliyon.2020.e04087

56. Dhangar, K., Kumar, M.: Tricks and tracks in removal of emerging contaminants from the wastewater through hybrid treatment systems: a review. Sci. Total Environ. 140320 (2020). https://doi.org/10.1016/j.scitotenv.2020.140320
57. Ji, J., Kakade, A., Yu, Z., Khan, A., Liu, P., Li, X.: Anaerobic membrane bioreactors for treatment of emerging contaminants: a review. J. Environ. Manage. **270**, 110913 (2020). https://doi.org/10.1016/j.jenvman.2020.110913
58. Ahmed, M.J., Hameed, B.H.: Removal of emerging pharmaceutical contaminants by adsorption in a fixed-bed column: a review. Ecotox. Environ. Safe **149**, 257–266 (2018). https://doi.org/10.1016/j.ecoenv.2017.12.012
59. Zhao, L., Deng, J., Sun, P., Liu, J., Ji, Y., Nakada, N., Qiao, Z., Tanaka, H., Yang, Y.: Nanomaterials for treating emerging contaminants in water by adsorption and photocatalysis: Systematic review and bibliometric analysis. Sci. Total Environ. **627**, 1253–1263 (2018). https://doi.org/10.1016/j.scitotenv.2018.02.006
60. Zhou, X., Zeng, Z., Zeng, G., Lai, C., Xiao, R., Liu, S., Huang, D., Qin, L., Liu, X., Li, B., Yi, H., Wang, Z.: Persulfate activation by swine bone char-derived hierarchical porous carbon: multiple mechanism system for organic pollutant degradation in aqueous. Chem. Eng. J. 123091 (2019). https://doi.org/10.1016/j.cej.2019.123091
61. Ali, S., Thakur, S.K., Sarkar, A., Shekhar, S.: Worldwide contamination of water by fluoride. Environ. Chem. Lett. **14**(3), 291–315 (2016). https://doi.org/10.1007/s10311-016-0563-5
62. Yousefi, M., Ghalehaskar, S., Asghari, F.B., Ghaderpoury, A., Dehghani, M.H., Ghaderpoori, M., Mohammadi, A.A.: Distribution of fluoride contamination in drinking water resources and health risk assessment using geographic information system, northwest Iran. Regul. Toxicol. Pharm. 104408 (2019). https://doi.org/10.1016/j.yrtph.2019.104408
63. Alkurdi, S.S.A., Al-Juboori, R.A., Bundschuh, J., Hamawand, I.: Bone char as a green sorbent for removing health threatening fluoride from drinking water. Environ. Int. **127**, 704–719 (2019). https://doi.org/10.1016/j.envint.2019.03.065
64. Walia, T., Abu Fanas, S., Akbar, M., Eddin, J., Adnan, M.: Estimation of fluoride concentration in drinking water and common beverages in United Arab Emirates (UAE). Saudi Dent. J. **29**(3), 117–122 (2017). https://doi.org/10.1016/j.sdentj.2017.04.002
65. Qasemi, M., Afsharnia, M., Zarei, A., Farhang, M., Allahdadi, M.: Non-carcinogenic risk assessment to human health due to intake of fluoride in the groundwater in rural areas of Gonabad and Bajestan, Iran: a case study. Hum. Ecol. Risk Assess. 1–12 (2018). https://doi.org/10.1080/10807039.2018.1461553
66. Aghaei, M., Derakhshani, R., Raoof, M., Dehghani, M., Mahvi, A.H.: Effect of fluoride in drinking water on birth height and weight: an ecological study in Kerman Province, Zarand County. Iran. Fluoride **48**, 160–168 (2015)
67. Borysewicz-Lewicka, M., Opydo-Szymaczek, J.: Fluoride in Polish drinking water and the possible risk of dental fluorosis. Pol. J. Environ. Stud. **25**(1), 9–15 (2016). https://doi.org/10.15244/pjoes/60352
68. Roy, S., Dass, G.: Fluoride contamination in drinking water—A review. Resour. Environ. **3**(3), 53–58 (2012). https://doi.org/10.5923/j.re.20130303.02
69. Vocciante, M., Trofa, M., Rodriguez-Estupiñan, P., Giraldo, L., D'Auria, T., Moreno-Pirajan, J.C., Erto, A.: A rigorous procedure for the design of adsorption units for the removal of cadmium and nickel from process wastewaters. J. Cleaner Prod. **77**, 35–46 (2014). https://doi.org/10.1016/j.jclepro.2013.12.001
70. Fletcher, H.R., Smith, D.W., Pivonka, P.: Modeling the sorption of fluoride onto alumina. J. Environ. Eng. **132**(2), 229–246 (2006). https://doi.org/10.1061/(asce)0733-9372(2006)132:2(229)
71. Tripathy, S.S., Bersillon, J.L., Gopal, K.: Removal of fluoride from drinking water by adsorption onto alum-impregnated activated alumina. Sep. Purif. Technol. **50**(3), 310–317 (2006). https://doi.org/10.1016/j.seppur.2005.11.036
72. Turner, B.D., Binning, P., Stipp, S.L.S.: Fluoride removal by calcite: evidence for fluorite precipitation and surface adsorption. Environ. Sci. Technol. **39**(24), 9561–9568 (2005). https://doi.org/10.1021/es0505090

73. Kumar, E., Bhatnagar, A., Ji, M., Jung, W., Lee, S.H., Kim, S.J., Lee, G., Song, H., Choi, J.Y., Yang, J.S., Jeon, B.H.: Defluoridation from aqueous solutions by granular ferric hydroxide (GFH). Water Res. **43**(2), 490–498 (2009). https://doi.org/10.1016/j.watres.2008.10.031
74. Biswas, K., Debnath, S., Ghosh, U.C.: Physicochemical aspects on fluoride adsorption for removal from water by synthetic hydrous Iron(III)–Chromium(III) mixed oxide. Sep. Sci. Technol. **45**(4), 472–485 (2010). https://doi.org/10.1080/01496390903526667
75. Deng, S., Liu, H., Zhou, W., Huang, J., Yu, G.: Mn–Ce oxide as a high-capacity adsorbent for fluoride removal from water. J. Hazard. Mater. **186**(2–3), 1360–1366 (2011). https://doi.org/10.1016/j.jhazmat.2010.12.024
76. Samatya, S., Yüksel, Ü., Yüksel, M., Kabay, N.: Removal of Fluoride from water by metal ions (Al^{3+}, La^{3+} and ZrO^{2+}) loaded natural zeolite. Sep. Sci. Technol. **42**(9), 2033–2047 (2007). https://doi.org/10.1080/01496390701310421
77. Onyango, M.S., Kojima, Y., Kumar, A., Kuchar, D., Kubota, M., Matsuda, H.: Uptake of fluoride by Al^{3+} pretreated low-silica synthetic zeolites: adsorption equilibrium and rate studies. Sep. Sci. Technol. **41**(4), 683–704 (2006). https://doi.org/10.1080/01496390500527019
78. Lai, Y., Yang, K., Yang, C., Tian, Z., Guo, W., Li, J.: Thermodynamics and kinetics of fluoride removal from simulated zinc sulfate solution by La(III)-modified zeolite. Trans. Nonferrous Met. Soc. **28**(4), 783–793 (2018). https://doi.org/10.1016/s1003-6326(18)64711-9
79. Zhang, J., Xie, S., Ho, Y.S.: Removal of fluoride ions from aqueous solution using modified attapulgite as adsorbent. J. Hazard. Mater. **165**(1–3), 218–222 (2009). https://doi.org/10.1016/j.jhazmat.2008.09.098
80. Thakre, D., Rayalu, S., Kawade, R., Meshram, S., Subrt, J., Labhsetwar, N.: Magnesium incorporated bentonite clay for defluoridation of drinking water. J. Hazard. Mater. **180**(1–3), 122–130 (2010). https://doi.org/10.1016/j.jhazmat.2010.04.001
81. Gogoi, P.K., Baruah, R.: Fluoride removal from water by adsorption on acid activated kaolinite clay. Indian J. Chem. Technol. **15**, 500–503 (2008)
82. Maiti, A., Basu, J.K., De, S.: Chemical treated laterite as promising fluoride adsorbent for aqueous system and kinetic modeling. Desalin **265**(1–3), 28–36 (2011). https://doi.org/10.1016/j.desal.2010.07.026
83. Bhatnagar, M., Bhatnagar, A., Jha, S.: Interactive biosorption by microalgal biomass as a tool for fluoride removal. Biotechnol. Lett. **24**(13), 1079–1081 (2002). https://doi.org/10.1023/a:1016086631017
84. Thakre, D., Jagtap, S., Bansiwal, A., Labhsetwar, N., Rayalu, S.: Synthesis of La-incorporated chitosan beads for fluoride removal from water. J. Fluorine Chem. **131**(3), 373–377 (2010). https://doi.org/10.1016/j.jfluchem.2009.11.024
85. Parmar, H.S., Patel, J.B., Sudhakar, P., Koshy, V.J.: Removal of fluoride from water with powdered corn cobs. J. Environ. Sci. Eng. **48**, 135–138 (2006)
86. Alagumuthu, G., Rajan, M.: Equilibrium and kinetics of adsorption of fluoride onto zirconium impregnated cashew nut shell carbon. Chem. Eng. J. **158**(3), 451–457 (2010). https://doi.org/10.1016/j.cej.2010.01.017
87. Bombuwala Dewage, N., Liyanage, A.S., Pittman, C.U., Mohan, D., Mlsna, T.: Fast nitrate and fluoride adsorption and magnetic separation from water on α-Fe_2O_3 and Fe_3O_4 dispersed on Douglas fir biochar. Bioresour. Technol. **263**, 258–265 (2018). https://doi.org/10.1016/j.biortech.2018.05.001
88. Wang, J., Chen, N., Li, M., Feng, C.: Efficient removal of fluoride using polypyrrole-modified biochar derived from slow pyrolysis of pomelo peel: sorption capacity and mechanism. J. Polym. Environ. **26**(4), 1559–1572 (2017). https://doi.org/10.1007/s10924-017-1061-y
89. Medellin-Castillo, N.A., Leyva-Ramos, R., Ocampo-Perez, R., Garcia de la Cruz, R.F., Aragon-Piña, A., Martinez-Rosales, J.M., Guerrero-Coronado, R.M., Fuentes-Rubio, L.: Adsorption of Fluoride from Water Solution on Bone Char. Ind. Eng. Chem. Res. **46**(26), 9205–9212 (2007). https://doi.org/10.1021/ie070023n
90. Ismail, Z.Z., AbdelKareem, H.N.: Sustainable approach for recycling waste lamb and chicken bones for fluoride removal from water followed by reusing fluoride-bearing waste in concrete. Waste Manage. **45**, 66–75 (2015). https://doi.org/10.1016/j.wasman.2015.06.039

91. Delgadillo-Velasco, L., Hernandez-Montoya, V., Cervantes, F.J., Montes-Moran, M.A., Lira-Berlanga, D.: Bone char with antibacterial properties for fluoride removal: preparation, characterization and water treatment. J. Environ. Manage. **201**, 277–285 (2017). https://doi.org/10.1016/j.jenvman.2017.06.038
92. Organisation for Economic Co-operation and Development/Food and Agriculture Organization of the United Nations (OECD/FAO): Agricultural Outlook 2018–2027. https://www.fao.org/3/i9166en/I9166EN.pdf (2018). Accessed 14 July 2020
93. Hyder, A.H.M.G., Begum, S.A., Egiebor, N.O.: Adsorption isotherm and kinetic studies of hexavalent chromium removal from aqueous solution onto bone char. J. Environ. Chem. Eng. **3**(2), 1329–1336 (2015). https://doi.org/10.1016/j.jece.2014.12.005
94. Ishikawa, K., Miyamoto, Y., Tsuchiya, A., Hayashi, K., Tsuru, K., Ohe, G.: Physical and histological comparison of hydroxyapatite, carbonate apatite, and β-tricalcium phosphate bone substitutes. Materials **11**(10), 1993 (2018). https://doi.org/10.3390/ma11101993
95. Mendoza-Castillo, D.I., Bonilla-Petriciolet, A., Jauregui-Rincon, J.: On the importance of surface chemistry and composition of Bone char for the sorption of heavy metals from aqueous solution. Desalin. Water Treat. 1–12 (2014). https://doi.org/10.1080/19443994.2014.888684
96. Reynel-Avila, H.E., Mendoza-Castillo, D.I., Bonilla-Petriciolet, A.: Relevance of anionic dye properties on water decolorization performance using bone char: adsorption kinetics, isotherms and breakthrough curves. J. Mol. Liq. **219**, 425–434 (2016). https://doi.org/10.1016/j.molliq.2016.03.051
97. Aranda Usón, A., Lopez-Sabiron, A.M., Ferreira, G., Llera Sastresa, E.: Uses of alternative fuels and raw materials in the cement industry as sustainable waste management options. Renew. Sustain. Energy Rev. **23**, 242–260 (2013). https://doi.org/10.1016/j.rser.2013.02.024
98. Naliaka, B.: Evaluation of bone char defluoridation of water using adsorption isotherms and the bed depth service time (BDST) Model. Dissertation, University of Nairobi (2016)
99. Nigri, E.M., Bhatnagar, A., Rocha, S.D.F.: Thermal regeneration process of bone char used in the fluoride removal from aqueous solution. J. Cleaner Prod. **142**, 3558–3570 (2017). https://doi.org/10.1016/j.jclepro.2016.10.112
100. Rojas-Mayorga, C., Bonilla-Petriciolet, A., Sanchez-Ruiz, F.J., Moreno-Perez, J., Reynel-Avila, H.E., Aguayo-Villarreal, I.A., Mendoza-Castillo, D.I.: Breakthrough curve modeling of liquid-phase adsorption of fluoride ions on aluminum-doped bone char using micro-columns: Effectiveness of data fitting approaches. J. Mol. Liq. **208**, 114–121 (2015). https://doi.org/10.1016/j.molliq.2015.04.045
101. Wang, M., Liu, Y., Yao, Y., Han, L., Liu, X.: Comparative evaluation of bone chars derived from bovine parts: Physicochemical properties and copper sorption behavior. Sci. Total Environ. 134470 (2019). https://doi.org/10.1016/j.scitotenv.2019.134470
102. Shahid, M.K., Kim, J.Y., Choi, Y.G.: Synthesis of bone char from cattle bones and its application for fluoride removal from the contaminated water. Groundwater Sustain. Dev. (2018). https://doi.org/10.1016/j.gsd.2018.12.003
103. Jia, P., Tan, H., Liu, K., Gao, W.: Adsorption behavior of methylene blue by bone char. Int. J. Mod. Phys. B **31**(16–19), 1744099 (2017). https://doi.org/10.1142/s0217979217440994
104. Hashemi, S., Rezaee, A., Nikodel, M., Ganjidost, H., Mousavi, S.M.: Equilibrium and kinetic studies of the adsorption of sodium dodecyl sulfate from aqueous solution using bone char. React. Kinet. Mech. Catal. **109**(2), 433–446 (2013). https://doi.org/10.1007/s11144-013-0559-0
105. Rezaee, A., Ghanizadeh, G., Behzadiyannejad, G., Yazdanbakhsh, A., Siyadat, S.D.: Adsorption of endotoxin from aqueous solution using bone char. Bull. Environ. Contam. Tox. **82**(6), 732–737 (2009). https://doi.org/10.1007/s00128-009-9690-z
106. Herath, H., Kawakami, T., Tafu, M.: The extremely high adsorption capacity of fluoride by chicken bone char (CBC) in defluoridation of drinking water in relation to its finer particle size for better human health. Healthcare **6**(4), 123 (2018). https://doi.org/10.3390/healthcare6040123
107. Côrtes, L.N., Druzian, S.P., Streit, A.F.M., Sant'anna Cadaval Junior, T.R., Collazzo, G.C., Dotto, G.L.: Preparation of carbonaceous materials from pyrolysis of chicken bones and its

application for fuchsine adsorption. Environ. Sci. Pollut. Res. (2018). https://doi.org/10.1007/s11356-018-3679-2
108. Martins, J.I., Órfão, J.J.M., Soares, O.S.G.P.: Sorption of copper, nickel and cadmium on bone char. Prot. Met. Phys. Chem. Surf. **53**(4), 618–627 (2017). https://doi.org/10.1134/s2070205117040153
109. Cheung, C., Porter, J., Mckay, G.: Sorption kinetic analysis for the removal of cadmium ions from effluents using bone char. Water Res. **35**(3), 605–612 (2001). https://doi.org/10.1016/s0043-1354(00)00306-7
110. Chen, Y.N., Chai, L.Y., Shu, Y.D.: Study of arsenic(V) adsorption on bone char from aqueous solution. J. Hazard. Mater. **160**(1), 168–172 (2008). https://doi.org/10.1016/j.jhazmat.2008.02.120
111. Reynel-Avila, H.E., Mendoza-Castillo, D.I., Bonilla-Petriciolet, A., Silvestre-Albero, J.: Assessment of naproxen adsorption on bone char in aqueous solutions using batch and fixed-bed processes. J. Mol. Liq. **209**, 187–195 (2015). https://doi.org/10.1016/j.molliq.2015.05.013
112. Liu, J., Huang, X., Liu, J., Wang, W., Zhang, W., Dong, F.: Adsorption of arsenic(V) on bone char: batch, column and modeling studies. Environ. Earth Sci. **72**(6), 2081–2090 (2014). https://doi.org/10.1007/s12665-014-3116-x
113. Ferreira Mendes, K., de Oliveira Freguglia, R.M., Barbosa Martins, B.A., de Carvalho Dias, R., Floriano Pimpinato, R., Luiz Tornisielo, V.: Cow bonechar for pesticide removal from drinking water. Sch. J. Agric. Vet. Sci. **4**(11), 504–512 (2017). https://doi.org/10.21276/sjavs.2017.4.11.11
114. Rojas-Mayorga, C., Bonilla-Petriciolet, A., Aguayo-Villarreal, I.A., Hernandez-Montoya, V., Moreno-Virgen, M.R., Tovar-Gomez, R., Montes-Moran, M.A.: Optimization of pyrolysis conditions and adsorption properties of bone char for fluoride removal from water. J. Anal. Appl. Pyrolysis **104**, 10–18 (2013). https://doi.org/10.1016/j.jaap.2013.09.018
115. Piccirillo, C., Moreira, I.S., Novais, R.M., Fernandes, A.J.S., Pullar, R.C., Castro, P.M.L.: Biphasic apatite-carbon materials derived from pyrolysed fish bones for effective adsorption of persistent pollutants and heavy metals. J. Environ. Chem. Eng. **5**(5), 4884–4894 (2017). https://doi.org/10.1016/j.jece.2017.09.010
116. Brunson, L.R., Sabatini, D.A.: An evaluation of fish bone char as an appropriate arsenic and fluoride removal technology for emerging regions. Environ. Eng. Sci. **26**(12), 1777–1784 (2009). https://doi.org/10.1089/ees.2009.0222
117. Iriarte-Velasco, U., Sierra, I., Cepeda, E.A., Bravo, R., Ayastuy, J.L.: Methylene blue adsorption by chemically activated waste pork bones. Color Technol. **131**(4), 322–332 (2015). https://doi.org/10.1111/cote.12160
118. Pan, X., Wang, J., Zhang, D.: Sorption of cobalt to bone char: Kinetics, competitive sorption and mechanism. Desalin **249**(2), 609–614 (2009). https://doi.org/10.1016/j.desal.2009.01.027
119. Dawlet, A., Talip, D., Mi, H.Y.: MaLiKeZhaTi: Removal of mercury from aqueous solution using sheep bone charcoal. Procedia Environ. Sci. **18**, 800–808 (2013). https://doi.org/10.1016/j.proenv.2013.04.108
120. Kawasaki, N., Ogata, F., Tominaga, H., Yamaguchi, I.: Removal of fluoride ion by bone char produced from animal biomass. J. Oleo Sci. **58**(10), 529–535 (2009). https://doi.org/10.5650/jos.58.529
121. Figueiredo, M., Fernando, A., Martins, G., Freitas, J., Judas, F., Figueiredo, H.: Effect of the calcination temperature on the composition and microstructure of hydroxyapatite derived from human and animal bone. Ceram. Int. **36**(8), 2383–2393 (2010). https://doi.org/10.1016/j.ceramint.2010.07.016
122. Medellin-Castillo, N.A., Padilla-Ortega, E., Tovar-Garcia, L.D., Leyva-Ramos, R., Ocampo-Perez, R., Carrasco-Marin, F., Berber-Mendoza, M.S.: Removal of fluoride from aqueous solution using acid and thermally treated bone char. Adsorption **22**(7), 951–961 (2016). https://doi.org/10.1007/s10450-016-9802-0
123. Rojas-Mayorga, C.K., Silvestre-Albero, J., Aguayo-Villarreal, I.A., Mendoza-Castillo, D.I., Bonilla-Petriciolet, A.: A new synthesis route for bone chars using CO_2 atmosphere and their

application as fluoride adsorbents. Microp. Mesopor. Mater. **209**, 38–44 (2015). https://doi.org/10.1016/j.micromeso.2014.09.002
124. Alkurdi, S.S.A., Al-Juboori, R.A., Bundschuh, J., Bowtell, L., McKnight, S.: Effect of pyrolysis conditions on bone char characterization and its ability for arsenic and fluoride removal. Environ. Pollut. (2020). https://doi.org/10.1016/j.envpol.2020.114221
125. Patel, S., Han, J., Qiu, W., Gao, W.: Synthesis and characterisation of mesoporous bone char obtained by pyrolysis of animal bones, for environmental application. J. Environ. Chem. Eng. **3**(4), 2368–2377 (2015). https://doi.org/10.1016/j.jece.2015.07.031
126. Kader, A.A., Aly, A., Girgis, B.: Bone char decolorisation efficiency. A laboratory study over four consecutive cycles. Int. Sugar J. **98**(1174), 542–554 (1996)
127. Wei, S., Zhang, H., Huang, Y., Wang, W., Xia, Y., Yu, Z.: Pig bone derived hierarchical porous carbon and its enhanced cycling performance of lithium–sulfur batteries. Energy Environ. Sci. **4**(3), 736 (2011). https://doi.org/10.1039/c0ee00505c
128. Goodman, P.A., Li, H., Gao, Y., Lu, Y.F., Stenger-Smith, J.D., Redepenning, J.: Preparation and characterization of high surface area, high porosity carbon monoliths from pyrolyzed bovine bone and their performance as supercapacitor electrodes. Carbon **55**, 291–298 (2013). https://doi.org/10.1016/j.carbon.2012.12.066
129. Bhattacharjya, D., Yu, J.S.: Activated carbon made from cow dung as electrode material for electrochemical double layer capacitor. J. Power Sources **262**, 224–231 (2014). https://doi.org/10.1016/j.jpowsour.2014.03.143
130. Isaacs-Paez, E.D., Medellin-Castillo, N., Manriquez-Guerrero, F., Cercado, B.: Characterization of bone char and carbon xerogel as sustainable alternative bioelectrodes for bioelectrochemical systems. Waste Biomass Valorization (2019). https://doi.org/10.1007/s12649-019-00817-4
131. Obadiah, A., Swaroopa, G.A., Kumar, S.V., Jeganathan, K.R., Ramasubbu, A.: Biodiesel production from Palm oil using calcined waste animal bone as catalyst. Bioresour. Technol. **116**, 512–516 (2012). https://doi.org/10.1016/j.biortech.2012.03.112
132. Smith, S.M., Oopathum, C., Weeramongkhonlert, V., Smith, C.B., Chaveanghong, S., Ketwong, P., Boonyuen, S.: Transesterification of soybean oil using bovine bone waste as new catalyst. Bioresour. Technol. **143**, 686–690 (2013). https://doi.org/10.1016/j.biortech.2013.06.087
133. Chakraborty, R., Bepari, S., Banerjee, A.: Application of calcined waste fish (*Labeo rohita*) scale as low-cost heterogeneous catalyst for biodiesel synthesis. Bioresour. Technol. **102**(3), 3610–3618 (2011). https://doi.org/10.1016/j.biortech.2010.10.123
134. Farooq, M., Ramli, A., Naeem, A.: Biodiesel production from low FFA waste cooking oil using heterogeneous catalyst derived from chicken bones. Renew. Energy **76**, 362–368 (2015). https://doi.org/10.1016/j.renene.2014.11.042
135. Zhou, X., Zeng, Z., Zeng, G., Lai, C., Xiao, R., Liu, S. … Wang, Z.: Insight into the mechanism of persulfate activated by bone char: Unraveling the role of functional structure of biochar. Chem. Eng. J. **401**, 126127 (2020). https://doi.org/10.1016/j.cej.2020.126127
136. Mardani Ghahfarrokhi, A., Moshiri, P., Ghiaci, M.: Studies on calcined cow bone and pyrolyzed wood, suitable supports for immobilizing hybrid nano particles of Co–Mn as new catalysts for oxidation of 2,6-diisopropyl naphthalene. Appl. Catal. A **456**, 51–58 (2013). https://doi.org/10.1016/j.apcata.2013.02.013
137. Jia, P., Tan, H., Liu, K., Gao, W.: Enhanced photocatalytic performance of ZnO/bone char composites. Mater. Lett. **205**, 233–235 (2017). https://doi.org/10.1016/j.matlet.2017.06.099
138. Abe, I., Iwasaki, S., Tokimoto, T., Kawasaki, N., Nakamura, T., Tanada, S.: Adsorption of fluoride ions onto carbonaceous materials. J. Colloid Interface Sci. **275**(1), 35–39 (2004). https://doi.org/10.1016/j.jcis.2003.12.031
139. Leyva-Ramos, R., Rivera-Utrilla, J., Medellin-Castillo, N.A., Sanchez-Polo, M.: Kinetic modeling of fluoride adsorption from aqueous solution onto bone char. Chem. Eng. J. **158**(3), 458–467 (2010). https://doi.org/10.1016/j.cej.2010.01.019
140. Ma, W., Ya, F., Wang, R., Zhao, Y.: Fluoride removal from drinking water by adsorption using bone char as a biosorbent. Int. J. Environ. Technol. Manage. **9**(1), 59 (2008). https://doi.org/10.1504/ijetm.2008.017860

141. Dahi, E., Bregnhøj, H.: Significance of oxygen in processing of bone char for defluoridation of water. In: Dahi, E., Bregnhøj, H. (eds.) Proceedings of the 1st International Workshop on Fluorosis and Defluoridation of Water, Tanzania (1995)
142. Onyango, M.S., Matsuda, H.: Chapter 1 fluoride removal from water using adsorption technique. In: Tressaud, A. (ed.) Fluorine and the Environment—Agrochemicals, Archaeology, Green Chemistry & Water, 1st edn., pp. 1–48. Elsevier, B.V (2006)
143. Ferreira Brum, R., Marinuchi, R., Martins Gomes, J., Ostroski, I.C., Arroyo, P.A., Granhen Tavares, C.R., Barros, M.A.S.D.: Equilibrium studies of Mn(II), Pb(II) and Cr(III) in bone char. Chem. Eng. Trans. **21**, 721–726 (2010). https://doi.org/10.3303/CET1021121
144. Nigri, E.M., Cechinel, M.A.P., Mayer, D.A., Mazur, L.P., Loureiro, J.M., Rocha, S.D.F., Vilar, V.J.P.: Cow bones char as a green sorbent for fluorides removal from aqueous solutions: batch and fixed-bed studies. Environ. Sci. Pollut. Res. Int. **24**(3), 2364–2380 (2016). https://doi.org/10.1007/s11356-016-7816-5
145. Fang, J.L., Stingley, R.L., Beland, F.A., Harrouk, W., Lumpkins, D.L., Howard, P.: Occurrence, efficacy, metabolism, and toxicity of triclosan. J. Environ. Sci. Health **28**(3), 147–171 (2010). https://doi.org/10.1080/10590501.2010.504978
146. Webb, S., Gaw, S., Marsden, I.D., McRae, N.K.: Biomarker responses in New Zealand green-lipped mussels Perna canaliculus exposed to microplastics and triclosan. Ecotox. Environ. Safe. **201**, 110871 (2020). https://doi.org/10.1016/j.ecoenv.2020.110871
147. Montaseri, H., Forbes, P.B.C.: A review of monitoring methods for triclosan and its occurrence in aquatic environments. Trac. Trends Anal. Chem. **85**, 221–231 (2016). https://doi.org/10.1016/j.trac.2016.09.010

Current Advances in Characterization of Nano-porous Materials: Pore Size Distribution and Surface Area

Kaihang Shi, Erik E. Santiso, and Keith E. Gubbins

Abstract Methods for characterizing the most important textural properties of porous materials, particularly pore size distribution (PSD) and specific surface area (SSA), using theoretical models to analyze standard experimental physisorption isotherms are reviewed. For PSD analysis, we explain the basics of the classical methods, including those based on the Kelvin equation and adsorption potential theory, as well as the modern molecular methods based on the density functional theory (DFT) and molecular simulation. For SSA analysis, the discussion is focused on kinetic theory, and primarily on the application of the Brunauer-Emmett-Teller (BET) method for characterization of microporous materials. Current advances and extensions of model-based methods are discussed, including those related to machine learning techniques. We reflect on the limitations of the current state-of-the-art methods and point out some possible directions for future studies.

1 Introduction

The demand for accurate and fast characterization of porous materials has increased dramatically in recent years, due to the wide application of these materials to solve engineering and global environmental problems (for example, design of supercapacitors [1], hydrogen storage [2], and carbon capture [3]). The adsorption capacity [4, 5], phase equilibrium [6–8], and transport [9] of confined phases in porous materials are directly linked to their structure properties, such as specific surface area (SSA) and pore size distribution (PSD). Knowledge and control of the PSD and surface area of porous materials is of utmost importance in many applications, including medical and dental implants [10, 11], filtration [12, 13], catalysis [14–17], next-generation

K. Shi · E. E. Santiso · K. E. Gubbins
Department of Chemical and Biomolecular Engineering, North Carolina State University, Raleigh, NC 27695, USA

K. Shi (✉)
Department of Chemical and Biological Engineering, Northwestern University, Evanston, IL 60208, USA
e-mail: kshi@northwestern.edu

J. C. Moreno-Piraján et al. (eds.), *Porous Materials*, Engineering Materials,
https://doi.org/10.1007/978-3-030-65991-2_12

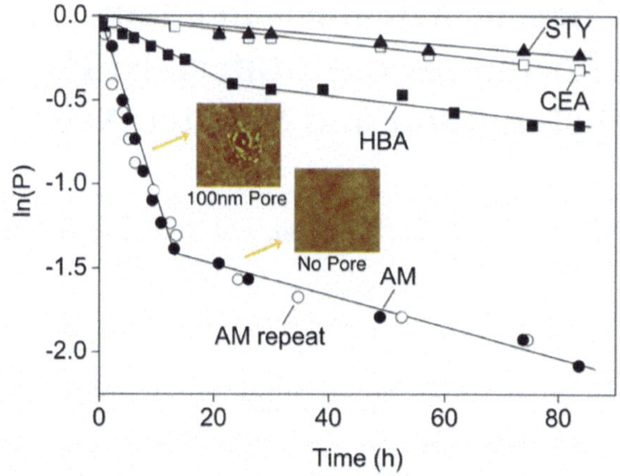

Fig. 1 Crystallization of aspirin on porous materials with different pore size distributions. $P(t)$ is the probability of no crystal forming at time (t). The pore sizes have a dramatic effect on the crystallization rate. Reprinted with permission from Ref. [25]. Copyright (2011) American Chemical Society

high-pressure manufacturing [18–22], geological applications (including storage of gas, oil, and water in reservoirs [23]), crystallization control (see an example in Fig. 1) [24, 25], among others. For this reason, the development of accurate, high-throughput techniques to determine PSD and SSA can have an impact in many different fields.

There is a long history of the development of both experimental and theoretical methods to obtain the textural properties of porous materials. Experimental methods for measuring the PSD include small-angle X-ray (SAXS) and neutron scattering (SANS), gas adsorption, mercury porosimetry, microscopy, and X-ray microtomography, among many others [26–28]. The experimental measurement of the SSA is often limited to the gas sorption method. Physisorption experiments are the most used techniques to understand the thermodynamic state of confined phases and the structure of porous materials. Physisorption has the advantage of being a convenient, non-destructive, and reversible method. Hybrid methods, combining physisorption experiments and theoretical models based on classical thermodynamics and statistical mechanics [29, 30], are able to extract much critical structural information of the porous materials, including the PSD and SSA, and they are performed as a routine analysis nowadays.

This book chapter focuses primarily on theory- and simulation-based hybrid approaches for the PSD and SSA characterization of porous materials, including some of the recent advances and extensions of the traditional approaches. The chapter is organized as follows: Sect. 2 describes methods for the determination of the PSD, including classical methods based on the Kelvin equation and the adsorption potential theory, as well as more modern methods based on the classical density functional theory (DFT) and grand canonical Monte Carlo (GCMC) simulation. Section 3 discusses methods to determine the SSA, including the widely used Brunauer-Emmett-Teller (BET) method, and discusses newer extensions of the BET method as well as modern approaches based on machine learning techniques. Finally, Sect. 4 provides some concluding remarks and outlines some possible future developments.

2 Pore Size Distribution

2.1 Classical Methods

Mesopore Analysis The adsorption behavior in mesopores (2–50 nm) depends not only on the solid-fluid interactions, but also on the fluid-fluid interactions, leading to the occurrence of pore condensation and multiplayer adsorption. Before the advent of statistical mechanical methods, mesopore size characterization generally made use of the correlation between the pore condensation pressure and the pore size, established by the Kelvin equation [31]. The Kelvin equation relates the shift of pore condensation pressure, P, relative to the bulk coexistence pressure, P_0, to the macroscopic properties:

$$\ln \frac{P}{P_0} = -\frac{2\gamma \bar{V}}{rRT} \tag{1}$$

where P is the actual vapor pressure (i.e., the pore condensation pressure, $P < P_0$), γ is the liquid/vapor surface tension, $\bar{V}$ is the molar volume of the liquid, r is the radius of the droplet (or pore radius), R is the gas constant and T is the temperature. However, Eq. (1) does not consider the pre-adsorbed layers near the pore wall before the pore condensation happens. The original Kelvin equation was later modified to correct this problem [32]. In the modified Kelvin equation, the variable r in Eq. (1) is replaced by $\left(r_p - t_c\right)$, where r_p and t_c are the actual pore radius and the critical thickness of the adsorbed films at which pore condensation occurs. The modified Kelvin equation serves as the basis for many methods developed for pore size analysis, among which are included the Barrett-Joyner-Halenda (BJH) method [33], which is the most widely-used method for mesopore size analysis. The typical procedure to obtain a PSD profile is to build a work-table, including the columns of the following parameters: relative pressure, the corresponding Kelvin radius [r, from Eq. (1)] and the thickness of the pre-adsorbed layers (t_c, by Halsey equation [34, 35] for example), the pore radius (r_p) at each relative pressure, the mean value of Kelvin radius and pore radius ($\bar{r}_p$), and the change of pre-adsorbed film thickness and of liquid volume (converted from experimental volumetric gas physisorption data). The mean values and changes of quantities are calculated between two consecutive pressure conditions. The pore volume at a certain $\bar{r}_p$ can be calculated based on these quantities. An example calculation of the PSD based on the modified Kelvin equation is demonstrated in Ref. [27]. All methods that are based on the classical Kelvin equation make three main assumptions:

1. The porous material is composed of a collection of well-defined pores (cylindrical or slit-shaped), and the solid-fluid interactions are negligible for pore condensation in the inner core.
2. The thermodynamic properties of the confined phase (liquid-like) are the same as those of the corresponding bulk phase.

3. The thickness of the pre-adsorbed multi-layers can be estimated by the statistical thickness of layers on non-porous materials which share similar surface chemistry (i.e., similar BET *C* value, see Sect. 3.1) to the porous sample.

The first assumption sets the limitation that the method can only be applied to large mesopores and macropores (see Fig. 2), and caution should always be exercised when applying this method to calculate the PSD of materials having small and connected pores. Gelb and Gubbins [36] showed that the BJH method yields a qualitatively similar PSD profile compared to the exact geometric PSD in porous glasses with complex pore shape and connectivity, but the BJH method underestimates the pore size in a systematic way for small mesopores (~ 2 to 4 nm, consistent with the data presented in Fig. 2). The second assumption leads to deviations in small pores, where the confined fluid is structured layer-by-layer. In addition, the surface tension of the curved interface at the nanoscale is no longer a constant, as in the planar case. According to the Tolman equation [37], the surface tension of a droplet depends on the radius of the curvature [38]. The third assumption makes it easy to correlate the thickness of the layer with pressure, but the effect of adsorbent surface curvature on the adsorption amount (and thus on the thickness of the layer) has been omitted. The third assumption is also the basic approximation adopted by many other classical

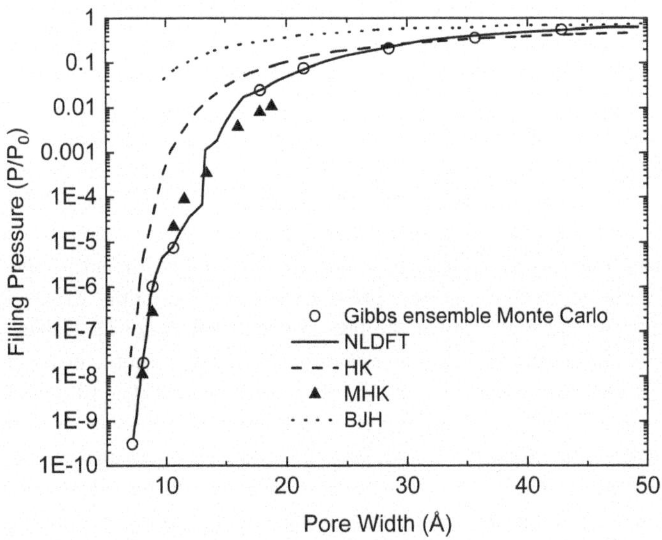

Fig. 2 Pore filling correlation predicted by Gibbs ensemble Monte Carlo (GEMC) simulation, non-local density functional theory (NLDFT), the original Horvath-Kawazoe (HK) method, modified Horvath-Kawazoe (MHK) method and the Barrett-Joyner-Halenda (BJH) method for nitrogen adsorption in carbon slit pores at 77 K. The HK method gives better agreement with the (exact) GEMC simulations than the modified Kelvin equation (BJH) in the micropore range, but performs more poorly in mesopores. The MHK method, using a more realistic 10-4-3 potential, gives better agreement with the NLDFT and GEMC simulations in micropores. The MHK data are from Ref. [48] and the rest are from Ref. [49, 50]

methods for the PSD analysis that involve predicting the layer thickness in pores. Nguyen and Do [39] showed that the PSDs calculated by the modified Kelvin equation in conjunction with an enhanced film thickness equation that depends on both pressure and pore size are comparable with those from the density functional theory.

Due to their theoretical simplicity and convenient implementation, methods based on the Kelvin equation (mainly the BJH method) are still widely used nowadays for analyzing the PSD of mesoporous materials. Efforts have been made to adapt the Kelvin equation to pores of finite length [40, 41], and to calibrate the performance of the BJH method for specific systems, such as for MCM-41 [42], CO_2 in activated carbon [43] and alumina samples having large and irregular slit-shaped mesopores (>10 nm) [44]. A general improvement of the BJH method has been made recently by introducing a correction term in the algorithm [45, 46]. The value of the correction term was chosen to fit the experimental isotherm data, and the PSD obtained by the proposed method is in favorable comparison with that from density functional theory [46]. The BJH method was originally developed for a single pore shape; it has been extended to analyze dual pore size distribution for coexistence of slit-shaped and cylindrical pores [47].

Micropore Analysis For micropores (< 2 nm), the Kelvin equation is no longer valid. Here we introduce several popular methods that have been widely used in the community to determine the PSD of microporous materials.

The micropore analysis (MP) method by Mikhail, Brunauer and Bodor [51] allows one to obtain the surface area, pore volume and PSD of the micropores from a single isotherm. The MP method can be considered as an extension of the *t*-method [52, 53], where the adsorbed liquid volume, V_{liq}, is plotted against the statistical thickness of the adsorbed film, t, which is determined from adsorbed films on nonporous materials having a similar BET C value, as a function of pressure. Once the $V_{liq} - t$ plot is made, the slope of the curve evaluated at a point between two thicknesses, t_i and t_{i+1}, is considered as the surface area of the pore that has not been filled:

$$S_{i+1} = \left.\frac{dV_{liq}}{dt}\right|_{r_h} \tag{2}$$

where the hydraulic radius is $r_h = \frac{(t_i+t_{i+1})}{2}$. The difference, $(S_i - S_{i+1})$, is the surface area for pores with hydraulic radii between t_i and t_{i+1} that has become filled. The pore volume corresponding to the hydraulic radius r_h can then be calculated by

$$V_i = (S_i - S_{i+1})r_h \tag{3}$$

The MP method is applicable to both slit-shaped and cylindrical pores, and the value $2r_h$ is the pore width for slit-shaped pores or the pore radius for cylindrical pores. Because the statistical thickness of the film gives the actual measure of the pore size in the MP method, it is not hard to imagine that the lack of statistical thickness data for certain adsorbents, and the approximation of the film thickness in pores to that on a non-porous material, will make the results from the MP method questionable.

The Dubinin-Radushkevich (DR) method provides another way to measure the micropore PSD, based on Polanyi's adsorption potential theory [54]. In Polanyi's theory, the adsorption potential A at a certain distance from the surface corresponds to the molar free energy to compress the vapor from pressure P, to the liquid state at the saturation pressure, P_0:

$$A = RT \ln \frac{P_0}{P} \tag{4}$$

The DR method postulates that the experimental micropore volume occupied by the condensed liquid, V_{liq}, follows a Gaussian distribution in terms of this adsorption potential,

$$V_{liq} = V_0 \exp\left[-\left(\frac{A}{\beta E_0}\right)^2\right] \tag{5}$$

where V_0 is the total micropore volume; β is the affinity coefficient defined as the characteristic adsorption energy of an adsorbate, E, to the reference value, E_0, where benzene was used as the reference adsorbate [55]. To extend the DR equation to heterogeneous surfaces, the exponent of the term, $\frac{A}{(\beta E_0)}$, in Eq. (5), is simply replaced by n, leading to the Dubinin-Astakhov (DA) equation [56]. The value of exponent n ranges from 2 to 6 in practice. The DR and DA formalisms have been compared in terms of adsorption potential distribution for pillared clays, zeolites and activated carbons [57]. The reference characteristic adsorption energy, E_0, has a one-to-one mapping to the pore size [55]. Therefore, for a real microporous material composed of a collection of different micropores, the PSD can be obtained by solving the integral equation [58]:

$$V_{liq} = \int f(E_0) \exp\left[-\left(\frac{A}{\beta E_0}\right)^2\right] dE_0 \tag{6}$$

where $f(E_0)$ is the PSD function in terms of the reference characteristic adsorption energy. It can be converted to the normal PSD by using the relation between the pore size and E_0 [55]. To solve Eq. (6), $f(E_0)$ is usually assumed to be the Gaussian distribution [58]. It has been shown that, without restricting $f(E_0)$ to a particular shape, the DR method is able to capture the bimodal PSD of chars [59]. The DR method in principle is not applicable to low pressures (Henry's law region). Sun [60] has modified the DR method to make it flexible enough to model the N_2 isotherm over a wide range of pressure.

Horvath and Kawazoe (HK) [61] introduced a simple PSD analysis that considers pore shape and molecular adsorbate-adsorbent interactions. They obtained a direct relation between the pore filling pressure and the physical pore width (defined as the distance between the nuclei of the two opposing pore surfaces for a slit-shaped pore) by equating the $(-A)$ in Eq. (4) to a uniform potential field. The uniform (unweighted) potential field was obtained by spatially averaging the local 10-4 potential over the entire pore. Both adsorbate-adsorbent and adsorbate-adsorbate interactions were included in the potential field, although the adsorbate-adsorbate part was not handled correctly [48]. For nitrogen adsorbed in a microporous carbon material at 77.4 K, the pore filling correlation is given by [61]

$$\ln\left(\frac{P}{P_0}\right) = \frac{62.38}{H - 0.64}\left[\frac{1.895 \times 10^{-3}}{(H - 0.32)^3} - \frac{2.7087 \times 10^{-7}}{(H - 0.32)^9} - 0.05014\right] \quad (7)$$

where the physical pore width, H, is in units of nanometers. Thus, for a given physical pore width, a unique pore filling pressure, P, can be found (see Fig. 2). The HK method assumes the pore filling mechanism is discontinuous, that is, for each pressure, P, only pores that have physical pore width equal to or smaller than the corresponding physical pore width (calculated by Eq. (7), for example) are filled. The assumed discontinuous mechanism is a good approximation for small micropores but has been shown to be a source of error in the characterization of large micropores [62]. If we plot the experimental volume of adsorbed gas in its liquid state at a certain relative pressure versus the corresponding internal pore width, $w = H - \sigma_{ss}$, where σ_{ss} is the diameter of the adsorbent atom, we then have a $V_{liq} - w$ plot, i.e., a cumulative pore volume curve. The first-order gradient of the plot, $\frac{dV_{liq}}{dw}$, is the PSD. The HK method was originally developed for micropores having slit-shaped pores. It has been extended to cylindrical pores by Saito and Foley [63], and to spherical pores by Cheng and Yang [64]. Lastoskie and co-authors have modified the HK method by replacing the original unweighted potential field with a density-weighted integral [65] and by using a more realistic adsorbate-adsorbent potential (10-4-3 Steele potential [66]) and potential parameters [48]. It turned out that the latter modification, with the original unweighted scheme (here the adsorbate-adsorbate interaction was not included), yields a pore filling correlation in good agreement with the exact ones from the density functional theory and molecular simulation [48] (see Fig. 2 for MHK data). Although the HK method improves on the DR method by taking a realistic adsorbate-adsorbent potential and the pore geometry into account, it still makes the assumption that the thermodynamic properties of the confined phase are similar to those of the bulk phase, which is incorrect, as revealed by modern molecular methods based on statistical mechanics.

2.2 *Modern Molecular Methods: Density Functional Theory and Molecular Simulations*

The development of the statistical mechanical methods, including classical density functional theory (recently reviewed by Landers et al. [67]) and molecular simulations, has advanced the understanding of the fluids confined in pores, and leads to a universal approach that, in principle, enables the accurate characterization of

PSD over the complete pore range (micro-, meso- and macro-pores). The molecular methods have been considered superior to the classical methods, and they have been recommended as the standard ones for routine PSD analysis. They are available in commercial software for easy use by the experimentalists.

Classical Density Functional Theory (DFT) The principle behind DFT is that, for an open system, where molecules can freely exchange with a reservoir, the grand potential functional, Ω, can be written as a functional of local fluid density, $\rho(\boldsymbol{r})$, at position **r**:

$$\Omega[\rho(\boldsymbol{r})] = F[\rho(\boldsymbol{r})] - \int d\boldsymbol{r}\rho(\boldsymbol{r})[\mu - V_{ext}(\boldsymbol{r})] \quad (8)$$

where F is the intrinsic Helmholtz free energy functional, μ is the chemical potential of adsorbate molecules, and $V_{ext}(\boldsymbol{r})$ is the spatially-varying external potential exerted by the pore walls. The integration is carried out over the entire system. We can obtain the equilibrium density profile $\rho_{eq}(\boldsymbol{r})$ by minimizing the functional derivative of the grand potential with respect to the fluid density profile, i.e.,

$$\left.\frac{\delta\Omega[\rho(\boldsymbol{r})]}{\delta\rho(\boldsymbol{r})}\right|_{\rho(\boldsymbol{r})=\rho_{eq}(\boldsymbol{r})} = 0 \quad (9)$$

When more than one minimum is present, the density profile that gives the lowest grand potential is the stable branch. Readers are referred to Hansen and McDonald [29] for detailed derivations of the DFT framework. Once the grand potential at equilibrium is known, other thermodynamic properties of the system are readily available, and the adsorbed amount in the pores (i.e., isotherm) can be calculated from the equilibrium density profile. The PSD is calculated by minimizing the following expression:

$$\left[n(P) - \int_{w_{\min}}^{w_{\max}} f(w)K(w,P)dw\right]^2 + \lambda \int_{w_{\min}}^{w_{\max}} \left[f''(w)\right]^2 dw \quad (10)$$

where $n(P)$ is the experimental specific adsorption or desorption isotherm at pressure P, the internal pore width is w, $f(w)$ is the pore size distribution function and $K(w, P)$ is the so-called kernel function which is the isotherm, predicted from either DFT or GCMC simulation at a specific pore size w. The integration in Eq. (10) is from the minimum pore width to the maximum pore width. Simply minimizing the first term in Eq. (10) for $f(w)$, however, is an ill-posed inverse problem, and a regularization term (second term in Eq. (10)) should be added, where λ is the regularization parameter. A stable numerical method for solving for $f(w)$ is available in the SAIEUS program [68, 69]. Seaton, Walton and Quirke [70] were the first to apply DFT to predict the PSD. Their method is based on the local version of DFT, which assumes that the specific free energy at a point **r** depends only on the local density at that point, and so neglects effects of strong density gradients near the pore walls. Nevertheless, the local DFT provides significant improvement over the classical methods for PSD in both the mesopore and micropore range. However, it fails to reproduce the strong oscillations characteristic of the fluid density profile at a solid-fluid interface for small pores. Lastoskie et al. [50], and independently Olivier et al. [71], applied the non-local DFT (NLDFT) to the PSD analysis. Compared to the local version, the non-local version evaluates the excess Helmholtz free energy in the hard sphere term using a non-local density approximation [72, 73], that accounted for strong oscillations in the density profile near the pore walls. The NLDFT leads to quantitatively accurate fluid structure in nanopores compared to the local DFT, as has been confirmed by molecular simulations [74]. The consistency and validity of the NLDFT has further been confirmed by comparing the calculated PSD to experimental X-ray diffraction (XRD) measurements for porous materials with well-defined pore geometry [75, 76] (see Fig. 3). The PSD calculation using NLDFT generally makes the following main assumptions:

1. The pore wall is homogeneous and smooth [50, 77, 78]. For example, the 10-4-3 Steele potential derived from a homogeneous graphite surfaces is usually adopted for modeling a carbon slit-shaped pore [50].
2. Pores in the material have a single geometry and single-level surface heterogeneity.

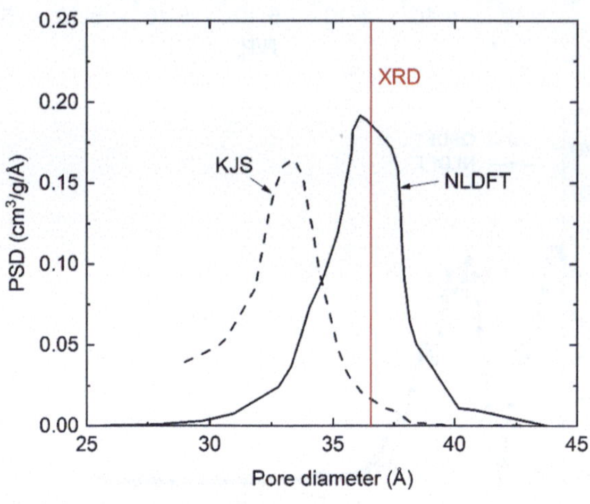

Fig. 3 Pore size distribution of MCM-41 sample AM-1 by Kruk-Jaroniec-Sayari (KJS) [42] method and NLDFT methods. KJS is a modified method based on the BJH method, and it was calibrated for MCM-41 materials. KJS method was found to be quite successful for MCM-41 materials in the pore size range of 2–6.5 nm, but for sample AM-1 here, it underpredicts the pore size by about 3 Å. The calculated PSD by the NLDFT method is in perfect agreement with the pore size from the XRD measurement. Adapted with permission from Ref. [76]. Copyright (2000) American Chemical Society

3. The pore size information can be fully extracted from a single-component adsorption or desorption isotherm.

It has been known for a long time that the first assumption leads to two artifacts: a typical S-shaped deviation from the experimental isotherm, due to layering transitions on the (assumed in the model) smooth surface, and a minimum in the PSD plot at a pore width of about 10 Å (see Fig. 4). In fact, nearly all real surfaces exhibit some degree of defects through geometric curvature, heteroatoms in the surface structure and chemical groups attached to the surface. Efforts have been devoted to eliminating

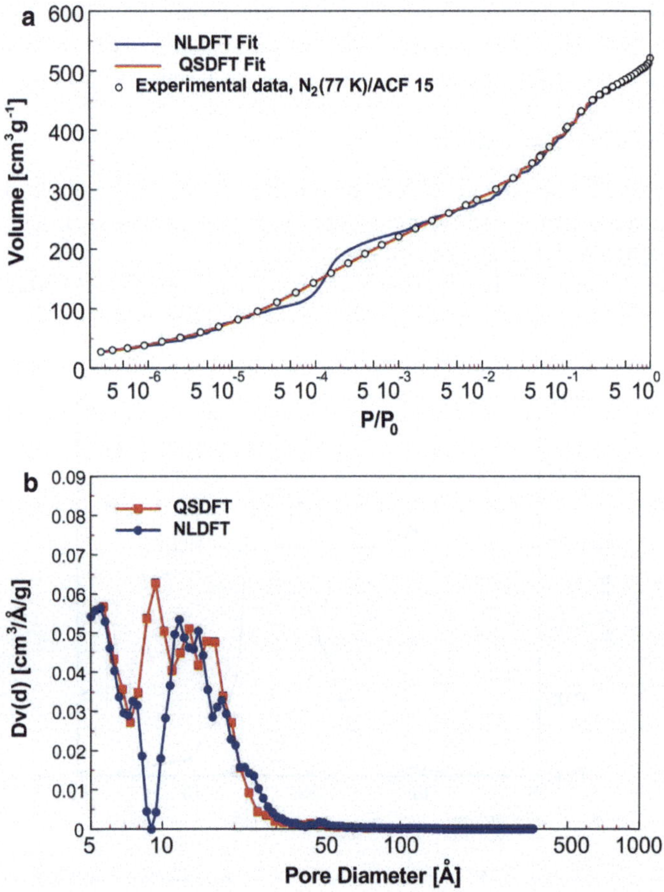

Fig. 4 Comparison of the NLDFT and QSDFT methods for nitrogen adsorption for activated carbon fiber ACF-15. **a** Experimental adsorption isotherm in comparison with the fitted isotherm by NLDFT and QSDFT methods. **b** PSD calculated by the NLDFT and QSDFT methods. The NLDFT method produces artifacts of a S-shaped isotherm and a minimum in the PSD plot at a pore width of about 10 Å. These two artifacts can be attributed to the homogeneous surface model used in the NLDFT method. QSDFT eliminates these two artifacts and presents a better fit to the experimental isotherm data. Adapted from Ref. [82], Copyright (2009), with permission from Elsevier

these two artifacts by introducing heterogeneities to the pore wall potential in the DFT framework. Examples are the introduction of an external potential accounting for variable wall thickness [79], variable surface density [80–82], incorporating pore edge effects [83, 84], and also models that incorporate a periodic function to the fluid-wall potential in a direction parallel to the pore surface [85–89]. Among these modifications, two notable versions are quenched solid density functional theory (QSDFT) and 2D-NLDFT. In QSDFT [81, 82], the geometrically heterogeneous pore wall is represented explicitly by a one-dimensional density profile of carbon atoms controlled by a single roughness parameter, while in 2D-NLDFT, by introducing periodic functions into the pore wall potential, both energetic and geometric heterogeneities of the pore wall have been accounted for either in a separate [85] or combined way [90]. Both methods eliminate the two artifacts of the traditional NLDFT method for activated carbon materials [82, 85] (see Fig. 4 for an example of QSDFT). When characterizing the same carbon materials, the QSDFT and 2D-NLDFT methods give similar PSD results [91]. This might be because both methods adjust the potential parameters for the carbon pore model by fitting the theoretical adsorption isotherm to the experimental data of the reference Cabot BP-280 [92]. For silica adsorbents, QSDFT calculates the parameters from the XRD measurement [93], while 2D-NLDFT fits potential parameters to the data of reference silica LiChrospher-Si-1000 [88]. It would be good to test both methods on the same silica materials and see if they give consistent results. In addition to modifying the pore wall model in the theory, Kupgan et al. [94] showed that, by tuning the regularization parameter, λ, in Eq. (10), the performance of NLDFT methods can also be improved for calculating the PSD of amorphous microporous materials. They proposed a smooth-shift method to obtain the regularization parameter in comparison to the conventional L-curve method [95]. The smooth-shifted PSD agrees well with the exact geometric PSD from molecular simulation. It is not clear, however, if the smooth-shifted method can eliminate the artifact of S-shaped adsorption isotherm from NLDFT.

The second assumption in NLDFT (single pore geometry and single-level surface heterogeneity) is not bad for a general PSD analysis. Allowing for the variation of the pore geometry and surface heterogeneity, however, adds more flexibility and reality to the theoretical model, leading to a better fitting to the experimental isotherm and better representation of the materials. Thommes et al. [96] developed a set of hybrid NLDFT kernels allowing for different pore geometries for hierarchically structured, micro-mesoporous silica materials. The hybrid kernels use a cylindrical pore model for micropores and small mesopores, and a spherical pore model for large mesopores where hysteresis occurs. The authors found that the hybrid approach is able to capture the full PSD in the complete micro- and meso-pore range, and the calculated PSD agreed with the results from independent SANS/SAXS measurement [96]. Gor et al. [97] were able to construct four different sets of hybrid kernels that use different pore geometry within different ranges of pore size, and these hybrid kernels have been successfully applied to characterize the micro-mesoporous carbons [97, 98]. Attempts to construct hybrid kernels allowing for different levels of surface heterogeneity have only been made recently. Lucena et al. [99] proposed a set of hybrid kernels composed of simulated isotherms on surfaces having different levels of etching [100], covering from a homogeneous surface to a highly heterogenous surface. They have shown that employing the hybrid kernels for the PSD calculations reveals new structural details of the carbon porous materials (see Fig. 5). However, the use of the hybrid kernels requires a better understanding of the materials a priori, and so far no rational strategy has been proposed for choosing a suitable collection of kernels for the PSD calculations.

The third assumption (PSD can be fully extracted from a single-component isotherm) is the safest one and has been validated by comparing the calculated PSD with independent experimental measurements [67, 88, 96]. Although the pure adsorption/desorption kernels of N_2 at 77 K or Ar at 87 K are dominant in standard PSD calculations, it becomes problematic when analyzing fine micropores where the diffusion of N_2 is very slow [101], making the experimental measurement of equilibrium adsorption difficult to achieve. Recently, a dual gas analysis of PSD with the 2D-NLDFT method was proposed, where the adsorption data of N_2 & CO_2 [102, 103] or O_2 & H_2 (for minimal interaction of gas quadrupole moment with surface

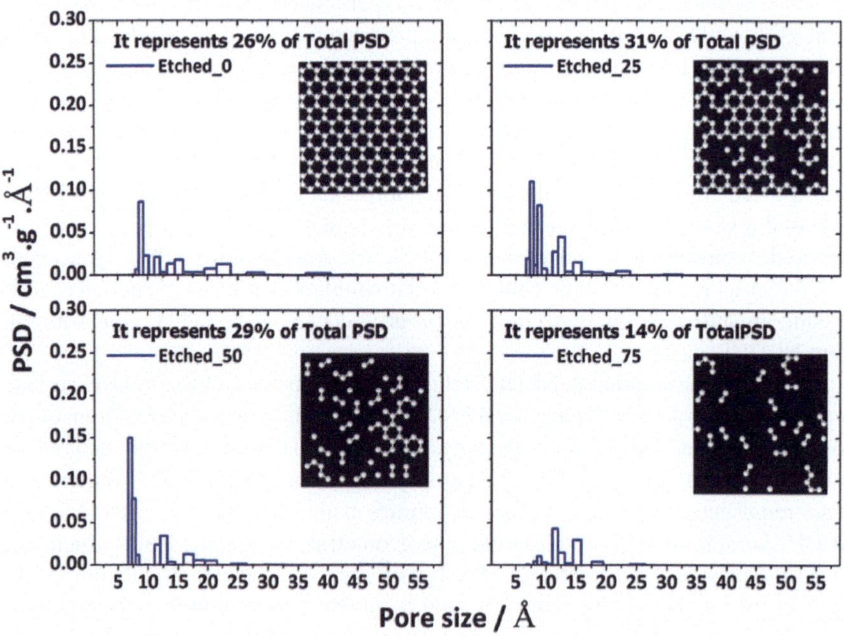

Fig. 5 Pore size distribution (PSD) by the hybrid kernels representing different levels of surface heterogeneity for PC58 sample. 26% of the total pore volume belong to pores having homogeneous surfaces, 31% of the total pore volume belong to pores having surfaces of 25% etching level, 29% of the total pore volume belong to pores having surfaces of 50% etching level, and 14% of the total pore volume belong to pores having surfaces of 75% etching level. Using the hybrid kernels reveal new information about the materials. Reprinted from Ref. [99]. Copyright (2017), with permission from Elsevier

polar sites) [104] were processed simultaneously to get a high-resolution PSD in full micro- and meso-pore range.

Monte Carlo Simulation GCMC simulation [105] is another method to get the kernels for PSD analysis. Unlike DFT, where the equations are numerically evaluated, the confined system in GCMC can be explicitly set up in a 3D simulation box with specified intermolecular force fields (see Fig. 6). The temperature, chemical potential and volume of the simulation box are pre-specified constants. During the simulation, the adsorbate molecules in the simulation box are in free exchange with a "ghost" bulk reservoir whose chemical potential is equal to the pre-set value (see Fig. 6). When enough exchanges (i.e., insertion or deletion move of molecules in the box) have been made, the simulation box is in chemical equilibrium with the "ghost" reservoir. Meanwhile, translational, and rotational (for non-spherical molecules) moves are also attempted randomly for adsorbate molecules in the simulation box to make sure the system eventually reaches the thermal equilibrium (assuming the structure of adsorbent is fixed). The Metropolis scheme is usually employed to determine if an attempted move should be accepted or declined. Once enough moves have been made for the simulation to reach equilibrium, the adsorption amount in the modeled

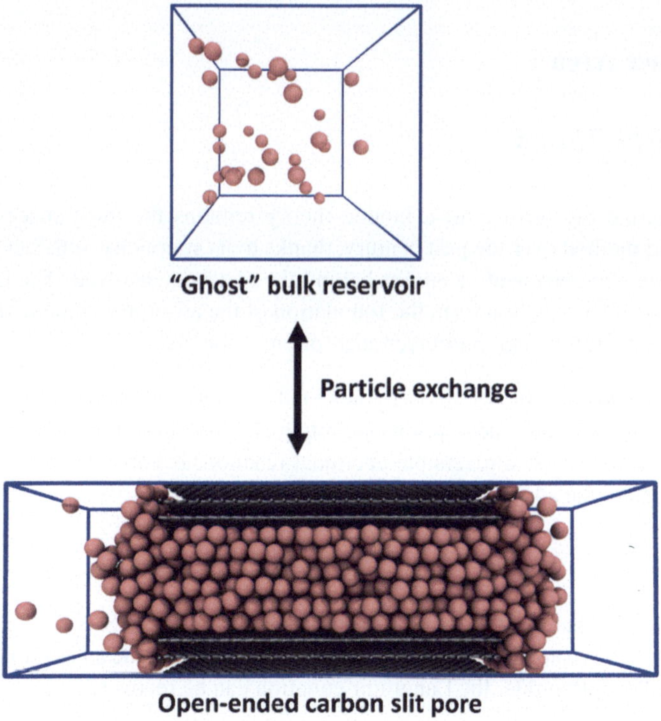

Fig. 6 Schematic of grand canonical Monte Carlo simulation. The argon molecules (pink particle) are in free exchange between the "ghost" bulk reservoir (top) and the simulation box (bottom). When chemical equilibrium is reached, the chemical potentials in both systems are equal

nanopore can then be directly calculated from the average over all equilibrium configurations of the molecules. In contrast to DFT-based methods, where some approximations are made (e.g., mean field approximation), the results from the GCMC simulation are considered to be exact for the model system (model for the intermolecular forces, pore geometry, etc.). In addition, the pore geometry and atomic surface structure can be modeled as realistically as desired in the GCMC simulation [99, 106–108]. However, consistency of the PSDs obtained from GCMC-based kernels and DFT-based kernels can be reached if suitable intermolecular parameters are chosen [109]. As of now, DFT-based kernels are still dominant over the GCMC-based kernels, mainly due to the fast execution of DFT calculations (minutes) compared to GCMC simulations (days). To explore the atomic details of the surface structure, or to access adsorption of complex molecules, this time expense must be paid, although some efforts to address this have been made on the DFT side recently by combining QSDFT and statistical associating fluid theory [110]. With increasing computational power and better understating of the surface structure at the microscopic scale, the GCMC method is expected to become more widely used in constructing the kernels for PSD determination.

3 Surface Area

3.1 Kinetic Theory

To characterize the surface area, kinetic theory remains the most successful and widely used method over the past century, thanks to its simplicity, satisfactory accuracy, and the clear physical meanings behind the concepts involved. The Langmuir equation and BET equation form the foundation of the adsorption kinetic theory for monolayer adsorption and multilayer adsorption, respectively.

Langmuir Equation In 1918, Langmuir [111] proposed the first quantitative description of monolayer adsorption (including chemisorption) by viewing adsorption onto the surface as a reversible chemical reaction. If only one type of surface site is available, the adsorption reaction can be written as:

$$A + S \rightleftarrows \text{AS}$$

where A is adsorbate, S is an unoccupied surface site, and AS is the surface site occupied by adsorbate through chemical forces. By equating the adsorption reaction rate to the desorption rate, the Langmuir equation can be recovered:

$$n = n_m \frac{KP}{1 + KP} \tag{11}$$

where n is the experimental specific adsorbed amount (e.g., in units of mol/g) at the partial pressure of the adsorbate, P; n_m is the monolayer saturation capacity, and K is the adsorption equilibrium constant. To obtain the surface area of the material, Eq. (11) is fitted to the experimental adsorption isotherm. The SSA, a_s, is calculated by

$$a_s = n_m N_A \sigma_m \tag{12}$$

where N_A is Avogadro's number and σ_m is the molecular cross-sectional area. The application of Langmuir's equation should, in principle, be limited to the monolayer adsorption on non-porous materials, but it is often applied to the general adsorption isotherm of Type I (see Fig. 7) which also represents adsorption behavior in microporous materials. Without a deep understanding of the pore topology and pore filling mechanism in microporous materials, a good fit of the Langmuir equation to the experimental isotherm should only be interpreted as the mathematical effectiveness

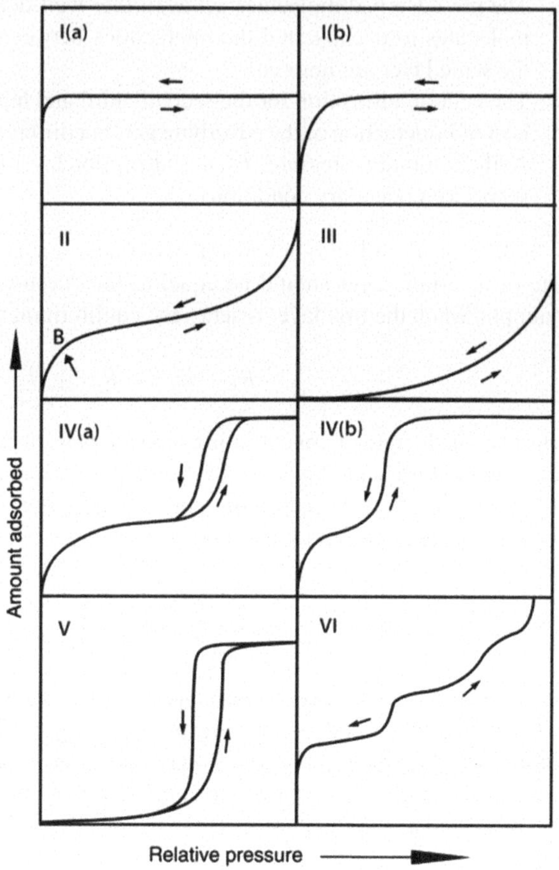

Fig. 7 International Union of Pure and Applied Chemistry (IUPAC) classification of physisorption isotherms. Reprinted with permission from Ref. [26]. Copyright (2015) IUPAC & De Gruyter

of the model, and the surface area obtained usually does not reflect the true surface area. The Langmuir equation assumes the surface is homogeneous. For adsorption on heterogeneous surfaces, several semi-empirical equations have been proposed based on the Langmuir equation and the Freundlich isotherm, such as the Sips equation [112] and the Toth equation [113]. There is no preference for one empirical model over the other, and the choice of a particular equation usually depends on their fitting quality to the experimental isotherm data of Type I shape.

Brunauer-Emmett-Teller (BET) Equation In reality, almost all monolayer adsorption on an open surface will be followed by the stacking of second, third and higher layers as the gas pressure is increased. In 1938, Brunauer, Emmett and Teller [114] extended the Langmuir formulation to a multilayer adsorption scenario, by considering the formation of higher layers on top of the first adsorbed layer as a dynamic process. The BET theory is based on the following assumptions:

1. The surface is atomically homogeneous, and adsorption occurs only on well-defined surface sites (i.e., one adsorbate molecule per surface site).
2. The pre-adsorbed molecules act as further well-defined adsorption sites for gas molecules to occupy, and the interactions between the adsorbate molecules in the same layer are ignored.
3. The heat of adsorption for the second, third and higher layers is the same as the heat of liquefaction of the adsorbing gas, but different from that for the first layer.
4. At the saturation pressure, P_0, the adsorption layer becomes infinitely thick (this serves as a boundary condition).

Following a similar argument as that of Langmuir, at equilibrium, the "formation" rate of a certain layer should be equal to the "destruction" rate of that layer. For example, when the first layer reaches the equilibrium, we have

$$R_1 + R_{-2} = R_{-1} + R_2 \tag{13}$$

where R_1 is the rate of condensation onto the bare surface, R_{-2} is the rate of evaporation from the second layer, R_{-1} is the rate of evaporation from the first layer, and R_2 is the rate of condensation onto the second layer. The rate is directly related to the surface area covered by the layer. For example,

$$\begin{aligned} R_1 &= k_1 P \theta_0 \\ R_{-2} &= k_{-2} \theta_2 \end{aligned} \tag{14}$$

where k_1 and k_{-2} are the rate constants, θ_0 and θ_2 are the empty surface area and the area covered by the second layer, respectively. Figure 8 presents a diagram for BET multilayer adsorption. By writing equations similar to Eq. (13) for all other adsorbed layers and applying the boundary condition (point 4 listed above), the BET equation can be eventually reached as [114]:

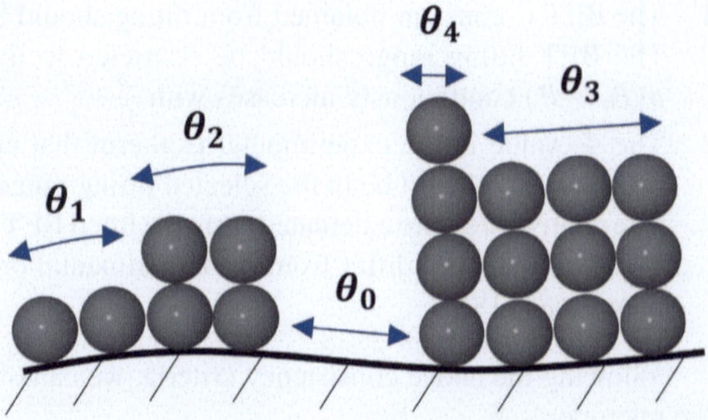

Fig. 8 BET model of multilayer adsorption. The total surface area is the sum of the empty area (θ_0), the area covered by the first adsorbed layer (θ_1), the area covered by the second adsorbed layer (θ_2), and so on

$$\frac{1}{n\left(\frac{P_0}{P}-1\right)} = \frac{C-1}{n_m C} \times \frac{P}{P_0} + \frac{1}{n_m C} \tag{15}$$

where C is the BET constant, which is directly related to the adsorbate-adsorbent interaction strength. Equation (15) can also be rigorously derived from statistical mechanics [115, 116]. By plotting experimental adsorption data as $\frac{P}{[n(P_0-P)]}$ versus relative pressure, $\frac{P}{P_0}$ ("the BET plot"), a linear correlation can usually be found as $\frac{P}{P_0}$ ranges from 0.05 to 0.3 for isotherms of Type II, IV and VI (see Fig. 7). By fitting such linear data with Eq. (15), the specific monolayer capacity, n_m, can then be determined, which leads to the estimation of the specific BET surface area of the materials [using Eq. (12)]. Because argon molecules do not have a quadrupole moment, and are less sensitive to the surface chemistry, IUPAC (2015) [26] recommends using argon at 87 K (its normal boiling point) as the standard adsorption condition for characterization.

3.2 *Current Advances in Characterization of Microporous Materials by the BET Method and Beyond*

BET Method Most of the porous materials that present excellent performance in real applications (such as gas storage and separation) contain micropores. Because of its restrictive assumptions, the Langmuir equation is rarely used in practice. Although the BET equation was also derived from several simplified assumptions (including the assumption that the adsorption is on open surfaces), scientists still found it convenient and useful for characterizing microporous materials. The most common issue associated with BET analysis of microporous materials is that multiple linear regions can be found in the BET plot for fitting, which leads to inconsistency of the calculations. To enhance the reproducibility, Rouquerol et al. [117] proposed four consistency criteria, which were later recommended by IUPAC (2015) [26], to locate a linear region objectively:

1. The BET C constant obtained from fitting should be positive to be meaningful.
2. The BET fitting range should be restricted to the region where the quantity $n(P_0 - P)$ continuously increases with $\frac{P}{P_0}$.
3. The $\frac{P}{P_0}$ value in the experimental isotherm that corresponds to the monolayer capacity, n_m, should be in the selected fitting range.
4. The relative pressure calculated from the fitted BET equation [Eq. (15)] by setting $n = n_m$ should not differ from the experimental one (from criterion 3) by more than 20% [118].

Following the above consistency criteria, we can obtain the SSA that belongs to the micropores, a_{micro}, by

$$a_{micro} = a_{BET} - a_{ext} \tag{16}$$

where a_{BET} is the specific BET surface area; a_{ext} is the specific surface area contributed from external (i.e., non-microporous) surfaces, which can be calculated with the help of the t-method [52, 53] or the α_s-method [119]. Walton and Snurr [120] have shown that these consistency criteria lead to a better BET surface area characterization for six microporous metal organic frameworks (MOFs), with the calculated values being in good agreement with the geometric (accessible) surface area calculated directly from the MOF atomic structure. It has recently been confirmed that, based on the calculation of true monolayer surface area [121] from molecular data for over 200 MOFs, the BET theory along with these consistency criteria can, in general, produce reasonable surface areas for MOFs with a high volumetric percentage of small micropores (pore diameter < 10 Å and surface area < 1500 m^2/g) [122]. Although the consistency criteria help standardize the BET analysis, they do not always lead to the correct identification of the pressure range where monolayer formation occurs, especially for porous materials having a combination of large micropores (10 Å < pore diameter < 20 Å) and mesopores (pore diameter > 20 Å) [121, 122]. The adsorption isotherm of these porous materials shows a characteristic step-wise pattern, and the consistency criteria tend to choose the pressure range where micropore pore filling occurs, thus usually overestimating the surface area. To locate the correct region for monolayer formation, Sinha et al. [122] proposed to use the excess sorption work (ESW) method together with the BET equation (termed as BET+ESW method). The ESW method [123, 124] was designed to determine the monolayer capacity directly from the adsorption isotherm without referring to any adsorption model. By plotting the excess sorption work, $\phi = nRT\ln\left(\frac{P}{P_0}\right)$, versus the adsorption loading, n, the monolayer capacity is expected to be the loading at the first local minimum in the $(\phi - n)$ plot (i.e., ESW plot). Although the ESW method alone usually underestimates the actual surface area [122], it can predict a rough pressure value that corresponds to the monolayer formation. Therefore, in addition to the first and second consistency criteria listed above, a new third criterion was suggested that the BET linear region should be chosen such that it includes the pressure that corresponds to the first minimum in the ESW plot (see Fig. 9 for an

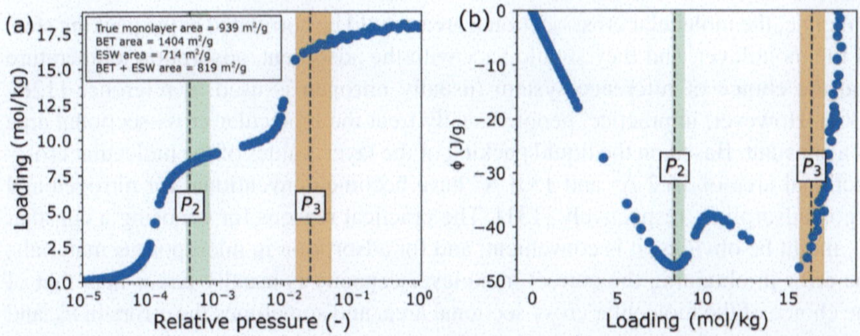

Fig. 9 The BET+ESW method for estimating the surface area of a carbon nanotube with diameter of 16.9 Å. **a** Adsorption isotherm. **b** The ESW plot. The ochre region is the BET fitting range determined from original four consistency criteria, and the green region is the new BET range determined by BET+ESW method. The first local minimum in the ESW plot corresponds to the pressure P_2, and the new BET fitting range is chosen to include this pressure. The BET+ESW method gives a specific surface area of 819 m^2/g, that is closer to the true surface area (939 m^2/g) than other methods. Reprinted with permission from Ref. [122]. Copyright (2019) American Chemical Society

example). The BET+ESW method has been shown to improve the surface area estimation for porous materials that contain large micropores and mesopores, however, it fails when no clear minimum is present in the ESW plot. In some cases, the micropores are so small that the first minimum in the ESW plot corresponds to micropore filling that finishes before the monolayer formation. The result is that the BET+ESW method tends to suggest the wrong BET fitting range, while the original consistency criteria work better. Therefore, the BET+ESW method cannot guarantee the correct estimation of the true monolayer surface area, but can serve as a supplement to the standard BET analysis to prevent overestimation [122]. It should be noticed that, when monolayer formation and pore filling become indistinguishable, even if all consistency criteria are satisfied, the BET related methods are unable to predict the correct surface area. In such cases, the BET plot only exhibits one linear region where the monolayer formation and pore filling occurs simultaneously, and the monolayer capacity determined from the BET method now also includes the non-monolayer amount, thus overestimating the surface area. It has been shown that this kind of pore indistinguishability happens for slit-shaped pores that can accommodate three adsorption layers (pore size of about 10 Å) [125]. Theories that are beyond the BET method, or additional adsorption data, are required to effectively separate the monolayer capacity from the pore filling.

According to Eq. (12), in addition to the monolayer capacity, the other source of error in the BET method is the molecular cross-sectional area, σ_m. As shown by Brunauer and coworkers in their 1937 [126] and 1938 [114] papers, and somewhat later in the monograph by Brunauer [127], the close packing of the adsorbate molecules on the substrate surface results in a surface area per molecule that is 7–19% smaller than expected, based on the density of the bulk liquid for these same gases. In principle, the molecular cross-sectional area should be correlated to the packing state of the monolayer, and they should vary with the adsorbent, adsorption temperature and the choice of reference system (usually nitrogen is used as reference [128–130]). However, in practice, people usually treat the molecular cross-sectional area as a constant. Based on the liquid packing of the layer, values of the molecular cross-sectional area of 16.2 Å^2 and 13.8 Å^2 have become conventional for nitrogen and argon adsorption, respectively [131]. The practical reasons for choosing a constant σ_m might be obvious: it is convenient; and for adsorption in microporous materials, the error in obtaining the correct monolayer capacity is usually larger than that of the choice of the molecular cross-sectional area, and sometimes the errors in n_m and σ_m may compensate each other. Nevertheless, we should still keep in mind that the choice of molecular cross-sectional area also matters in surface area calculations.

Beyond the BET Equation Since the original work of Langmuir [111] and Brunauer et al. [114], no major breakthrough has been made in the field of adsorption theory. This is partly because the BET theory is convenient and simple, and it has already been proved suitable for characterizing both porous and non-porous materials [132]. Most of the real porous materials contain hierarchical pores and pore connectivity, making the development of a general adsorption theory very difficult. Recently, advances in machine learning (ML) techniques [133] have opened up a new route to estimating the SSA from experimental data; in principle, such data do not necessarily have to be the physisorption isotherms. Figure 10 shows a general supervised ML workflow. In the case of predicting the surface area, a set of *features* that can describe the materials is first designed. Features need to be experimentally measurable, representative, and informative. A set of training data are then prepared, and they include representative materials, whose surface area (*label*) has been accurately pre-calculated from either simulation or direct measurement. The ML model (e.g., artificial neural networks

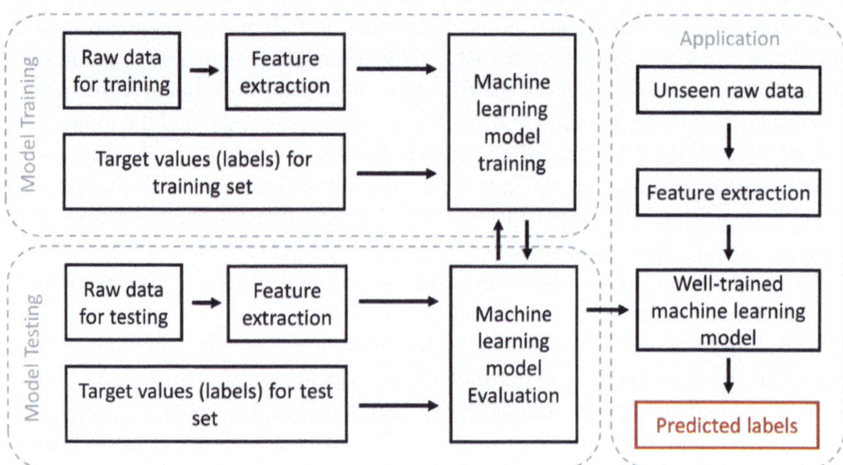

Fig. 10 A general supervised machine learning workflow

[133]) is subsequently trained using the training set with *features* as input and *labels* as the output target. If the ML model is properly trained without overfitting or underfitting, it can then be used to predict the surface area of other unseen materials by simply feeding the *features* of those materials into the well-trained ML model. The advantage of the ML method is clear: it enables prediction of the surface area through this large-scale regression algorithm without knowing the underlying physics of adsorption. This, however, can also be considered a disadvantage, as most ML models operate as "black boxes" that do not provide physical insight on the process. Attempts to use ML methods for surface area predictions have only been made very recently. Datar et al. [134] considered the mean loading in various pressure regions of the isotherm as the *feature*. They trained the least absolute shrinkage and selection operator (LASSO) model [135], using a training set of more than 300 MOFs with diverse structures and pre-calculated true monolayer surface area. They have shown that the ML model outperforms the BET method in the entire surface area region (see Fig. 11). In addition to extracting the materials' structure information from the normal adsorption isotherm, the micro-computed tomography (micro-CT) X-ray images can also serve as the input data [136]. With the power of convolutional neural networks [137], Alqahtani et al. [136] were able to predict the porosity, SSA, and average pore size for each input image. Although the physisorption isotherm still remains one of the most accessible sources of data for many laboratories, the ML methods open up a lot of possibilities to extract information from various sources of data, thus providing a more flexible, general and fast way to predict the properties of porous materials.

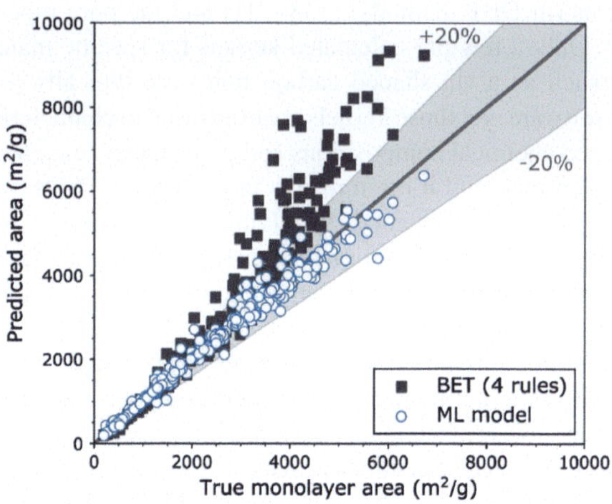

Fig. 11 Comparison between the predicted surface area by the BET method (satisfying the four consistency criteria of Rouquerol et al. [117]) and by the machine learning (ML) model. The ML model offers overall excellent prediction over a wide range of areas in agreement with the true surface area from molecular simulations. Adapted with permission from Ref. [134]. Copyright (2020) American Chemical Society

4 Concluding Remarks and Future Perspective

In this chapter, we have reviewed the characterization of the PSD and the SSA of porous materials by model-based hybrid methods. These methods extract the structural information of the materials from the experimental physisorption isotherms.

For PSD analysis, we have first introduced the classical methods, including those based on the Kelvin equation for mesopore analysis, and those for micropore analysis (MP method, DR/DA method, and HK method). The basics of the modern molecular methods (DFT and GCMC simulation) based on statistical mechanics are also explained. The modern molecular methods are superior to the classical methods in the complete pore size range, and they are recommended for routine PSD analysis.

For SSA analysis, we have focused on the kinetic theory (Langmuir and BET equations), and the application of the BET method for characterization of microporous materials. With the four consistency criteria proposed by Rouquerol et al. [117] and the recently proposed BET+ESW method, the BET equation is able to determine the surface area of the porous materials having micro- and mesopores. The accuracy of the BET surface area, however, should be cautiously examined with the understanding of the pore structure (e.g., PSD, pore geometry).

Some current challenges for the PSD and surface area analysis, and possible future directions are summarized as follows:

1. The current state-of-the-art methods for rapid characterization of the PSD are those based on statistical mechanics, namely DFT or GCMC simulations. However, more experimental evidence is needed to refine the effective solid-fluid potentials (in DFT methods) [138–141] and the pore models (in GCMC simulations). Only a few pre-calculated kernels for specific materials and pore geometries (such as a slit-shaped carbon pore) are typically available in the commercial software, yet those kernels are often used to characterize novel materials with varied chemical compositions and pore shapes (e.g., MOFs), which is in principle incorrect, and it often results in misleading PSDs even if the fit to the experimental isotherm is good. Resolving this discrepancy would require a rational strategy to combine a set of kernels representing the surface chemistry and pore geometry of the materials, with algorithms for fast generation of the custom kernels.
2. Due to the simplified assumptions, it is impossible to apply the BET equation to any materials with hierarchical pore structures. Especially, the BET equation will fail when monolayer formation and pore filling occur simultaneously. Recent developments in ML-based methods provide a new route for fast characterization of surface area. We anticipate that new, general ML models with the ability to predict the surface area of materials over the entire range of pore sizes will be developed in the near future. This involves the development of advanced ML algorithms and design of suitable *feature* set. It should be noted that, however, the ML model is usually trained with the simulation data (i.e., simulated isotherm and geometric surface area). When applying those simulation-data-trained models to

predict surface area in practice, the fidelity of materials' models in simulations should be carefully checked, so that the established correlation between *features* and *labels* can be safely translated from simulated world to the real world.

3. Some materials are not rigid and have a flexible skeleton. The deformation of the structure during the adsorption process [142] can affect the surface area and PSD analysis of the material. Future characterization tools should consider this effect to provide a more complete description of porous materials.

Acknowledgements It is our pleasure to dedicate this paper to Professor F. ('Paco') Rodriguez-Reinoso, in memory of his friendship and many significant contributions to the field of physical adsorption. KS thanks Professor Sebastiao Lucena, Mr. Archit Datar and Mr. Shivam Parashar for helpful discussions.

References

1. Kondrat, S., Pérez, C.R., Presser, V., Gogotsi, Y., Kornyshev, A.A.: Energy Environ. Sci. **5**, 6474 (2012)
2. Sdanghi, G., Canevesi, R.L.S., Celzard, A., Thommes, M., Fierro, V.: J. Carbon Res. **6**, 46 (2020)
3. Sumida, K., Rogow, D.L., Mason, J.A., McDonald, T.M., Bloch, E.D., Herm, Z.R., Bae, T.-H., Long, J.R.: Chem. Rev. **112**, 724 (2012)
4. Bobbitt, N.S., Chen, J., Snurr, R.Q.: J. Phys. Chem. C **120**, 27328 (2016)
5. Suresh Kumar, P., Korving, L., Keesman, K.J., van Loosdrecht, M.C.M., Witkamp, G.-J.: Chem. Eng. J. **358**, 160 (2019)
6. Gelb, L.D., Gubbins, K.E., Radhakrishnan, R., Sliwinska-Bartkowiak, M.: Rep. Prog. Phys. **62**, 1573 (1999)
7. Luo, S., Lutkenhaus, J.L., Nasrabadi, H.: Fluid Phase Equilib. **487**, 8 (2019)
8. Alba-Simionesco, C., Coasne, B., Dosseh, G., Dudziak, G., Gubbins, K.E., Radhakrishnan, R., Sliwinska-Bartkowiak, M.: J. Phys.: Condens. Matter **18**, R15 (2006)
9. Lian, C., Su, H., Li, C., Liu, H., Wu, J.: ACS Nano **13**, 8185 (2019)
10. Worthington, K.S., Wiley, L.A., Mullins, R.F., Tucker, B.A., Nuxoll, E.: J. Biomed. Mater. Res. Part B Appl. Biomater. **104**, 1602 (2016)
11. Li, G., Wang, L., Pan, W., Yang, F., Jiang, W., Wu, X., Kong, X., Dai, K., Hao, Y.: Sci. Rep. **6**, 34072 (2016)
12. Chang, J.S., Vigneswaran, S., Kandasamy, J.K., Tsai, L.J.: Sep. Sci. Technol. **43**, 1771 (2008)
13. Yang, H., Min, X., Xu, S., Bender, J., Wang, Y., Sustain, A.C.S.: Chem. Eng. **8**, 2531 (2020)
14. Santiso, E.E., George, A.M., Turner, C.H., Kostov, M.K., Gubbins, K.E., Buongiorno-Nardelli, M., Sliwinska-Bartkowiak, M.: Appl. Surf. Sci. **252**, 766 (2005)
15. Iwamoto, M., Tanaka, Y., Sawamura, N., Namba, S.: J. Am. Chem. Soc. **125**, 13032 (2003)
16. Santiso, E.E., Buongiorno Nardelli, M., Gubbins, K.E.: J. Chem. Phys. **128**, 034704 (2008)
17. Gubbins, K.E., Liu, Y.-C., Moore, J.D., Palmer, J.C.: Phys. Chem. Chem. Phys. **13**, 58 (2011)
18. Urita, K., Shiga, Y., Fujimori, T., Iiyama, T., Hattori, Y., Kanoh, H., Ohba, T., Tanaka, H., Yudasaka, M., Iijima, S., Moriguchi, I., Okino, F., Endo, M., Kaneko, K.: J. Am. Chem. Soc. **133**, 10344 (2011)
19. Fujimori, T., Morelos-Gómez, A., Zhu, Z., Muramatsu, H., Futamura, R., Urita, K., Terrones, M., Hayashi, T., Endo, M., Young Hong, S., Chul Choi, Y., Tománek, D., Kaneko, K.: Nat. Commun. **4**, 2162 (2013)

20. Vasu, K.S., Prestat, E., Abraham, J., Dix, J., Kashtiban, R.J., Beheshtian, J., Sloan, J., Carbone, P., Neek-Amal, M., Haigh, S.J., Geim, A.K., Nair, R.R.: Nat. Commun. **7**, 12168 (2016)
21. Gubbins, K.E., Gu, K., Huang, L., Long, Y., Mansell, J.M., Santiso, E.E., Shi, K., Śliwińska-Bartkowiak, M., Srivastava, D.: Engineering **4**, 311 (2018)
22. Srivastava, D., Turner, C.H., Santiso, E.E., Gubbins, K.E.: J. Phys. Chem. B **122**, 3604 (2018)
23. Anovitz, L.M., Cole, D.R.: Rev. Mineral. Geochem. **80**, 61 (2015)
24. Santiso, E.E.: Mol. Simul. **40**, 664 (2014)
25. Diao, Y., Myerson, A.S., Hatton, T.A., Trout, B.L.: Langmuir **27**, 5324 (2011)
26. Thommes, M., Kaneko, K., Neimark, A.V., Olivier, J.P., Rodriguez-Reinoso, F., Rouquerol, J., Sing, K.S.W.: Pure Appl. Chem. **87**, 1051 (2015)
27. Lowell, S., Shields, J.E., Thomas, M.A., Thommes, M.: Characterization of Porous Solids and Powders: Surface Area, Pore Size and Density. Springer, Netherlands, Dordrecht (2004)
28. Espinal, L.: Characterization of Materials. Wiley, Hoboken (2012)
29. Hansen, J.-P., McDonald, I.R.: Theory of Simple Liquids, 4th edn. Academic Press, Oxford (2013)
30. Allen, M.P., Tildesley, D.J.: Computer Simulation of Liquids, 2nd edn. Oxford University Press (2017)
31. Thomson, W.: Lon. Edinb. Dublin Philos. Mag. J. Sci. **42**, 448 (1871)
32. Cohan, L.H.: J. Am. Chem. Soc. **60**, 433 (1938)
33. Barrett, E.P., Joyner, L.G., Halenda, P.P.: J. Am. Chem. Soc. **73**, 373 (1951)
34. Halsey, G.: J. Chem. Phys. **16**, 931 (1948)
35. Dollimore, D., Heal, G.R.: J. Appl. Chem. **14**, 109 (1964)
36. Gelb, L.D., Gubbins, K.E.: Langmuir **15**, 305 (1999)
37. Tolman, R.C.: J. Chem. Phys. **17**, 333 (1949)
38. Santiso, E., Firoozabadi, A.: AIChE J. **52**, 311 (2006)
39. Nguyen, C., Do, D.D.: Langmuir **15**, 3608 (1999)
40. Malijevský, A., Parry, A.O., Pospíšil, M.: Phys. Rev. E **96**, 020801 (2017)
41. Malijevský, A., Parry, A.O.: Phys. Rev. Lett. **120**, 135701 (2018)
42. Kruk, M., Jaroniec, M., Sayari, A.: Langmuir **13**, 6267 (1997)
43. Yin, G., Liu, Q., Liu, Z., Wu, W.: Fuel Process. Technol. **174**, 118 (2018)
44. Huang, B., Bartholomew, C.H., Woodfield, B.F.: Microporous Mesoporous Mater. **184**, 112 (2014)
45. Villarroel Rocha, J., Barrera, D., Sapag, K.: Top. Catal. **54**, 121 (2011)
46. Villarroel-Rocha, J., Barrera, D., Sapag, K.: Microporous Mesoporous Mater. **200**, 68 (2014)
47. Liu, J., Li, P., Sun, Z., Lu, Z., Du, Z., Liang, H., Lu, D.: Fuel **210**, 446 (2017)
48. Dombrowski, R.J., Lastoskie, C.M., Hyduke, D.R.: Colloids Surf., A: Physicochem. Eng. Asp. **187–188**, 23 (2001)
49. Lastoskie, C.M., Quirke, N., Gubbins, K.E.: Stud. Surf. Sci. Catal., 745–775 (1997)
50. Lastoskie, C., Gubbins, K.E., Quirke, N.: J. Phys. Chem. **97**, 4786 (1993)
51. Mikhail, R.S., Brunauer, S., Bodor, E.: J. Colloid Interface Sci. **26**, 45 (1968)
52. Lippens, B.C., de Boer, J.H.: J. Catal. **4**, 319 (1965)
53. de Boer, J.H., Lippens, B.C., Linsen, B.G., Broekhoff, J.C.P., van den Heuvel, A., Osinga, T.J.: J. Colloid Interface Sci. **21**, 405 (1966)
54. Polanyi, M.: Verh. Dtsch. Phys. Ges. **16**, 1012 (1914)
55. Dubinin, M., Stoeckli, H.: J. Colloid Interface Sci. **75**, 34 (1980)
56. Dubinin, M.M., Astakhov, V.A.: Adv. Chem., 69–85 (1971)
57. Gil, A., Grange, P.: Colloids Surf., A Physicochem. Eng. Asp. **113**, 39 (1996)
58. Stoeckli, H.: J. Colloid Interface Sci. **59**, 184 (1977)
59. Pinto, M.L., Mestre, A.S., Carvalho, A.P., Pires, J.: Ind. Eng. Chem. Res. **49**, 4726 (2010)
60. Sun, J.: Carbon N. Y. **40**, 1051 (2002)
61. Horváth, G., Kawazoe, K.: J. Chem. Eng. Japan **16**, 470 (1983)
62. Valladares, D.L., Rodríguez Reinoso, F., Zgrablich, G.: Carbon N. Y. **36**, 1491 (1998)
63. Saito, A., Foley, H.C.: AIChE J. **37**, 429 (1991)
64. Cheng, L.S., Yang, R.T.: Chem. Eng. Sci. **49**, 2599 (1994)

65. Lastoskie, C.M.: Stud. Surf. Sci. Catal., 475–484 (2000)
66. Steele, W.A.: Surf. Sci. **36**, 317 (1973)
67. Landers, J., Gor, G.Y., Neimark, A.V.: Colloids Surf., A Physicochem. Eng. Asp. **437**, 3 (2013)
68. Jagiello, J.: Langmuir **10**, 2778 (1994)
69. Micromeritics, http://www.nldft.com/download/ (n.d.)
70. Seaton, N.A., Walton, J.P.R.B., Quirke, N.: Carbon N. Y. **27**, 853 (1989)
71. Olivier, J.P., Conklin, W.B., Szombathely, M. v.: Stud. Surf. Sci. Catal., 81–89 (1994)
72. Tarazona, P.: Phys. Rev. A **31**, 2672 (1985)
73. Tarazona, P., Marconi, U.M.B., Evans, R.: Mol. Phys. **60**, 573 (1987)
74. Lastoskie, C., Gubbins, K.E., Quirke, N.: Langmuir **9**, 2693 (1993)
75. Neimark, A.V., Ravikovitch, P.I., Grün, M., Schüth, F., Unger, K.K.: J. Colloid Interface Sci. **207**, 159 (1998)
76. Sonwane, C.G., Bhatia, S.K.: J. Phys. Chem. B **104**, 9099 (2000)
77. Ravikovitch, P.I., Haller, G.L., Neimark, A.V.: Adv. Colloid Interface Sci. **76–77**, 203 (1998)
78. Ravikovitch, P.I., Neimark, A.V.: Langmuir **18**, 1550 (2002)
79. Nguyen, T.X., Bhatia, S.K.: J. Phys. Chem. B **108**, 14032 (2004)
80. Ustinov, E.A., Do, D.D., Fenelonov, V.B.: Carbon N. Y. **44**, 653 (2006)
81. Ravikovitch, P.I., Neimark, A.V.: Langmuir **22**, 11171 (2006)
82. Neimark, A.V., Lin, Y., Ravikovitch, P.I., Thommes, M.: Carbon N. Y. **47**, 1617 (2009)
83. Jagiello, J., Olivier, J.P.: J. Phys. Chem. C **113**, 19382 (2009)
84. Jagiello, J., Kenvin, J., Olivier, J.P., Lupini, A.R., Contescu, C.I.: Adsorpt. Sci. Technol. **29**, 769 (2011)
85. Jagiello, J., Olivier, J.P.: Carbon N. Y. **55**, 70 (2013)
86. Röcken, P., Somoza, A., Tarazona, P., Findenegg, G.: J. Chem. Phys. **108**, 8689 (1998)
87. Huerta, A., Pizio, O., Bryk, P., Sokolowski, S.: Mol. Phys. **98**, 1859 (2000)
88. Jagiello, J., Jaroniec, M.: J. Colloid Interface Sci. **532**, 588 (2018)
89. Jagiello, J., Kyotani, T., Nishihara, H.: Carbon N. Y. **169**, 205 (2020)
90. Jagiello, J., Olivier, J.P.: Adsorption **19**, 777 (2013)
91. Puziy, A.M., Poddubnaya, O.I., Gawdzik, B., Sobiesiak, M.: Adsorption **22**, 459 (2016)
92. Kruk, M., Jaroniec, M., Gadkaree, K.P.: J. Colloid Interface Sci. **192**, 250 (1997)
93. Miyasaka, K., Neimark, A.V., Terasaki, O.: J. Phys. Chem. C **113**, 791 (2009)
94. Kupgan, G., Liyana-Arachchi, T.P., Colina, C.M.: Langmuir **33**, 11138 (2017)
95. Hansen, P.C., O'Leary, D.P.: SIAM J. Sci. Comput. **14**, 1487 (1993)
96. Thommes, M., Smarsly, B., Groenewolt, M., Ravikovitch, P.I., Neimark, A.V.: Langmuir **22**, 756 (2006)
97. Gor, G.Y., Thommes, M., Cychosz, K.A., Neimark, A.V.: Carbon N. Y. **50**, 1583 (2012)
98. Cychosz, K.A., Guo, X., Fan, W., Cimino, R., Gor, G.Y., Tsapatsis, M., Neimark, A.V., Thommes, M.: Langmuir **28**, 12647 (2012)
99. Lucena, S.M., Oliveira, J.C., Gonçalves, D.V., Silvino, P.F.: Carbon N. Y. **119**, 378 (2017)
100. Lucena, S.M., Paiva, C.A.S., Silvino, P.F., Azevedo, D.C., Cavalcante, C.L.: Carbon N. Y. **48**, 2554 (2010)
101. Rodríguez-Reinoso, F., López-González, J.d.D., Berenguer, C.: Carbon N. Y. **22**, 13 (1984)
102. Jagiello, J., Ania, C., Parra, J.B., Cook, C.: Carbon N. Y. **91**, 330 (2015)
103. Jagiello, J., Kenvin, J., Celzard, A., Fierro, V.: Carbon N. Y. **144**, 206 (2019)
104. Jagiello, J., Kenvin, J., Ania, C.O., Parra, J.B., Celzard, A., Fierro, V.: Carbon N. Y. **160**, 164 (2020)
105. Frenkel, D., Smit, B.: Understanding Molecular Simulation: From Algorithms to Applications, 2nd edn. Academic Press, San Diego (2002)
106. Kowalczyk, P., Gauden, P.A., Furmaniak, S., Terzyk, A.P., Wiśniewski, M., Ilnicka, A., Łukaszewicz, J., Burian, A., Włoch, J., Neimark, A.V.: Carbon N. Y. **111**, 358 (2017)
107. Alexandre de Oliveira, J., Soares Maia, D., Toso, J., Lopez, R., Azevedo, D., Zgrablich, G.: Colloids Surf., A Physicochem. Eng. Asp. **437**, 69 (2013)
108. de Oliveira, J.C.A., López, R.H., Toso, J.P., Lucena, S.M.P., Cavalcante, C.L., Azevedo, D.C.S., Zgrablich, G.: Adsorption **17**, 845 (2011)

109. Ravikovitch, P.I., Vishnyakov, A., Russo, R., Neimark, A.V.: Langmuir **16**, 2311 (2000)
110. Sermoud, V.M., Barbosa, G.D., Barreto, A.G., Tavares, F.W.: Fluid Phase Equilib. **521**, 112700 (2020)
111. Langmuir, I.: J. Am. Chem. Soc. **40**, 1361 (1918)
112. Sips, R.: J. Chem. Phys. **16**, 490 (1948)
113. Do, D.D.: Adsorption Analysis: Equilibria and Kinetics. Published by Imperial College Press and Distributed by World Scientific Publishing Co. (1998)
114. Brunauer, S., Emmett, P.H., Teller, E.: J. Am. Chem. Soc. **60**, 309 (1938)
115. Hill, T.L.: J. Chem. Phys. **14**, 263 (1946)
116. Keii, T.: J. Chem. Phys. **22**, 1617 (1954)
117. Rouquerol, J., Llewellyn, P., Rouquerol, F.: Stud. Surf. Sci. Catal., 49–56 (2007)
118. Llewellyn, P., Maurin, G., Rouquerol, J.: Adsorption by Powders Porous Solids, pp. 565–610. Elsevier (2014)
119. Kaneko, K., Ishii, C., Ruike, M., Kuwabara, H.: Carbon N. Y. **30**, 1075 (1992)
120. Walton, K.S., Snurr, R.Q.: J. Am. Chem. Soc. **129**, 8552 (2007)
121. Gómez-Gualdrón, D.A., Moghadam, P.Z., Hupp, J.T., Farha, O.K., Snurr, R.Q.: J. Am. Chem. Soc. **138**, 215 (2016)
122. Sinha, P., Datar, A., Jeong, C., Deng, X., Chung, Y.G., Lin, L.-C.: J. Phys. Chem. C **123**, 20195 (2019)
123. Adolphs, J., Setzer, M.J.: J. Colloid Interface Sci. **180**, 70 (1996)
124. Adolphs, J.: Appl. Surf. Sci. **253**, 5645 (2007)
125. de Lange, M.F., Lin, L.-C., Gascon, J., Vlugt, T.J.H., Kapteijn, F.: Langmuir **32**, 12664 (2016)
126. Brunauer, S., Emmett, P.H.: J. Am. Chem. Soc. **59**, 2682 (1937)
127. Brunauer, S.: The Adsorption of Gases and Vapours, vol. 1. Physical Adsorption. Oxford University Press (1943)
128. Rouquerol, J., Partyka, S., Rouquerol, F.: J. Chem. Soc. Faraday Trans. 1 Phys. Chem. Condens. Phases **73**, 306 (1977)
129. Gardner, L., Kruk, M., Jaroniec, M.: J. Phys. Chem. B **105**, 12516 (2001)
130. Lowell, S., Shields, J., Charalambous, G., Manzione, J.: J. Colloid Interface Sci. **86**, 191 (1982)
131. McClellan, A., Harnsberger, H.: J. Colloid Interface Sci. **23**, 577 (1967)
132. Ambroz, F., Macdonald, T.J., Martis, V., Parkin, I.P.: Small Methods **2**, 1800173 (2018)
133. Géron, A.: Hands-on Machine Learning with Scikit-Learn, Keras, and TensorFlow: Concepts, Tools, and Techniques to Build Intelligent Systems, 2nd Editio. O'Reilly Media, Inc. (2019)
134. Datar, A., Chung, Y.G., Lin, L.-C.: J. Phys. Chem. Lett. **11**, 5412 (2020)
135. Tibshirani, R., Stat, J.R.: Soc. Ser. B **58**, 267 (1996)
136. Alqahtani, N., Alzubaidi, F., Armstrong, R.T., Swietojanski, P., Mostaghimi, P.: J. Pet. Sci. Eng. **184**, 106514 (2020)
137. Krizhevsky, A., Sutskever, I., Hinton, G.E.: Advances in Neural Information Processing Systems, vol. 25, pp. 1097–1105 (Pereira, F., Burges, C.J.C., Bottou, L., Weinberger, K.Q., eds.). Curran Associates, Inc. (2012)
138. Shi, K., Santiso, E.E., Gubbins, K.E.: Langmuir **35**, 5975 (2019)
139. Shi, K., Santiso, E.E., Gubbins, K.E.: Langmuir **36**, 1822 (2020)
140. Forte, E., Haslam, A.J., Jackson, G., Müller, E.A.: Phys. Chem. Chem. Phys. **16**, 19165 (2014)
141. Ravipati, S., Galindo, A., Jackson, G., Haslam, A.J.: Phys. Chem. Chem. Phys. **21**, 25558 (2019)
142. Balzer, C., Waag, A.M., Gehret, S., Reichenauer, G., Putz, F., Hüsing, N., Paris, O., Bernstein, N., Gor, G.Y., Neimark, A.V.: Langmuir **33**, 5592 (2017)

Organoclays. Fundamentals and Applications for Removing Toxic Pollutants from Water Solution

Roberto Leyva-Ramos, Araceli Jacobo-Azuara, and Jesus Ivan Martínez-Costa

Abstract Organoclay is a novel adsorbent synthesized by adsorbing a surfactant on the natural clay surface. The surfactant adsorbed modified the nature of the surface of the clay. The surfactant can be adsorbed on the external surface of the clay and within the interlayer spacing of the clay. The arrangement of surfactant molecules adsorbed on the external and interlayer spacing has been designated varying from monolayer to bilayer structures. The adsorption capacity of the organoclay is highly dependent upon the properties of the precursor clay, the surfactant loading of the organoclay, and the physicochemical properties of the surfactant, such as the size. Besides, the physicochemical properties of the pollutant considerably affect the adsorption capacity of the organoclays. The organoclays have successfully been applied for removing organic compounds, metallic cations and anions, inorganic anions and pharmaceutical compounds from water solution. The adsorption capacity of the organoclay can be influenced by the operating conditions such as temperature and solution pH. The degree and tendency of the variation depend on the interactions between the organoclay and the contaminant in the solution.

1 Introduction

The pollution of water resources is a serious environmental problem that directly and indirectly affects the various ecosystems and human health. The contamination sources of aqueous systems can be natural and anthropogenic. The former are

R. Leyva-Ramos (✉) · J. I. Martínez-Costa
Centro de Investigación y Estudios de Posgrado, FCQ, Universidad Autonoma de San Luis Potosi, Av. Dr. M. Nava No 6, Zona Universitaria, 78210 San Luis Potosi, SLP, Mexico
e-mail: rlr@uaslp.mx

J. I. Martínez-Costa
e-mail: ji_mcosta@yahoo.com.mx

A. Jacobo-Azuara
División de Ciencias Naturales y Exactas, Departamento de Química, Universidad de Guanajuato, Col. Noria Alta S/N, 36050 Guanajuato, GTO, Mexico
e-mail: aazuara@ugto.mx

J. C. Moreno-Piraján et al. (eds.), *Porous Materials*, Engineering Materials,
https://doi.org/10.1007/978-3-030-65991-2_13

natural processes such as volcanic eruptions, forest fires, hurricanes and tornadoes. The primary sources of pollution are of anthropogenic origin, including pollution from residual discharges of human activities such as agriculture, livestock, metal-mechanical and metal finishing industries, and mining activities, among others [1].

The increase in industrial and domestic activities due to population growth has caused the overexploitation of aquatic resources and the contamination of surface and underground water supply sources. Therefore, the levels of various organic and inorganic toxic pollutants have been continually increased in aqueous effluents, causing a severe environmental problem since many of these compounds are toxic, carcinogenic, and mutagenic and can cause adverse effects on human beings, flora and fauna.

Many of the pollutants present in water are organic molecules such as pharmaceuticals, herbicides, solvents, fire retardants and plasticizers [2]. These compounds can remain in surface and groundwater for long periods as they have complex molecular structures that are difficult for microorganisms to biodegrade.

Currently, there is a growing concern in the detection and removal of Emerging Pollutants (EPs) from water resources [3]. The EPs are organic substances released into the aquatic environment, although no regulations are presently established [4]. The pharmaceutical and personal care products (PPCPs) are a class of CEs present in water due to excessive use and inappropriate disposal. In general, PPCPs refer to a large number of organic substances widely consumed by society and include the following compounds: (i) pharmaceutical products of very diverse nature prescribed for the treatment and prevention of diseases in humans and animals; (ii) cosmetics (creams, perfumes, make-up); (iii) household products (degreasers, window cleaners, detergents) [5, 6].

The occurrence of pharmaceuticals in water resources is attributed to effluents from wastewater treatment plants, municipal wastewaters, water discharges from hospitals, and inadequate disposal from drug manufacturers [7–10]. The pharmaceuticals most frequently found in effluents from wastewater treatment plants are antibiotics, antacids, antidepressants, analgesics, antipyretics, steroids, anti-inflammatories, β-blockers, tranquilizers and stimulants [11–13]. Recently, 45 out of 477 emerging pollutants analyzed in European surface waters compounds presented a potential risk to the environment [14].

Among the most toxic inorganic water pollutants are heavy metals [Cd(II), Pb(II), Cr(III) and Cr(VI)] and some anions of As(V), As(III), and fluoride (F-) [15]. On the other hand, the removal of these inorganic pollutants from aqueous solutions is critical because their presence, even at low concentrations, poses a serious risk to the environment and human health.

Therefore, efficient treatment processes are necessary for the removal of emerging and inorganic pollutants. In recent years, adsorption has been applied successfully for removing toxic contaminants, both organic and inorganic, present in aqueous solution. Usually, adsorption is an efficient and economical method to remove pollutants of different kinds present in aqueous solutions. Adsorption has several advantages, such as the great variety of adsorbents available to remove contaminants and the operation easiness of the process.

2 Natural Clays

Natural clays are composed mainly of hydrated aluminum silicates, so that they are composed mainly of silicon, aluminum, oxygen and hydrogen [16]. Many of the clay properties depend primarily on their structure. Clays are characterized by having a structure based upon the stacking of tetrahedral (T) and octahedral (O) sheets. Tetrahedral sheets are formed by linking the tetrahedral groups of $(SiO_4)^{4-}$ and $(AlO_4)^{5-}$, which have a silicon or aluminum atom surrounded by four oxygens [17, 18]. The $(SiO_4)^{4-}$ and $(AlO_4)^{5-}$ are bound through the vertices sharing oxygen, forming complete sheets. Octahedral groups have the composition $[AlO_n\ (OH)_m]$ and $[MgO_n\ (OH)_m]^-$ and comply with the relationship $n + m = 6$. The octahedral groups are joined by sharing oxygens and hydroxyls located on the same edge of the octahedron with the oxygen and hydroxyl of the edge of another octahedron [19] and in this way, the octahedral sheets are constituted. The structure of clays is based on the ordered stacking of layers formed by tetrahedra and octahedra, joined together through oxygen and hydroxyls [17, 18].

One of the main characteristics of clays is the excess of negative charge caused by isomorphic substitutions. The silicon, Si^{4+}, present in the center of the tetrahedrons, is replaced by Al^{3+} or Fe^{3+} atoms, and the Mg^{2+} or Fe^{2+} can replace the aluminum, Al^{3+} located in the center of the octahedron. The excess negative charge is commonly balanced by the exchangeable cations Na^+, K^+, Ca^{2+}, and Mg^{2+}. The presence of these cations makes the clays hydrophilic because water molecules are bound to these exchangeable cations. Clays are known as phyllosilicates because of their layered structure. This structure can be formed by two sheets: tetrahedral and octahedral called bilamellar or T:O; or by three sheets, one octahedral and two tetrahedral, called trilamellar or T:O:T [17, 18].

Some of the most important physicochemical properties of clays are surface area, cation exchange capacity, adsorption capacity, hydration, swelling, plasticity and thixotropy. The natural clays have been extensively applied for the adsorption of cationic metals, and their adsorption capacity depends on the nature of the cationic metal, type of natural clay and operating conditions. Some of the natural clays commonly used are bentonite, sepiolite and vermiculite.

The cation exchange capacity of natural clays can be ascertained by comparing the adsorption isotherms of Cd(II) on bentonite (Bent), vermiculite (Verm) and sepiolite (Sepio), as presented in Fig. 1 [20]. At an equilibrium concentration of Cd(II) of 300 mg/L, the mass of Cd(II) adsorbed on Verm, Bent and Sepio were 32.25, 30.12 and 23.06 mg/g, respectively. Therefore, the adsorption capacity of the natural clays towards Cd(II) decreased in the following order: Verm > Bent > Sep. Natural clays are cation exchangers, so their adsorption capacity can be attributed mainly to the cation exchange capacity (CEC). The CEC of Verm, Bent and Sepio were determined

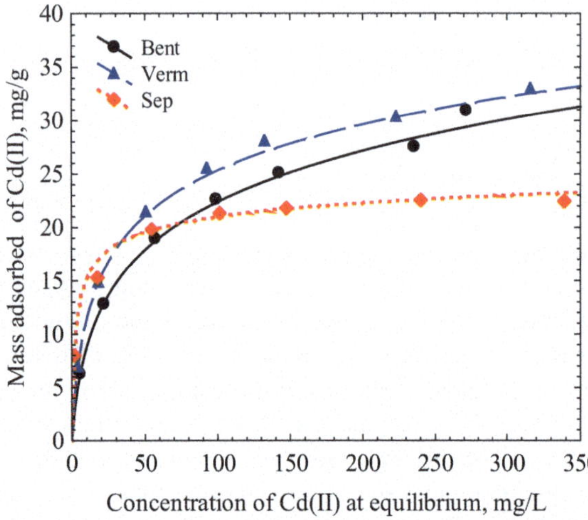

Fig. 1 Adsorption isotherms of Cd(II) on bentonite, vermiculite and sepiolite at pH = 7, T = 25 °C and I = 0.01 N. The lines were predicted with Prausnitz-Radke isotherm model [20]

to be 40.2, 76.6, 42.8 meq/100 g [21, 22], respectively. As expected, the adsorption capacity of Bent and Sepio decreased in the same order as the CEC. However, Verm presented the highest adsorption capacity but the lowest CEC. This trend can be attributed to the fact that adsorption of Cd(II) on natural clays can be attributed to the cation exchange mechanism, electrostatic attraction between the negatively charged surface and the cation, and the accessibility to the cationic sites.

3 Modification of Natural Clays

The capacity of natural clays for adsorbing non-ionic organic compounds and metal anions present in aqueous solution is very low because their surface is hydrophilic, and their surface charge is negative [23]. Modifying the character of the surface of natural clays is an excellent alternative to increase the adsorption capacity and thus favor the adsorption of certain compounds present in an aqueous solution that have little affinity for natural clays.

The most commonly used methods to modify natural clays are chemical and thermal activation, pillaring and adsorption of cationic surfactants. The first two have been widely used to obtain clays for specific applications since it is possible to improve the acidity level and, at the same time, modify the textural and surface properties. Clay pillaring is a process that uses swelling and cationic exchange properties to replace the exchangeable cations present in the clay structure with aluminum polyoxocations, which act as pillars to expand the interlaminar space and keep its sheets separate and therefore, generating a micro-mesoporous structure [24]. Finally, in recent years cationic surfactants have been used to modify the surface characteristics of clays. The surfactant molecule adsorbs on the cationic or negatively charged sites of natural clay, and the surface nature of the clay changes from cationic to anionic due to the quaternary ammonium group of the surfactant, providing a positive charge to the clay surface, and on the other, the surface nature changes from hydrophilic to organophilic and hydrophobic [25]. Hence, the organoclay has a greater capacity for adsorbing non-ionic organic compounds and metallic anions in an aqueous solution.

The cationic surfactants used to synthesize organoclays are typically composed of a quaternary ammonium and an organophilic group or chain and have the chemical formula of $(CH_3)_3NR^+$, where R can be an alkyl chain or benzyl group. The alkyl is generally composed of 1 to 18 carbons [26–28]. Dodecyltrimethylammmonium bromide (DDTMA), Hexadecyltrimethylammmonium bromide (HDTMA), Octadecyltrimethylammonium bromide (ODTMA) and Benzyltrimethylammmonium bromide (BTMA) have been commonly employed to prepared organoclays [28–31]. In these cationic surfactants, the polar or hydrophobic group is the quaternary ammonium, and the nonpolar or organophilic group is the hexadecyl, dodecyl and benzyl groups.

4 Adsorption of Cationic Surfactants on Natural Clays

The adsorption of the surfactant on the natural clay surface is an essential step in preparing an organoclay. The adsorption of cationic surfactants on natural clays has been studied in various works, and the predominant mechanisms are cation exchange and hydrophobic bonding [32]. The adsorption capacity of a clay depends on the characteristics of the surfactant and natural clay, as well as the amount or loading of surfactant adsorbed on the organoclay [31].

The adsorption equilibrium data of DDTMA and BTMA on vermiculite are presented in Fig. 2, and the organoclays are designated as OVDDTMA and OVBTMA, correspondingly. The shape of both isotherms is known as the bimodal

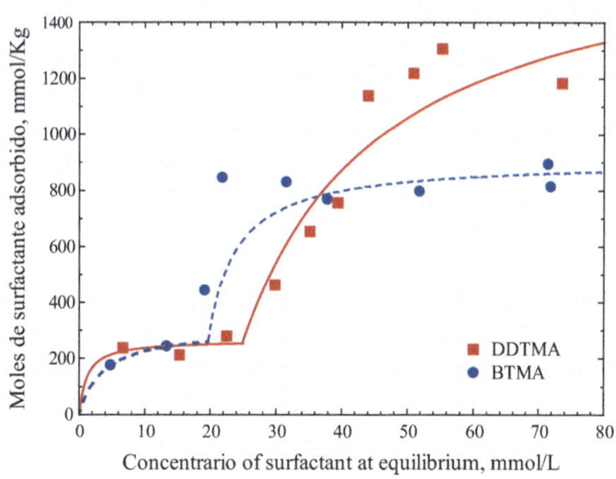

Fig. 2 Adsorption isotherm of DDTMA and BTMA on vermiculite at pH = 7 and T = 25 °C. The lines are the predictions of the bimodal Langmuir isotherm [34]

isotherm, and both isotherms displayed two shoulders. At concentrations of DDTMA below 10 mM, a monolayer of DDTMA was adsorbed on the surface of vermiculite. This behavior revealed that the surfactant molecules formed a monolayer on the surface of natural clay at low concentrations of surfactant at equilibrium. A bilayer of surfactant is formed at high concentrations of DDTMA at equilibrium and above the critical micelle concentration of DDTMA, which is 14 mM [33]. Thus, the surfactant formed a micelle in solution and adsorbed as micelle on the surface of the clay.

The bimodal behavior was also observed for the adsorption of HDTMA, DDTMA and BTMA on bentonite and HDTMA on vermiculite [31, 34] and myristylpyridinium bromide on bentonite [35]. The adsorption equilibrium data of the surfactant on natural clays can be interpreted using the modified bimodal Langmuir isotherm [36], which can be expressed by the following equations:

$$q_e = \frac{q_{m1} K_{L1} C_e}{1 + K_{L1} C_e} \quad C_e \leq C_{e1} \tag{1}$$

$$q_e = \frac{q_{m1} K_{L1} C_e}{1 + K_{L1} C_e} + \frac{q_{m2} K_{L2} (C_e - C_{e1})}{1 + K_{L2} (C_e - C_{e1})} \quad C_e > C_{e1} \tag{2}$$

where C_{e1} is the concentration of surfactant at the end of the first layer, mmol/L; C_e is the surfactant concentration at equilibrium, mmol/L; K_{L1} and K_{L2} are the Langmuir isotherm parameter for the first and second layers, respectively, L/mmol; q_e is the uptake of surfactant adsorbed on the clay, mmol/kg; q_{m1} is the maximum uptake of surfactant adsorbed for the first layer, mmol/kg; and q_{m2} is the maximum uptake of surfactant adsorbed for the second layer, mmol/kg. The bimodal model parameters for the adsorption of DDTMA and BTMA on vermiculite are given in Table 1.

The adsorption of the surfactant in the first layer took place on the cationic sites of the natural clay and can be attributed to ion exchange and electrostatic attraction. On the other hand, in the second layer, the surfactant is adsorbed by interactions between the surfactant molecules already adsorbed and the surfactant molecules in the solution. The van der Waals interactions occurred between alkyl chains of DDTMA molecules.

In general, a typical isotherm of a surfactant on clays can be subdivided into four regions by plotting on a log-log scale [37]. Each region is identified by the change in the slope of the isotherm. These four regions can be seen in the isotherm of HDTMA on vermiculite, as depicted in Fig. 3, and the three concentrations limiting these regions are designated as C1, C2 and C3.

Table 1 Bimodal isotherm parameters for the adsorption on DDTMA and BTMA on vermiculite at pH = 7 and T = 25 °C [34]

Surfactant	C_{e1} (mmol/L)	K_{L1} (L/mmol)	K_{L2} (L/mmol)	q_{m1} (mmol/kg)	q_{m2} (mmol/kg)
DDTMA	25.0	0.888	0.046	266.4	1484.7
BTMA	20.0	0.277	0.222	309.4	600.5

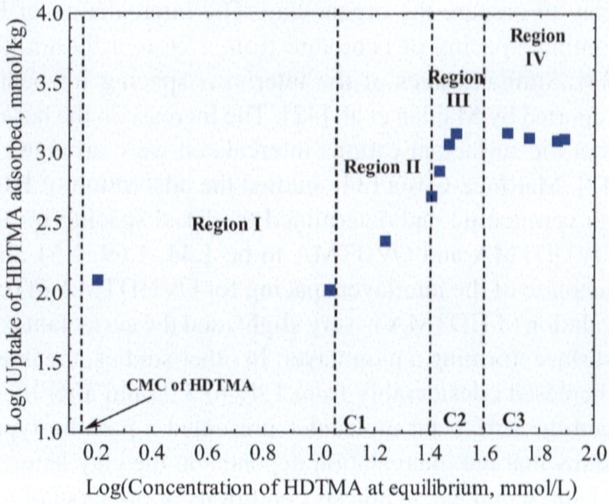

Fig. 3 Schematic illustration of the different regions in the adsorption of the HDTMA on vermiculite [34]

In region I, the shape of the surfactant adsorption isotherm varies depending on the type of exchangeable cations initially present in the clay. At very low concentrations, the cationic surfactant monomers are electrostatically adsorbed and randomly distributed on the surface, and nonpolar side interactions do not occur between the alkyl chains of the adsorbed surfactant molecules. In this region, the surfactant adsorption mechanism is mainly by cation exchange [32] and exhibits an asymptotic behavior.

For concentrations above C1, the surfactant begins to be adsorbed by hydrophobic bonding between the alkyl chains of the surfactant. The mass of surfactant adsorbed increased linearly in region II, as shown in Fig. 3. This sudden increase in the mass of surfactant adsorbed was due to lateral interactions between adsorbed monomers, resulting in surfactant aggregation on the surface [38].

An inflection point can be noted in the adsorption isotherm at C2. The amount of surfactant adsorbed increases sharply when the equilibrium concentration is between C2 and C3 (Region III) and then becomes asymptotic when the equilibrium concentration is higher than C3 (Region IV). The surfactant molecules are packed densely in the interlayer spacing at high concentrations, increasing the lateral interactions between the adsorbed surfactant molecules and resulting in a higher uptake of surfactant.

The surfactant can be adsorbed on the external surface and the interlayer spacing of the clays. In other words, the surfactants can intercalate within the interlayer spacing. The intercalation process depends on the nature and molecular size of the surfactant and the physicochemical properties of the clay, such as the charge density of the clay layer and cation exchange capacity. The organoclays would be more organophilic by increasing the length of the surfactant alkyl chain and the cationic surfactant/clay ratio, and the interlaminar spacing would be greater [39, 40].

The intercalation of the surfactant molecules can be corroborated by the widening of the interlayer or basal spacing reported during the modification of the natural clay to prepare the organoclay. The intercalation of HDTMA increased the interlaminar spacing of bentonite from 1.26 to 1.99 nm [41] and 1.58 to 2.01 nm [31, 34]. Similar values of the interlayer spacing for bentonite and OBHDTMA were reported by Majdan et al. [42]. The increase in the basal spacing of 0.73 nm revealed that the surfactant cations intercalated were arranged in the form of a bilayer [32, 42]. Martínez-Costa [34] studied the adsorption of HDTMA, DDTMA and BTMA on vermiculite and determined the basal spacing for the vermiculite, OVHDTMA, OVDDTMA and OVBTMA to be 1.44, 1.69, 1.51 and 1.43 nm, respectively. The increase of the interlayer spacing for OVHDTMA of 0.25 nm revealed that the intercalation of HDTMA is very slight, and the surfactant cations may be lying flat on the surface, forming a monolayer. In other studies, the interlayer spacing of vermiculite increased considerably from 1.49 to 5.23 nm after the adsorption of HDTMA [43], and the surfactant molecules presented a paraffin-type arrangement. These results show that the intercalation depends on the clay nature, physicochemical properties of surfactant and synthesis conditions of organoclay.

The surfactant arrangement in the surface and within the interlayer space of the clay sheets depends on the packing density, temperature and length of the alkyl chain [44, 45]. The configurations of the surfactant adsorbed vary between monolayer and bilayer structures. The arrangements of the surfactant molecules have been designated as monolayers, bilayers, pseudo-trimolecular layers and paraffin-type with different tilting angles of the alkyl chains depending on the amount of surfactant adsorbed [46, 47]. These arrangements of the surfactant molecules are illustrated in Fig. 4.

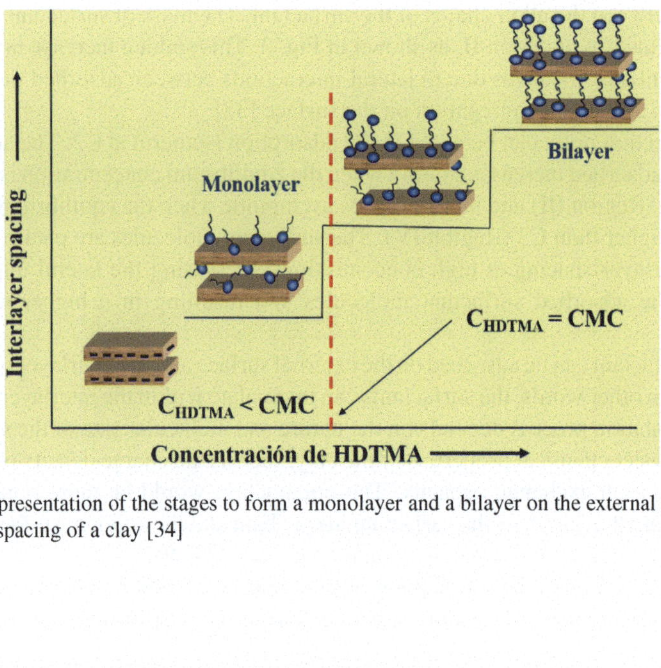

Fig. 4 Representation of the stages to form a monolayer and a bilayer on the external surface and interlayer spacing of a clay [34]

5 Surface Charge of Organoclays

Figure 5 shows the zeta potential of natural vermiculite and organovermiculite in the pH range of 2–13. The zeta potential of organovermiculite and vermiculite was reduced by increasing the pH from 2 to 11, but the surface of the organovermiculite was always positive, while that of vermiculite was always negative. The isoelectric point of the organovermiculite was $pH_{IEP} = 12.2$, while the pH_{IEP} of the vermiculite was slightly lower than pH = 2. The adsorption of the HDTMA surfactant on the vermiculite drastically modified the nature and surface charge. Similar results were reported for the adsorption of HDTMA on bentonite [31, 48].

The zeta potential of organoclay is dependent upon the loading or uptake of surfactant adsorbed because the surfactant modifies the charge of the surface of the clay. The effect of the uptake of HDTMA upon de zeta potential of OVHTMA is depicted in Fig. 6. As expected, the zeta potential increased considerably by raising

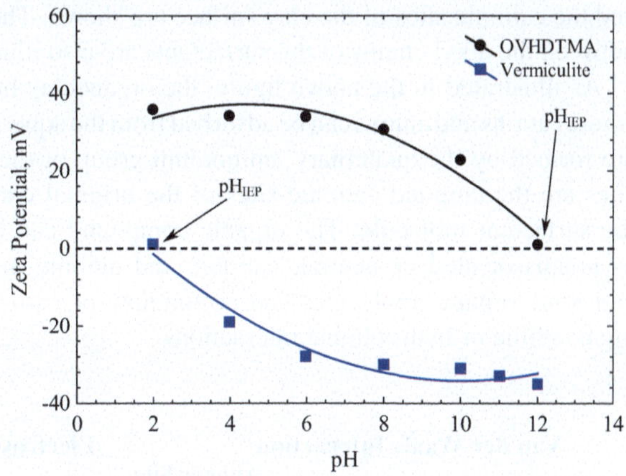

Fig. 5 Zeta potential distribution and pH_{IEP} of natural vermiculite and OVHDTMA. The lines represent the polynomial fitting of the data [34]

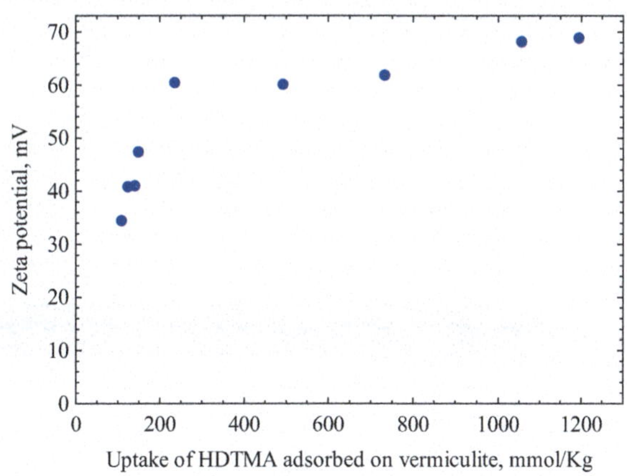

Fig. 6 Effect of uptake of HDTMA adsorbed on the zeta potential of OVHDTMA at T = 25 °C and pH = 7 [34]

the uptake of HDTMA adsorbed because the surface charge of the organoclay is due to the surfactant adsorbed on the clay surface.

6 Adsorption Mechanisms of Pollutants on Organoclay

The organoclays have been applied for adsorbing organic compounds, metallic cations and anions, and inorganic anions from water solution. The adsorption mechanism of a pollutant on organoclays is dependent upon the type of pollutant and the physicochemical characteristics of the organoclay.

To better understand the adsorption of any pollutant on organoclays is important to know the organoclay structure. Figure 4 shows that the different configurations of the organoclay depending upon the surfactant loading. An illustration of an organoclay with a high surfactant loading is schematically depicted in Fig. 7. The electrostatic and ion exchange interactions between the quaternary ammonium group of the surfactant and the cationic sites of the clay surface are shown. The Van der Waals interactions between the alkyl chains of the surfactants are also illustrated.

As illustrated in the above figure, the organoclay has cationic and anionic sites where cations and anions can be adsorbed from the aqueous solution. The anionic sites are formed by the quaternary ammonium group of the surfactant, and the cationic sites are the emptied cationic sites of the original clay that were no occupied by the surfactant molecule. The organic compound can be present in water solution as nondissociated or neutral, cationic and anionic molecules. The undissociated and ionic organic molecules can accumulate or sorb in the organic phase due to organophilic or hydrophobic interactions.

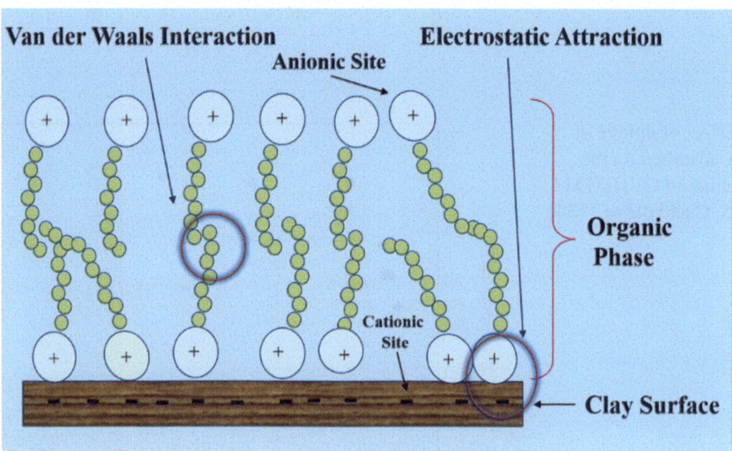

Fig. 7 Schematic illustation of the interactions and active sites of organoclay

7 Adsorption of Pollutants on Organoclays

7.1 *Adsorption of Organic Compounds on Organoclays from Water Solution*

The adsorption of organic compounds on organoclays has been extensively studied. The organic compounds included anionic and cationic dyes, non-ionic hydrophobic organic compounds, chlorinated phenolic and pharmaceutical compounds, pesticides and herbicides [25, 47, 49].

In several works [28, 34, 50], it has been shown that the adsorption capacity of organoclays towards organic compounds is dependent on the size of the surfactant molecule. The capacities of the organobentonites OBODTMA, OBHDTMA and OBDDTMA for adsorbing α-naphthylamine, β-naphthylamine, naphthol, nitrobenzene and aniline exhibited the following decreasing order OBODTMA > OBHDMA > OBDDTMA, so the adsorption capacity diminished, reducing the length of the octadecyl (OD), hexadecyl (HD) and dodecyl (DD) chains [28]. The adsorption isotherms of α-naphthylamine, β-naphthylamine and naphthol were satisfactorily represented by the Freundlich isotherm. The adsorption capacity of OBODTMA increased drastically at low concentrations, indicating a high affinity of the organobentonite for these pollutants. The adsorption isotherms of nitrobenzene and aniline presented a linear behavior suggesting that the main adsorption mechanism is partitioning of both pollutants between the organic phase of the organoclay and the water solution due to hydrophobic or organophilic interactions.

Additionally, Lee et al. [50] investigated the adsorption of naphthalene and phenanthrene on smectite modified with the surfactants tetramethylammonium bromide (TMA), HDTMA and DDTMA. The organosmectites synthesized by Lee et al. [50] were designated as OSTMA, OSHDTMA and OSDDTMA. The capacity of the organosmectites for adsorbing naphthalene decreased in the following order: OSHDTMA > OSDDTMA > OSTMA. Martínez-Costa [34] studied the adsorption of pyrogallol on the organobentonites OBHDTMA, OBDDTMA and OBBTMA, and the adsorption capacity of organobentonites towards pyrogallol decreased in the following order OBHDTMA > OBDDTMA ≈ OBBTMA. Furthermore, the adsorption capacities of OBHDTMA towards pyrogallol and OSHDTMA and OSDDTMA towards naphthalene increased almost linearly with the surfactant loading of the organoclay. The above results demonstrated that the adsorption capacity of the organoclays is dependent upon the size of the surfactant and surfactant loading of the organoclay because the volume of the organic phase of the organoclay increased with the size and loading of the surfactant.

The adsorption capacity of the organobentonite OBHDTMA towards 2,4-dichlorophenoxyacetic acid (2,4-D) at $pH = 4$ and phenol at $pH = 5$ was analyzed by Jacobo-Azuara et al. [41] and found that the capacity of OBHDTMA for adsorbing phenol was higher than that for 2,4-D at the equilibrium concentrations higher than 3 mmol/L. However, the opposite behavior occurred for concentrations less than

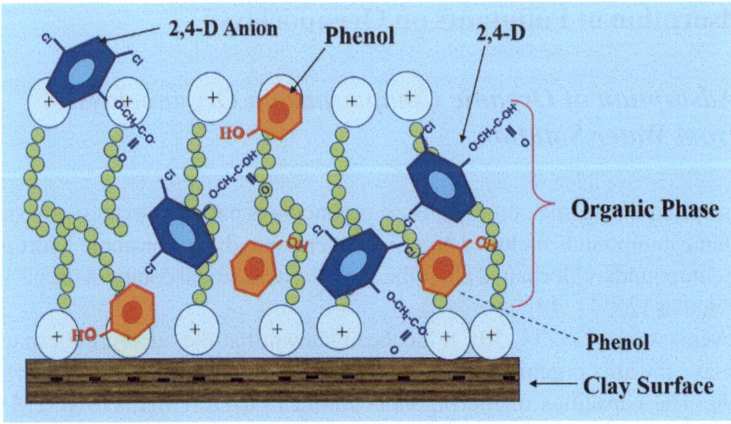

Fig. 8 Illustration of the sorption of 2,4-D (pH = 4) and phenol (pH = 5) on organobentonite OBHDTMA

3 mmol/L. The adsorption equilibrium data of phenol on OBHDTMA was interpreted by the Langmuir isotherm, whereas that bimodal Langmuir isotherm fitted the data for 2,4-D. Accordingly to the speciation diagrams, the phenol in water solution at pH = 5 is predominantly present as the protonated or undissociated species, and the 2,4-D is present as the protonated species (80%) and the deprotonated species or 2,4-D anion (20%) [51]. The adsorption of phenol and 2,4-D on organobentonite is schematically illustrated in Fig. 8.

As shown in the above figure, the 2,4-D anion was sorbed in the anionic sites, and the protonated species of 2,4-D and phenol were sorbed in the organic phase of the organoclay due to organophilic and hydrophobic interactions.

The effect of the substituent groups in the adsorption of phenolic compounds on organoclay was studied by determining the adsorption isotherms of phenol, hydroquinone and pyrogallol on OVDDTMA from water solution [34], and the adsorption isotherms of these phenolic compounds are graphed in Fig. 9. Pyrogallol, hydroquinone and phenol have three, two and one OH groups, respectively, and the values of the Octanol/Water Partition Coefficient, $LogK_{OW}$, are 0.21, 0.88 and 1.46 [52], respectively. The adsorption capacity of OVDDTMA decreased in the following order pyrogallol > hydroquinone > phenol, and the values of $LogK_{OW}$ increased in the same order. Hence, the higher adsorption capacity of OVDDTMA can be attributed to the larger hydrophobicity of the phenolic compound. In other words, the affinity of the phenolic compound towards the organic phase decreases in the same order as the adsorption capacity.

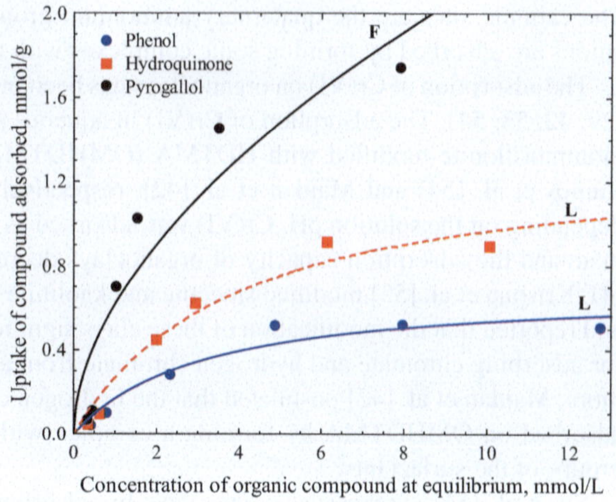

Fig. 9 Adsorption isotherms of hydroquinone, phenol and pyrogallol on OVDDTMA at pH = 7 and T = 25 °C. The lines represent the predictions of the Langmuir (L) and Freundlich (F) isotherms [34]

7.2 *Adsorption of Anions on Organoclays from Water Solution*

Several anions are present in water solutions, and some of them are very toxic and represent a health risk to human health. Among the very toxic anions are arsenate (AsO_4^{3-}), chromate (CrO_4^{2-}), hydrogen chromate ($HCrO_4^-$) and perchlorate (ClO_4^-), and the less toxic are nitrate (NO_3^-) and phosphate (PO_4^{3-}). These anions can be effectively removed from water solution by adsorption on organoclays. The adsorption mechanism of the anions on the organoclay is illustrated in Fig. 10. The anions can adsorb on the cationic sites of the organoclay as shown for AsO_4^{3-} and $HCrO_4^-$.

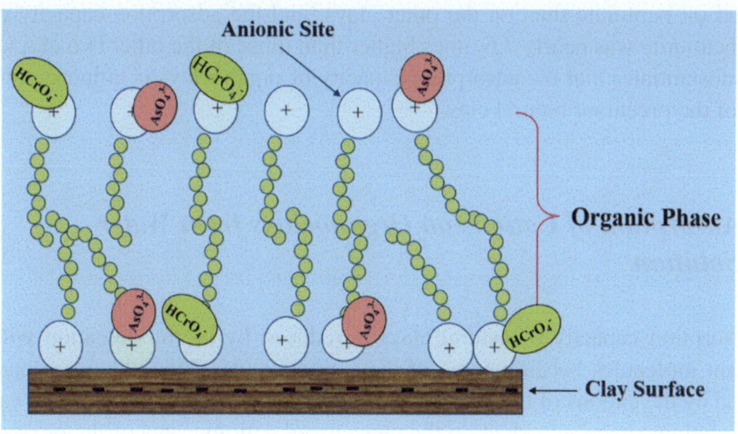

Fig. 10 Illustration of the sorption of arsenate (AsO_4^{3-}) and hydrogen chromate ($HCrO_4^-$) on organoclay

The cationic sites are the quaternary ammonium groups of the surfactant and the anions are adsorbed by forming some complexes with the cationic sites.

The adsorption of Cr(VI) on organoclays has been investigated by several authors [29, 42, 53, 54]. The adsorption of Cr(VI) in aqueous solution on OBHDTMA and montmorillonite modified with HDTMA (OMHDTMA) was analyzed by Leyva-Ramos et al. [54] and Majdan et al. [42], respectively. These authors found that depending on the solution pH, Cr(VI) was adsorbed as hydrogen chromate or chromate and the adsorption capacity of organoclays decreased by raising the solution pH. Krishna et al. [55] modified smectite and kaolinite with the surfactant HDTMA and reported that the modification of these clays significantly enhanced the capacity for adsorbing chromate and hydrogen chromate from an aqueous solution. Furthermore, Majdan et al. [42] postulated that the hydrogen chromate or chromate anions adsorbed on OMHDTMA by forming a complex with the quaternary ammonium groups of the surfactants.

Su et al. [56] synthesized organoclays by adsorbing octadecyltrimethylammonium chloride (OTMA), dioctadecyldimethylammonium chloride (DODMA) and octadecyl benzyldimethylammonium chloride (OBDMA) on bentonite, and the organobentonites were denominated as SMB1, SMB2 and SMB3, respectively. These authors studied the adsorption of As(III) and As(V) on these organobentonites and the decreasing order of adsorption capacity towards As(V) was SMB3 > SMB1 > SMB2 and that towards As(III) was SMB2 > SMB1 > SMB3, corroborating that the adsorption capacity of the organoclays is dependent on the characteristics of the surfactant and the anionic species present in solution.

Bentonite, kaolinite and halloysite were modified with HDTMA, and the organoclays were tested for removing nitrate [57], and the results indicate that the adsorption capacity decreased as follows organobentonite > organohalloysite ≈ organokaolinite. This trend was explained considering that the CEC of the bentonite was about 6.8 times those of halloysite and kaolinite, so much more HDTMA was adsorbed on bentonite than on the other clays, and the adsorption capacity of the organobentonite was nearly 7.6 times higher than those of the other two clays. This result substantiates that the adsorption capacity of organoclays is influenced by the nature of the precursor natural clay.

7.3 *Adsorption of Cations on Organoclays from Water Solution*

The adsorption capacity of natural clays is reduced by the modification with the surfactant molecules because some of the cationic sites of the clay were partially occupied by the cations of the surfactant during the modification. Hence, the cations in the solution can be adsorbed on the unoccupied cationic sites of the organoclay. This adsorption mechanism is illustrated in Fig. 11 for the adsorption of Cd^{2+} and Pb^{2+} from the water solution.

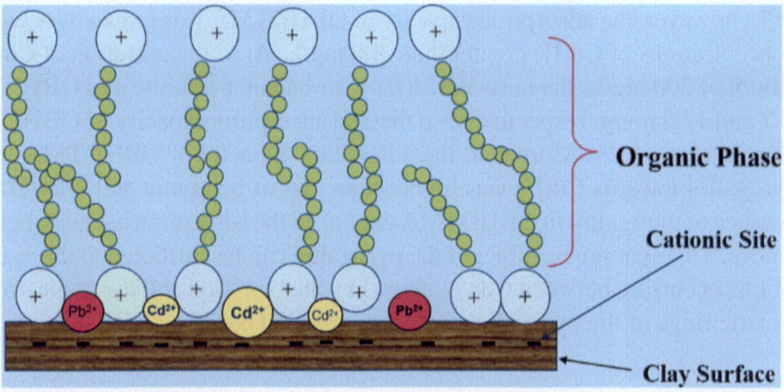

Fig. 11 Schematic representation of the adsorption mechanism of cations on organoclay

Recently, it has been reported that the adsorption of metal cations on organoclays from water solutions can be increased considerably if the organoclays were previously loaded with an organic compound. Martínez-Costa et al. [58] noted that the adsorption capacity of OBHDTMA towards Cd(II) was enhanced by sorbing diclofenac (DCF) and the adsorption capacity increased with the loading of DCF adsorbed. The enhancement of the adsorption capacity was up to 50%, depending on the loading of DCF sorbed [59].

The adsorption of Cd(II) on bentonite and OBHDTMA was evaluated by Martinez-Costa [34], and the experimental data and the adsorption isotherms are depicted in Fig. 12. As seen in this figure, the adsorption capacity of OBHDTMA was only slightly affected for the equilibrium concentrations of Cd(II) less than

Fig. 12 Adsorption isotherm of Cd(II) on bentonite, OBHDTMA and OBHDTMA loaded with pyrogallol at pH = 7 and T = 25 °C. The lines represent the predictions of the Freundlich (F) isotherm [34]

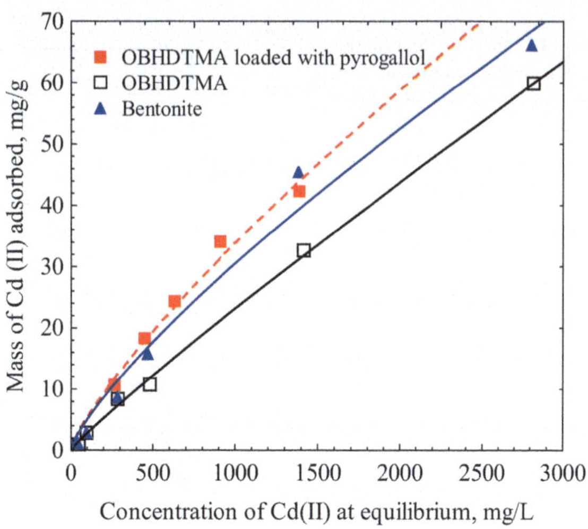

300 mg/L; however, the adsorption capacity of OBHDTMA was reduced for equilibrium concentrations of Cd(II) greater than 300 mg/L. At a concentration of Cd(II) at equilibrium of 500 mg/L, the mass of Cd(II) adsorbed on bentonite and OBHDTMA was 17.7 and 12.2 mg/g, respectively, so that the adsorption capacity of OBHDTMA was reduced about 31%. Moreover, the adsorption capacity of OBHDTMA loaded with pyrogallol towards Cd(II) was higher than that of bentonite and OBHDTMA. The presence of pyrogallol in OBHDTMA enhanced the adsorption capacity between 40 and 60%. This synergistic effect of the pyrogallol can be attributed to the π-cation interactions occurring between Cd^{2+} cations in water solution and the π-electrons of the aromatic rings of the pyrogallol sorbed on the OBHDTMA [58].

8 Effect of the Operating Conditions on the Adsorption Capacity of Organoclays

8.1 Effect of PH on the Adsorption Capacity of Organoclays

The effect of the pH on the adsorption capacity of an organoclay can be affected by the interaction between the species of pollutants present in the solution and the surface characteristics of the organoclay. Depending upon the nature of the pollutant, the pollutants in water solutions can be present as nondissociated, anionic and cationic species.

The effect of pH on the adsorption capacity of the organobentonite towards phenol is illustrated in Fig. 13, and the adsorption capacity remained essentially constant when the pH was increased from 5 to 7. In contrast, the adsorption capacity decreased

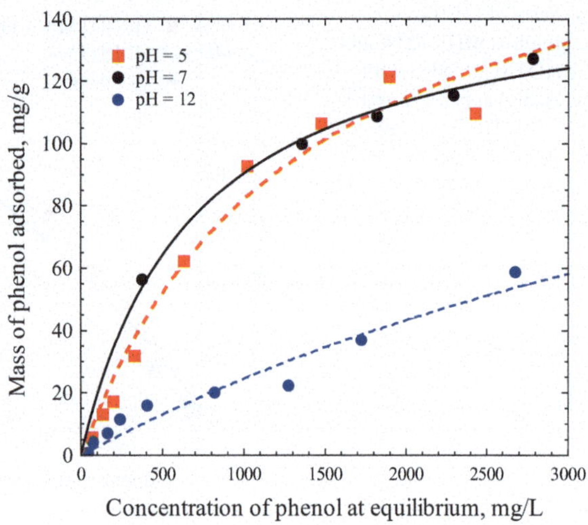

Fig. 13 Effect of pH on the adsorption isotherm of phenol on organobentonite at T = 25 °C. The lines represent the prediction of the Langmuir isotherm [60]

considerably by increasing the pH from 7 to 12. At an equilibrium concentration of 1700 mg/L, the maximum adsorption capacity was 110 mg/g at pH of 5 and 7 and 40 mg/g at a pH of 12. Thus, the adsorption capacity decreased 2.7 times when the pH was reduced from 5 to 12.

This behavior can be explained by recalling that phenol is mainly present as the nondissociated species in the pH range 5–7 and as the anionic species or phenolate at pH above 10 [41]. The point of zero charge of the organobentonite was determined to be $pH_{PZC} = 10$ [60], showing that the surface charge of the organobentonite is positive for $pH < pH_{PZC}$ and negative for $pH > pH_{PZC}$. Hence, the adsorption of phenol on the organobentonite can be attributed to organophilic and hydrophobic interactions in the pH range 5–7. Furthermore, the electrostatic repulsion between the negatively charged surface of the organobentonite and the phenolate at $pH = 12$ resulted in a reduction of the adsorption capacity.

Similar behavior was observed for the adsorption of 1-naphthalene sulphonic acid (NSA) and methyl blue (MTB) onto organobentonite in the pH range of 5–12 [61]. In the pH range 5–12, MTB and MTB are present as anionic species so that the electrostatic attraction favored the adsorption capacity for pH below pH_{PZC}. The adsorption capacity of organobentonite towards MTB and NSA exhibit a minimum at $pH = 12$ because of electrostatic repulsion.

An opposite behavior had been noticed for the adsorption of methylene blue (MB) because the capacity of organobentonite for adsorbing MB was reduced by increasing the pH solution from 3 to 11, and the maximum adsorption capacity was exhibited at $pH = 11$ [61]. It is very well documented that undissociated (MB°) and the cationic species (MB^{+}) can coexist at $pH = 3.8$, and the MB^{+} is the predominant species for pH higher than 5. At $pH = 11$, the enhancement of the adsorption capacity can be rationalized by considering that electrostatic attraction occurred between the MB^{+} and the negatively charged surface of the organobentonite.

8.2 *Effect of Temperature on the Adsorption Capacity of Organoclays*

At $pH = 5$, the adsorption isotherms of phenol on organobentonite at 15, 25 and 35 °C at $pH = 5$ are depicted in Fig. 14. As shown in this figure, there was no significant difference in the adsorption capacity of organobentonite when the temperature was raised from 25 to 35 °C, but a decrease in the adsorption capacity was observed when the temperature was reduced from 25 to 15 °C. At an equilibrium concentration of 1500 mg/L, the mass of phenol adsorbed was 115 mg/g for a temperature of 25 and 35 °C and 70 mg/g at the temperature of 15 °C. Thus, the mass of phenol adsorbed on organobentonite decreased 1.6 times by decreasing temperature from 25 to 15 °C.

The dependence of the adsorption capacity of organoclay on temperature was also analyzed for the adsorption of diclofenac [58, 59] and pyrogallol [31, 34] on organobentonite. The results demonstrated that the adsorption capacity towards both

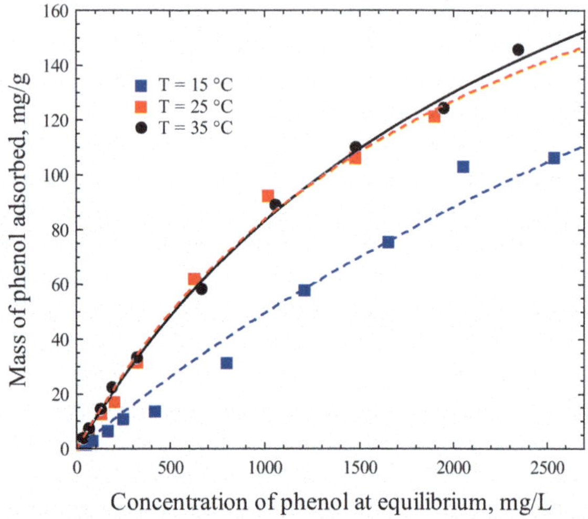

Fig. 14 Effect of the temperature on the adsorption isotherm of phenol on organobentonite at pH = 5. The lines represent the Langmuir isotherm [60]

organic pollutants increased by raising the temperature from 15 to 35 °C so that the adsorption of both organic pollutants was endothermic.

The isosteric heat of adsorption of these pollutants on organobentonite was estimated by the following equation [62]:

$$(\Delta H_{ads})_q = \frac{RLn\frac{C_{Ae1}}{C_{Ae2}}}{\frac{1}{T_1} - \frac{1}{T_2}} \tag{3}$$

where C_{Ae1} and C_{Ae2} are the equilibrium concentrations of pollutant at the temperatures of T_1 and T_2, respectively, both at the same uptake of pollutant, mmol/L; R is the universal gas law constant, 8.314 J/mol K; and $(\Delta H_{ads})_q$ is the isosteric heat of adsorption of pollutant on organoclay, J/mol.

The $(\Delta H_{ads})_q$ of pyrogallol and DCF on organobentonite were estimated to be 36.5 kJ/mol and 11.5 kJ/mol for an uptake of pollutant adsorbed of 98.1 mg/g and 114 mg/g, correspondingly [34, 59]. These values corroborated that the adsorption was endothermic, and the adsorption was physical because the $(\Delta H_{ads})_q$ was less than 40 kJ/mol [63].

As pointed out above, the adsorption capacity is dependent on the nature of the pollutant, and the effect of temperature on the adsorption capacity can also be affected. The adsorption isotherm of the antibiotic trimethoprim (TMP) from aqueous solution onto organobentonite is graphed in Fig. 15. As seen in this figure, the adsorption capacity of the organobentonite towards trimethoprim (TMP) decreased by increasing the temperature from 15 to 35 °C. This trend is contradictory to the behavior argued above. Thus, the adsorption of TMP on organobentonite is exothermic. It is important to point out that the adsorption capacity of the organobentonite towards TMP is very low compared to that towards pyrogallol and DCF.

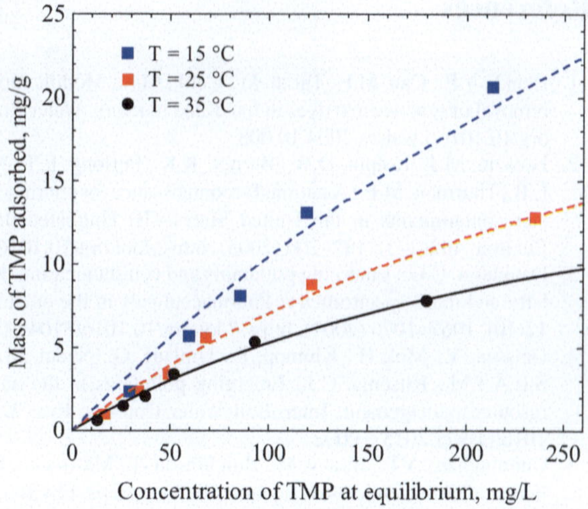

Fig. 15 Effect of temperature on the adsorption isotherm of TMP on organobentonite at pH = 7. The lines represent the Langmuir isotherm [59]

9 Conclusions

The modification of the nature of the clay surface represents a novel alternative for the preparation of adsorbents for removing toxic pollutants from water solutions. The synthesis of organoclays is based upon the adsorption of surfactants on the clay surface, and the characteristics of the organoclay are highly dependent on the synthesis conditions.

The type and physicochemical properties of the surfactant play a significant role in the adsorption capacity of the organoclay. The adsorption capacity of the organoclay is highly affected by the type and molecular size of the surfactant and loading or uptake of surfactant adsorbed. Besides, the surfactant can be adsorbed on the external surface and the interlayer spacing of the natural clay. The intercalation of the surfactants within the interlayer spacing happens in higher quantity the longer the length of the alkyl chain of the surfactant. It is crucial to understand the different arrangements or configurations of the surfactant adsorbed, and these configurations can be monolayers, bilayers, pseudo-trimolecular layers and paraffin-type layers and can be formed on the external surface and interlayer spacing.

The organoclays can be effectively applied for removing different types of organic pollutants such as anionic and cationic dyes, non-ionic hydrophobic organic compounds, chlorinated phenolic and pharmaceutical compounds, pesticides and herbicides. The organoclays exhibited a reasonably high adsorption capacity towards very toxic metallic and inorganic anions such as chromate, hydrogen chromate, arsenate and arsenite.

The adsorption mechanisms of the pollutants on the organoclay depend on the operating conditions such as solution pH and temperature, and the types of pollutant species present in the water solution. In general, the adsorption mechanisms include electrostatic attraction, ion exchange and organophilic and hydrophobic interactions.

References

1. Benkli, Y.E., Can, M.F., Turan, M., Celik, M.S.: Modification of organo-zeolite surface for the removal of reactive azo dyes in fixed-bed reactors. Water Res. **39**, 487–493 (2005). https://doi.org/10.1016/j.watres.2004.10.008
2. Focazio, M.J., Kolpin, D.W., Barnes, K.K., Furlong, E.T., Meyer, M.T., Zaugg, S.D., Barber, L.B., Thurman, M.E.: A national reconnaissance for pharmaceuticals and other organic wastewater contaminants in the United States—II) Untreated drinking water sources. Sci. Total Environ. **402**(2–3), 192–200 (2008). https://doi.org/10.1016/j.scitotenv.2008.02.021
3. Daughton, C.G.: Emerging pollutants and communicating the science of environmental chemistry and mass spectrometry: Pharmaceuticals in the environment. J. Am. Soc. Mass Spectr. **12**(10), 1067–1076 (2001). https://doi.org/10.1016/S1044-0305(01)00287-2
4. Geissen, V., Mol, H., Klumpp, E., Umlauf, G., Nadal, M., van der Ploeg, M., van de Zee, S.E.A.T.M., Ritsema, C.J.: Emerging pollutants in the environment: a challenge for water resource management. Inter. Soil Water Conserv. Res. **3**, 57–65 (2005). https://doi.org/10.1016/j.iswcr.2015.03.002
5. Cunningham, V.L., Buzby, M., Hutchinson, T., Mastrocco, F., Parke, N., Roden, N.: Effects of human pharmaceuticals on aquatic life: next steps. Environ. Sci. Technol. **40**(11), 3456–3462 (2006). https://doi.org/10.1021/es063017b
6. Daughton, C.G., Ternes, T.A.: Pharmaceuticals and personal care products in the environment: agents of subtle change? Environ. Health Perspect. **107**(6), 907–938 (1999). https://doi.org/10.1289/ehp.99107s6907
7. Gros, M., Petrovic, M., Ginebreda, A., Barceló, D.: Removal of pharmaceuticals during wastewater treatment and environmental risk assessment using hazard indexes. Environ. Int. **36**, 15–26 (2010). https://doi.org/10.1016/j.envint.2009.09.002
8. Loos, R., Gawlik, B.M., Locoro, G., Rimaviciute, E., Contini, S., Bidoglio, G.: EU-wide survey of polar organic persistent pollutants in European river waters. Environ. Pollut. **157**, 561–568 (2009). https://doi.org/10.1016/j.envpol.2008.09.020
9. Larsson, D.G.J.: Pollution from drug manufacturing: review and perspectives. Philos. Trans. R Soc. Lond. B **369**, 1–7 (2014). https://doi.org/10.1098/rstb.2013.0571
10. Loraine, G.A., Pettigrove, M.E.: Seasonal variations in concentrations of pharmaceuticals and personal care products in drinking water and reclaimed wastewater in southern California. Environ. Sci. Technol. **40**(3), 687–695 (2006). https://doi.org/10.1021/es051380x
11. Daughton, C.G.: Pharmaceuticals and personal care products in the environment: overarching issues and overview. In: Daughton, C.G., Jones-Lepp, T.L. (eds.) ACS Symposium Series 791, pp. 2–38. American Chemical Society, Washington, DC (2001). https://doi.org/10.1021/bk-2001-0791.ch001
12. Hirsch, R., Ternes, T., Haberer, K., Kratz, K.L.: Occurrence of antibiotics in the aquatic environment. Sci. Total Environ. **225**(1–2), 109–118 (1999). https://doi.org/10.1016/S0048-9697(98)00337-4
13. Gogoi, A., Mazumder, P., Tyagi, V.K., Chamind, G.G.T., An, A.K., Kumar, M.: Occurrence and fate of emerging contaminants in water environment: a review. Groundwater Sustain. Dev. **6**, 169–180 (2018). https://doi.org/10.1016/j.gsd.2017.12.009
14. Zhou, S., Di Paolo, C., Wu, X., Shao, Y., Seiler, T.-B., Hollert, H.: Optimization of screening-level risk assessment and priority selection of emerging pollutants—the case of pharmaceuticals in European surface waters. Environ. Int. **128**, 1–10 (2019). https://doi.org/10.1016/j.envint.2019.04.034
15. Clifford, D., Subramonian, S., Sorg, T.J.: Water treatment processes. III. Removing dissolved inorganic contaminants from water. Environ. Sci. Technol. **20**(11), 1072–1080 (1986). https://doi.org/10.1021/es00153a001
16. Domínguez, J.M., Schifter, I.: Las arcillas: El Barro Noble. Fondo de Cultura Económica, Mexico (1992)
17. Velde, B.: Introducion of Clay Minerals. Chemistry, Origin, Uses and Environmental Significance. Chapman & Hall, London, UK (1992)

18. Meunier, A.: Clays. Springer-Verlag, Berlin Heidelbergh, Germany (2005)
19. García, R.E., Suárez, B.M.: Las Arcillas: Propiedades y Usos. Textos Universitarios, España (2001)
20. Coronado-Oyarvide, P.: Equilibrio y Velocidad de Adsorción de Cadmio(II) sobre Arcillas Naturales y Clinoptilolita. M. Sc. Thesis, Universidad Autónoma de San Luis Potosí, San Luis Potosi, Mexico (2015)
21. Padilla-Ortega, E.: Adsorción de Metales Pesados en Solución Acuosa sobre Arcillas Naturales. M. Sc. Thesis, Universidad Autónoma de San Luis Potosí, San Luis Potosi, Mexico (2008)
22. Padilla-Ortega, E.: Adsorción de Metales Pesados en Solución Acuosa sobre Arcillas Naturales, Modificadas y Bio-nanocomposites. Ph.D. Thesis, Universidad Autónoma de San Luis Potosí, San Luis Potosi, Mexico (2013)
23. Shen, Y.H.: Preparations of organobentonite using nonionic surfactants. Chemosphere **44**, 989–995 (2001). https://doi.org/10.1016/S0045-6535(00)00564-6
24. Sapag, K., Mendioroz, S.: Synthesis and characterization of micro-mesoporous solids: pillared clays. Colloid Surface A **187–188**, 141–149 (2001). https://doi.org/10.1016/S0927-7757(01)00617-3
25. Mortland, M.M., Shaobal, S., Boyd, S.N.A.: Clay-organic complexes as adsorbents for phenol and chlorophenols. Clays Clay Miner. **34**, 581–585 (1986). https://doi.org/10.1346/CCMN.1986.0340512
26. Lin, S.H., Cheng, M.J.: Adsorption of phenol and m-chlorophenol on organobentonites and repeated thermal regeneration. Waste Manage. **22**, 595–603 (2002). https://doi.org/10.1016/S0956-053X(01)00029-0
27. Vidal, N.C., Volzone, C.: Analysis of tetramethylammonium–montmorillonite and retention of toluene from aqueous solution. Appl. Clay Sci. **45**, 227–231 (2009). https://doi.org/10.1016/j.clay.2009.06.013
28. Zhu, L., Li, Y., Zhang, J.: Sorption of organobentonites to some organic pollutants in water. Environ. Sci. Technol. **31**, 1407–1410 (1997). https://doi.org/10.1021/es960641n
29. Haggerty, G.M., Bowman, R.S.: Sorption of chromate and other inorganic anions by organo-zeolite. Environ. Sci. Technol. **28**, 452–458 (1994). https://doi.org/10.1021/es00052a017
30. Kim, Y.S., Song, D.I., Jeon, Y.W., Choi, S.J.: Adsorption of organic phenols onto hexadecyltrimethylammonium treated montmorillonite. Sep. Sci. Technol. **31**(20), 2815–2830 (1996). https://doi.org/10.1080/01496399608000829
31. Martínez-Costa, J.I., Leyva-Ramos, R.: Effect of surfactant loading and type upon the sorption capacity of organobentonite towards pyrogallol. Colloid Surface A **520**, 676–685 (2017). https://doi.org/10.1016/j.colsurfa.2017.02.033
32. Xu, S., Boyd, S.A.: Alternative model for cationic surfactant adsorption by layer silicates. Environ. Sci. Technol. **29**, 3022–3028 (1995). https://doi.org/10.1021/es00012a020
33. Ismadji, S., Soetaredjo, F.E., Ayucitra, A.: Clay Materials for Environmental Remediation. Springer, Berlin, Heidelberg, Germany (2015)
34. Martínez-Costa, J.I.: Adsorción de Contaminantes Orgánicos Tóxicos en Solución Acuosa sobre Organoarcillas Preparadas con Diferentes Surfactantes Catiónicos. M.Sc., Thesis, Universidad Autónoma de San Luis Potosí, San Luis Potosi, Mexico (2012)
35. Chen, B., Zhu, L., Zhu, J., Xing, B.: Configurations of the bentonite sorbed myristyl pyridinium cation and their influences on the uptake of organic compounds. Environ. Sci. Technol. **39**, 6093–6100 (2005). https://doi.org/10.1021/es0502674
36. Guo, Z., Zheng, S., Zheng, Z., Jiang, F., Hu, W., Ni, L.: Selective adsorption of p-chloronitrobenzene from aqueous mixture of p-chloronitrobenzene and o-chloronitrobenzene using HZSM-5 zeolite. Water Res. **39**, 1174–1182 (2005). https://doi.org/10.1016/j.watres.2004.12.031
37. Fuerstenau, D.W.: Equilibrium and nonequilibrium phenomena associated with the adsorption of ionic surfactants at solid–water interfaces. J. Colloid Interf. Sci. **256**(1), 79–90 (2002). https://doi.org/10.1006/jcis.2001.8139
38. Wang, W., Kwak, J.C.T.: Adsorption at the alumina-water interface from mixed surfactant solutions. Colloid Surface A **156**, 95–110 (1999). https://doi.org/10.1016/S0927-7757(99)00062-X

39. Kooli, F., Liu, Y., Alshahateet, S., Messali, M., Bergaya, F.: Reaction of acid activated montmorillonites with hexadecyl trimethylammonium bromide solution. App. Clay Sci. **43**(3–4), 357–363 (2009). https://doi.org/10.1016/j.clay.2008.10.006
40. Xi, Y., Ding, Z., He, H., Frost, R.L.: Structure of organoclays-an X-ray diffraction and thermogravimetric analysis study. J. Colloid Interf. Sci. **277**, 116–120 (2004). https://doi.org/10.1016/j.jcis.2004.04.053
41. Jacobo-Azuara, A., Leyva-Ramos, R., Padilla-Ortega, E., Aragón-Piña, A., Guerrero-Coronado, R.M., Mendoza-Barron, J.: Removal of toxic pollutants from water solutions by adsorption on organobentonite. Ads. Sci. Technol. **24**(8), 687–699 (2006). https://doi.org/10.1260/026361706781355000
42. Majdan, M., Maryuk, O., Pikus, S., Olszewska, E., Kwiatkowski, R., Skrzypek, H.: Equilibrium, FTIR, scanning electron microscopy and small wide angle X-ray scattering of chromates adsorption on modified bentonite. J. Molecular Struct. **740**, 203–211 (2005). https://doi.org/10.1016/j.molstruc.2005.01.044
43. Yu, X., Wei, C., Ke, L., Hu, Y., Xie, X., Wu, H.: Development of organovermiculite-based adsorbent for removing anionic dye from aqueous solution. J. Hazard. Mater. **180**, 499–507 (2010). https://doi.org/10.1016/j.jhazmat.2010.04.059
44. De Paiva, L.B., Morales, A.R., Diaz, F.R.V.: Organoclays: properties, preparation and applications. App. Clay Sci. **42**(1–2), 8–24 (2008). https://doi.org/10.1016/j.clay.2008.02.006
45. Wang, L., Liu, J., Exarhos, G., Flanigan, K., Bordia, R.: Conformation heterogeneity and mobility of surfactant molecules in intercalated clay minerals studied by solid-state NMR. J. Phys. Chem. B **104**(13), 2810–2816 (2000). https://doi.org/10.1021/jp993058c
46. Lagaly, G.: Interaction of alkylamines with different types of layered compounds. Solid State Ionics **22**, 43–51 (1986). https://doi.org/10.1016/0167-2738(86)90057-3
47. Park, Y., Ayoko, G.A., Frost, R.L.: Application of organoclays for the adsorption of recalcitrant organic molecules from aqueous media. J. Colloid Interf. Sci. **354**, 292–305 (2011). https://doi.org/10.1016/j.jcis.2010.09.068
48. Wang, C.-C., Juang, L.-C., Lee, C.-K., Hsu, T.-C., Lee, J.-F., Chao, H.-P.: Effects of exchanged surfactant cations on the pore structure and adsorption characteristics of montmorillonite. J. Colloid Interf. Sci. **280**, 27–35 (2004). https://doi.org/10.1016/j.jcis.2004.07.009
49. Lee, S.M., Tiwari, D.: Organo and inorgano-organo-modified clays in the remediation of aqueous solutions: an overview. App. Clay Sci. **59–60**, 84–102 (2012). https://doi.org/10.1016/j.clay.2012.02.006
50. Lee, S.Y., Kim, S.J., Chung, S.Y., Jeong, C.H.: Sorption of hydrophobic organic compounds onto organoclays. Chemosphere **55**, 781–785 (2004). https://doi.org/10.1016/j.chemosphere.2003.11.007
51. Díaz-Flores, P.E., Leyva-Ramos, R., Rangel-Mendez, J.R., Michel-Ortiz, M., Guerrero-Coronado, R.M., Mendoza-Barron, J.: Adsorption of 2,4 dichlorophenoxyacetic acid from aqueous solution on activated carbon cloth. J. Environ. Eng. Manage. **16**(4), 249–257 (2006)
52. Wang, X., Yu, J., Wang, Y., Wang, L.: Mechanism-based quantitative structure–activity relationships for the inhibition of substituted phenols on germination rate of Cucumis sativus. Chemosphere **46**(2), 241–250 (2002). https://doi.org/10.1016/S0045-6535(01)00082-0
53. Li, Z., Bowman, R.S.: Sorption of chromate and PCE by surfactant-modified clay minerals. Environ. Eng. Sci. **15**, 237–245 (1998). https://doi.org/10.1089/ees.1998.15.237
54. Leyva-Ramos, R., Jacobo-Azuara, A., Torres-Rivera, O.L., Guerrero-Coronado, R.M., Berber-Mendoza, M.S., Alonso-Davila, P.: Adsorption of chromium (VI) from water solution onto organobentonite. J. Environ. Eng. Manage. **18**(5), 313–319 (2008)
55. Krishna, B.S., Murty, D.S.R., Prakash, B.S.J.: Surfactant-modified clay as adsorbent for chromate. Appl. Clay Sci. **20**, 65–71 (2001). https://doi.org/10.1016/S0169-1317(01)00039-4
56. Su, J., Huang, H.-G., Jin, X.-Y., Lu, X.-Q., Chen, Z.-L.: Synthesis, characterization and kinetic of a surfactant-modified bentonite used to remove As(III) and As(V) from aqueous solution. J. Hazard. Mater. **185**, 63–70 (2011). https://doi.org/10.1016/j.jhazmat.2010.08.122
57. Xi, Y., Mallavarapu, M., Naidu, R.: Preparation, characterization of surfactants modified clay minerals and nitrate adsorption. App. Clay Sci. **48**(1–2), 92–96 (2010). https://doi.org/10.1016/j.clay.2009.11.047

58. Martínez-Costa, J.I., Leyva-Ramos, R., Padilla-Ortega, E.: Sorption of diclofenac from aqueous solution on an organobentonite and adsorption of cadmium on organobentonite saturated with diclofenac. Clays Clay Miner. **66**(6), 515–528 (2018). https://doi.org/10.1346/CCMN.2018.064119
59. Martínez-Costa, J.I.: (2017) Eliminación de Compuestos Orgánicos en Solución Acuosa por medio de Sorción sobre Organoarcillas y Procesos Avanzados de Oxidación. Ph. D. Thesis, Universidad Autónoma de San Luis Potosí, San Luis Potosi, Mexico (2017)
60. Jacobo-Azuara, A.: Adsorción de Compuestos Tóxicos en solución Acuosa sobre Organobentonita y Zeolita Modificada con Surfactante, Ph.D., Thesis, Universidad Autónoma de San Luis Potosí, San Luis Potosi, Mexico (2007)
61. Ocampo-Pérez, R., Leyva-Ramos, R., Padilla-Ortega, E.: Equilibrium and kinetic adsorption of organic compounds onto organobentonite: application of a surface diffusion model. Ad. Sci. Technol. **29**(10), 1007–1024 (2011). https://doi.org/10.1260/0263-6174.29.10.1007
62. Leyva-Ramos, R.: Effect of temperature and pH on the adsorption of an anionic detergent on activated carbon. J. Chem. Technol. Biotechnol. **45**, 231–240 (1989). https://doi.org/10.1002/jctb.280450308
63. Froment, G.F., Bischoff, K.B.: Chemical reactor analysis and design, 2nd edn. Wiley, New York (1990)

The manufacturer's authorised representative in the EU is Springer
Nature Customer Service Centre GmbH, Europaplatz 3, 69115 Heidelberg,
Germany. If you have any concerns regarding our products, please
contact ProductSafety@springernature.com

Printed and bound by CPI Group (UK) Ltd, Croydon, CR0 4YY

30/09/2026

02225508-0014